Modern Oceanography

Modern Oceanography

Editor: Jeremy Harper

R CALLISTO
REFERENCE

www.callistoreference.com

Callisto Reference,
118-35 Queens Blvd., Suite 400,
Forest Hills, NY 11375, USA

Visit us on the World Wide Web at:
www.callistoreference.com

ISBN: 978-1-64116-060-5 (Hardback)

Cataloging-in-Publication Data

Modern oceanography / edited by Jeremy Harper.
    p. cm.
Includes bibliographical references and index.
ISBN 978-1-64116-060-5
1. Oceanography. 2. Marine sciences. I. Harper, Jeremy.
GC11.2 .M63 2019
551.46--dc23

Table of Contents

Preface

Oceanography is the study of oceans, their physical and chemical properties. It is a branch of Earth science. It delves into the varied aspects of oceans such as ocean currents, waves dynamics, geology of the sea floor among many others. It combines the principles of biology, hydrology, physics, climatology, chemistry, etc. to better understand the process of ocean ecosystems. The geological past of the oceans is studied under the domain of paleoceanography. This book presents detailed analysis of all crucial concepts and theories related to the advances made in this field of study. From theories to research to practical applications, chapters related to all contemporary topics of relevance to this field have been included in this book. It will serve as an ideal reference text for students, academicians and professionals associated with the field of oceanography at various levels.

All of the data presented henceforth, was collaborated in the wake of recent advancements in the field. The aim of this book is to present the diversified developments from across the globe in a comprehensible manner. The opinions expressed in each chapter belong solely to the contributing authors. Their interpretations of the topics are the integral part of this book, which I have carefully compiled for a better understanding of the readers.

At the end, I would like to thank all those who dedicated their time and efforts for the successful completion of this book. I also wish to convey my gratitude towards my friends and family who supported me at every step.

Editor

Changes in extreme regional sea level under global warming

Changes in extreme regional sea level under global warming

Changes in extreme regional sea level under global warming

Changes in extreme regional sea level under global warming

Changes in extreme regional sea level under global warming

uncertainty. In Slangen et al. (2012) and Bordbar et al. (2015), the spread due to decadal-to-centennial variability is considered by looking at ensemble simulations using CMIP3 and CMIP5 climate models, respectively. It was shown that the (CMIP5) ensemble spread of the projected DSL is of the same order of magnitude as the globally averaged sea level rise (Bordbar et al., 2015). Several regions were identified where the forced sea level change signal is relatively strong with respect to the internal variability, e.g., the Indo-Pacific part of the Southern Ocean and the eastern equatorial Pacific, and hence may be detected earlier (Bordbar et al., 2015).

However, in all these model studies the strongest component of oceanic internal variability, i.e., that due to ocean meso-scale eddies, was not represented. Rectification processes due to eddies can lead to strong changes in mean ocean surface flows and their response to atmospheric forcing, in particular in the Southern Ocean (Böning et al., 2008). In strongly eddying ocean models even new modes of low-frequency variability may appear, such as the multidecadal Southern Ocean Mode (Le Bars et al., 2016). Using the eddy-permitting (about $1/4°$ horizontal resolution) version of the MIROC3.2 model, Suzuki et al. (2005) showed that representing ocean eddies provides a more detailed projection of regional sea level changes under the IPCC SRES-A1B scenario and that eddies are strongly involved in regional sea level extremes. In addition, as demonstrated by Firing and Merrifield (2004) from observational data, a high background sea level superposed on the sea level change due to an arriving ocean eddy can lead to extreme local sea levels.

Eddies can also have a strong effect on the sensitivity of the AMOC to freshwater forcing. The study of Weijer et al. (2012) indicates that the AMOC in the strongly eddying (about 0.1° horizontal resolution) version of the Parallel Ocean Program (POP) model version is more sensitive to freshwater perturbations than the non-eddying version of the same model. Climate model studies on the projections of the AMOC with non-eddying ocean components show only an AMOC decline of 22 to 40 % over the period 2000–2100, depending on the IPCC scenario (Weaver et al., 2013). Only 2 (out of 30) of these models project a substantial decrease of the AMOC under the RCP8.5 scenario until year 2100, and no model shows an abrupt transition after the 21st century (Weaver et al., 2013). However, in high-resolution ocean models strong variations in the AMOC strength lead to changes in ocean currents and eddy pathways, which induce an additional contribution to the variability in DSL and hence affect extreme DSL values (Brunnabend et al., 2014).

It is important to assess the role of eddies in projections of future regional sea level changes, in particular on the DSL extremes. In this paper, we study a scenario of future DSL change using the high-resolution version of POP as in Brunnabend et al. (2014), but now forced with atmospheric fields from a coupled climate model that evolved under the SRES-A1B scenario. We focus on the changes in the probability density function of regional (and more local) DSL val-

ues and 10-year return extreme values over the period 2000–2100, computed using the generalized extreme value theory (Coles, 2001), and compare these results to those obtained from a similarly forced non-eddying version of POP.

2 Ocean model

The high-resolution version of the POP (http://www.cesm.ucar.edu/models/ccsm4.0/pop/) used has a spatial resolution of 0.1° horizontally and 42 depth levels of which the thickness varies from 10 m near the surface to 250 m near the ocean bottom (Maltrud et al., 2008). The high spatial resolution captures the processes leading to meso-scale ocean eddies and provides a more detailed representation of the western boundary currents. Specific details about the high-resolution model setup, such as the treatment of the bottom topography, sea ice and river runoff, are described in Weijer et al. (2012). The high-resolution model was optimized for use on the Cartesius supercomputer in Amsterdam (www.surfsara.nl) and about 3 model years are simulated per 24 h using about 1000 cores.

The POP simulation was initialized from a 75-year spin-up simulation (Maltrud et al., 2008) under the CORE-I climatology dataset (Large and Yeager, 2004) as atmospheric forcing. This initial condition is indicated here as the year 1950. Under a freshwater flux which is diagnosed from the last 5 years of this spin-up, the model displays only a very small drift over a 200-year control simulation (Le Bars et al., 2016). Here, the model was forced with monthly mean atmospheric forcing fluxes over the period 1950–2100, which were derived from simulations with the ECHAM5-OM1 model within the ESSENCE (Sterl et al., 2008) project (see www.knmi.nl/~sterl/Essence/). The used forcing fields are 10 m wind speed, downward flux of short-wave and longwave radiation, 2 m temperature, humidity, precipitation, runoff, and the surface wind stress field. The atmospheric forcing fields are given on a global $1° × 1°$ grid and are interpolated to the curvilinear POP model grid. The outgoing heat and freshwater fluxes are computed within the model using bulk formulae. There is an initial adjustment after the switch in forcing in 1950, for example measured by the change in the AMOC strength, which lasts for about a decade.

Over the years 1950–2000, the POP model was forced by the ensemble mean atmospheric fields from the ESSENCE project that take observed concentrations of greenhouse gases and anthropogenic aerosols into account. Over the years 2001–2100, POP was forced with atmospheric forcing fields obtained from the ECHAM5-OM1 model according to the SRES A1B scenario (from an arbitrarily chosen ensemble member of the ESSENCE project; Sterl et al., 2008). We focus on ensemble member no. 021 for which the high-resolution simulation is denoted by R_{021}. Two additional simulations are performed using the forcing from the

ESSENCE ensemble members nos. 029 and 033 to address the robustness of the change of extreme DSL values.

In addition to this high-resolution simulation, a similarly forced simulation is performed with a low-resolution POP version, indicated in the following by R_{021}^{low}. This non-eddying version has an average $1.0°$ horizontal resolution and 40 vertical levels (Weijer et al., 2012). The Gent and McWilliams (1990) scheme is used to represent eddy-driven tracer transports. Such a scheme is not needed in the strongly eddying version as these tracer transports are explicitly resolved.

The POP model directly computes the DSL, which can be decomposed into a mass redistribution term and a steric contribution. Because the freshwater flux is included into the model as a virtual salt flux and the global mean of precipitation, evaporation and river runoff is zero, no mass-induced global mean sea level changes can be represented. Due to the applied Boussinesq approximation, global mean steric sea level variations are not accounted for explicitly during this study, but this spatially independent contribution was computed from the model output (Greatbatch, 1994).

To demonstrate the performance of both versions of the POP model, we compare the DSL over the years 1993–2012 (computed from monthly means) with observations derived from altimetry. The altimeter products were produced by Ssalto/Duacs and distributed by Aviso, with support from CNES (http://www.aviso.altimetry.fr/duacs/). No salinity or heat restoring is applied as even a weak salinity restoring artificially constrains the AMOC. The model is also configured with no weak restoring of the global sea surface salinity field. However, as the POP model does not include a thermodynamic–dynamic sea-ice component, a prescribed climatological flux of heat and salt is included in sea-ice regions. These fluxes are the same for the control and hosing simulations and are an order of magnitude smaller than the mean fluxes. The sea-ice regions are indicated by white areas and are not considered in the analyses below. The mean DSL of simulation R_{021} over the years 1993–2012 agrees well with observations, both for the mean (compare Fig. 1a and c) and the standard deviation (compare Fig. 1b and d). This shows that the model adequately determines the mean ocean circulation, including the western boundary currents and also represents the eddy-induced variability. Differences with respect to observations appear due to the general overestimation of the modeled variability during this time period, which may be due to the prescribed low resolution of the atmospheric forcing and the lack of feedback of the atmosphere on the ocean variability. Differences in variability may also occur due to the higher horizontal resolution of the ocean model ($0.1°$) than the altimetry dataset used ($0.25°$) as more small-scale features can be resolved. Regional differences in variability in the South Atlantic are caused by a too regular Agulhas ring formation rate in POP compared to observations (Le Bars et al., 2014).

In contrast, results for the low-resolution simulation R_{021}^{low} shown in Fig. 1e (mean) and Fig. 1f (standard deviation) indicate that only the mean DSL change is reasonably well captured. The variability captured in the model is mainly related to the seasonal cycle and internal variability is weak, in particular in the regions of the western boundary currents (similar to many other non-eddying ocean model results; Bordbar et al., 2015). This weak variability also has consequences for the mean flow through the lack of representation of rectification processes causing, for example, too small DSL values in the Agulhas and the Gulf Stream regions.

3 Future dynamic sea level changes

In the results below, all long-term changes are computed by taking the difference between values over the last 20 years (2081–2100) and the first 20 years (2001–2020) of the model simulations. Monthly mean data are used for the analysis of changes in the mean and standard deviation (Sect. 3.1), while daily data are used in the extreme value analyses (Sect. 3.2).

3.1 Mean and standard deviation

In the POP simulation R_{021}, global mean steric height increases by about $2.2\,\text{mm}\,\text{yr}^{-1}$ from year 2000 to 2100. As this signal is homogeneous over the Earth, it is not considered in the results below. Largest changes in mean DSL between the periods 2081–2100 and 2001–2020 occur in the North Atlantic (Fig. 2a), in particular near the western part of this basin (Fig. 2c). There is a mean DSL decrease in the Atlantic and Pacific parts of the Southern Ocean, while mean DSL increases in the Indian part of the Southern Ocean. The mean DSL increases in the eastern part of the North Atlantic basin and decreases in the center of the subpolar gyre. Large changes in DSL variability occur in the Agulhas retroflection region and near Drake Passage (Fig. 2b). The DSL variability decreases in the western North Atlantic, in the center of the subpolar gyre and slightly along the western boundary of the North Atlantic while it substantially increases in the eastern Atlantic (Fig. 2d). However, the separation of DSL change in the North Atlantic into steric height change and change in regional ocean mass show that the change is mainly caused by regional steric height changes (Fig. 3a, b). These regional steric height changes and the positive mass redistribution that increases DSL near the North American coast (Fig. 3c, d) correspond well with the pattern found by the study of Yin et al. (2009).

In the POP simulation R_{021}^{low}, global mean steric height varies only by a few centimeters over the period 2000 to 2100 and again is not considered further. Regional steric height changes and the redistribution of ocean mass towards the North American coast are also visible in the low-resolution results. In addition, the small dipole pattern visible in the North Atlantic is caused by the reduced strength and the shift

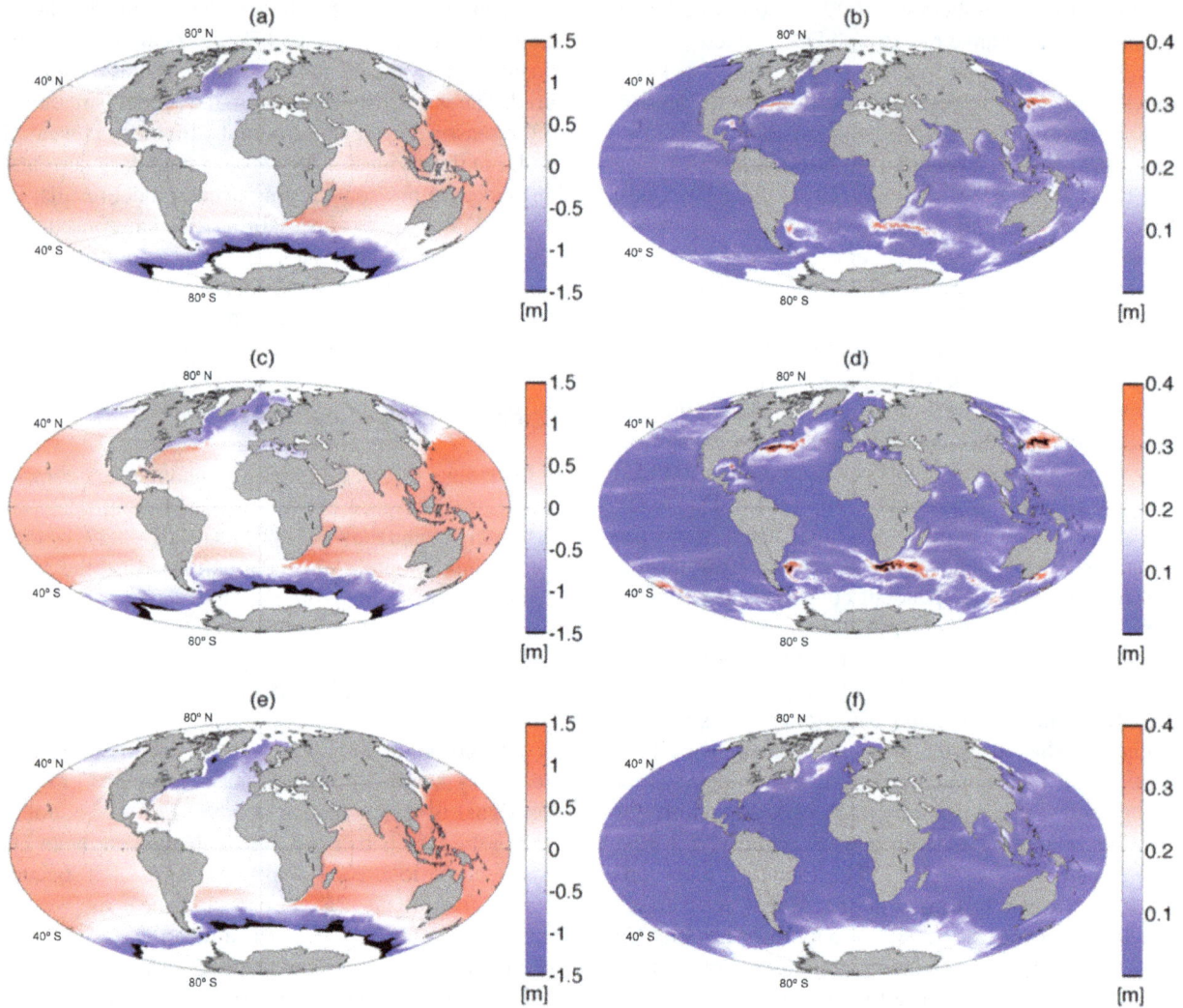

Figure 1. Mean sea surface height (SSH in meters) **(a–c)** and its standard deviation **(b–d)** over the years 1993–2012. **(a)** and **(b)** are derived from altimetry and **(c)** and **(d)** of the high-resolution simulation R_{021}. Panels **(e)** and **(f)** show the mean SSH and the standard deviation for the low-resolution simulation R_{021}^{low}, respectively.

in ocean currents, which are discussed later in this section. Regarding mean DSL patterns and amplitudes, the results of the low-resolution simulation (R_{021}^{low}), as shown in Fig. 2e, agree well with many other model studies using non-eddying ocean models (Landerer et al., 2007; Yin et al., 2009, 2010; Bordbar et al., 2015). At first sight, the results also look similar to those for the R_{021} simulation (compare Fig. 2a and e). However, when regional details are considered, the results are different. The Southern Ocean basin contrast (Indian versus Atlantic/Pacific) is much stronger in the R_{021} results. The DSL change in the Northern Atlantic is more dipolar in the North Atlantic than in the R_{021}^{low} results, with a large area south of Greenland with decreasing mean DSL. The change in DSL variability is, as expected, different in both models (compare Fig. 2b and f), in particular in western boundary current regions. In the North Atlantic, (compare Fig. 2d and h), the

changes in variability are less coherent in the Gulf Stream region and have larger amplitudes in the eastern part of the basin.

To explain the changes in DSL in the North Atlantic for the R_{021} simulation (Fig. 2c–d), the behavior of the AMOC is shown in Fig. 4. The maximum AMOC at 26° N decreases from about 20 Sv to about 5 Sv (red curve in Fig. 4a). The spatial pattern of the AMOC does not change, but the North Atlantic Deep Water shallows by about 1000 m (Fig. 4b–c). The maximum strength of the AMOC at 35° S decreases (blue curve in Fig. 4a) by more than 60 %. The decline in AMOC causes a rise in mean DSL of up to 0.4 m near the North American continent, mostly because of a redistribution of ocean mass towards these regions (see Fig. 1a). The reduction of the AMOC in the R_{021}^{low} simulation is only a few sverdrups, as in the ESSENCE ensemble (Sterl et al., 2008;

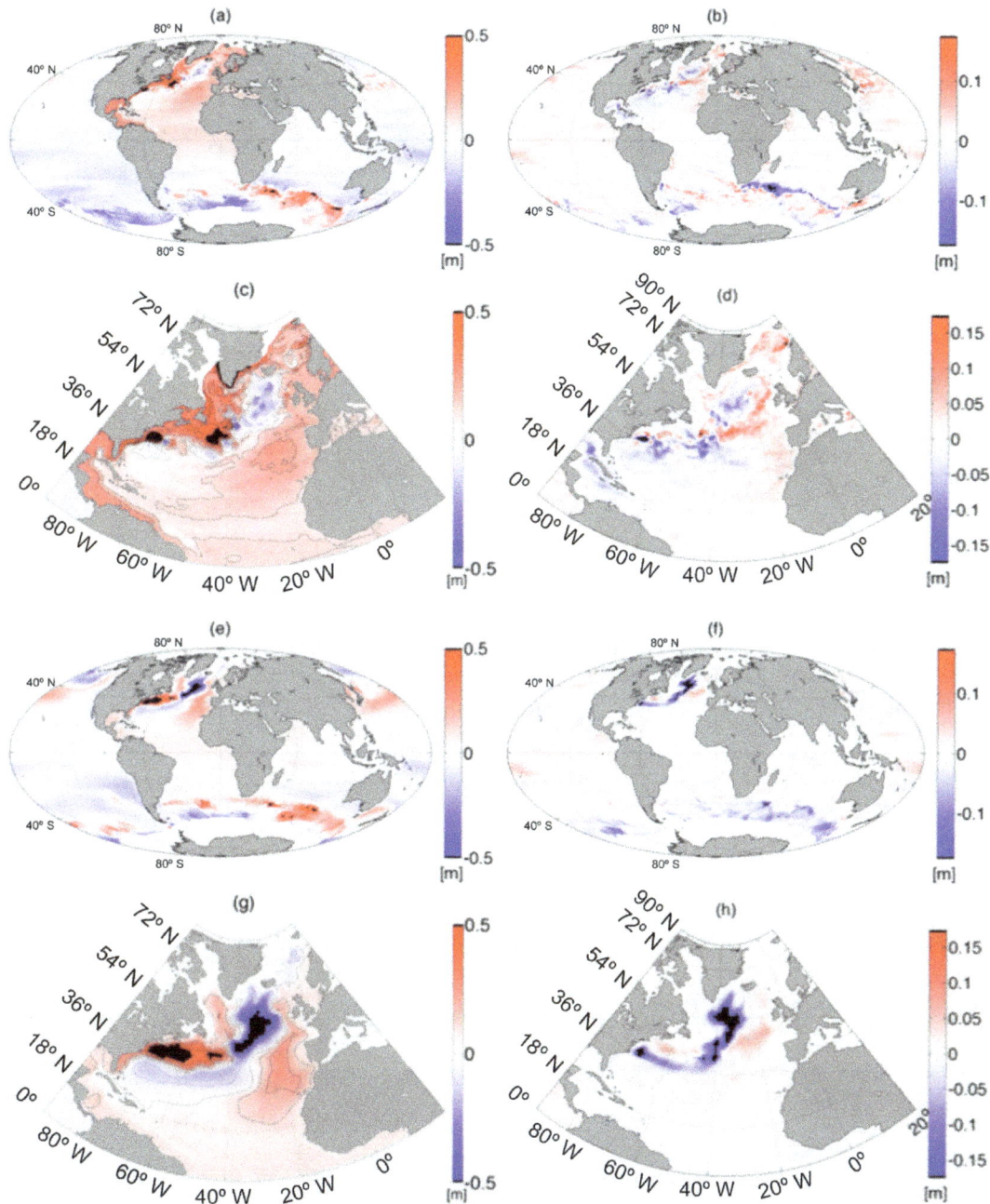

Figure 2. Change in the (a) mean and (b) standard deviation of modeled DSL in meters between the periods 2081–2100 and 2001–2020 for the R_{021} simulation. The panels (c) and (d) are magnifications of (a) and (b) for the North Atlantic region. (e–h) Same as (a–d) but for the R_{021}^{low} simulation.

Van Oldenborgh et al., 2009). The strong variations at 26° N in the R_{021}^{low} simulation are very likely due to an adjustment as a consequence of the change in forcing. At 26° N the AMOC also measures the Gulf Stream in the model, which can intensify due to a change in buoyancy gradient.

The DSL change in the Southern Ocean between the periods 2081–2100 and 2001–2020 (Fig. 2a) is caused by a southward shift of the westerly winds. In addition, the west-

erly wind stress strengthens by about 0.03 Pa (Fig. 6b). The increase in zonal momentum flux accelerates the Antarctic Circumpolar Current and increases the northward Ekman transport that changes the slope of the isopycnal surfaces in the South Atlantic (Yin et al., 2010). These effects cause changes in the water mass properties leading to steric contraction in the Southern Ocean and steric expansion in the

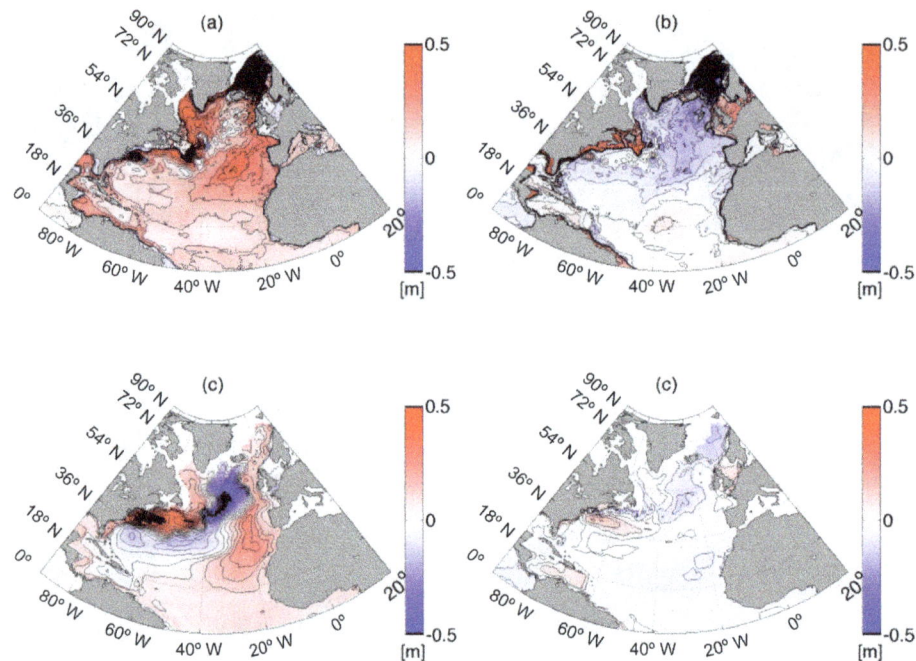

Figure 3. Change in **(a)** modeled mean steric height in meters, and change in **(b)** modeled mean ocean bottom pressure change in meters of equivalent water height between the periods 2081–2100 and 2001–2020 for the R_{021} simulation. The panels **(c)** and **(d)** are the same as **(a)** and **(b)** but for the R_{021}^{low} simulation.

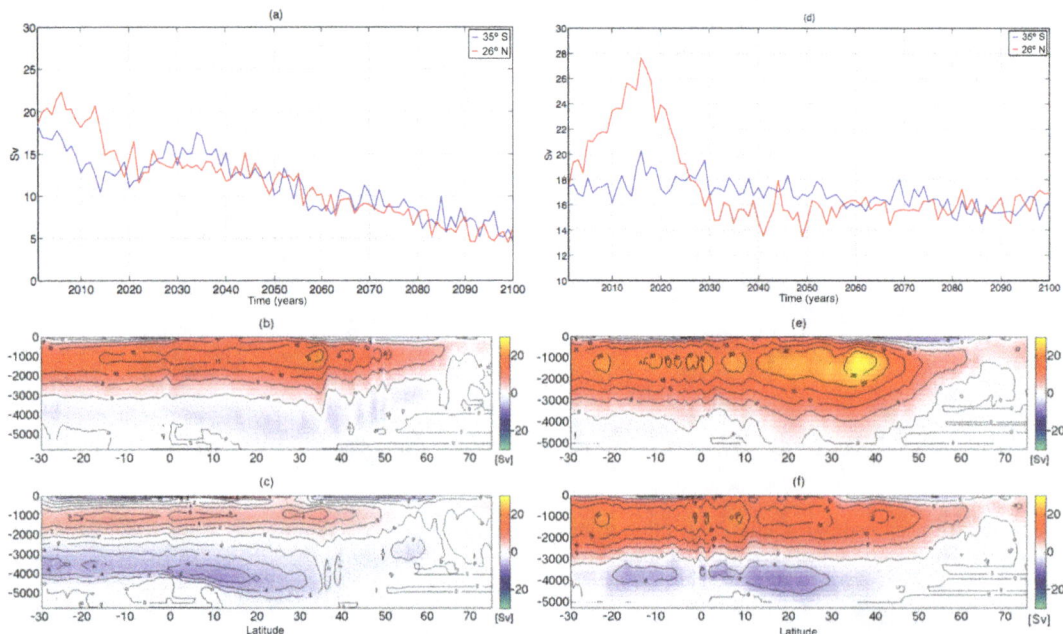

Figure 4. (a, d) Maximum AMOC strength at 35° S (blue) and 26° N (red) over the period 2000–2100 of **(a–c)** R_{021} and **(d–f)** R_{021}^{low}; **(b, e)** AMOC streamfunction (mean of years 2001–2020); **(c, f)** same as **(b)** and **(e)** but over the period 2081–2100.

region of the Agulhas return current (Yin et al., 2010), explaining the results in Fig. 2a.

The reduction of the AMOC is also associated with a northward shift of the latitude separation of the Gulf Stream.

This result has also been found in the non-eddying model studies (Landerer et al., 2007; Saba et al., 2016) and previous strongly eddying model studies (Brunnabend et al., 2014). In addition, eastward shifts of the path of the Gulf Stream and

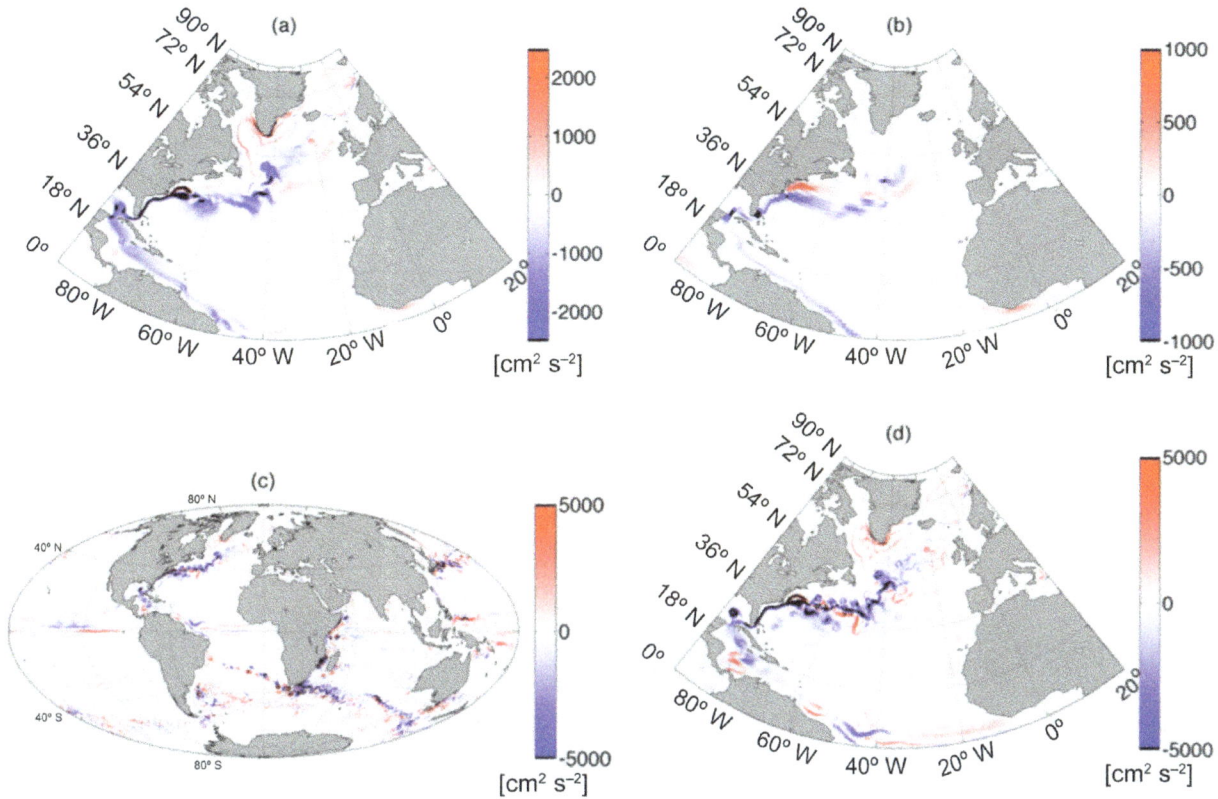

Figure 5. Difference of horizontal surface kinetic energy (energy flux per unit area) in $cm^2\,s^{-2}$ of the simulations **(a)** R_{021} and **(b)** R_{021}^{low} in the North Atlantic (mean of years 2081–2100 minus mean of years 2001–2020). **(c)** shows the difference in eddy kinetic energy (EKE) of the years 2090 and 2010 of R_{021}. Before computing EKE, the mean KE of the years 2080–2100 and 2000–2020 has been subtracted. **(d)** is the same as **(c)** showing only the North Atlantic.

North Atlantic Current occur. This is shown more clearly by the change in surface mean kinetic energy (Fig. 5a) which has decreased over most of the Gulf Stream path in the R_{021} simulation. Figure 5c and d show the change of the eddy kinetic energy (EKE) of year 2090 with respect to 2010. The changes in the mean current path redirect eddies and lead to higher variability in the eastern Atlantic while in the subpolar region the variability is reduced. In the R_{021}^{low} simulation, similar shifts in the current system in the North Atlantic occur (Fig. 5b). However, the amplitude of the kinetic energy changes is much smaller compared to the R_{021} simulation, in particular in the Labrador Sea and in the Caribbean Sea.

In the R_{021} simulation, the global mean sea surface temperature (SST) rises by about 2 °C over the period 2000–2100 (Fig. 6a). Almost all ocean regions experience a warming, and near the east coast of North America there is a warming of up to 4 °C as also shown by Saba et al. (2016). However, in the Southern Ocean, SST remains almost unchanged over large regions. This can be explained by the atmospheric forcing fields associated with the SRES-A1B scenario as they lead to changes in the radiative forcing between atmosphere and ocean. In addition, SST decreases by more than 3 °C in the subpolar gyre region of the North Atlantic. This cooling

is related to changes in deepwater formation, as discussed by Weijer et al. (2012), associated with a decrease of the AMOC strength and the shift in the currents that reduce the heat transport to the northern polar regions, which leads to thermal contraction and a negative DSL change (Fig. 2a). The dipole pattern of SST changes and the corresponding changes in DSL are robust fingerprints of AMOC weakening and are consistent with most low-resolution coupled model projections (e.g., Drijfhout et al., 2012; Lorbacher et al., 2010; Danabasoglu et al., 2012; Drinkwater et al., 2014 and others).

The reduction of the AMOC also decreases the ocean–atmosphere temperature difference in the subpolar Atlantic region and hence leads to a reduction in the net ocean–atmosphere surface heat flux, i.e., a reduced heat loss to the atmosphere (Fig. 6c; positive values: flux into the ocean). However, this heat gain is not strong enough to compensate for the cooling caused by the reduced AMOC strength and the shift in current. The overall cooling in the subpolar gyre region in the North Atlantic tends to strengthen the AMOC, but it cannot compensate for the influence of the general warming in the upper ocean. Furthermore, the cooling in this region leads to reduced evaporation resulting in a further freshening of the upper ocean (Fig. 6d) in a region where

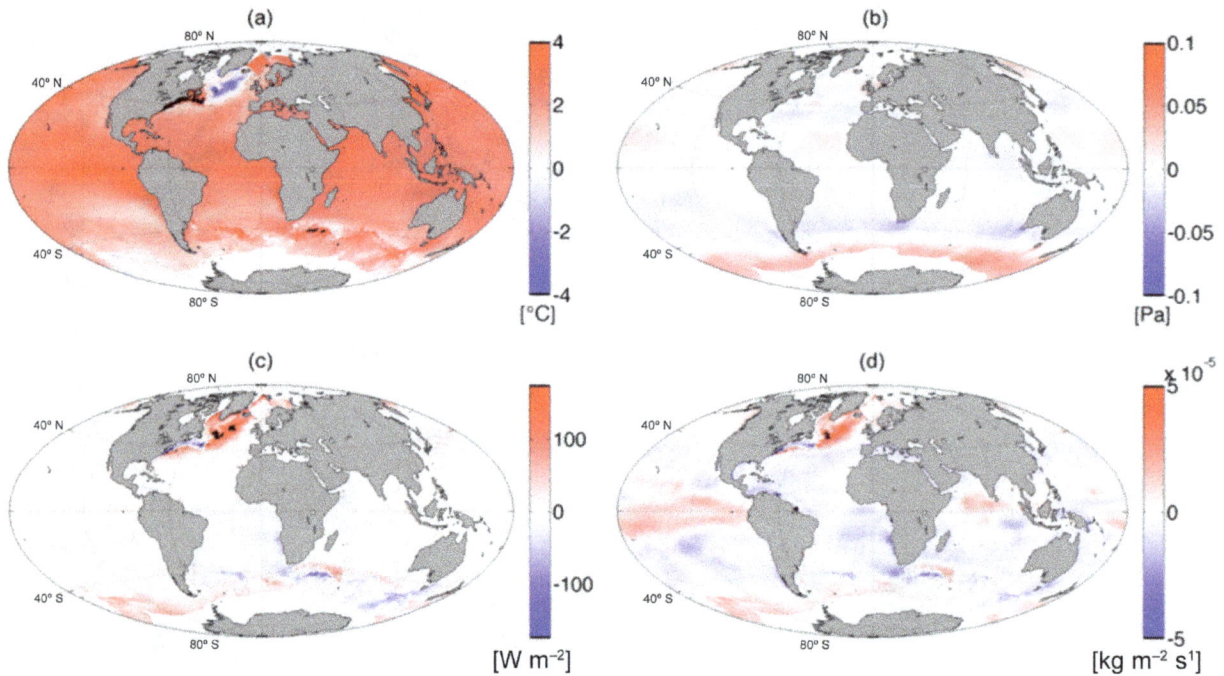

Figure 6. Change in (**a**) sea surface temperature (°C), (**b**) zonal wind stress (Pa), (**c**) surface heat flux (W m^{-2}), and (**d**) surface freshwater flux (kg m^{-2} s^{-1}) for the R_{021} simulation; again the mean over the last 20 years (2081–2100) minus that over the first 20 years (2001–2020) is shown. (positive values mean a flux into the ocean).

the AMOC is particularly sensitive to freshwater anomalies (Smith and Gregory, 2009; Weijer et al., 2012). The reduced heat loss and the additional freshening cause a further slow-down of the AMOC. The changes in surface fluxes for the simulation R_{021}^{low} (not shown) are very similar as they are derived from the same atmospheric forcing fields and are only slightly differently affected by the ocean fields, compared to the R_{021} simulation. Because the mechanism of deepwater formation is very different in the low-resolution model, the AMOC responds more mildly to changes in surface forcing than that in the high-resolution model (Weijer et al., 2012).

3.2   Regional probability density function and extremes

To determine an estimate of the probability density function (PDF) of DSL we show histograms of modeled daily-mean DSL data over two 20-year periods (2001–2020 and 2081–2100). To remove variations on long timescales, all signals with frequencies lower than 550 days are first filtered out of these DSL time series. This leaves the seasonal and annual signals in the DSL time series and hence changes on these timescales also lead to changes in the PDFs and the DSL extremes. The PDFs are computed for three different regions in the North Atlantic, i.e., in the region of the subpolar gyre, near the US east coast and near the European coast (as shown in Fig. 7) using the daily-mean maximum value (over the region) in each of the regions from the daily-mean time series. The PDFs for three specific locations near the

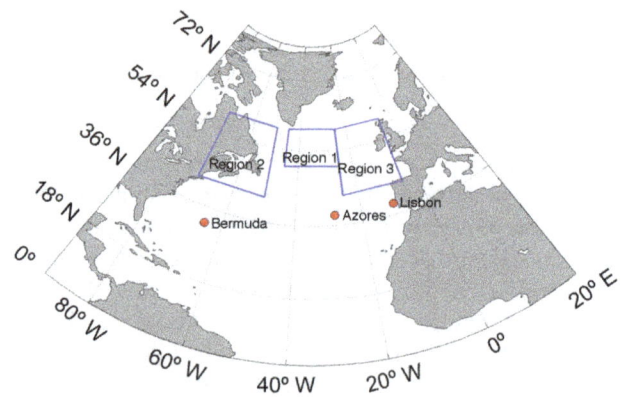

Figure 7. Regions in the North Atlantic (region of the subpolar gyre, near the US east coast and near the European coast) and locations (near Lisbon, Azores, and Bermuda) used for determining the PDFs and for the extreme value analysis.

Azores, Bermuda, and Lisbon are also computed by using the monthly maximum value (at that location) from the daily time series. A generalized extreme value (GEV) distribution function has been fitted to the PDFs using the maximum-likelihood method (Coles, 2001). It describes the behavior of the extremes using the location, scale and shape parameter in Coles (2001) and is computed in the same way as in Brunnabend et al. (2014).

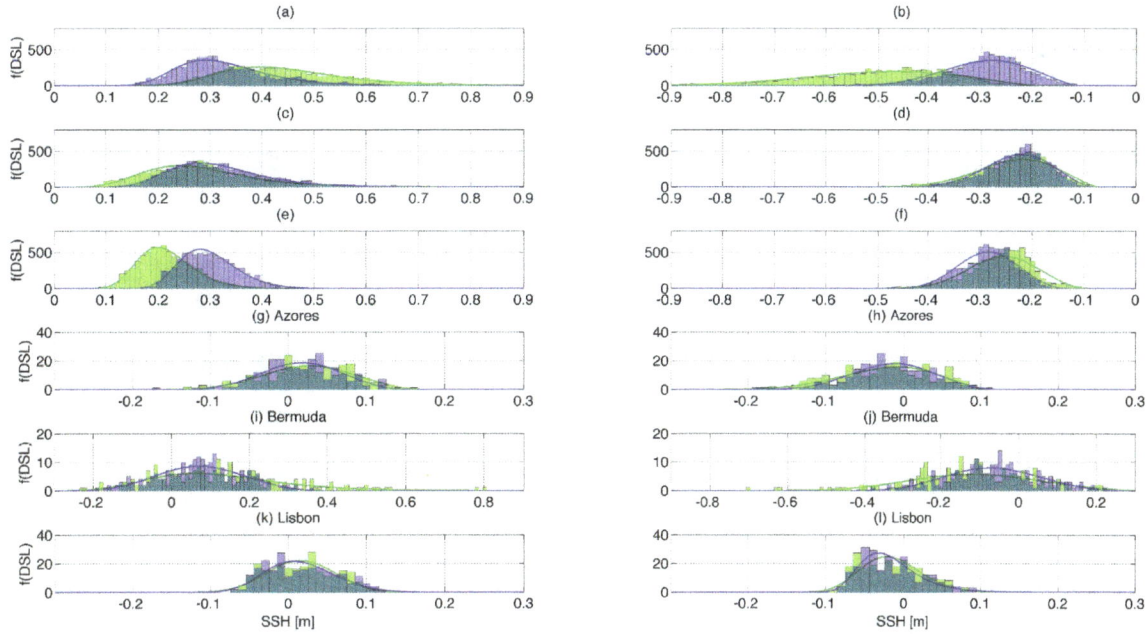

Figure 8. (a, c, e) Estimated probability density function (PDF) of daily regional maximum DSL of simulation R_{021} and **(b, d, f)** of the daily regional minimum DSL in the three different regions in the North Atlantic shown in Fig. 7 (**a** region of the subpolar gyre, **b** near the US east coast, and **c** near the European coast). In each plot, a maximum daily value over the region is identified after all variability with frequencies lower than 550 days has been filtered out. **(g–l)** Same but for the locations indicated in Fig. 7 and using **(g, i, k)** monthly maximum local DSL values and **(h, j, l)** monthly minimum local DSL values derived from daily mean time series. The green histogram is the PDF for the first 20 years (2001–2020) and the blue histogram that for the last 20 years (2081–2100). The green and blue lines are the GEV distribution function fitted to the corresponding green and blue histogram, respectively.

The changes in each PDF for the R_{021} simulation for the different regions and locations are plotted in Fig. 8 with the blue histogram being the future PDF. The variance of DSL decreases in mid-Atlantic region 1 (see Fig. 2b, d), which is seen by the shift of the PDF to the left (Fig. 8a). This also leads to a reduction of the highest DSL extremes by more than 10 cm. In region 2 (western North Atlantic), DSL is mainly driven by mean changes due to steric effects and the mass redistribution and hence the PDF shifts to the right (Fig. 8c). In the eastern North Atlantic (region 3), the variance of the DSL increases (see Fig. 2b, d) due to the changes in the pathways of eddies causing the changes in EKE (Fig. 5c, d). This leads to a rightward shift of the PDF by about 10 cm in this region (Fig. 8e). The PDF of minimum DSL in region 2 and 3 (Fig. 8d, f) shifts left indicating an intensification of eddy activity affecting the sea level change in these regions. In region 1 (Fig. 8a), the PDF of minimum DSL shifts right as the intensity of the eddy activity decreases in this region.

Changes in the pathways of eddies are also important when considering local DSL extremes. The Azores are located in a region of slightly decreased variability (Fig. 1b, d) due to reduced eddy kinetic energy in this region, shifting the PDF slightly to the left (Fig. 8g). Near Bermuda the shift in the ocean currents leads to lower probabilities of higher sea level extremes. (Fig. 8i). The most interesting result, however, is shown in Fig. 8k for the coast near Lisbon. Due to the shift in the Gulf Stream and North Atlantic Current one would expect increased probabilities for high DSL values in this region. However, because these currents are not only shifted but also reduced in strength almost no changes in DSL extremes can be identified (Fig. 8k). As the influence of ocean eddies decreases when reaching coastal region, no clear signal in the eddy intensity change can be identified at the three coastal locations (Fig. 8g–l).

The changes in the PDFs for the R_{021}^{low} simulation show quite a different behavior than those in the R_{021} simulation for most regions and locations. While the relative shift in the mean is comparable for both models in the regions 1 and 2 (Fig. 9a, b), the amplitude is much smaller for R_{021}^{low}. For region 3 (Fig. 9c), the PDF has bimodal characteristics and hardly changes under climate change, in contrast to the change in the R_{021} simulation (Fig. 8c). The PDF change for the Azores is the opposite (Fig. 9d) in both models due to the fact that the eastward shift in the Gulf Stream has no influence on ocean eddy paths in the R_{021}^{low} simulation (Fig. 5b). The PDFs of the other two locations (Fig. 9e, f) show the same behavior as in the R_{021} simulation.

From the fit of parameters in generalized extreme value (GEV) distributions, the extreme DSL values for a return

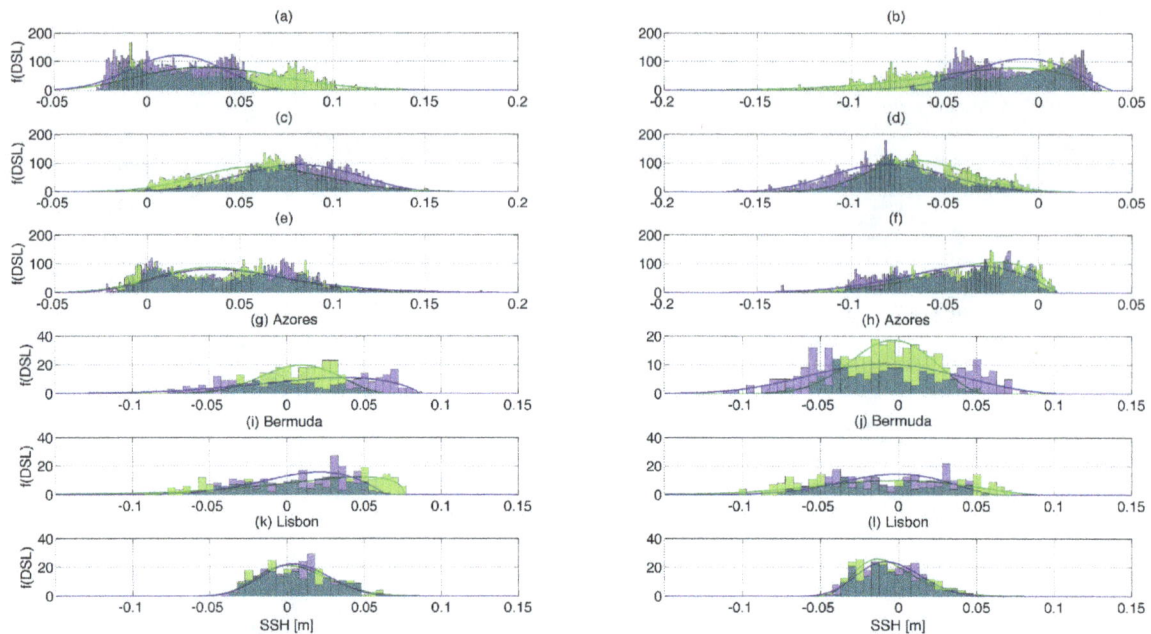

Figure 9. (a, c, e) Estimated probability density function (PDF) of daily regional maximum DSL of simulation R_{021}^{low} and **(b, d, f)** of the daily regional minimum DSL in the three different regions in the North Atlantic shown in Fig. 7 (**a** region of the subpolar gyre, **b** near the US east coast, and **c** near the European coast). In each plot, a maximum daily value over the region is identified after all variability with frequencies lower than 550 days has been filtered out. **(g–l)** Same but for the locations indicated in Fig. 7 and using **(g, i, k)** monthly maximum local DSL values and **(h, j, l)** monthly minimum local DSL values derived from daily mean time series. The green histogram is the PDF for the first 20 years (2001–2020) and the blue histogram that for the last 20 years (2081–2100). The green and blue lines are the GEV distribution function fitted to the corresponding green and blue histogram, respectively.

time of 120 months (10 years) over the period 2001–2020 and their changes over the different 20-year periods (2081–2100 and 2001–2020) of the R_{021} simulation can be determined (Fig. 10). Over the period 2000–2020 higher extreme sea levels occur in regions of high variability, i.e., in regions of the major current systems such as the Gulf Stream and the Agulhas Current (Fig. 10a, c). Therefore, the regional pattern of changes in extreme sea levels for a return time of 10 years (Fig. 10b, d) reflects the changes in sea level variability as shown in Fig. 2b and d. Sea level extremes can increase by 50 cm near Tasmania. Furthermore, in the northern and eastern North Atlantic, sea level extremes with a 10-year return time will increase by up to 20 cm. A comparison of the PDFs and the DSL extremes (for the 10-year return time) using a 550-day filter and a 180-day filter (not shown) indicates that the changes in DSL extremes are dominated by the change in short-term variability caused mainly by the shift in the ocean currents changing the eddy pathways (Fig. 5c, d).

To show that the mechanisms leading to extreme sea level change under the SRES-A1B scenario are robust, Fig. 10e–h show the change of extreme DSL values for a 10-year return time of two additional high-resolution simulations forced by the ensemble members 029 and 033. The similar pattern in the change of the extreme DSL values indicates similar

changes in behavior of the AMOC, ocean circulation, and DSL as in the R_{021} simulation.

Changes in extreme sea level values are shown in Fig. 11 for the R_{021}^{low} simulation. The amplitude of these extremes is much smaller, in particular in western boundary current regions (Fig. 11a) and in the Gulf Stream region (Fig. 11c). The low-resolution ocean model simulation leads to different extreme sea level projections in the northern North Atlantic (in particular, in the Labrador Sea and Barents Sea) than for the R_{021} simulation. The sign of the change in sea level extremes is also different in the Caribbean Sea. This shows the importance of including an explicit representation of eddy processes into an ocean model when looking at regional projections of DSL.

4 Summary and discussion

In this paper, we considered future dynamic sea level (DSL) changes using a strongly eddying ocean model forced by atmospheric fields according to an SRES A1B scenario. The results show that changes in local and regional PDFs (between the periods 2001–2020 and 2081–2100) are mainly due to changes in DSL variability on short timescales and therefore related to changes in the ocean eddy field. This can be deduced from both the changes in the eddy kinetic energy

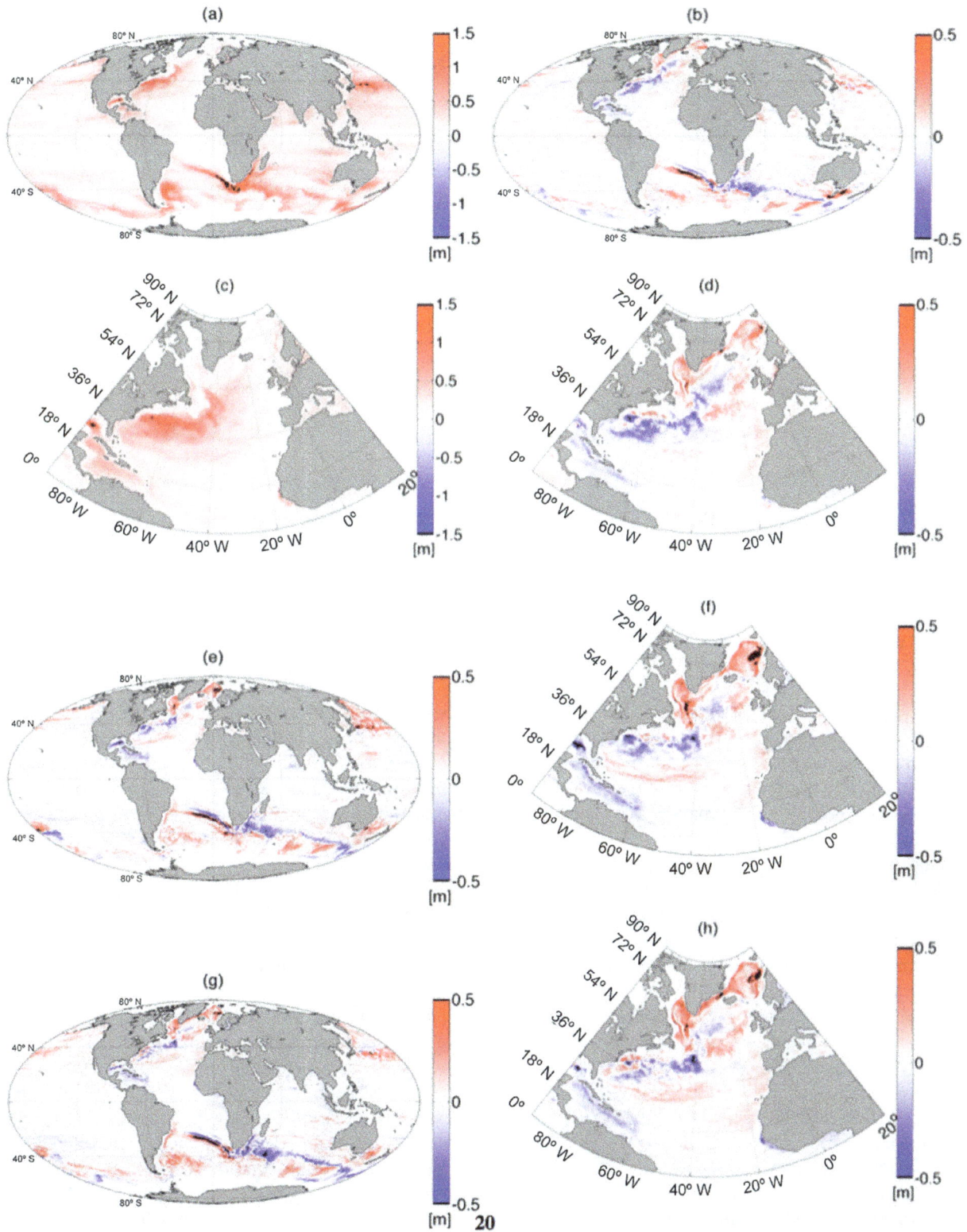

Figure 10. Extreme DSL values in meters for a 10-year return time of simulation R_{021} for **(a)** the first 20 years (2001–2020) and **(b)** the differences between the period 2081–2100 and 2001–2020. All signals with frequencies lower than 550 days have been filtered out. The panels **(c)** and **(d)** are magnifications of **(a)** and **(b)** for the North Atlantic region. **(e, f)** and **(g, h)** are the differences between the period 2081–2100 and 2001–2020 for two additional simulations forced by ensemble members 029 and 033, respectively.

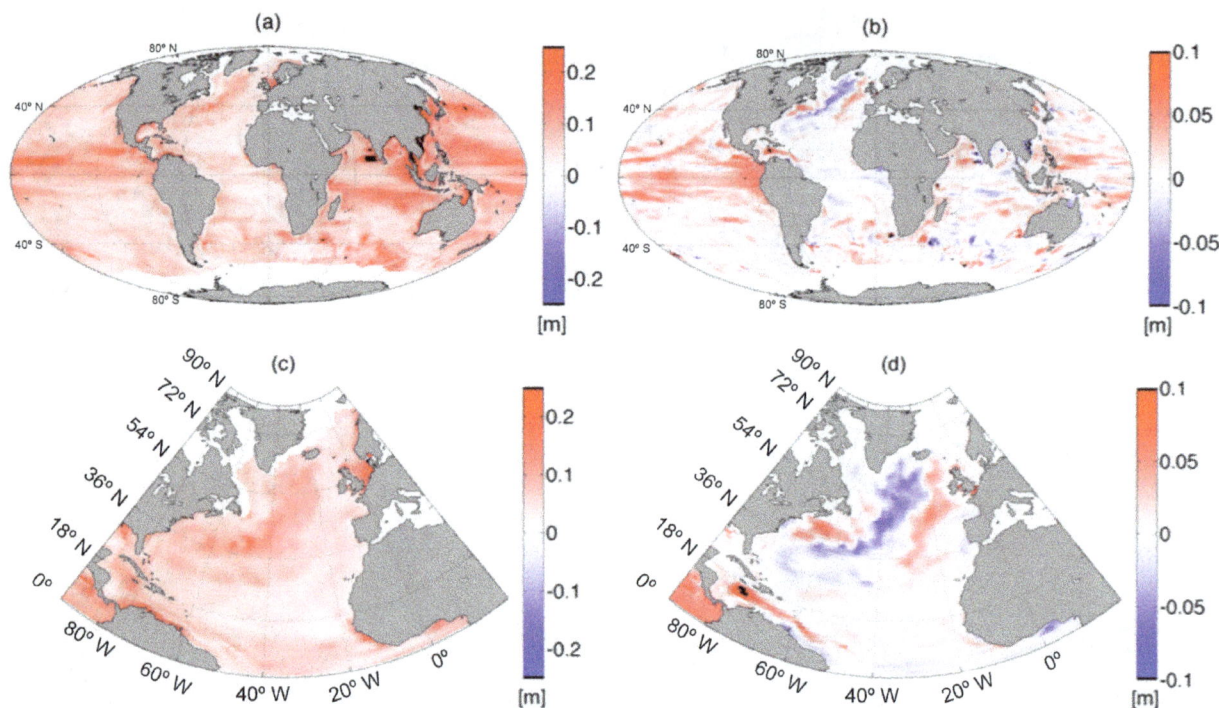

Figure 11. Extreme DSL values in meters for a 10-year return time of simulation R_{021}^{low} for **(a)** the first 20 years (2001–2020) and **(b)** the differences between the period 2081–2100 and 2001–2020. All signals with frequencies lower than 550 days have been filtered out. The panels **(c)** and **(d)** are magnifications of **(a)** and **(b)** for the North Atlantic region.

of the ocean surface velocity field and from a comparison of DSL changes in a non-eddying version of the same model. In the high-resolution model simulation, the changes in eddy pathways are caused by a strong decrease of the AMOC with simultaneous eastward shifts in the path of the Gulf Stream and the North Atlantic Current.

Our main result is that the patterns of 10-year return time DSL extremes (as shown in Fig. 10) are determined by changes in the ocean eddy field (Suzuki et al., 2005; Brunnabend et al., 2014). In the POP model, eddies can come within 100 km of the coast and their maximum sea surface signal is often strongly correlated with that at the coast. In some regions of the globe these extreme DSL values can be up to 0.5 m, which are of the same order of magnitude as the mean DSL change. This shows the importance of internal ocean variability for regional extreme sea levels, not only on longer timescales (Bordbar et al., 2015) but also on shorter timescales (Firing and Merrifield, 2004). These findings agree well with the study of Kanzow et al. (2009), where it has been shown that the influence of eddies on SSH variability is strongly reduced near ocean boundaries but may still be several centimeters.

Low-resolution ocean–climate models are not capable of accurately representing these changes in extreme sea levels. Some low-resolution model studies do capture a shift in ocean currents in the case of a declining AMOC (Landerer et al., 2007; Yin et al., 2010, 2009; Pardaens et al., 2011;

Kienert and Rahmstorf, 2012). However, the model resolution does not resolve DSL variability caused by ocean eddies, as the parameterization of eddies in these models only affects the heat and salt transport in the models. Although the use of an eddy-permitting ocean–climate model (with a 0.25° horizontal resolution) already indicated the importance of resolving ocean eddies to accurately estimate future sea level variability (Suzuki et al., 2005), the western boundary currents usually do not have a correct separation behavior in these models.

There are several caveats in this model study which may modify the results quantitatively but which do not affect the main message of this paper that strongly eddying models are important for regional future sea level change projections. First, the AMOC in the R_{021} POP model simulation appears to be quite sensitive to freshwater anomalies, and hence the scenario here may be quite an extreme one. Second, the use of an ocean-only model with mixed boundary conditions, restoring conditions below sea-ice regions, and atmospheric forcing fields from a climate model restricts the capabilities of the model in simulating the coupled ocean–atmosphere interactions occurring in reality. However, it is expected that shifts in the ocean eddy fields would also occur in coupled models with strongly eddying ocean model components. Third, the model does not simulate many other processes causing regional and coastal sea levels changes (e.g., glacial isostatic adjustment (GIA), gravity). Many of these processes would only affect the mean DSL values and not their variability. Hence, as a first approximation, these sea level changes can be added to the mean DSL values (Slangen et al., 2012, 2014) determined here. Finally, although we show robustness using a small ensemble it would be better to use a larger ensemble of simulations (Bordbar et al., 2015) to determine the effect of ocean initial conditions and to have better statistics on the extreme DSL values. The latter is still hardly feasible with the current computational capabilities.

We conclude from the results that when developing plans for adapting to future changing sea level, not only mean regional changes should be considered, although they may be substantial. Also the changes in variability should be accounted for, as with higher variability the probability of sea level extremes may increase. This in particular holds for the North Atlantic region, where many areas are vulnerable to sea level rise.

Acknowledgement. This study was supported by the Netherlands eScience Center (NLeSC) through the eSALSA (An eScience Approach to determine future Sea-level chAnges) project. The simulations were performed on the Cartesius supercomputer at SURFsara (https://www.surfsara.nl) through the project SH-243-13. This work was also partially funded by the Dutch national research program COMMIT. The altimeter products were produced by Ssalto/Duacs and distributed by Aviso, with support from CNES (http://www.aviso.altimetry.fr/duacs/). We thank the two anonymous reviewers for their constructive comments that improved the manuscript.

Edited by: M. Hecht

References

Böning, C. W., Dispert, A., Visbeck, M., Rintoul, S. R., and Schwarzkopf, F. U.: The response of the Antarctic Circumpolar Current to recent climate change, Nat. Geosci., 1, 864–869, doi:10.1038/ngeo362, 2008.

Bordbar, M. H., Martin, T., Latif, M., and Park, W.: Effects of long-term variability on projections of twenty-first-century dynamic sea level, Nat. Clim. Change, 5, 343–347, doi:10.1038/NCLIMATE2569, 2015.

Bouttes, N., Gregory, J. M., Kuhlbrodt, T., and Smith, R. S.: The drivers of projected North Atlantic sea level change, Clim. Dynam., 43, 5, 1531–1544, doi:10.1007/s00382-013-1973-8, 2013.

Brunnabend, S.-E., Dijkstra, H. A., Kliphuis, M. A., van Werkhoven, B., Bal, H. E., Seinstra, F., Maassen, J., and van Meersbergen, M.: Changes in extreme regional sea surface height due to an abrupt weakening of the Atlantic meridional overturning circulation, Ocean Sci., 10, 881–891, doi:10.5194/os-10-881-2014, 2014.

Church, J. A. and White, N. J.: Sea-Level Rise from the Late 19th to the Early 21st Century, Surv. Geophys., 32, 585–602, doi:10.1007/s10712-011-9119-1, 2011.

Church, J. A., Clark, P. U., Cazenave, A., Gregory, J. M., Jevrejeva, S., Levermann, A., Merrifield, M. A., Milne, G. A., Nerem, R. S., Nunn, P. D., Payne, A. J., Pfeffer, W. T., Stammer, D., and Unnikrishnan, A. S.: Sea Level Change, in: The Scientific Basis, Contribution of Working Group I to the Fifth Assessment Report of the Intergovernmental Panel on Climate Change, edited by: Stocker, T. F., Qin, D., Plattner, G.-K., Tignor, M., Allen, S. K., Boschung, J., Nauels, A., Xia, Y., Bex, V., and Midgley, P. M., Cambridge University Press, Cambridge, New York, 1137–1216, 2013.

Coles, S.: An Introduction to Statistical Modeling of Extreme Values, Springer-Verlag London Ltd., ISBN: 1-85233-459-2, 2001.

Danabasoglu, G., Yeager, S. G., Kwon, Y.-O., Tribbia, J. J., Phillips, A. S., and Hurrell, J. W.: Variability of the Atlantic Meridional Overturning Circulation in CCSM4, J. climate, 25, 5153–5172, doi:10.1175/JCLI-D-11-00463.1, 2012.

Drijfhout, S., Oldenborgh, G. J. V., and Cimatoribus, A.: Is a Decline of AMOC Causing the Warming Hole above the North Atlantic in Observed and Modeled Warming Patterns?, J. Climate, 25, 8373–8379, doi:10.1175/JCLI-D-12-00490.1, 2012.

Drinkwater, K. F., Miles, M., Medhaug, I., Ottera, O. H., T.Kristiansen, Sundby, S., and Gao, Y.: The Atlantic Multidecadal Oscillation: Its manifestations and impacts with special emphasis on the Atlantic region north of 60° N, J. Marine Syst., 133, 117–130, doi:10.1016/j.jmarsys.2013.11.001, 2014.

Firing, Y. L. and Merrifield, M. A.: Extreme sea level events at Hawaii: Influence of mesoscale eddies, Geophys. Res. Lett., 31, L24306 doi:10.1029/2004GL021539, 2004.

Gent, P. R. and McWilliams, J. C.: Isopycnal mixing in ocean circulation models, J. Phys. Oceanogr., 20, 150–155, 1990.

Greatbatch, R. J.: A note on the representation of steric sea level in models that conserve volume rather than mass, J. Geophys. Res., 99, 12767–12771, 1994.

Kanzow, T., Johnson, H. L., Marshall, D. P., Cunningham, S. A., Hirschi, J.-M., Bryden, H. L., and Johns, W. E.: Besinwide Integration Volume Transports in an Eddy-Filled Ocean, J. Phys. Oceanogr., 39, 3091–3110, doi:10.1175/2009JPO4185.1, 2009.

Kienert, H. and Rahmstorf, S.: On the relation between Meridional Overturning Circulation and sea-level gradients in the Atlantic, Earth Syst. Dynam., 3, 109–120, doi:10.5194/esd-3-109-2012, 2012.

Landerer, F. W., Jungclaus, J. H., and Marotzke, J.: Regional Dynamic and Steric Sea Level Change in Response to the IPCC-A1B Scenario, J. Phys. Oceanogr., 37, 296–312, doi:10.1175/JPO3013.1, 2007.

Large, W. G. and Yeager, S. G.: Diurnal to decadal global forcing for ocean and sea-ice models; the datasets and flux climatologies, NCAR Technical Note TN 460 STR, available at: http://www.clivar.org/organization/wgomd/resources/core/core-i (last access: 8 October 2008), 2004.

Le Bars, D., Durgadoo, J. V., Dijkstra, H. A., Biastoch, A., and De Ruijter, W. P. M.: An observed 20-year time series of Agulhas leakage, Ocean Sci., 10, 601–609, doi:10.5194/os-10-601-2014, 2014.

Le Bars, D., Viebahn, J. P., and Dijkstra, H. A.: A Southern Ocean mode of multidecadal variability, Geophys. Res. Lett., 43, 2102–2110, doi:10.1002/2016GL068177, 2016.

Lorbacher, K., Dengg, J., Böning, C. W., and Biastoch, A.: Regional Patterns of Sea Level Change Related to Interannual Variability and Multidecadal Trends in the Atlantic Meridional Overturning Circulation, J. Climate, 23, 4243–4254, doi:10.1175/2010JCLI3341.1, 2010.

Maltrud, M., Bryan, F., Hecht, M., Hunke, E., Ivanova, D., McClean, J., and Peacock, S.: Global Ocean Modeling in the Eddying Regime Using POP, CLIVAR Exchange, 44, 5–8, 2008.

Pardaens, A. K., Gregory, J. M., and Lowe, J. A.: A model study of factors influencing projected changes in regional sea level over the twenty-first century, Clim. Dynam., 36, 2015–2033, doi:10.1007/s00382-009-0738-x, 2011.

Rhein, M., Rintoul, S., Aoki, S., Campos, E., Chambers, D., Feely, R. A., Gulev, S., Johnson, G. C., Josey, S., Kostianoy, A., Mauritzen, C., Roemmich, D., Talley, L. D., and Wang, F.: Observations: Ocean, in: The Scientific Basis, Contribution of Working Group I to the Fifth Assessment Report of the Intergovernmental Panel on Climate Change, edited by: Stocker, T. F., Qin, D., Plattner, G.-K., Tignor, M., Allen, S. K., Boschung, J., Nauels, A., Xia, Y., Bex, V., and Midgley, P. M., Cambridge University Press, Cambridge, New York, 255–316, 2013.

Saba, V. S., Griffies, S. M., Anderson, W. G., Winton, M., Alexander, M. A., Delworth, T. L., Hare, J. A., Harrison, M. J., Rosati, A., Vecchi, G. A., and Zhang, R.: Enhanced warming of the Northwest Atlantic Ocean under climate change, J. Geophys. Res.-Oceans, 121, 118–132, doi:10.1002/2015JC011346, 2016.

Slangen, A., Katsman, C. A., van de Wal, R. S. W., Vermeersen, L. L. A., and Riva, R. E. M.: Towards regional projections of twenty-first century sea-level change based on IPCC SRES scenarios, Clim. Dynam., 38, 1191–1209, doi:10.1007/s00382-011-1057-6, 2012.

Slangen, A., Carson, M., Katsman, C. A., van de Wal, R. S. W., Koehl, A., Vermeersen, L. L. A., and Stammer, D.: Projecting twenty-first century regional sea-level changes, Clim. Change, 124, 317–332, doi:10.1007/s10584-014-1080-9, 2014.

Smith, R. S. and Gregory, J. M.: A study of the sensitivity of ocean overturning circulation and climate to freshwater input in different regions of the North Atlantic, Geophys. Res. Lett., 26, L15701 doi:10.1029/2009GL038607, 2009.

Sterl, A., Severijns, C., van Oldenborgh, G. J., Dijkstra, H., Hazeleger, W., van den Broeke, M., Burgers, G., van den Hurk, B., van Leeuwen, P., and van Velthoven, P.: When can we expect extremely high surface temperatures?, Geophys. Res. Lett., 35, L14703 doi:10.1029/2008GL034071, 2008.

Suzuki, T., Hasumi, H., Sakamoto, T. T., Nishimura, T., Abe-Ouchi, A., Segawa, T., Okada, N., Oka, A., and Emori, S.: Projection of future sea level and its variability in a high-resolution climate model: Ocean processes and Greenland and Antarctic ice-melt contributions, Geophys. Res. Lett., 32, L19706 doi:10.1029/2005GL023677, 2005.

van Oldenborgh, G. J., te Raa, L. A., Dijkstra, H. A., and Philip, S. Y.: Frequency- or amplitude-dependent effects of the Atlantic meridional overturning on the tropical Pacific Ocean, Ocean Sci., 5, 293–301, doi:10.5194/os-5-293-2009, 2009.

Watson, C. S., White, N. J., Church, J. A., King, M. A., and Burgette, R. J.: Unabated global mean sea-level rise over the satellite altimeter era, Nat. Clim. Change, 5, 565–569, doi:10.1038/NCLIMATE2635, 2015.

Weaver, A. J., Sedlacek, J., Eby, M., Alexander, K., Crespin, E., Fichefet, T., Philippon-Berthier, G., Joos, F., Kawamiya, M., Matsumoto, K., andK. Tachiiri, M. S., Tokos, K., Yoshimori, M., and Zickfeld, K.: Stability of the Atlantic meridional overturning circulation: A model intercomparison, Geophys. Res. Lett., 39, L20709 doi:10.1029/2012GL053763, 2013.

Weijer, W., Maltrud, M. E., Hecht, M. W., Dijkstra, H. A., and Kliphuis, M. A.: Response of the Atlantic Ocean circulation to Greenland Ice Sheet melting in a strongly-eddying ocean model, Geophys. Res. Lett., 39, L09606, doi:10.1029/2012GL051611, 2012.

Yin, J., Schlesinger, M. E., and Stouffer, R. J.: Model projections of rapid sea-level rise on the northeast coast of the United States, Nat. Geosci., 2, 262–266, doi:10.1038/ngeo462, 2009.

Yin, J., Griffies, S. M., and Stouffer, R. J.: Spatial Variability of Sea Level Rise in Twenty-First Century Projections, J. Climate, 23, 4585–4607, doi:10.1175/2010JCLI3533.1, 2010.

2

Measuring pH variability using an experimental sensor on an underwater glider

Michael P. Hemming[1,2], **Jan Kaiser**[1], **Karen J. Heywood**[1], **Dorothee C.E. Bakker**[1], **Jacqueline Boutin**[2],
Kiminori Shitashima[3], **Gareth Lee**[1], **Oliver Legge**[1], **and Reiner Onken**[4]

[1]Centre for Ocean and Atmospheric Sciences, School of Environmental Sciences, University of East Anglia, Norwich
Research Park, Norwich NR4 7TJ, UK
[2]Laboratoire d'Océanographie et du Climat, 4 Place Jussieu, 75005 Paris, France
[3]Tokyo University of Marine Science and Technology, 4-5-7 Konan, Minato, Tokyo 108-0075, Japan
[4]Helmholtz-Zentrum Geesthacht, Max-Planck-Straße 1, 21502 Geesthacht, Germany

Correspondence to: Michael P. Hemming (m.hemming@uea.ac.uk)

Abstract. Autonomous underwater gliders offer the capability of measuring oceanic parameters continuously at high resolution in both vertical and horizontal planes, with timescales that can extend to many months. An experimental ion-sensitive field-effect transistor (ISFET) sensor measuring pH on the total scale was attached to a glider during the REP14-MED experiment in June 2014 in the Sardinian Sea in the northwestern Mediterranean. During the deployment, pH was sampled at depths of up to 1000 m along an 80 km transect over a period of 12 days. Water samples were collected from a nearby ship and analysed for dissolved inorganic carbon concentration and total alkalinity to derive the pH for validating the ISFET sensor measurements. The vertical resolution of the pH sensor was good (1 to 2 m), but stability was poor and the sensor drifted in a non-monotonous fashion. In order to remove the sensor drift, a depth-constant time-varying offset was applied throughout the water column for each dive, reducing the spread of the data by approximately two-thirds. Furthermore, the ISFET sensor required temperature- and pressure-based corrections, which were achieved using linear regression. Correcting for this decreased the apparent sensor pH variability by a further 13 to 31 %. Sunlight caused an apparent sensor pH decrease of up to 0.1 in surface waters around local noon, highlighting the importance of shielding the sensor from light in future deployments. The corrected pH from the ISFET sensor is presented along with potential temperature, salinity, potential density anomalies (σ_θ), and dissolved oxygen concentrations ($c(O_2)$) measured by the glider, providing insights into the physical and biogeochemical variability in the Sardinian Sea. The pH maxima were identified close to the depth of the summer chlorophyll maximum, where high $c(O_2)$ values were also found. Longitudinal pH variations at depth ($\sigma_\theta > 28.8 \, \mathrm{kg\,m^{-3}}$) highlighted the variability of water masses in the Sardinian Sea. Higher pH was observed where salinity was > 38.65, and lower pH was found where salinity ranged between 38.3 and 38.65. The higher pH was associated with saltier Levantine Intermediate Water, and it is possible that the lower pH was related to the remineralisation of organic matter. Furthermore, shoaling isopycnals closer to shore coinciding with low pH and $c(O_2)$, high salinity, alkalinity, dissolved inorganic carbon concentrations, and chlorophyll fluorescence waters may be indicative of upwelling.

1 Introduction

It is estimated that one-third of the anthropogenic carbon dioxide emitted between 2004 to 2013 was absorbed by the oceans (Le Quéré et al., 2015). Normally unreactive in the atmosphere, carbon dioxide dissolved in seawater ($CO_2(aq)$) takes part in a number of chemical reactions. In particular, carbonic acid (H_2CO_3) forms as a result of ($CO_2(aq)$) reacting with water, which dissociates into bicarbonate (HCO_3^-) and carbonate (CO_3^{2-}). These seawater carbonate species are referred to as dissolved inorganic carbon concentrations

(c(DIC), with "c" representing a concentration) with HCO_3^- accounting for 90 % of c(DIC). During the dissociation of carbonate species, hydrogen ions (H^+) are released. The pH is calculated as the negative decadal logarithm of the activity (commonly referred to as a concentration) of these H^+ ions. Thus the pH in the ocean is directly related to the activity of H^+ ions in the water (Zeebe and Wolf-Gladrow, 2001).

The pH varies on timescales spanning less than 1 day (Hofmann et al., 2011) to many years (Rhein et al., 2013). Since before the industrial revolution (year 1760), the global surface ocean pH has fallen from 8.21 to 8.10 (corresponding to a 30 % increase in H^+ ion activity) as a result of the atmospheric CO_2 mole fraction increasing by more than $100\,\mu mol\,mol^{-1}$ (Doney et al., 2009; Fabry et al., 2008). Future projections of anthropogenic carbon dioxide emissions suggest that ocean uptake of CO_2 will continue for many decades, thus contributing to long-term ocean acidification (Rhein et al., 2013). This may have a significant effect on marine organisms, such as calcifying phytoplankton (e.g. coccolithophores) and corals (e.g. scleractinian), dependent on the solubility state of calcium carbonate (Doney et al., 2009).

Since 1989, it has been possible to measure pH to an accuracy of 10^{-3} using a spectrophotometric approach (Byrne and Breland, 1989). Although there have been some advances in adapting this approach to measure pH autonomously in situ (Martz et al., 2003; Aßmann et al., 2011), spectrophotometry is largely used for shipboard measurements, as it requires the use of indicator dye and a means to measure spectrophotometric blanks, which is challenging outside of a laboratory (Martz et al., 2010).

A limited number of hydrographic surveys have been undertaken, and stations offering long-term time series of pH are available (Rhein et al., 2013). However, there is a drive to improve spatial and temporal data coverage via autonomous means, similar to what was experienced for temperature and salinity with Argo floats 16 years ago (Roemmich et al., 2003). There is demand to develop a reliable autonomous sensor with a precision and accuracy of 10^{-3} that is affordable to the scientific community (Johnson et al., 2016).

The Mediterranean Sea comprises just 0.8 % of the global oceanic surface, but it is regarded as an important sink for anthropogenic carbon due to its physical and biogeochemical characteristics (Álvarez et al., 2014). Between 1995 and 2012, surface c(DIC) increased by $3\,\mu mol\,kg^{-1}\,a^{-1}$ in the northwest Mediterranean Sea, which is consistent with a rise in temperature of $0.06\,°C\,a^{-1}$ and a decrease in pH of $0.003\,a^{-1}$ (Yao et al., 2016). In contrast, the pH in the neighbouring North Atlantic Ocean decreased by just $0.0017\,a^{-1}$ associated with an increase in c(DIC) of around $1.4\,\mu mol\,kg^{-1}\,a^{-1}$ and a temperature rise of $0.01\,°C\,a^{-1}$ (Bates et al., 2012). The greater potential of the Mediterranean Sea to store anthropogenic carbon can be explained by its higher alkalinity, warmer temperatures, and thus lower Revelle factor (Álvarez et al., 2014; Touratier and Goyet,

2011) when compared with other oceans, such as the North Atlantic.

The pH in the Mediterranean Sea is typically higher than most other oceanic regions (Álvarez et al., 2014). The pH on the total scale normalised to $25\,°C$ ($pH_{T,25}$) collected by ship between 1998 and 1999 within the northwestern Mediterranean Sea varied between 7.92 and 8.04 at the surface and between 7.9 and 7.93 at depths greater than 100 m (Copin-Montégut and Bégovic, 2002). When considering the Mediterranean Sea as a whole, $pH_{T,25}$ obtained by ship in April 2011 varied between 7.98 and 8.02 at the surface and between 7.88 and 7.96 at greater depths (Álvarez et al., 2014). The peak-to-peak amplitude of the annual pH cycle in the northwest Mediterranean Sea is typically 0.1 with maxima and minima in spring and summer, respectively (Yao et al., 2016).

Measurements of pH with higher temporal resolution, such as those measured by in situ sensors, can vary greatly depending on their location and depth. Hofmann et al. (2011) presented the results of 15 deployments using SeaFET pH sensors (Sea-Bird Scientific) close to the surface at a number of locations worldwide. They found that pH could vary by as much as 1.1 in extreme environments, such as those obtained close to volcanic CO_2 vents off the coast of Italy, or as little as 0.02 in open ocean areas, such as in the temperate eastern Pacific Ocean, over a time period of 30 days. Hofmann et al. (2011) were able to capture diel cycles in pH, with the most consistent variations found in coral reef locations. The pH was at a maximum in the early evening and at a minimum in the morning with amplitudes between 0.1 and 0.25; this is similar in range to other studies based in subtropical estuaries (Yates et al., 2007).

Autonomous underwater gliders offer the possibility to observe the oceanic system with a greater level of detail on both temporal and spatial scales when compared with ship measurements (Eriksen et al., 2001). A low consumption of battery power and a great degree of manoeuvrability enable such vehicles to cover large areas and profile depths of up to 1000 m during missions that can last from weeks to months. They are suitable platforms for a range of sensors, measuring both physical and biogeochemical parameters (Piterbarg et al., 2014; Queste et al., 2012).

This paper is a contribution to the special issue "REP14-MED: A Glider Fleet Experiment in a Limited Marine Area". The main goal of this paper is to describe the trial of a novel ion-sensitive field-effect transistor (ISFET) pH sensor, which was attached to an autonomous underwater glider in the northwest Mediterranean Sea during the REP14-MED sea experiment. The secondary objective is to provide a method of correcting the pH measured by this sensor and to discuss the spatial and temporal variability observed. The experiment, the glider sensors including the ISFET sensor, and the method of validation are described in Sect. 2. The ship-based data are presented in Sect. 3.1, and a comparison between ship and glider measurements is made in Sect. 3.2.1.

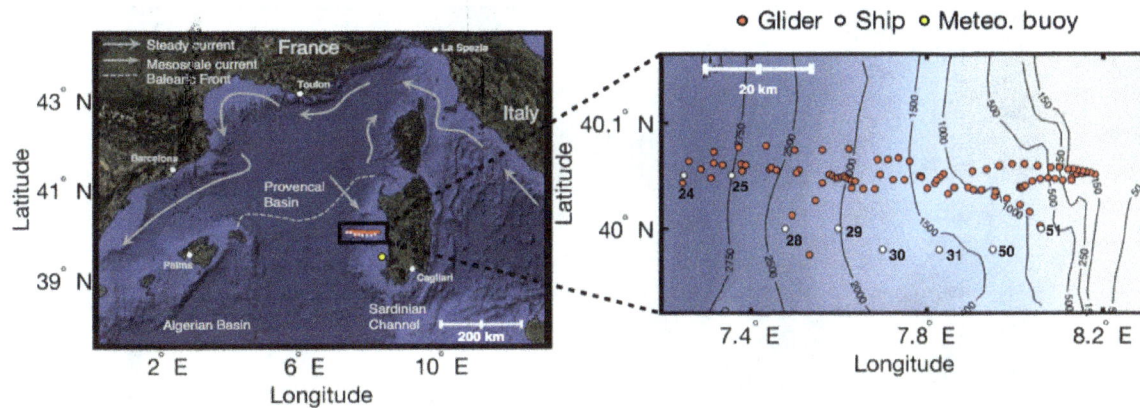

Figure 1. The locations of the 93 dives undertaken by the Seaglider (red markers), the eight ship CTD casts in which water samples were obtained (white markers), and meteorological buoy M1 (yellow marker) within the REP14-MED observational domain off the coast of Sardinia, Italy between 11 and 23 June 2014. GEBCO 1 min resolution bathymetry data (metres) were used (http://www.bodc.ac.uk/projects/international/gebco/), and surface circulation patterns were adapted from figures presented by Millot (1999).

The initial pH results and validation, the method of further correcting pH, and an artefactual light-induced effect are described in Sect. 3.2.2. Corrected pH measurements are analysed alongside other collected parameters in Sect. 3.2.3, and the conclusions are presented in Sect. 4.

2 Methodology

2.1 REP14-MED sea trial

This trial took place between 6 and 25 June 2014 in the northwest Mediterranean Sea off the coast of Sardinia, Italy (Fig. 1). This was part of the Environmental Knowledge and Operational Effectiveness (EKOE) research programme led by the North Atlantic Treaty Organisation (NATO) Centre for Maritime Research and Experimentation (CMRE) based in La Spezia, Italy. This was the fifth Recognised Environmental Picture (REP) trial, which was jointly conducted by two research vessels: the NRV *Alliance* and the RV *Planet*.

Eleven gliders with varying pressure tolerances were deployed during the trial, each making repeated west–east transects separated by roughly 0.13° of latitude from one another within the REP14-MED observational domain. One of these gliders was operated by the University of East Anglia (UEA): an iRobot Seaglider model 1KA (SN 537) with an ogive fairing. All gliders were deployed to meet the objectives of the trial, such as to improve ocean forecasting techniques (e.g. model validation and evaluation of forecasting skill), conduct a cost–benefit analysis of autonomous gliders, analyse mesoscale and sub-mesoscale features, and test new glider payloads. The latter objective was perhaps the most relevant to the deployment of the UEA glider. A more in-depth overview of the REP14-MED trial, its objectives, and the collected observational data is described by Onken et al. (2016).

The UEA glider completed a total of 126 dives between 11 and 23 June 2014. The first 24 dives did not record pH and the last 9 dives were very shallow, leaving 93 usable dives. Successive dives were approximately 2 to 4 km apart, descending to depths of up to 1000 m.

2.2 Glider sensors

Conductivity, pressure, and in situ temperature measurements were obtained by the glider using a Sea-Bird Scientific glider payload CTD sensor (Bellevue, WA, USA; Fig. 2). These measurements were then used to obtain potential temperature (θ) and practical salinity.

Dissolved oxygen concentrations ($c(O_2)$, where "c" refers to a concentration) were measured using an Aanderaa 4330 oxygen optode sensor (Aanderaa Data Instruments, Bergen, Norway) positioned towards the rear of the glider fairing (Fig. 2). The method of calibrating $c(O_2)$ closely followed that described by Binetti (2016), using the oxygen sensor engineering parameters TCPhase and CalPhase, which will be summarised here. The first step involved correcting $c(O_2)$ to account for the response time (τ) of the sensor, as the diffusion of oxygen across the silicon foil of the sensor is not an instantaneous process. Each oxygen sensor has a different τ, which depends on the structure, thickness, age, and usage of the foil (McNeil and D'Asaro, 2014), as well as the external environmental conditions such as temperature. An average τ of 17 s was obtained using the method outlined by Binetti (2016). After correcting the TCPhase for τ, glider TCPhase profiles were matched in time and space with pseudo-CalPhase profiles back-calculated from the measurements of $c(O_2)$ obtained by the ship Sea-Bird Scientific SBE 43 sensor (CTD package) using the manufacturer's set of optode calibration equations. The relationship between the glider TCPhase and the ship pseudo-CalPhase was established, and the calculated slope and offset coefficients were

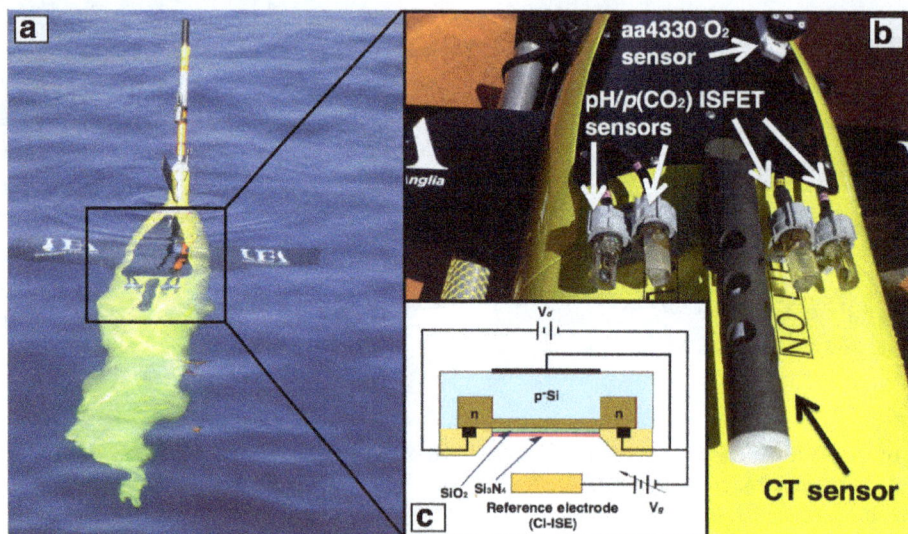

Figure 2. (a) Seaglider SN 537 during the deployment: **(b)** a close-up of the sensors and **(c)** a schematic diagram of the ISFET sensor adapted from Shitashima (2010).

used to correct the glider CalPhase, which is required for calibrating $c(O_2)$ measurements. A comparison between the ship $c(O_2)$ measurements and the calibrated glider $c(O_2)$ measurements is made in Sect. 3.2.1.

Glider variables have been processed using an open-source MATLAB-based toolbox (https://bitbucket.org/bastienqueste/uea-seaglider-toolbox/) in order to correct for differing timestamp allocations, sensor lags (Garau et al., 2011; Bittig et al., 2014), and to tune the hydrodynamical flight model (Frajka-Williams et al., 2011). Outliers outside of a specified range (e.g. 6 standard deviations) were flagged and not used for analysis, and glider profiles were smoothed using a lowess low-pass filter with a span of 5 data points (< 4 m range), which implements a local regression using weighted linear least squares and a first-order polynomial linear model. Individual profiles were inspected afterwards to ensure that potentially correct data points were not removed.

The ISFET pH sensor used in this study (Fig. 2) was custom built by a working group led by Kiminori Shitashima at the Tokyo University of Marine Science and Technology (previously the University of Kyushu), and it is not commercially available. The ISFET unit was housed in acrylic resin material. The ISFET unit and the reference chlorine ion-selective electrode (Cl-ISE) were moulded with epoxy resin in the custom-built housing. The ISFET pH unit was stand-alone, meaning that the sensor was not integrated into any of the onboard glider electronics. The power source of the sensor was 10.5 V supplied by three 3.5 V Li-ion batteries.

The glider also carried another ISFET pH sensor that was integrated into the glider electronics (Fig. 2) and two $p(CO_2)$ sensors (Shitashima, 2010), one stand-alone and one inte-

grated. The data retrieved from the integrated pH sensor and the $p(CO_2)$ sensors could not be used due to quality issues. We think the regular on/off cycling of power to the integrated dual pH–$p(CO_2)$ sensor between sampling did not allow it to function properly. In future, we would suggest the addition of backup batteries to supply power to the sensor between sampling. The cause of the problem with the stand-alone $p(CO_2)$ unit is unclear.

To measure pH, the activity of H^+ ions is determined using the interface potential between the Cl-ISE and the semi-conducting ion-sensing transistor coated with silicon dioxide (SiO_2) and silicon nitride (Si_3N_4). The ISFET pH sensor was previously found to have a response time of a few seconds with an accuracy of 0.005 pH and suitable temperature and pressure sensitivities (Shitashima et al., 2002, 2013; Shitashima, 2010). Before deploying the sensor, the ISFET and Cl-ISE were conditioned (as recommended by Bresnahan et al., 2014 and Takeshita et al., 2014) in a bucket of local sea surface water with a salinity of 38.05. However, due to time constraints, conditioning took place over just 1 hour rather than weeks as specified by Bresnahan et al. (2014) and Takeshita et al. (2014). During the deployment, pH measurements were obtained vertically every 1 to 2 m.

The measurements obtained by the ISFET pH sensor were converted from raw output counts to pH on a total scale using a two-point calibration with 2-aminopyridine (AMP) and a 2-amino-2-hydroxymethil-1, 3-propanediol (TRIS) buffer solution before and after the deployment of the glider. The same buffer solutions (Wako Pure Chemical Industries, Ltd., Osaka, Japan) created in synthetic seawater ($S = 35$, ionic strength of around 0.7 M) were used before and after deployment. These buffer solutions had a pH of 6.79 ± 0.03 (AMP) and 8.09 ± 0.03 (TRIS). The pH uncertainty of the buffer solutions takes into account the effect of changing air temperature, ranging between 30.5 and 33.3 °C during pre-calibration and between 27.5 and 28 °C during post-calibration. A linear fit using the raw output measured from these buffer solutions was used to convert the raw counts to pH (Shitashima et al., 2002). A drift was observed between the pH of these buffer solutions before and after the deployment, which was corrected for. As the ISFET sensor was previously described as having a pressure-resistant performance and good temperature characteristics for oceanographic use (Shitashima et al., 2002; Shitashima, 2010), no compensations for temperature and pressure were performed on the ISFET measurements at this stage. The ISFET pH sensor has a salinity sensitivity of $\partial \mathrm{pH} / \partial S = 0.011$, which was taken into account. The effect of biofouling on the ISFET pH measurements, as well as on all other glider measurements, was ruled out after a post-deployment inspection of the sensors indicated no problems.

2.3 Ship-based measurements

As the in situ ISFET pH sensor was under trial, some form of validation of the results was required. In total, 124 water samples were collected from Niskin bottles sampled at 12 depths (down to 1000 m) using a CTD rosette platform at eight locations (eight casts, numbered 24–51) close to the path of the glider (Fig. 1). Water samples were collected between 05:19 local time (LT, UTC+2) on 9 June and 16:58 LT on 11 June. The glider ISFET pH sensor started operating at 16:36 LT on 11 June. Overall, the measurements obtained by the glider and the CTD overlapped better in space than in time (Fig. 3).

When collecting carbon samples, water was drawn into 250 mL borosilicate glass bottles from Niskin bottles on the CTD rosette using tygon tubing. Bottles were rinsed twice before filling and were overflowed for 20 s, allowing the bottle volume to be flushed twice. Each sample was poisoned with 50 µL of saturated mercuric chloride and then sealed using greased stoppers, secured with elastic bands, and stored in the dark (Dickson et al., 2007). The total alkalinity (A_T) and the c(DIC) of each water sample was measured in the laboratory using a Marianda Versatile INstrument for the Determination of Titration Alkalinity (VIN-DTA 3C; www.marianda.com). The c(DIC) was measured by coulometry (Johnson et al., 1985) following standard operating procedure SOP 2, and A_T was measured by potentiometric titration (Mintrop et al., 2000) following SOP 3b, both described by Dickson et al. (2007). During the analytical process, 21 bottles of certified reference material (CRM; batch 107) supplied by the Scripps Institution of Oceanography (San Diego, CA, USA) were run through the instrument to keep track of stability and to calibrate the instrument. For each day in the lab, one CRM was used before and after the samples were processed. A total of 19 concurrent replicate-depth water samples were collected with around 2 to 3 replicates per CTD cast. Calculating the mean standard deviation of these replicate samples enabled a measurement of the instrument precision. The mean standard deviations of the c(DIC) and A_T replicates were 1.7 µmol kg^{-1} and 1.4 µmol kg^{-1}, respectively. This corresponds to a pH uncertainty of 0.003 for c(DIC) and A_T, resulting in a combined uncertainty of 0.009.

Once A_T and c(DIC) were known, pH could be derived using the CO2SYS programme (Van Heuven et al., 2011). This calculation has an estimated pH probable error of around 0.006 due to uncertainty in the dissociation constants pK_1 and pK_2 (Millero, 1995). Temperature and salinity were obtained from the Sea-Bird CTD sensor on the ship rosette sampler, and the seawater equilibrium constants presented by Mehrbach et al. (1973) were used as refitted by Dickson and Millero (1987), which has been recommended by previous studies (e.g. the CARINA data synthesis project) for the Mediterranean Sea (Álvarez et al., 2014; Key et al., 2010). The sulfate constant described by Dickson (1990) and the parameterisation of total borate presented by Uppström (1974) were used. More information on the equilibrium constants used in CO2SYS and other available carbonate system packages is described by Orr et al. (2015). The pH derived from the water samples collected by ship and the glider-retrieved ISFET pH are both on the total pH scale (as described by Dickson, 1984) at in situ temperature and will from now on be referred to as pH$_\mathrm{s}$ and pH$_\mathrm{g}$, respectively.

Standard deviation ranges will from this point be listed when referring to variability in measurements, such as pH$_\mathrm{s}$ and pH$_\mathrm{g}$. To obtain these standard deviation ranges, data points for a given variable were sorted into 10 m depth bins down to a maximum depth of 1000 m. The standard deviation was then calculated for each bin.

3 Results and discussion

3.1 Ship-based data

Measurements obtained by the ship CTD package provide an overview of the temporal and spatial variability for the time at which the water samples used to derive pH$_\mathrm{s}$ were collected (Fig. 4). The θ gradient was strong in the top 100 m of the water column due to limited vertical mixing with a maximum between 19 and 23 °C found in the upper 10 m of the water column, decreasing to between 13 and 14 °C at depths greater than 100 m. The salinity was low in the top 100 m, increasing to a maximum at around 400 to 600 m. These fresher waters in the top 100 m are likely modified Atlantic water (MAW), typically having a salinity between 38 and 38.3 in the northwest Mediterranean Sea (Millot, 1999). These waters enter from the Atlantic Ocean through the Strait of Gibraltar, flowing along the North African coast. Some water makes its way northward and follows the shelf back west towards the Atlantic Ocean (Rivaro et al., 2010; Millot, 1999). At deeper depths, warmer saltier waters were found

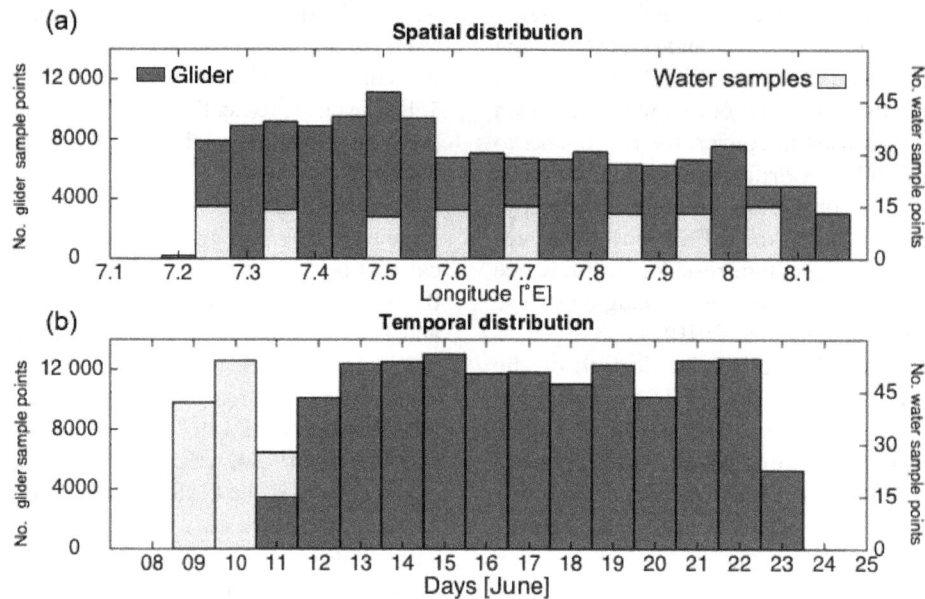

Figure 3. Histograms showing the **(a)** spatial and **(b)** temporal distribution of samples collected by the glider (dark grey) and by CTD water bottle sampling (light grey). The y axis on the left is for the sum of the glider samples, whilst the y axis on the right is for the sum of the water samples.

east of 7.5° E, which is likely to be Levantine Intermediate Water (LIW) identified in the western Mediterranean Sea by a salinity range of 38.45 to 38.65 and θ between 13.07 and 13.88 °C (Rivaro et al., 2010), typically found at depths between 200 and 800 m close to the shelf slope (Millot, 1999). The $c(O_2)$ maxima were found at depths between 20 and 90 m. The Mediterranean Sea on the whole is considered to be oligotrophic (Álvarez et al., 2014). However, a deep chlorophyll maximum (DCM) is common at these depths when waters are thermally stratified (Estrada, 1996). There is a build-up of actively growing biomass with greater cell pigment content as a result of photoacclimation due to increased concentrations of nitrate, phosphate, and silicate, as well as sufficient levels of light at these depths (Estrada, 1996). It is likely that this high $c(O_2)$ was related to the DCM, further evidenced by the high chlorophyll fluorescence layer observed at 60 to 100 m of depth, particularly in the east.

The objective of deriving pH_s using A_T and $c(DIC)$ was to make a comparison with the pH_g measured by the ISFET sensor. The $c(DIC)$ and A_T were greatest at depths below 250 m with lower values seen closer to the surface (Fig. 5a–b), which is typical of the northwest Mediterranean Sea (Copin-Montégut and Bégovic, 2002; Álvarez et al., 2014). The higher values of A_T and $c(DIC)$ at depth and in the east support the notion that this is LIW, as this water mass has previously been identified as having an A_T of around 2590 µmol kg^{-1} and a $c(DIC)$ of roughly 2330 µmol kg^{-1} (Álvarez et al., 2014), coinciding with the warmer saltier waters. The mean $c(DIC)$ and A_T (averages over eight casts) have standard deviations of 6.1 to 11.9 and 5.9 to

10.6 µmol kg^{-1}, respectively, for the top 150 m of the water column and 1.7 to 3.9 µmol kg^{-1} and 3.7 to 7.6 µmol kg^{-1} for deeper waters. The pH_s had a maximum of 8.14 between 50 and 70 m of depth (Fig. 5c). The mean pH_s values have standard deviations of 0.004 to 0.011 within the top 150 m and 0.006 to 0.017 deeper than this. A proportion of these standard deviations can be explained by the instrumental error associated with the analysis of $c(DIC)$ and A_T discussed in Sect. 2.3.

3.2 Glider data

3.2.1 Temperature, salinity, and oxygen validation

Since the sensors were calibrated before deployment, it was expected that the measurements from the glider would match those from the CTD because any discrepancies between datasets would indicate possible instrumental or methodological issues with the glider measurements. The mean profiles of θ, salinity, and $c(O_2)$ collected by the glider and by ship (Fig. 6) agreed well. The values obtained by both ship and glider were mostly within 1 standard deviation of one another. The mean θ and salinity retrieved during the eight ship pH_s casts differed from the binned mean calculated using all available REP14-MED ship casts at depths between 100 and 500 m. However, this is likely related to temporal or spatial variability as mean θ and salinity were within the range of all available glider measurements. Furthermore, differences of roughly 0.1 °C, 0.02, and 1.5 µmol kg^{-1} can be seen for θ, salinity, and $c(O_2)$, respectively, between the binned mean profile of CTD measurements and the binned mean profile

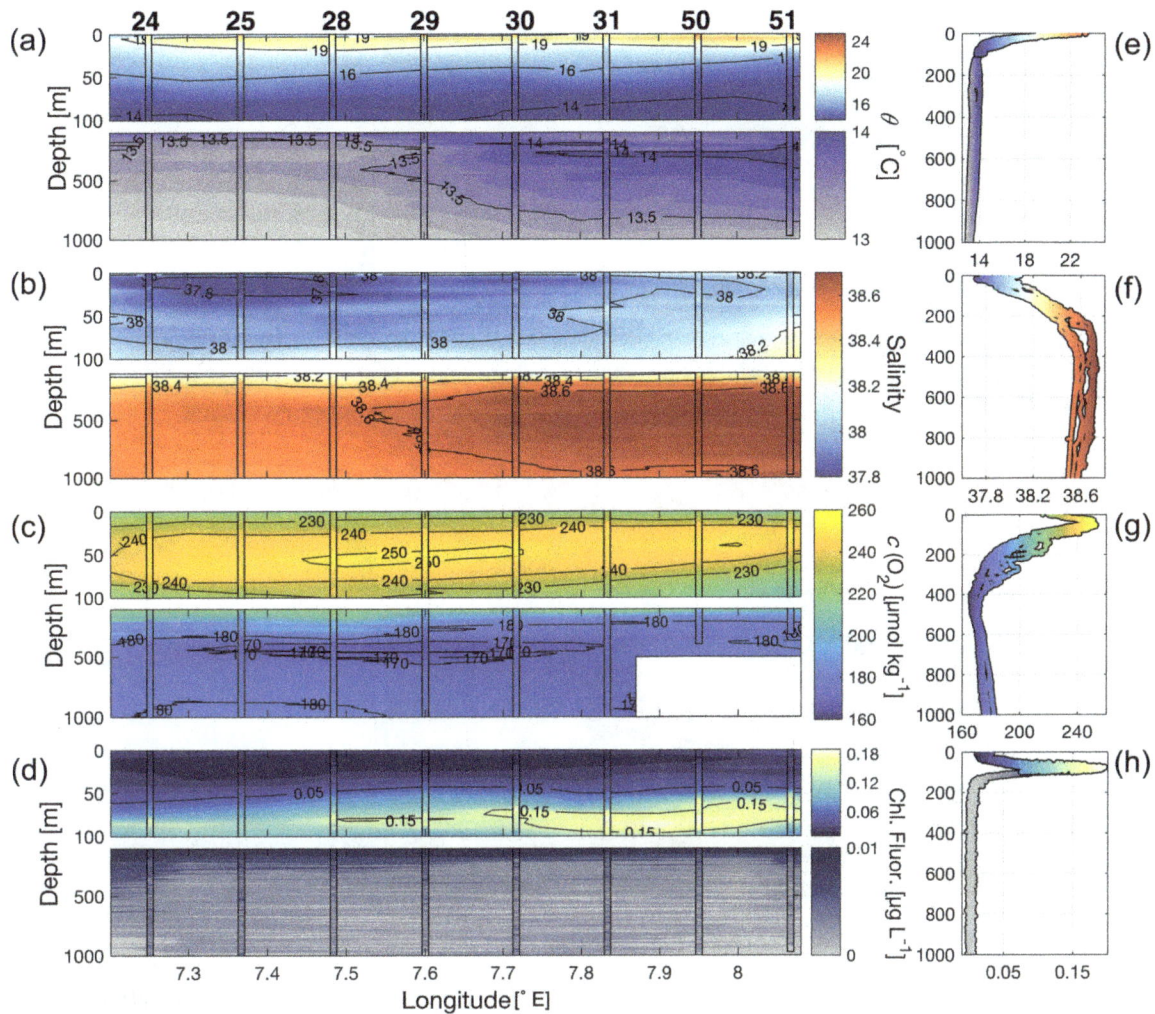

Figure 4. Transects of optimally interpolated (**a**) potential temperature (θ), (**b**) salinity, (**c**) dissolved oxygen concentrations ($c(O_2)$), and (**d**) chlorophyll fluorescence, along with their depth profiles (**e–h**) obtained by ship. These parameters were sorted into $0.1°$ longitude $\times$ 5 m bins, and the radius of influence used for optimal interpolation was $0.2°$ longitude $\times$ 20 m. The data retrieved during the eight CTD casts (displayed in Fig. 1) used for optimal interpolation are superimposed on the interpolated fields.

of glider measurements at depths greater than 500 m. These differences in θ, salinity, and $c(O_2)$ are related to the different spatial distribution of the two datasets, as the glider measured predominantly at $40°$ N where deep, cooler, and fresher waters were observed in the west (Fig. 4a–b) that are uncommon in other areas of the observational domain (Knoll et al., 2015b).

3.2.2 ISFET pH validation

The mean pH_g and pH_s agreed best between 60 and 250 m (Fig. 7), although pH_g variability was a lot higher than for pH_s. Larger differences between these profiles can be seen above and below this depth range, with a pH_g 0.1 higher at the surface and roughly 0.07 lower between 950 and 1000 m when compared with pH_s. The pH_s maximum at approximately 50 to 70 m of depth was not apparent in the pH_g pro-

file, with the highest pH_g seen at the surface. The standard deviations for pH_g were large, ranging between 0.044 and 0.114 in the top 150 m of the water column and between 0.027 and 0.053 at other points in the water column. Comparing all pH_g dive profiles obtained during the mission suggests a great degree of temporal and spatial variability, with pH ranging from 8.02 to 8.28 at the surface and between 7.97 and 8.13 at 800 m of depth.

A diel cycle in pH_g anomalies (calculated by subtracting the all-time mean from the hourly means within a given depth interval) was found predominantly at depths shallower than 20 m (Fig. 8b). Lower pH was found between 09:00 and 18:00 LT, decreasing by > 0.1 between 12:00 and 14:00 LT. Contrastingly, potential temperature, salinity, and $c(O_2)$ anomalies (calculated in the same way as pH_g anomalies) did not have strong diel cycles (Fig. 8c–e), suggesting

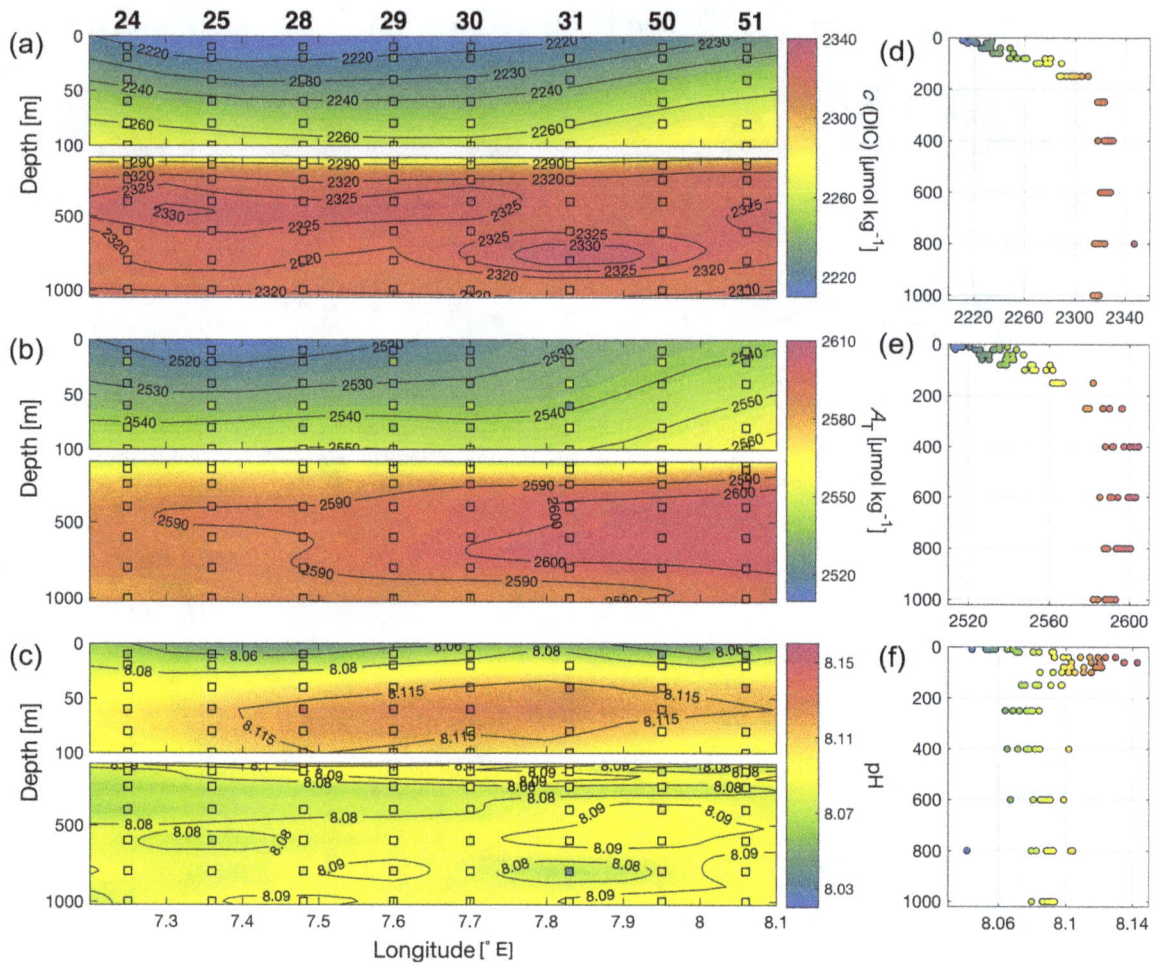

Figure 5. Optimally interpolated fields of (a) dissolved inorganic carbon (c(DIC)), (b) total alkalinity (A_T), and (c) pH derived using c(DIC) and A_T are displayed along with their depth profiles (d–f). These parameters were sorted into 0.1° longitude × 20 m bins, and the radius of influence used for optimal interpolation was 0.3° longitude × 200 m for c(DIC) and (A_T) and 0.3° longitude × 80 m for pH. The water sample values retrieved during the eight CTD casts (displayed in Fig. 1) used for optimal interpolation are superimposed on the interpolated fields as squares.

that the decrease in pH was not caused by changing environmental conditions. Particularly, one might expect c(O_2) to have a similar pattern to pH if it were related to photosynthesis or respiration due to variations in p(CO_2) (Cornwall et al., 2013; Copin-Montégut and Bégovic, 2002). However, c(O_2) remained relatively constant throughout the day at all depth ranges, implying that the level of biological activity in the Sardinian Sea did not change on average throughout the day and hence would not have caused this reduction in pH_g.

The decrease in pH_g coincided with increased levels of solar irradiance (Fig. 8a) recorded at meteorological buoy M1 (Fig. 1) during the day at the surface; hence it was likely a light-induced instrumental artefact. The effect of light on the voltage output of FET-based sensors using SiO_2- and Si_3N_4-sensitive layers is known (Wlodarski et al., 1986), as the presence of photons can excite electrons in the valence band of the semiconductor material, creating holes and allowing

the flow of electrons to the conduction band. This increases the voltage threshold to falsely measure higher hydrogen ion activity, leading to lower apparent pH (Liao et al., 1999).

The effect of light on our sensor was investigated further by exposing two ISFET pH sensors to artificial light whilst placed in reference buffer solutions (TRIS and AMP) under laboratory conditions. The results (not shown here) confirmed that our ISFET sensor is affected by light. The light-induced offset depended on the strength and type of the light source and which sensor was being used. The offset remained relatively constant whilst the light was turned on. Maximum pH offsets of -0.7 (-6×10^6 counts) and -0.15 (-3×10^6 counts) were found when the LED and halogen lights were used, respectively.

There were not enough dives for a robust light correction, and an irradiance-measuring sensor was not attached to the glider; hence data collected within the top 50 m be-

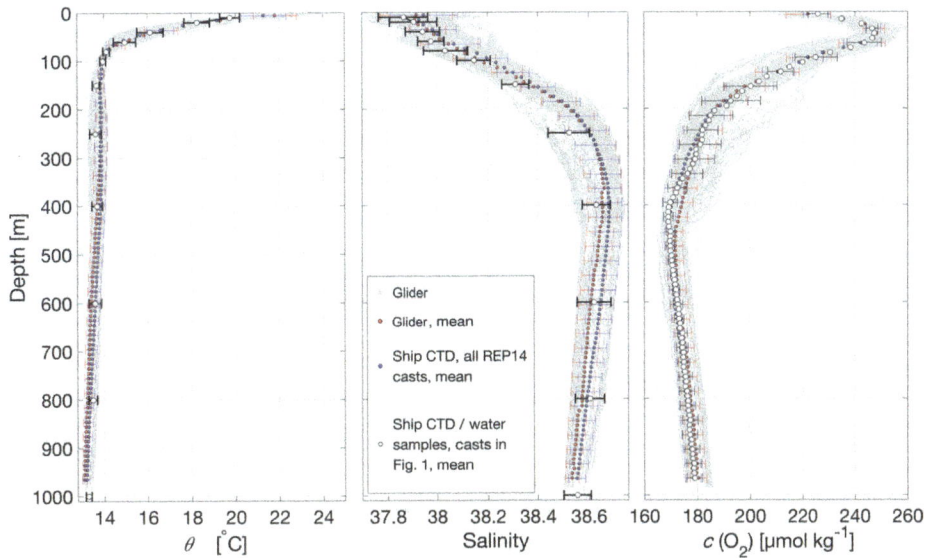

Figure 6. A comparison between the measurements retrieved by the glider and those obtained by the ship CTD package. The binned means (red) calculated using glider measurements (grey) are compared with the binned means of the CTD casts obtained from the entire REP14-MED observational domain (blue) and the binned mean values obtained from water samples (SBE oxygen optode sensor for dissolved oxygen concentrations ($c(O_2)$)) during the eight CTD casts in Fig. 1 (white) for **(a)** potential temperature (θ), **(b)** salinity, and **(c)** $c(O_2)$. Standard deviations (calculated for every 10 m bin) are displayed as error bars in this figure every 30 m for glider and CTD measurements and at every sampled depth for water samples.

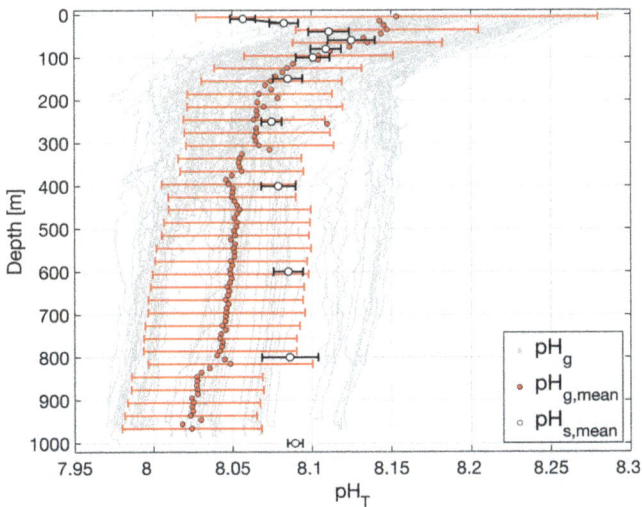

Figure 7. The pH obtained by the glider ISFET sensor (pH_g; grey) is compared with the depth-binned mean of these profiles (red) along with the corresponding standard deviation error bars (using 10 m bins) displayed in this figure every 30 m. The mean pH values obtained by the ship (pH_s) during the eight CTD casts displayed in Fig. 1 are shown (white), with their relevant standard deviations displayed as error bars. The mean pH_g values and their corresponding standard deviations were calculated using 10 m depth bins, whereas the mean pH_s values and standard deviations were calculated at each sampled depth.

tween 05:00 and 21:00 LT, representing roughly 5 % of all pH_g measurements, were not used in later analysis. In order to reduce this light effect on pH measurements in future, IS-FET sensors will have to be placed on the underside of the glider or equipped with a light shield.

Comparing pH_g to pH_s indicated that the range observed by the ISFET sensor was much larger. It could be argued that this difference in range is due to the differing temporal and spatial resolution between the glider and the ship measurements. However, comparing pH_g further with pH measurements in the literature on a similar timescale and spatial scale (Álvarez et al., 2014) suggests that this is not an issue with resolution. The $pH_{T,25}$ collected in the western Mediterranean Sea over a period of around 2 weeks (Álvarez et al., 2014), which is comparable in length to this trial, varied by roughly 0.02 at the surface and by around 0.08 at depths greater than 100 m. The range observed by the glider ISFET sensor was therefore 13 times larger at the surface and roughly 3 times larger at depths below 100 m. This difference in range cannot be explained by the high sampling frequency of the glider. Furthermore, the larger variations in pH_g were not a result of changing environmental conditions, as evidenced by the relatively stable $c(O_2)$, θ, and salinity measured by the glider, as discussed in Sects. 3.2.1 and 3.2.3.

It is likely that the ISFET pH measurements were not only related to the amount of hydrogen ion activity in the water, but also to the temperature and pressure that the sensor experienced, which was unexpected considering the sensor has previously shown good temperature and pressure character-

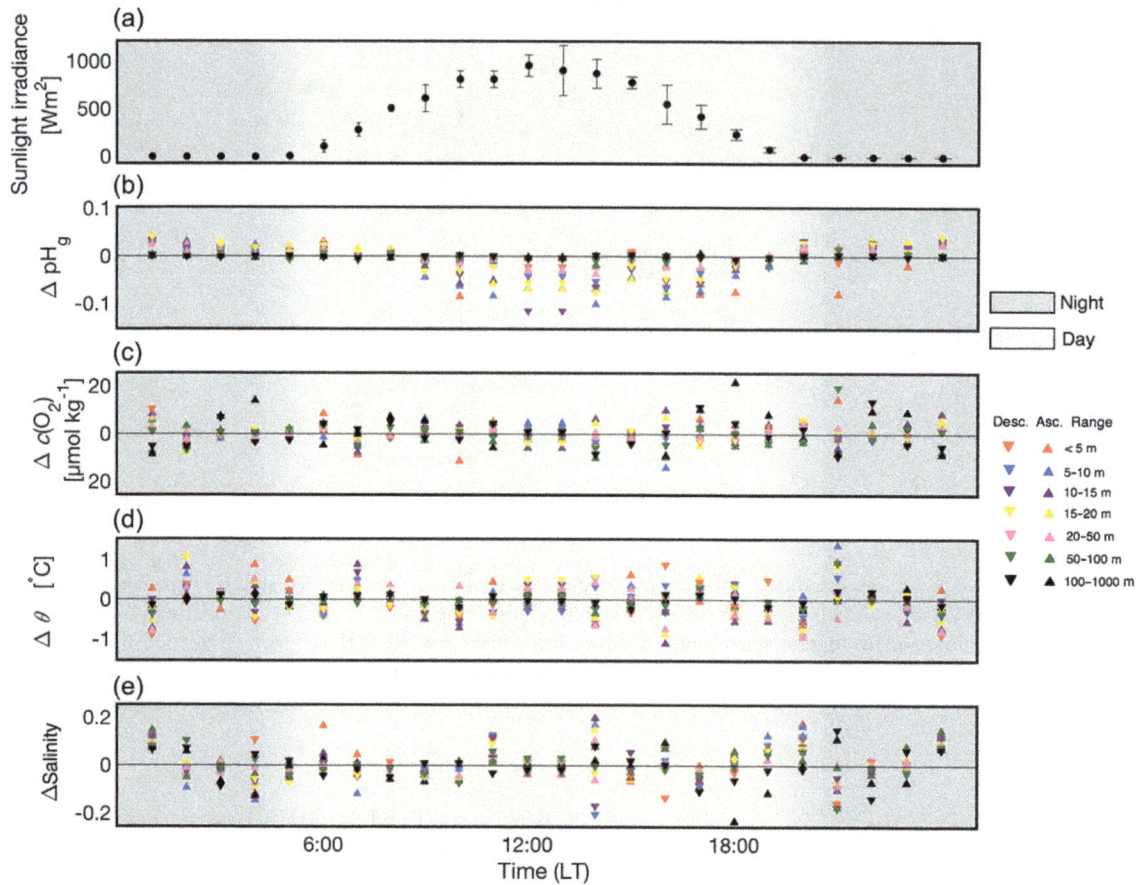

Figure 8. (a) Solar irradiance measured using a pyranometer on meteorological buoy M1 (Fig. 1). **(b)** Glider-retrieved pH (pH_g), **(c)** dissolved oxygen concentrations ($c(O_2)$), **(d)** potential temperature (θ), and **(e)** salinity average anomalies (calculated by subtracting the all-time mean from the hourly means within a given depth interval) for each hour of the day in local time (LT) for five near-surface depth ranges: < 5 m, 5–10, 10–15, 15–20, and 20–50 m, as well as two deeper depth ranges of 50–100 and 100–1000 m for both ascending (upward triangle) and descending (downward triangle) dive profiles. The grey shaded area represents the nighttime, whilst the lightly shaded area represents the daytime.

istics (Shitashima et al., 2002, 2013; Shitashima, 2010). Furthermore, comparing ISFET measurements with the pH_s and pH presented by Álvarez et al. (2014) suggests that the accuracy of the sensor was not as good as previously claimed (Shitashima et al., 2002). Therefore, it was necessary to correct the pH_g measurements for instrumental drift, temperature, and pressure.

The response of the ISFET sensor can be described by the Nernst equation (Eq. 1), which relates sensor voltage to hydrogen ion activity:

$$E = E^* - m_N \lg(a(H^+)a(Cl^-)),\qquad(1)$$

which incorporates the Nernst slope (Eq. 2),

$$m_N = RT\ln(10)/F,\qquad(2)$$

where T is temperature (K), R is the gas constant ($8.3145\,\mathrm{J\,K^{-1}\,mol^{-1}}$), F is the Faraday constant ($96\,485\,\mathrm{C\,mol^{-1}}$), $a(H^+)$ and $a(Cl^-)$ are the proton

and chloride ion activities, E is the measured voltage by the sensor (i.e. electromotive force), and E^* is representative of the two half-cells in the ISFET sensor forming a circuit (i.e. interface potential) (Martz et al., 2010). It is known that temperature and pressure have an effect on E^* (strong linear relationship) and that the Nernst slope is a function of temperature. Studies have also shown that it is possible for ISFET sensors to experience some form of hysteresis as a result of changing T and pressure (Martz et al., 2010; Bresnahan et al., 2014; Johnson et al., 2016).

The first step in correcting pH_g aimed to reduce the measured extent of variability to within the measured limits of pH_s. This in part removed the non-monotonous instrumental drift experienced by the sensor, which we think was likely due to the E^* between the two n-type silicon parts of the semiconductor being affected. A depth-constant time-varying offset correction (i.e. one offset value determined for each dive and applied to the entire profile) was applied

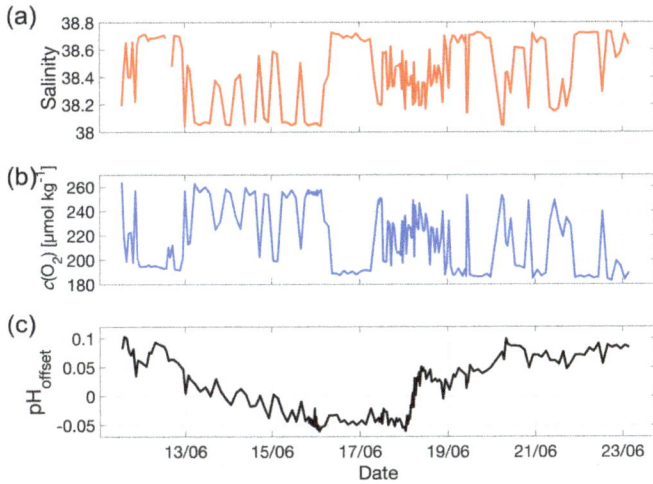

Figure 9. Salinity **(a)**, dissolved oxygen concentrations ($c(O_2)$) **(b)**, and the calculated pH offset values **(c)** as a function of time at the depth where the in situ temperature was $14 \pm 0.1\,°C$.

(Eq. 3) using the difference between the mean pH_s and each pH_g dive measurement where the in situ temperature was $14.0\,°C$, as water with this temperature was situated at a depth below the thermocline for most dives where the density gradient was weak:

$$pH_{Offset} = pH_s(T)_{mean} - pH_g(T) \quad \text{for} \quad T = 14 \pm 0.1\,°C. \quad (3)$$

The calculated offset values as a function of time were compared with salinity and $c(O_2)$ where the in situ temperature was constant at $14\,°C$ (Fig. 9). Variability in salinity and $c(O_2)$ with time were strongly related ($r^2 = 0.97$), whereas the relationship between pH offset values, salinity, and $c(O_2)$, were not (r^2 of around 0.2). Furthermore, the majority of the offset values were calculated below 100 m (Fig. 12d–f) where the density and pH gradients were weak. This indicated that our depth-constant time-varying offset correction decreased the apparent range of pH variability by an amount that was mostly associated with instrumental drift rather than physical and biogeochemical variability. Applying these offsets to the data decreased the range of pH measured by the IS-FET sensor by approximately two-thirds (Fig. 11), with new pH_g standard deviations ranging between 0.009 and 0.048 within the top 150 m and between 0.008 and 0.017 at greater depths.

After applying this offset correction, pH_g was further corrected for in situ temperature and pressure using linear regression models. The method is outlined below.

1. Calculate ΔpH (Eq. 4) as the difference between the mean pH_s and pH_g:

$$\Delta pH = pH_{s,mean} - pH_g. \quad (4)$$

2. Determine the line of best fit between ΔpH and in situ temperature in the top 100 m of the water column where

the temperature gradient was the strongest using linear regression.

3. Correct pH_g for in situ temperature for the entire water column using the slope (m) and intercept (c) coefficients of the best fit line in step 2 to obtain $pH_{g,tc}$, where "tc" stands for "temperature-corrected" values.

4. Calculate the difference between $pH_{g,tc}$ profiles and mean pH_s, producing ΔpH_{tc}, using an equation similar to Eq. (4).

5. Determine the line of best fit between ΔpH_{tc} and pressure for the lower 900 m of the water column using linear regression.

6. Correct $pH_{g,tc}$ for pressure for the entire water column using the coefficients m and c in a similar way to step 3 to obtain $pH_{g,tpc}$, where "tpc" stands for "temperature- and pressure-corrected" values.

The derived equation used for correcting pH_g is as follows:

$$pH_{g,tpc} = pH_g - 0.021t/°C + 4.5 \times 10^{-5} P/\text{dbar} + 0.261, \quad (5)$$

where t is in situ temperature and P is pressure. A good fit was found between pH_g and in situ temperature, and a reasonable fit was found with pressure (Fig. 10). The standard deviations of $pH_{g,tpc}$ ranged between 0.008 and 0.039 in the top 150 m of the water column and between 0.007 and 0.013 at greater depths, corresponding to a further decrease in apparent variability of 13 to 23 and 14 to 31 % respectively (Fig. 11).

Johnson et al. (2016) ran a series of temperature and pressure cycling experiments when testing an ISFET pH sensor based on the Honeywell Durafet ISFET. They found a temperature sensitivity of around $\partial pH / \partial T = -0.018\,K^{-1}$, which is similar to our calculated slope, and a pressure hysteresis of 0.5 mV (pH of around 0.01) at maximum compression (2000 dbar). This is equivalent to a pressure sensitivity of roughly $\partial pH / \partial P = 5 \times 10^{-6}\,\text{dbar}^{-1}$, which is an order of magnitude smaller than in this study. This difference in pressure sensitivity could be related to the different housing materials used, as Johnson et al. (2016) used polyether ether ketone (PEEK), whereas acrylic resin was used for our sensor.

Salinity covaries with temperature and pressure, and some of the salinity dependence of the offset between pH_s and pH_g might have been misattributed to the regression coefficients associated with temperature and pressure. The sensor characteristics should therefore be studied in detail under controlled laboratory conditions. However, for the purposes of calibrating the high-resolution poor-accuracy measurements (relative to pH_s) obtained from the ISFET pH sensor, the present empirical correction based on temperature and pressure appears to be sufficient to achieve a match within the pH repeatability of the discrete samples between 0.004 and 0.017.

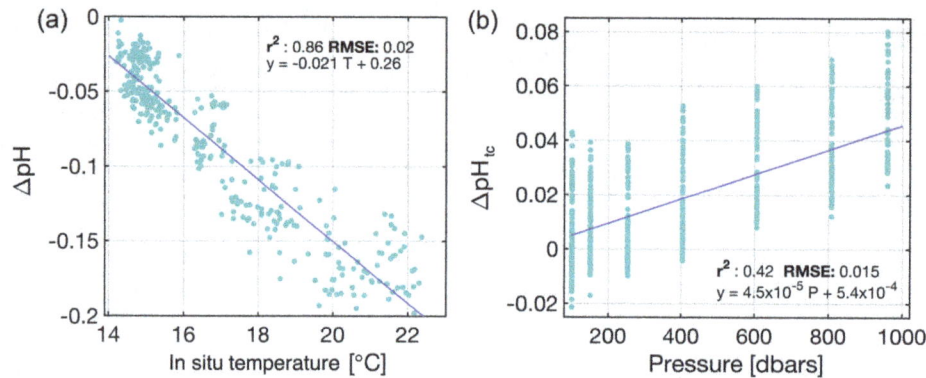

Figure 10. Linear regression fits are displayed for **(a)** ΔpH (the difference between the mean pH_s and pH_g corrected for drift) vs. in situ temperature in the top 100 m of the water column and **(b)** ΔpH corrected for in situ temperature (ΔpH_{tc}) vs. pressure between 100 and 1000 m using all available dives. The r^2, root mean square error (RMSE), and the equation of the line are displayed for each linear fit.

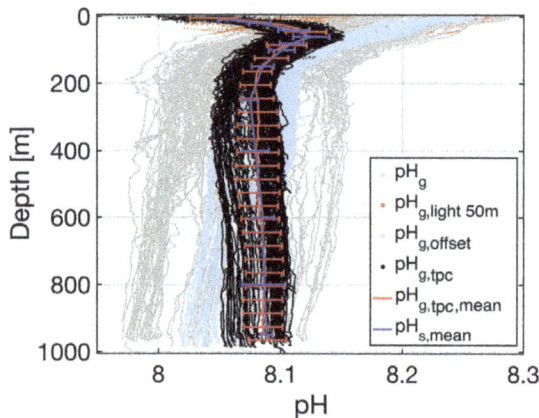

Figure 11. Profiles of glider-retrieved pH (pH_g) pre-correction (grey) with a depth-constant time-varying offset correction applied (light blue) and further corrected for in situ temperature and pressure (black). The pH_g measurements affected by light in the top 50 m of the water column (orange) and not used for the drift; temperature and pressure corrections are also shown. The depth-binned mean profile of drift, temperature, and pressure-corrected (tpc) pH_g (using 10 m bins) is shown in the foreground (red) along with the standard deviation ranges displayed every 40 m in this figure. The depth-binned mean profile of the pH measurements retrieved by ship (pH_s) is plotted for comparison (dark blue) with standard deviation ranges displayed at each sampled depth.

3.2.3 Coast to open ocean high-resolution hydrographic and biogeochemical variability

Spatial and temporal variability can be seen in $pH_{g,tpc}$ for three individual east–west transects using measurements obtained within different time periods (Fig. 12a–c). This pH variability is likely related to the air–sea exchange of carbon dioxide (weak), changes in temperature (indirectly), and biological activity (Yao et al., 2016). In the top 100 m, pH higher than 8.12 was found at depths ranging from 20 to 95 m, whereas lower pH values ranging from 8.06 to 8.09

were present closer to the surface at some locations (e.g. between 7.5 and 7.7° E and east of 8° E). The pH maxima were found at depths between 40 and 70 m where θ was around 15 °C (Fig. 12d–f) within the pycnocline (Fig. 12j–l). This band of high pH situated at 20 to 95 m of depth corresponded with a thick layer of $c(O_2)$-rich water at similar depths (Fig. 12m–o). The pH and $c(O_2)$ in the top 200 m of the water column more or less followed isopycnal surfaces at a range of points in time and space. For example, the slanted isopycnals closer to the coast (east of 7.95° E) associated with geostrophic shear corresponded with horizontal gradients in pH and $c(O_2)$. Below 100 m, $c(O_2)$ decreased to a minimum of $< 170\,\mu mol\,kg^{-1}$; although not spatially homogeneous, this corresponded with generally colder, saltier, and lower-pH waters.

All three east–west transects can be separated into two parts roughly either side of 7.7° E for depths greater than 100 m. Lower pH values between 8.05 and 8.1 were found in the western part, whereas a higher pH ranging from 8.07 to 8.12 was found in the eastern part, which was partially seen in the pH_s measurements (Fig. 5). The spatial variability in these eastern and western parts differed for each of the three time periods (times labelled in Fig. 12) with both the eastern high and western low pH patches changing in size vertically and horizontally, corresponding to spatial changes in θ and salinity. Furthermore, salinity, θ, and $c(O_2)$ were lower in the western part compared with the values found at similar depths in the eastern section (Fig. 12d–i, m–o).

In the top 100 m of the water column, the variability in pH and $c(O_2)$ is likely related to biological activity and air–sea gas exchange. As discussed in Sect. 3.1, a DCM within this depth range is common in the Mediterranean Sea when waters are thermally stratified and sufficient nutrients and light are available below the mixed layer (Estrada, 1996). High chlorophyll fluorescence was observed by the ship's sensor here (Fig. 4d). Enhanced $c(O_2)$ values at these depths are likely the by-product of photosynthesis, and the higher pH

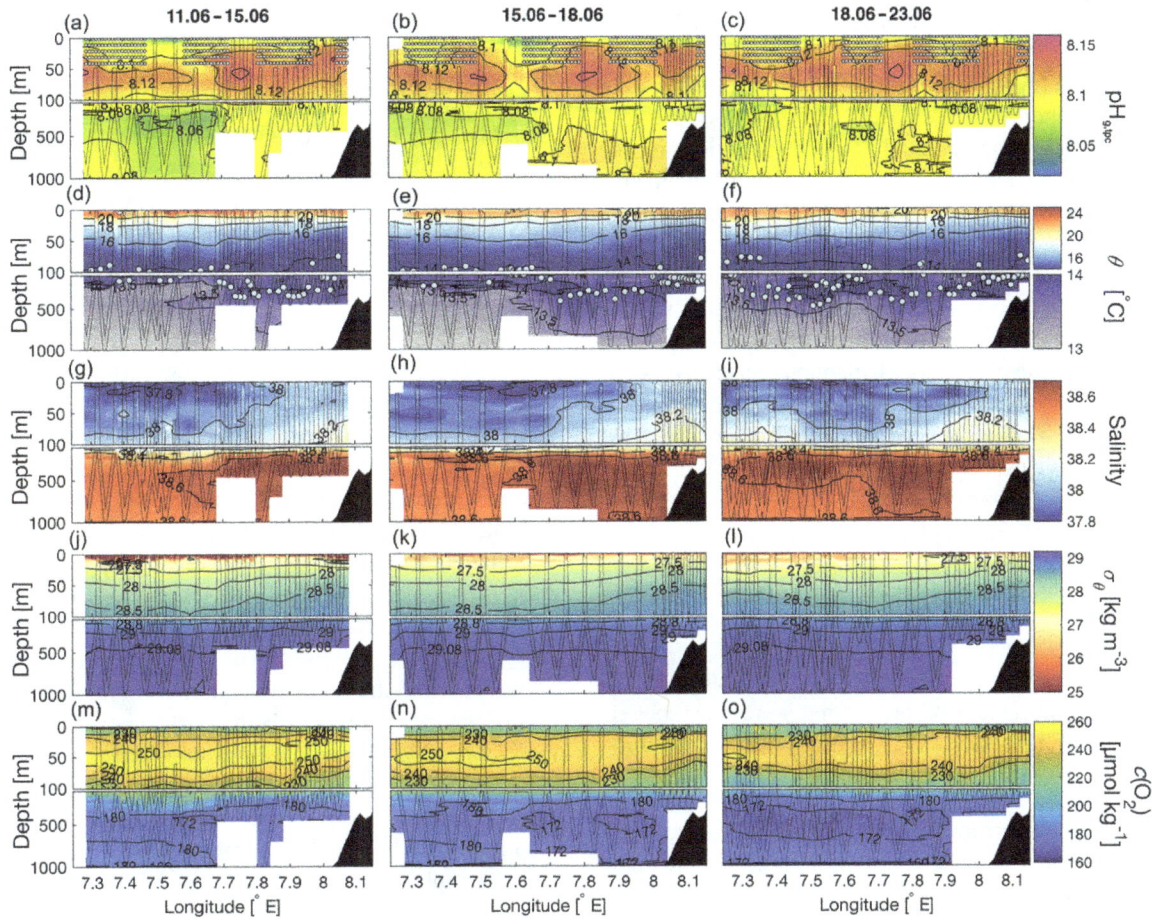

Figure 12. Objectively mapped transects of glider-retrieved **(a–c)** pH corrected for drift, temperature, and pressure (pH$_{g,tpc}$). **(d–f)** Potential temperature (θ), **(g–i)** salinity, **(j-l)** potential density anomalies (σ_θ), and **(m–o)** dissolved oxygen concentrations ($c(O_2)$) for three different time periods between 11 and 15, 15 and 18, and 18 and 23 June 2014. The spatial ranges of pH measurements affected by light and removed prior to corrections are represented by small white points **(a–c)**. The depth–longitude points at which pH offsets were calculated at a temperature of 14 °C are indicated by pale blue points **(d–f)**. Glider measurements were sorted into 0.04° longitude × 2 m bins, and the radius of influence used for optimal interpolation was 0.1° longitude × 10 m. Glider measurements used for optimal interpolation are superimposed on the interpolated fields for reference.

values are likely the result of changes in the carbon equilibrium due to the consumption of CO_2 (Cornwall et al., 2013; Rivaro et al., 2010; Copin-Montégut and Bégovic, 2002). A similar relationship between pH and primary production was described by Álvarez et al. (2014) in the western Mediterranean Sea. As discussed in Sect. 3.1, the fresher waters found in the top 100 m are likely MAW.

The difference in pH between the eastern and western parts at depths greater than 100 m depth highlighted the variability of water masses in this region. In particular, the higher pH found in the eastern part of the transect (east of 7.7° E), coinciding with high A_T and $c(DIC)$ (Fig. 5), was likely related to the flow of LIW, as described in Sect. 3.1. The LIW flows from the eastern Mediterranean basin (east of the Strait of Sicily) where pH is higher than in the western Mediterranean basin (Álvarez et al., 2014) towards the west along the continental shelf edge (Millot, 1999). This high pH found

in the eastern section of the glider transect may therefore be a remnant of these eastern Mediterranean waters. The low-pH low-$c(O_2)$ waters found deeper than 100 m result from increased respiration and remineralisation of organic matter (Lefèvre and Merlivat, 2012), coinciding with higher levels of $c(DIC)$ deeper than 200 m (Merlivat et al., 2015), which may have been more prominent in the western part of the transect (west of 7.7° E) leading to decreased levels of pH.

The pycnocline shallowed east of 7.7° E in the top 100 m of the water column during all three time periods (times labelled in Fig. 12), which corresponded with shoaling, high-salinity, low-pH, low-$c(O_2)$ waters and high $c(DIC)$, A_T, and chlorophyll fluorescence obtained by ship (Figs. 4 and 5). These features may be related to upwelling. Meteorological buoy M1 located south of the glider transect recorded an average surface wind direction of 198° towards the south-southwest, which would be favourable for coastal upwelling.

However, the mean wind speed was only $2\,\mathrm{m\,s^{-1}}$, which is weak. On the other hand, salinity maxima seen at depths of 200 to 700 m seem to suggest an intrusion of LIW westward. An intrusion of water away from the coast towards the open ocean has been shown to increase divergence in regions close to shore with strong alongshore currents (Roughan et al., 2005). Upwelling signatures at this longitudinal range along the Sardinian coast have been simulated, particularly in the summer, by Olita et al. (2013) using a hydrodynamic 3-D mesoscale-resolving numerical model. They suggest that a mixture of both current flow and wind preconditioned and enhanced upwelling in this region, which may have also been the case during our deployment. Furthermore, chlorophyll fluorescence (Fig. 4) obtained by ship was higher closer to shore, which is indicative of a greater abundance of biomass in the top 100 m, perhaps fuelled by upwelled nutrients (Porter et al., 2016; El Sayed et al., 1994).

4 Conclusions

Our trials of an experimental pH sensor in the Mediterranean Sea uncovered instrumental problems that were unexpected and will need to be addressed in future usage. These are summarised here.

1. The data retrieved from the dual pH–$p(CO_2)$ integrated sensor and from the $p(CO_2)$ unit of the stand-alone dual sensor could not be used due to quality issues. It is unclear why there was a problem with the measurements obtained by the stand-alone $p(CO_2)$ unit; however, we think the regular on/off cycling of power to the integrated dual pH–$p(CO_2)$ sensor between sampling did not allow it to function properly. In future, we would suggest the addition of backup batteries to supply electricity to the sensor between sampling.

2. The stand-alone pH sensor was subject to drift. This could be reduced by subtracting a depth-constant time-varying offset from each dive using the difference between pH_g and pH_s at a more dynamically stable depth, but such an approach is not generally recommended or valid. We think that a change in E^* between the two n-type silicon parts of the semiconductor might be the cause of the drift. To elucidate this drift further, in future two ISFET sensors should be tested in laboratory conditions within a bridge circuit to attempt to isolate possible factors contributing to drift. Focussing on the root cause of the sensor drift rather than correcting the pH data for drift after the deployment would be more beneficial to the long-term study of ISFET pH–$p(CO_2)$ sensors.

3. The sensor was apparently affected by temperature and pressure, but it is unclear to what extent the empirical relationship between in situ temperature and ΔpH in

the thermocline (top 100 m) and between pressure and ΔpH_{tc} in the deeper water (100–900 m) can be generalised.

4. The effect of light caused the sensor to measure lower levels of pH_g in surface waters. This effect is expected to be ubiquitous wherever the sensor nears the surface during daytime. In future, the sensor will have to be positioned on the underside of the glider or equipped with a light shield to limit the effect of the sun when close to the surface.

Despite the overall disappointing performance, we were able to demonstrate the potential use of the corrected glider pH measurements to uncover the biogeochemical variability associated with biological and physical mesoscale features. The pH_g corrected for drift, temperature, and pressure was compared temporally and spatially with the other physical and biogeochemical parameters obtained by the glider. This comparison indicated that the pH in the top 100 m of the water column was mostly related to biological activity where $c(O_2)$ was high. Below 100 m, low pH west of 7.7° E was likely linked to the remineralisation of organic matter, whilst east of this point, higher pH may have been transported from the eastern Mediterranean basin via LIW. Shoaling isopycnals east of 7.7° E closer to shore may have been indicative of upwelling, and possible upwelling signatures at the same location could be seen in salinity, θ, pH, $c(O_2)$, $c(DIC)$, A_T, and chlorophyll fluorescence.

Competing interests. The authors declare that they have no conflict of interest.

Acknowledgements. The authors would like to thank all the partners who helped make REP14-MED a success: the engineers, technicians, and scientists onboard the NRV *Alliance* and the NRV *Planet*, those on land responsible for the logistics of the experiment, and the UEA glider science team for piloting the glider. We thank Bastien Queste and Gillian Damerell for their help and support regarding the analysis of the glider data. Michael Hemming's PhD project is funded by the Defence Science and Technology Laboratory (DSTL, UK), in close cooperation with Direction Générale de l'Armement (DGA, France), with oversight provided by Tim Clarke and Carole Nahum. We thank the Natural Environment Research Council (NERC, UK) for providing financial support for the demonstration of the glider capability.

Edited by: J. Chiggiato

References

Álvarez, M., Sanleón-Bartolomé, H., Tanhua, T., Mintrop, L., Luchetta, A., Cantoni, C., Schroeder, K., and Civitarese, G.: The CO_2 system in the Mediterranean Sea: a basin wide perspective, Ocean Sci., 10, 69–92, doi:10.5194/os-10-69-2014, 2014.

Aßmann, S., Frank, C., and Körtzinger, A.: Spectrophotometric high-precision seawater pH determination for use in underway measuring systems, Ocean Sci., 7, 597–607, doi:10.5194/os-7-597-2011, 2011.

Bates, N. R., Best, M. H. P., Neely, K., Garley, R., Dickson, A. G., and Johnson, R. J.: Detecting anthropogenic carbon dioxide uptake and ocean acidification in the North Atlantic Ocean, Biogeosciences, 9, 2509–2522, doi:10.5194/bg-9-2509-2012, 2012.

Binetti, U.: Dissolved oxygen-based annual biological production from glider observations at the Porcupine Abyssal Plain (North Atlantic), Ph.D. thesis, University of East Anglia, 2016.

Bittig, H. C., Fiedler, B., Scholz, R., Krahmann, G., and Körtzinger, A.: Time response of oxygen optodes on profiling platforms and its dependence on flow speed and temperature, Limnol. Oceanogr. Methods, 12, 617–636, 2014.

Bresnahan, P. J., Martz, T. R., Takeshita, Y., Johnson, K. S., and LaShomb, M.: Best practices for autonomous measurement of seawater pH with the Honeywell Durafet, Methods Oceanogr., 9, 44–60, 2014.

Byrne, R. H. and Breland, J. A.: High precision multiwavelength pH determinations in seawater using cresol red, Deep Sea Res. A, 36, 803–810, 1989.

Copin-Montégut, C. and Bégovic, M.: Distributions of carbonate properties and oxygen along the water column (0–2000 m) in the central part of the NW Mediterranean Sea (Dyfamed site): influence of winter vertical mixing on air–sea CO_2 and O_2 exchanges, Deep Sea Res. II, 49, 2049–2066, 2002.

Cornwall, C. E., Hepburn, C. D., McGraw, C. M., Currie, K. I., Pilditch, C. A., Hunter, K. A., Boyd, P. W., and Hurd, C. L.: Diurnal fluctuations in seawater pH influence the response of a calcifying macroalga to ocean acidification, Proc. Roy. Soc. London B, 280, 20132201, 2013.

Dickson, A.: pH scales and proton-transfer reactions in saline media such as sea water, Geochem. Cosmochim. Acta, 48, 2299–2308, 1984.

Dickson, A. and Millero, F. J.: A comparison of the equilibrium constants for the dissociation of carbonic acid in seawater media, Deep Sea Res. A, 34, 1733–1743, 1987.

Dickson, A. G.: Thermodynamics of the dissociation of boric acid in synthetic seawater from 273.15 to 318.15 K, Deep Sea Res. A, 37, 755–766, 1990.

Dickson, A. G., Sabine, C. L., and Christian, J. R.: Guide to best practices for ocean CO_2 measurements, PICES Special Publication 3, 191 pp., 2007.

Doney, S. C., Fabry, V. J., Feely, R. A., and Kleypas, J. A.: Ocean acidification: the other CO_2 problem, Marine Sci., 1, 169–192, 2009.

El Sayed, M. A., Aminot, A., and Kerouel, R.: Nutrients and trace metals in the northwestern Mediterranean under coastal upwelling conditions, Cont. Shelf Res., 14, 507–530, 1994.

Eriksen, C. C., Osse, T. J., Light, R. D., Wen, T., Lehman, T. W., Sabin, P. L., Ballard, J. W., and Chiodi, A. M.: Seaglider: A long-range autonomous underwater vehicle for oceanographic research, Oceanic Engineering, IEEE Journal of, 26, 424–436, 2001.

Estrada, M.: Primary production in the northwestern Mediterranean, Scientia Marina, 60, 55–64, 1996.

Fabry, V. J., Seibel, B. A., Feely, R. A., and Orr, J. C.: Impacts of ocean acidification on marine fauna and ecosystem processes, ICES Journal of Marine Science: Journal du Conseil, 65, 414–432, 2008.

Frajka-Williams, E., Eriksen, C. C., Rhines, P. B., and Harcourt, R. R.: Determining vertical water velocities from Seaglider, J. Atmos. Ocean. Tech., 28, 1641–1656, 2011.

Garau, B., Ruiz, S., Zhang, W. G., Pascual, A., Heslop, E., Kerfoot, J., and Tintoré, J.: Thermal lag correction on Slocum CTD glider data, J. Atmos. Ocean. Tech., 28, 1065–1071, 2011.

Hofmann, G. E., Smith, J. E., Johnson, K. S., Send, U., Levin, L. A., Micheli, F., Paytan, A., Price, N. N., Peterson, B., Takeshita, Y., Matson, P. G., Crook, E. D., Kroeker, K. J., Gambi, M. C., Rivest, E. B., Frieder, C. A., Yu, P. C., and Martz, T. R.: High-frequency dynamics of ocean pH: a multi-ecosystem comparison, PloS one, 6, e28983, doi:10.1371/journal.pone.0028983, 2011.

Johnson, K. M., King, A. E., and Sieburth, J. M.: Coulometric TCO_2 analyses for marine studies; an introduction, Marine Chem., 16, 61–82, 1985.

Johnson, K. S., Jannasch, H. W., Coletti, L. J., Elrod, V. A., Martz, T. R., Takeshita, Y., Carlson, R. J., and Connery, J. G.: Deep-Sea DuraFET: A pressure tolerant pH sensor designed for global sensor networks, Anal. Chem., 88, 3249–3256, 2016.

Key, R. M., Tanhua, T., Olsen, A., Hoppema, M., Jutterström, S., Schirnick, C., van Heuven, S., Kozyr, A., Lin, X., Velo, A., Wallace, D. W. R., and Mintrop, L.: The CARINA data synthesis project: introduction and overview, Earth Syst. Sci. Data, 2, 105–121, doi:10.5194/essd-2-105-2010, 2010.

Knoll, M., Benecke, J., Russo, A., and Ampolo-Rella, M.: Comparison of CTD measurements obtained by NRV Alliance and RV Planet during REP14-MED, Technical Report WTD71-0083/2015 WB 34 pp., 2015b.

Le Quéré, C., Moriarty, R., Andrew, R. M., Peters, G. P., Ciais, P., Friedlingstein, P., Jones, S. D., Sitch, S., Tans, P., Arneth, A., Boden, T. A., Bopp, L., Bozec, Y., Canadell, J. G., Chini, L. P., Chevallier, F., Cosca, C. E., Harris, I., Hoppema, M., Houghton, R. A., House, J. I., Jain, A. K., Johannessen, T., Kato, E., Keeling, R. F., Kitidis, V., Klein Goldewijk, K., Koven, C., Landa, C. S., Landschützer, P., Lenton, A., Lima, I. D., Marland, G., Mathis, J. T., Metzl, N., Nojiri, Y., Olsen, A., Ono, T., Peng, S., Peters, W., Pfeil, B., Poulter, B., Raupach, M. R., Regnier, P., Rödenbeck, C., Saito, S., Salisbury, J. E., Schuster, U., Schwinger, J., Séférian, R., Segschneider, J., Steinhoff, T., Stocker, B. D., Sutton, A. J., Takahashi, T., Tilbrook, B., van der Werf, G. R., Viovy, N., Wang, Y.-P., Wanninkhof, R., Wiltshire, A., and Zeng, N.: Global carbon budget 2014, Earth Syst. Sci. Data, 7, 47–85, doi:10.5194/essd-7-47-2015, 2015.

Lefèvre, N. and Merlivat, L.: Carbon and oxygen net community production in the eastern tropical Atlantic estimated from a moored buoy, Global Biogeochem. Cy., 26, doi:10.1029/2010GB004018, 2012.

Liao, H.-K., Wu, C.-L., Chou, J.-C., Chung, W.-Y., Sun, T.-P., and Hsiung, S.-K.: Multi-structure ion sensitive field effect transistor with a metal light shield, Sensors and Actuators B: Chemical, 61, 1–5, 1999.

Martz, T. R., Carr, J. J., French, C. R., and DeGrandpre, M. D.: A submersible autonomous sensor for spectrophotometric pH measurements of natural waters, Anal. Chem., 75, 1844–1850, 2003.

Martz, T. R., Connery, J. G., and Johnson, K. S.: Testing the Honeywell Durafet® for seawater pH applications, Limnol. Oceanogr. Methods, 8, 172–184, 2010.

McNeil, C. L. and D'Asaro, E. A.: A calibration equation for oxygen optodes based on physical properties of the sensing foil, Limnol. Oceanogr. Methods, 12, 139–154, 2014.

Mehrbach, C., Culberson, C., Hawley, J., and Pytkowics, R.: Measurement of the apparent dissociation constants of carbonic acid in seawater at atmospheric pressure, Limnol. Oceanogr., 18, 40 pp., 1973.

Merlivat, L., Boutin, J., and Antoine, D.: Roles of biological and physical processes in driving seasonal air–sea CO_2 flux in the southern ocean: new insights from CARIOCA pCO_2, J. Marine Syst., 147, 9–20, 2015.

Millero, F. J.: Thermodynamics of the carbon dioxide system in the oceans, Geochim. Cosmochim. Acta, 59, 661–677, 1995.

Millot, C.: Circulation in the western Mediterranean Sea, J. Marine Syst., 20, 423–442, 1999.

Mintrop, L., Pérez, F. F., González-Dávila, M., Santana-Casiano, M., and Körtzinger, A.: Alkalinity determination by potentiometry: Intercalibration using three different methods, Ciencias Marinas, 26, 23–37, 2000.

NATO Centre for Maritime Research and Experimentation (CMRE): Geo Spatial Data Portal, available at: http://geos3.cmre.nato.int/portal, last access: May 2016.

Olita, A., Ribotti, A., Fazioli, L., Perilli, A., and Sorgente, R.: Surface circulation and upwelling in the Sardinia Sea: A numerical study, Cont. Shelf Res., 71, 95–108, 2013.

Onken, R., Fiekas, H.-V., Beguery, L., Borrione, I., Funk, A., Hemming, M., Heywood, K. J., Kaiser, J., Knoll, M., Poulain, P.-M., Queste, B., Russo, A., Shitashima, K., Siderius, M., and Thorp-Küsel, E.: High-Resolution Observations in the Western Mediterranean Sea: The REP14-MED Experiment, Ocean Sci. Discuss., doi:10.5194/os-2016-82, in review, 2016.

Orr, J. C., Epitalon, J.-M., and Gattuso, J.-P.: Comparison of ten packages that compute ocean carbonate chemistry, Biogeosciences, 12, 1483–1510, doi:10.5194/bg-12-1483-2015, 2015.

Piterbarg, L., Taillandier, V., and Griffa, A.: Investigating frontal variability from repeated glider transects in the Ligurian Current (North West Mediterranean Sea), J. Marine Syst., 129, 381–395, 2014.

Porter, M., Inall, M., Hopkins, J., Palmer, M., Dale, A., Aleynik, D., Barth, J., Mahaffey, C., and Smeed, D.: Glider observations of enhanced deep water upwelling at a shelf break canyon: A mechanism for cross-slope carbon and nutrient exchange, J. Geophys. Res.-Oceans, 121, 7575–7588, 2016.

Queste, B. Y., Heywood, K. J., Kaiser, J., Lee, G., Matthews, A., Schmidtko, S., Walker-Brown, C., and Woodward, S. W.: Deployments in extreme conditions: Pushing the boundaries of Seaglider capabilities, in: Autonomous Underwater Vehicles (AUV), 2012 IEEE/OES, 1–7, IEEE, 2012.

Rhein, M., Rintoul, S., Aoki, S., Campos, E., Chambers, D., Feely, R., Gulev, S., Johnson, G., Josey, S., Kostianoy, A., Mauritzen, C., Roemmich, D., Talley, L. D., and Wang, F.: Chapter 3: Observations: Ocean, Climate Change, 255–315, 2013.

Rivaro, P., Messa, R., Massolo, S., and Frache, R.: Distributions of carbonate properties along the water column in the Mediterranean Sea: Spatial and temporal variations, Marine Chem., 121, 236–245, 2010.

Roemmich, D. H., Davis, R. E., Riser, S. C., Owens, W. B., Molinari, R. L., Garzoli, S. L., and Johnson, G. C.: The argo project. global ocean observations for understanding and prediction of climate variability, Tech. rep., DTIC Document, 2003.

Roughan, M., Terrill, E. J., Largier, J. L., and Otero, M. P.: Observations of divergence and upwelling around Point Loma, California, J. Geophys. Res.-Oceans, 110, C04011, doi:10.1029/2004JC002662, 2005.

Shitashima, K.: Evolution of compact electrochemical in-situ pH-pCO_2 sensor using ISFET-pH electrode, in: Oceans 2010 MTS/IEEE Seattle, 1–4, 2010.

Shitashima, K., Kyo, M., Koike, Y., and Henmi, H.: Development of in situ pH sensor using ISFET, in: Underwater Technology, 2002. Proceedings of the 2002 International Symposium on, 106–108, IEEE, 2002.

Shitashima, K., Maeda, Y., and Ohsumi, T.: Development of detection and monitoring techniques of CO_2 leakage from seafloor in sub-seabed CO_2 storage, Appl. Geochem., 30, 114–124, 2013.

Takeshita, Y., Martz, T. R., Johnson, K. S., and Dickson, A. G.: Characterization of an ion sensitive field effect transistor and chloride ion selective electrodes for pH measurements in seawater, Anal. Chem., 86, 11189–11195, 2014.

Touratier, F. and Goyet, C.: Impact of the Eastern Mediterranean Transient on the distribution of anthropogenic CO_2 and first estimate of acidification for the Mediterranean Sea, Deep Sea Res. I, 58, 1–15, 2011.

Uppström, L. R.: The boron/chlorinity ratio of deep-sea water from the Pacific Ocean, in: Deep Sea Research and Oceanographic Abstracts, vol. 21, 161–162, Elsevier, 1974.

Van Heuven, S., Pierrot, D., Rae, J., Lewis, E., and Wallace, D.: MATLAB program developed for CO_2 system calculations, ORNL/CDIAC-105b, Carbon Dioxide Inf, Anal. Cent., Oak Ridge Natl. Lab., US DOE, Oak Ridge, Tennessee, 2011.

Wlodarski, W., Bergveld, P., and Voorthuyzen, J.: Threshold voltage variations in n-channel MOS transistors and MOSFET-based sensors due to optical radiation, Sensors and Actuators, 9, 313–321, 1986.

Yao, K. M., Marcou, O., Goyet, C., Guglielmi, V., Touratier, F., and Savy, J.-P.: Time variability of the north-western Mediterranean Sea pH over 1995–2011, Marine Environ. Res., 116, 51–60, 2016.

Yates, K. K., Dufore, C., Smiley, N., Jackson, C., and Halley, R. B.: Diurnal variation of oxygen and carbonate system parameters in Tampa Bay and Florida Bay, Marine Chem., 104, 110–124, 2007.

Zeebe, R. E. and Wolf-Gladrow, D. A.: CO_2 in seawater: equilibrium, kinetics, isotopes, Gulf Professional Publishing, 2001.

3

First year of practical experiences of the new Arctic AWIPEV-COSYNA cabled Underwater Observatory in Kongsfjorden, Spitsbergen

Philipp Fischer[1], **Max Schwanitz**[1], **Reiner Loth**[2], **Uwe Posner**[3], **Markus Brand**[1], and **Friedhelm Schröder**[4]

[1]Alfred-Wegener-Institut Helmholtz Centre for Polar and Marine Research, Centre for Scientific Diving at the Biological Station Helgoland, Kurpromenade 211, 27498 Helgoland, Germany
[2]loth-engineering GmbH, Lochmühle 1, 65527 Niedernhausen, Germany
[3]-4H-JENA engineering GmbH, Mühlenstr. 126, 07745 Jena, Germany
[4]Helmholtz-Zentrum Geesthacht, Institut für Material- und Küstenforschung,
Max-Planck-Straße 1, 21502 Geesthacht, Germany

Correspondence to: Philipp Fischer (philipp.fischer@awi.de)

Abstract. A combined year-round assessment of selected oceanographic data and a macrobiotic community assessment was performed from October 2013 to November 2014 in the littoral zone of the Kongsfjorden polar fjord system on the western coast of Svalbard (Norway). State of the art remote controlled cabled underwater observatory technology was used for daily vertical profiles of temperature, salinity, and turbidity together with a stereo-optical assessment of the macrobiotic community, including fish. The results reveal a distinct seasonal cycle in total species abundances, with a significantly higher total abundance and species richness during the polar winter when no light is available underwater compared to the summer months when 24 h light is available. During the winter months, a temporally highly segmented community was observed with respect to species occurrence, with single species dominating the winter community for restricted times. In contrast, the summer community showed an overall lower total abundance as well as a significantly lower number of species. The study clearly demonstrates the high potential of cable connected remote controlled digital sampling devices, especially in remote areas, such as polar fjord systems, with harsh environmental conditions and limited accessibility. A smart combination of such new digital "sampling" methods with classic sampling procedures can provide a possibility to significantly extend the sampling time and frequency, especially in remote and difficult to access areas. This can help to provide a sufficient data density and therefore statistical power for a sound scientific analysis without increasing the invasive sampling pressure in ecologically sensitive environments.

1 Introduction

Kongsfjorden (78°55′ N, 11°56′ E) on the western coast of Spitsbergen (Fig. 1) is described as one of the best studied polar fjord systems in the Arctic (Wiencke, 2004). The 20 km long ecosystem opens without a sill in a westerly direction toward the Fram straight (Hop et al., 2002) and is alternatively penetrated by warm saline Atlantic water masses from the West Spitsbergen Current, by cold less saline Arctic water from the East Spitsbergen Current, or a mixture of both (Cottier et al., 2005). This bi-modal hydrographic situation leads to a complex spatio-temporal pattern in the fjord hydrography with an occasionally more Atlantic and in other instances more Arctic characteristic with respect to the water masses, even in the inner fjord system (Svendsen et al., 2002). Due to an increased advection rate of warmer Atlantic water masses in the fjord systems over the last decade (Cottier et al., 2005), the first signs of an overall warming of the fjord system have been observed, with an overall decrease in seasonal ice coverage (Walczowski et al., 2012), signifi-

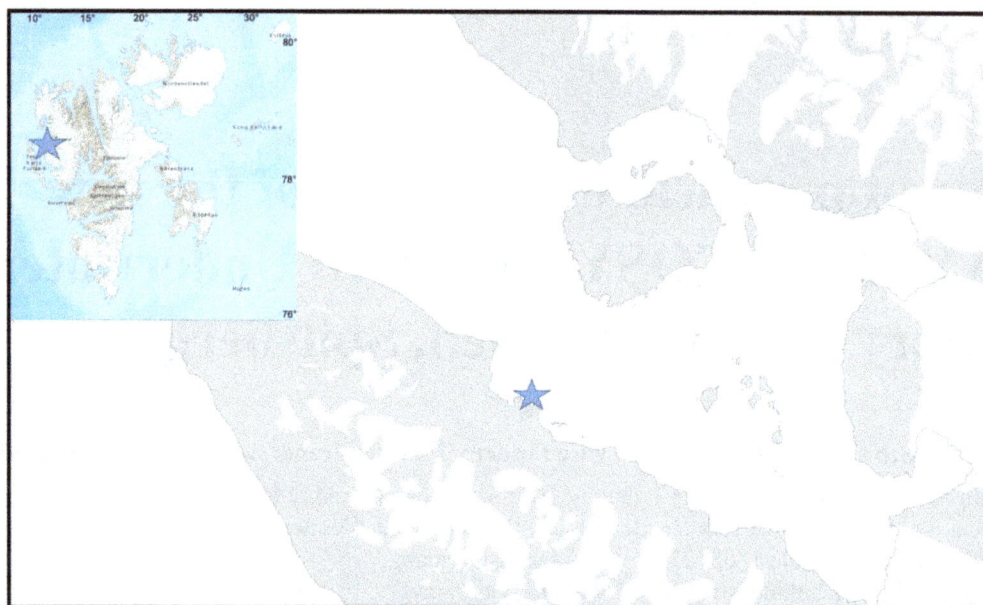

Figure 1. Spitzbergen with Kongsfjorden (★ in the small inlay panel in the upper left corner) and the location of NyÅlesund in Kongsfjorden (★). Source: Norwegian Polar Institute (2014), 2017.

cant changes in the phytoplankton community (Hegseth and Tverberg, 2013; Willis et al., 2006), changes in the depth distribution of macroalgae in the shallow waters (Bartsch et al., 2016) and in the macrozoobenthos community (Parr at al., 2015), as well as an increase in turbidity due to increased meltwater runoff from the glaciers (Peterson et al., 2002; Bartsch et al., 2016). Although Renaud et al. (2011) and Voronkov et al. (2013) recently started to study the food-chain length, trophic levels, and the main feeding groups in Kongsfjorden, our knowledge of the temporal and spatial dynamics of the higher trophic levels of the food web is still extremely limited (Stempniewicz et al., 2007). Therefore, important knowledge gaps such as a lack of quantitative data on production, abundance of key prey species, and the role of advection in the biological communities in the fjord still exist (Hop et al., 2002).

Such knowledge, however, is mandatory for a better understanding of this polar fjord system and potentially to use it as a model system for future Arctic change scenarios under the pressure of global warming. The most comprehensive review thus far of the occurrence and higher trophic level species in the Kongsfjorden ecosystem has been performed by Hop et al. (2002) and revealed approximately 34 zooplankton taxa, between 29 and 396 macrozoobenthos species, as well as approximately 30 fish species in the fjord system in total, depending on the type of substratum. Most of these data have been sampled during intense summer campaigns with ship-supported sampling methods or by occasional scuba diving operations at different sites of the fjord. Although these datasets are highly valuable, they are mainly restricted to the polar summer when light is available and sampling can be

performed on a regular basis. A systematic year-round assessment of the fjord community, especially of the shallow water habitats, which are well known as most important as spawning, hatching, and nursery grounds for juvenile specimens (Fischer and Eckmann, 1997a, b; Werner, 1977), is missing.

Thorough assessments especially of higher tropic levels such as fish and macroinvertebrates are demanding already in northern temperate non-polar waters because of the required logistics, methods, and manpower (Wehkamp and Fischer, 2013a, b, c). In Arctic waters with the even harsher conditions with respect to low winter temperatures, seasonally limited daylight availability and a partial or complete ice coverage, longer-term and year-round assessments especially in shallow coastal areas are almost completely lacking. Furthermore, in several hard bottom fjord systems, such as the Kongsfjorden system, the shallow water areas are relatively inaccessible by trawling with larger vessels due to a complex and highly structured benthic habitat, with a mixture of rocky bottom and ice-rafted pebbles and stones (Jørgenson and Gulliksen, 2001). Therefore, most available studies are temporally restricted to the summer months and the open or deeper water bodies.

In the present study, we present data from a 13-month (October 2013 to November 2014) long hydro-biological survey in the sublittoral zone of the Arctic Kongsfjorden at the southern shoreline close to the research village of NyÅlesund at UMT 8763953° N, 433992° E (Fig. 1). With a 2012 installed cabled underwater observatory (COSYNA@AWIPEV Underwater Observatory – subsequently called UWO), we continuously recorded the main hydrological parameters temperature, salinity, pH, Chl *a*, and turbidity and additionally made a quantitative analysis of the abundance, species occurrence, and (for selected species) length–frequency distribution of the fish and macroinvertebrate taxa. For the latter assessment, a stereo-optical macrobiota observatory called "RemOS1" (Remote Optical System) was used, specifically designed for long-term exposure and assessments of fish and macroinvertebrate communities in shallow water areas (Fischer et al., 2007b). Data acquisition was conducted year-round, remote controlled with a temporal resolution of 1 Hz for the hydrological data and with a stereoscopic imaging frequency of 30 min. Parallel to this study, classic fishing campaigns were performed in 2012, 2013, and 2014 in the months June/July and September in the same area with standard fyke nets to provide ground-truth data for the remotely sampled fish data. These fishing data are published in Brand and Fischer (2016) for the years 2012 and 2013. The data for 2014 will be published together with a comparative analysis of the results of the UWO elsewhere (M. Brand, personal communication, 2016).

The present study aims to demonstrate the high potential of remote controlled sensors to quantitatively assess not only hydrological data such as temperature, current, or plankton community with classical CTD (conductivity–temperature–depth) probes or VPRs (video plankton recorders), but also for the assessment of higher tropic levels such as macroinvertebrates and fish. To the best of our knowledge, there are only a small number of studies and observatories available worldwide that are trying to also assess higher trophic levels with remote controlled optical systems (Aguzzi et al., 2011; Buckland et al., 2005; Fischer et al., 2007b; Wehkamp and Fischer, 2014), and even fewer with regard to quantitative assessments with respect to a specimen's abundances and species-specific length–frequency analysis in an area. Because these technologies will certainly develop and improve over the next years, this study also discusses certain specific requirements and challenges for such systems, especially for shallow water Arctic areas.

2 Materials and methods

The UWO was built up in 2012 in the framework of COSYNA (Coastal Observing Systems of the Northern and Arctic Seas). The system comprises a land-based FerryBox system equipped with various hydrographic sensors (Table 1) receiving water from a remote controlled underwater pump station at 11 m water depth. Additionally, a cable connected (fibre-optic and 240 V power) underwater node (Fig. 2) was installed close to the pump station at a 11 m water depth providing power (48 V) and a network (TCP/IP 100 Mbit) connection to additional in situ sensors. To install or exchange sensor equipment at the node system by divers, the node is equipped with four underwater matable power/ethernet con-

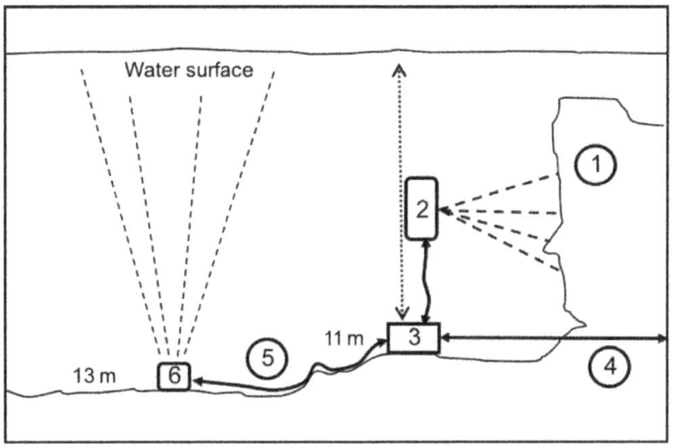

Figure 2. Sketch of the underwater installations with the underwater base station and the vertical profiling unit off NyÅlesund. Numbers refer to numbers in the sketch. (1) Steep wall (drop-off) with vertical zonated macrophyte coverage. (2) Vertical profiling sensor carrier with CTD and a stereo-optical imaging device (RemOs1) looking towards the wall. (3) Underwater node with wet-matable plugs. (4) Combined power/fibre-optic cable to land. (5) Combined power/rs232 cable from node to ADCP. (6) ADCP. For details on the single components, see the text.

nectors and two additional underwater matable power/rs232 connectors.

For the experiment described in this study, the node system was equipped with an upward looking ADCP positioned at 13–15 m water depth (depending on the tide cycle), a SBE38 temperature sensor positioned at 11–13 m water depth (depending on the tide cycle), and a vertical profiling sensor carrier. The profiling sensor carrier was fully remote controlled via the Internet and was operated year-round from October 2013 to November 2014 from Germany. It was equipped with a CTD for the assessment of the main hydrographical parameters and the RemOS1 stereo-optical camera system (Fischer, 2017; Fischer et al., 2007b; Wehkamp and Fischer, 2014) for macrobiota assessments. Using the stereo-optic sensor, we assessed the macrobiota, jellyfish, and fish community along the vertical depth profile from 11 m water depth to the surface with the sensors looking from a distance of about 2.5 m towards a steep wall that reached from 11 m of water depth to 3 m below the mean sea level (Fig. 2). The upper part of the wall was dominated by brown algae of the type of *Alaria esculenta*, the lower part by *Saccharina latissima* and the two red algal species *Phycodryis rubens* and *Ptilota gunneri*. Using the vertical profiling unit, we conducted a 1-year continuous stereo-optical survey of the fish and the macrozoobenthos community in five depth strata (11–9, 9–7, 7–5, 5–3, and 3 m from the water surface). The stereo-optical system and the CTD probe were remotely positioned every day between 11:00 and 13:00 h in one of the five depth layers, with the exact depth being calculated as the distance from the bottom. This means that the effective water depth

Table 1. Sensors attached to the COSYNA@AWIPEV UWO at UMT 8763953° N, 433992° E. The FerryBox has its water inlet at a fixed depth of 11 m below mean sea level (http://vannstand.no/index.php/nb/english-section/sea-level-data). The RemOs1 system is profiling from 11 m water depth to the surface (for further descriptions, see the text).

Sensor carrier	Sensor type	Water depth	Sensor unit manufacturer
FerryBox	Water temperature (°C)	11 m	SBE45
	Conductivity (ms m^{-1})/salinity (PSU)[1]		SBE45
	Oxygen (%)		Anderra
	Chl a (mg m^3)		Cyclops
	pH		Meinsberg
	Turbidity (FTU)		Seapoint
Underwater node	Current (ADCP Teledyne Workhorse 600 kHz)	13 m	Teledyne
Underwater node	Stereo-optical imaging system RemOs1	Profiling[2]	Fischer et al. (2007)
	Pressure (dbar)		
	Water temperature (°C)		
Underwater node	Conductivity (ms m^{-1})/salinity (PSU)[1]	Profiling[2]	Sea&Sun CTD90
	Oxygen (%)		
	Chl a (mg m^3)		
	Turbidity (FTU)		

[1] Calculated after actual UNESCO procedures. [2] Between 11 m water depth and the surface.

changed with the tide cycle for max. 1.5 m, but the system itself had a fixed position above the ground (1 m distance from the bottom for the depth stratum 11–9 m, 3 m distance for the depth stratum 9–7 m, 5 m distance for the depth stratum 5–7 m, 7 m distance for the depth stratum 3–5 m, and 9 m distance for the depth stratum 3–0 m). The daily target depths were selected randomly for each week such that all of the depth strata were sampled once per week for 24 h. Missing depths, e.g. because of system or connection problems to the underwater observatory, were repeated on the weekend. The system was positioned for 24 h at the selected depth stratum and made stereoscopic images every 30 min. In parallel, all other in situ and FerryBox sensors recorded with a frequency of 1 Hz. The image pairs and all the hydrographic data were transferred automatically via the Internet to Germany for further daily processing. All hydrographic data were automatically quality controlled by automated procedures, flagged as good, probably good, and bad, and stored at a central data server in Geesthacht, Germany, under an open-access policy at http://codm.hzg.de/codm/. For our study, only the data with the quality flags probably good and good were used. Based on these data, we analysed the temporal succession of the shallow water fish, jellyfish, and macrozoobenthos community in this kelp-dominated shallow water Arctic habitat in Kongsfjorden. Organisms on the stereoscopic images were analysed in a two-step procedure following the routines described in Wehkamp and Fischer (2014). The 48 stereoscopic image pairs of each day were first scanned manually for the presence of organisms. This scanning was performed with image analysis software that presented the left image of the stereoscopic pair for at least 5 s on a 21″ high-resolution computer screen. Only two persons did this basic analysis step over the entire year and thoroughly counterchecked their object findings. During this first step, all the specimens found on an image were counted and pre-classified into the categories fish, jellyfish, appendicularia, pelagic crustacean, benthic crustacean, pteropods, and chaetognats. Organisms that could not be classified into one of these categories were classified as "others". The analyser (the person who did the analysis) had the possibility of increasing or decreasing the image brightness or of enhancing the contrast by a single mouse click quickly. The possibility of such a rapid pre-processing of the 48 stereoscopic image pairs was revealed to be most important because 48 image pairs were produced every day year-round. This rapid assessment procedure allowed a first analysis of all the images per day within approximately 15 min, so that a quasi-online overview of the actual situation under water in the target area and of the functioning of the monitoring system was achieved within 24 h. With this procedure, problems of the system itself or with the data transfer could be detected fast and could be addressed and solved. With this daily rapid assessment routine, we could achieve an acceptable level of operational stability of the systems with less than 15 unplanned offline days over the entire sampling period of 13 months. Unplanned offline days occurred mainly due to failures in the land-based power support system. During such phases, the underwater part of the system was shut down to avoid hardware damage due to spontaneous and possibly critical voltage fluctuations.

In a second image analysis step, all the images where organisms were detected were rectified, which means that the geometry of the images was corrected to eliminate image distortions due to the lens of the camera. This correction was performed with the "stereo_gui" modified MATLAB routine (Wehkamp and Fischer, 2014). After this step, all the objects that were detected in the first image analysis step were measured (standard length in fish, carapax length in macrocrustacea, and max. dimension in all other organisms) and identified as precisely as possible, i.e. to species level in most fish species except for the two cod species *Boreogadus saida* and *Gadus morhua*, which were not distinguished properly on the images. Furthermore, amphipoda or appendicularia were only identified to the class level.

Because we had a clearly restricted water volume that was assessed by the camera system (volume between the camera and the vertical wall), we calculated the "catch per unit effort" of the system by summarizing all the individuals found on the images per 24 h and depth stratum. These CPUE $\times 24\,h^{-1}$ data were used as the basis for all further calculation. We did not recalculate these data on a defined water volume (which is possible) to avoid confounding calculations between benthic organisms living on the two-dimensional bottom or the surface of the algae and planktonic organisms living in the three-dimensional water column.

Length–frequency measurements on the three-dimensional-image pairs were performed pooled for each month for the cod species (mainly *Gadus morhua*), the common sea spiders (*Hyas araneus*), the two main jellyfish species (*Beroe* sp. and *Aglantha digitale*), the appendicularia, and the pteropods (*Clione limacina*). For these species, all the organisms were measured except for the month when more than 200 specimens occurred within 1 month. In this case, only 200 specimens were measured by randomly selecting over the day of the month.

3 Results

3.1 Habitat description

The Kongsfjorden shallow water ecosystem is characterized by large kelp beds of different species of macroalgae between 0 and approximately 12–15 m water depth (Bartsch et al., 2016). The site where the observatory has been set up is, therefore, characteristic of the fjord habitat and provides a highly diverse habitat with a steep wall completely covered with large macroalgae followed by a sandy to muddy slope that begins at approximately 11 m water depth at the base station of the observatory. The five depth layers covered by the stereo-optical camera system cover the typical vertical gradient of a littoral habitat with a surface near-pelagic habitat (depth range 0–2 m water depth (Fig. 3a), a typical litho-pelagic habitat close to the upper edge of the drop-off (2–4 m water depth (Fig. 3b), the upper drop-off edge between 4 and 6 m water depth) with dense horizontal and vertical macrophyte coverage (Fig. 3c), the vertical wall of the drop-off with overhanging structures and grotto-like crevices

(water depth 6–8 m, Fig. 3d) and, finally, the lower edge of the drop-off where the wall goes over in the typical benthic habitat with a gentle slope formed by sand and mud at a depth of around 11 m, decreasing further towards north to the centre of the fjord (Fig. 3e).

The observatory technology allows for daily vertical CTD profiles every noon at approximately 12:00 with a sampling frequency of 1 Hz at a constant profiling speed of 1.5 m per minute from approximately 10 m water depth (depending on the tide) to 1 m below the surface. The FerryBox unity additionally provides complementary hydrographic data from a fixed water depth of 11 m. Figure 4 shows the compiled data for water temperature (°C), salinity (PSU), and turbidity (FTU) from October 2013 to November 2014. The data reveal a distinct seasonal cycle in the water temperature, with the lowest values of approximately $-1.0\,°C$ in the winter months from October to April and the highest temperatures up to approximately $8\,°C$ during the summer months, May to September. Most interestingly, however, are the distinct short-term changes in water temperatures even within the individual seasons. These changes spanned ranges of up to $4\,°C$ within the shortest time periods of a few days both in the summer and in the winter. While the average water temperature, for example, during the middle of December to the end of January was between -0.5 and $+0.5\,°C$, the water temperatures then suddenly increased within a few days up to $3\,°C$ and stayed at this comparatively high level until the end of March, when it dropped again to approximately $0.5\,°C$. In May, the temperatures increased again and reached the highest values of up to $7.7\,°C$ in the surface layers, which indicates a distinct stratification during this time. In July to September, this stratification dissolved, and the water temperatures were almost equally distributed over the water column. Similar temporal patterns were observed also in salinity (Fig. 4), which indicates that the overall patterns in the water temperature in the shallow littoral zone of the fjord system were also significantly determined by a fast (within days) exchange of water masses that brought either colder and lower saline Arctic water or warmer higher saline water masses even to the shallow fjord areas.

Figure 4 shows the seasonal patterns in turbidity over the water columns. The data indicate that the overall turbidity significantly increased during the seasonal cycle, with higher values from July to September and low values during the rest of the year. However, Fig. 4 also shows a longer lasting local and distinct increase in turbidity close to the bottom in May and June. These high turbidity values during this time are confirmed by both systems, the vertical profiling in situ probe as well as the FerryBox unit.

3.2 Species community

Figure 5 (upper panel) shows the sum of individual organisms counted on the images per week for the months October 2013 to November 2014. The average values and stan-

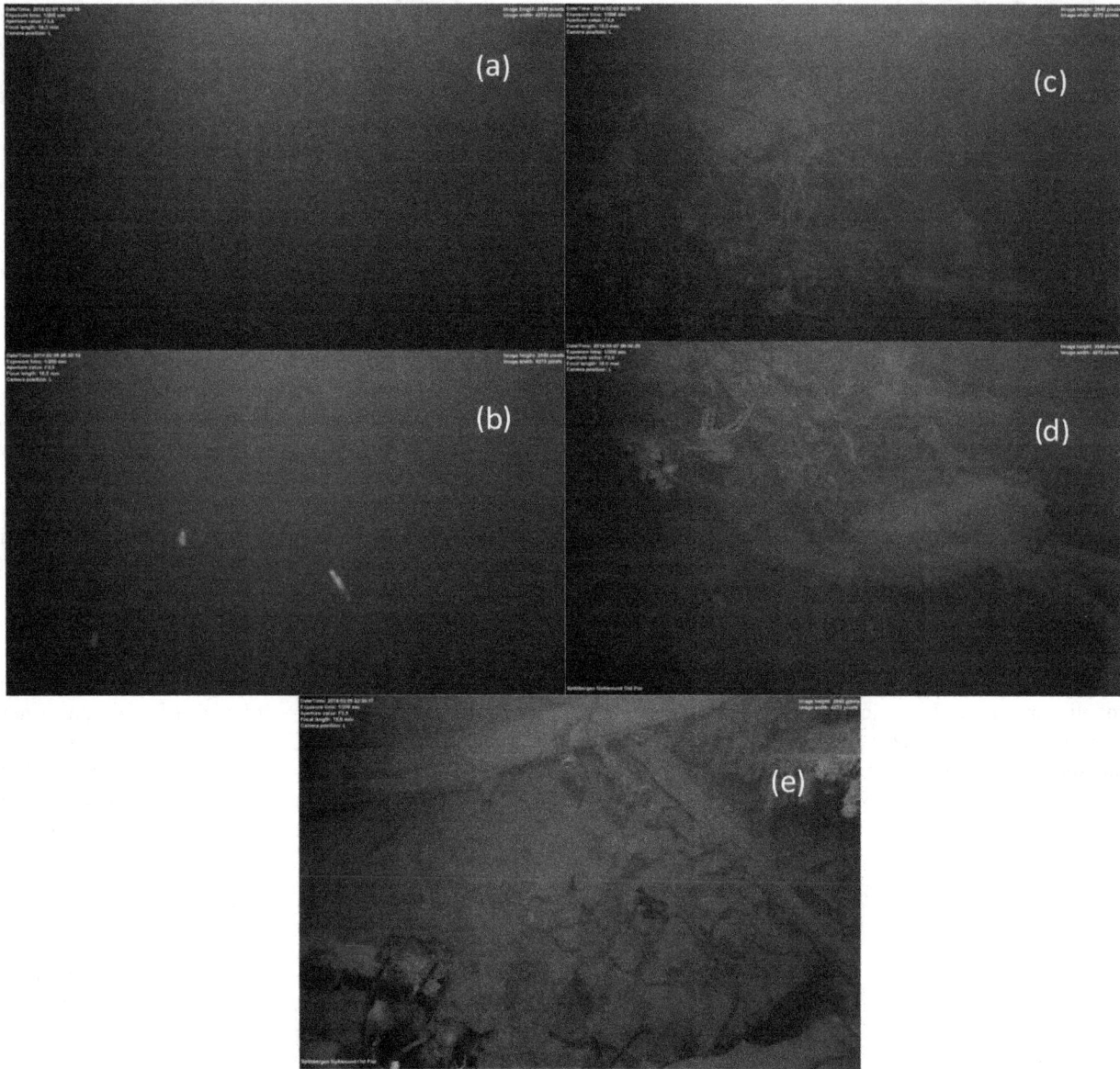

Figure 3. (a, b, c, d, e) View of the RemOs1 stereo-optical system in the five different depth strata. **(a)** Depth stratum 0–2 m, **(b)** depth stratum 2–4 m, **(c)** depth stratum 4–6 m, **(d)** depth stratum 6–8 m, and depth stratum **(e)** 8–11 m.

dard deviations per month were calculated based on four or five weekly CPUE values depending on how many weeks a month had. The analysis revealed a distinct seasonal cycle with high specimen abundances during the winter months from December to April, lowest values from May to July, and a second smaller peak in August and September. Figure 5 (lower panel) shows the same monthly abundance values but separated by groups of organisms. Ten different groups of organisms were identified over the year, namely, appendicularia, benthic crustacea, birds, chaetognaths, fish, jellyfish, molluscs, pelagic crustaceans, polychaets, and pteropods. From these groups, six occurred in higher abundances, at

least during a certain phase of the year (benthic crustacean, fish, jellyfish, appendicularia, chaetognaths, and pteropods).

During the winter–spring peak, benthic crustaceans had the highest share of the total species abundances, followed by jellyfish, pteropods, and fish (Fig. 5, lower panel). In contrast, the summer–autumn peak was almost completely formed by appendicularia and a smaller share of fish.

When analysing the winter–spring phase (December–March) and the summer–autumn phase (August–October) separately and in detail, a strong spatial separation of the winter–spring and summer–autumn communities emerged with respect to the position in the water column (Fig. 6). While the overall share of the winter–spring community was

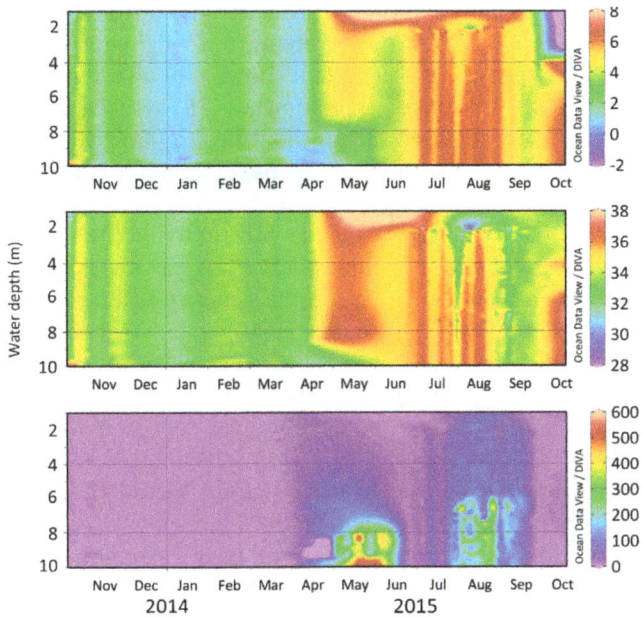

Figure 4. Temporal–spatial pattern in water temperature (°C – upper panel), salinity (PSU – central panel), and turbidity (FTU – lower panel) from October 2013 to October 2014 for the depth range 1 to 11 m based on daily vertical CTD profiles from 10 to 1 m and the FerryBox data from 11 m (fixed inlet).

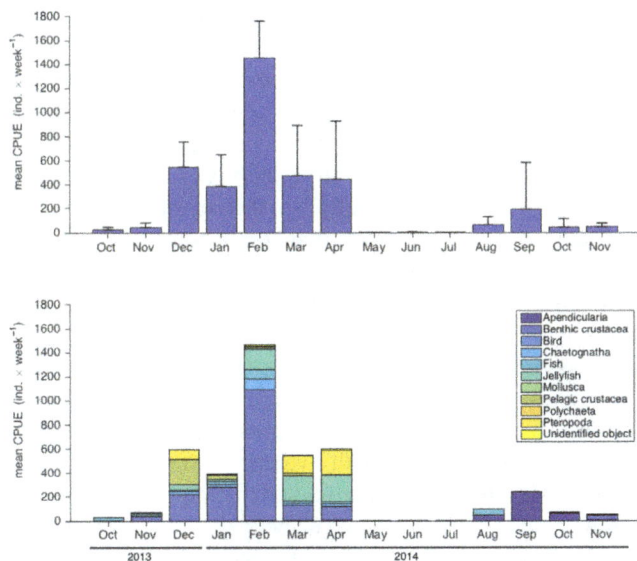

Figure 5. Seasonal cycle in total species abundance (upper panel) and species composition (lower panel) pooled per month of the year. For details with respect to "Catch per unit effort", see the text.

benthic or benthic-associated except for the jellyfish, this benthic-associated community was almost completely missing in the summer and autumn, except for a small share of fish.

Except for appendicularia, all of the other highly abundant species were identified to the species level if possible. Fig-

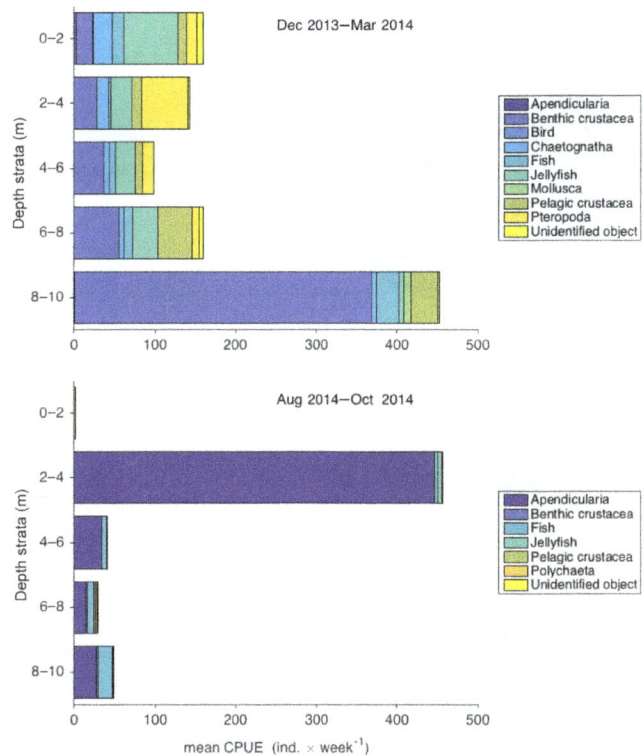

Figure 6. Vertical distribution of the different species groups over the water columns. For details with respect to "Catch per unit effort", see the text.

ure 7 shows the species composition of benthic crustaceans (upper panel), fish (middle panel), and jellyfish (lower panel). The analysis revealed that approximately 90 % of the benthic crustaceans identified over the year were made up of a single species, the great spider crab *Hyas araneus* (L.). In addition, hermit crabs (*Paguridae*) were also found occasionally as well as benthic living decapod crustaceans, which most probably belonged to the mysid species *Mysis oculata* (approximately 10 % share). *Hyas araneus*, however, clearly dominated the benthic decapod community, especially in the winter month of February, when a mass invasion of this species was observed in the area.

A similar uniform pattern was observed in fish (Fig. 7 – middle panel); 81 % of the fish on the images were classified as cod of either one of the two species *Gadus morhua* (L.) (50 %) or *Bodeogadus saida* (L.) (31 %). The differentiation of these two species, however, has to be perceived critically because it was based on coloration, which is especially problematic in young specimens. For all the subsequent analyses, we pooled these two fish species and summarized them under "Gadidae".

The most diverse groups over the year were the jellyfish (Fig. 7 – lower panel). A total of nine different species plus one class "unidentified" were found. Integrated over the year, the most dominant jellyfish species (57 %) belonged to the

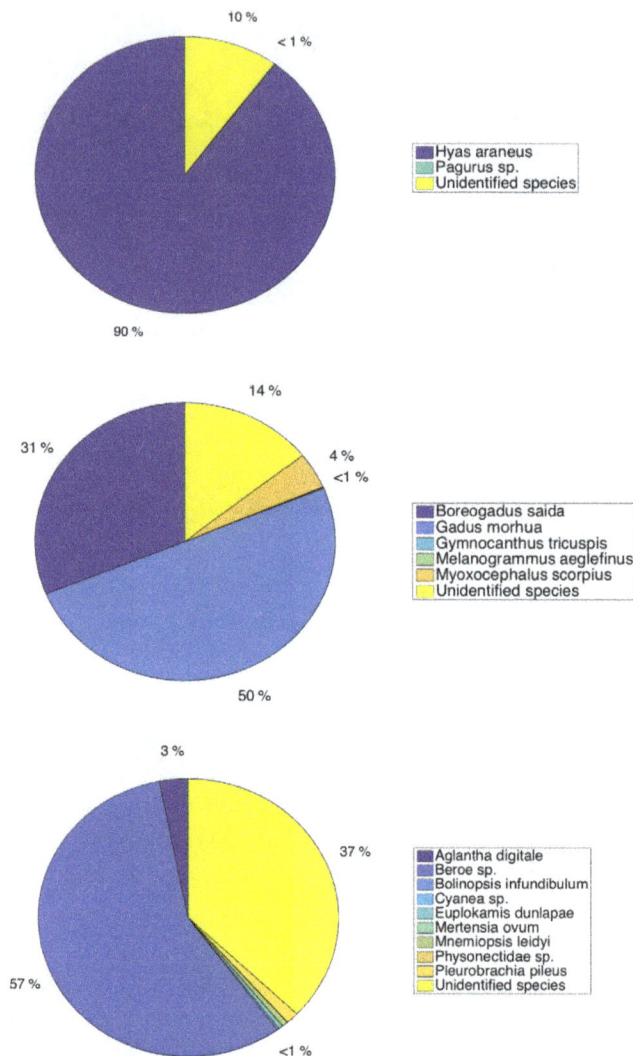

Figure 7. Percent distribution of the different species within the different biota groups. For details, see the text.

group *Beroe* sp., followed by *Aglantha digitale* (8 %) and *Pleurobrachia pileus* (5 %). All the other identified species (*Physonectidae* sp., *Mnemiopsis leidyi*, *Mertensia ovum*, *Euplocamis dunlapa*, *Cyanea* sp., *Bolinopsis iunfundibulum*, and *Aglantha digitale*) occurred in abundances with a total share of < 1 %. Unfortunately, 37 % of the jellyfish could not be clearly identified to the species level and, therefore, had to be left unidentified. These species most certainly did not belong to the above-mentioned identified species, which indicates that the jellyfish diversity in this area is even higher.

For the dominant species of the six major biota groups (benthic crustacean, fish, jellyfish, appendicularia, chaetognaths, and pteropods), the body sizes were measured for up to 200 randomly selected specimens per month (if available). In benthic crustaceans, the carapax length from the tip of the rostrum to the end of the telson (in a normal body position) was measured; for fish, the standard length; for jellyfish, the

largest body dimension (either longitudinal or transversal); and for chaetognaths and pteropods, the longitudinal body axes were measured. The system allowed for an accuracy in length measurements of approximately 3 % (Wehkamp and Fischer, 2014). Figures 8 to 10 show the size–frequency distributions of the six measured groups per month over the seasonal cycle from October 2013 to November 2014. As the most abundant species during the winter months, November to March, *Hyas araneus* showed an average carapax length of between 50 and 100 mm (Fig. 8 – upper panel) with no temporal trend over the months. However, in November and December 2013, larger animals with a carapax length of up to 180 mm also appeared in the area, which disappeared during the spring and re-appeared again 1 year later in November 2014.

In contrast, in the pooled species group "Gadidae", a clear increase in the average length over the months was observed (Fig. 8 – lower panel). Starting in November 2013, the young-of-the-year (YOY) cohort appeared in the area with an average standard length between 70 and 100 mm. This 2013 cohort stayed in the area until March 2014, when they reached an average length between 100 and 125 mm. After this time, no more cod was observed in the area over the spring and summer until then next YOY cohort appeared for a short time in higher abundances in August 2014 with an average standard length between 40 and 70 mm (mean $\pm$ SD = 65 $\pm$ 16 mm). After this time, no more YOY cod could be observed in the shallow area. Instead, larger cod of up to 300 mm were observed sporadically in the shallow waters (Fig. 8 – lower panel, September–October 2014).

All of the other species that occurred in higher abundances in the shallow areas around NyÅlesund belonged to the pelagic community. In jellyfish, the ctenophore *Beroe* sp. made up a major share of the planktonic community and appeared with higher abundances in the winter months, November to April, but with only a few specimens during the summer months. For *Beroe* sp., no temporal size distribution pattern was observed over the months (Fig. 9 – upper panel). The highest abundances were observed in February, with an average size in the longitudinal direction of 45 mm spanning from 10 to 75 mm with average values of 32 $\pm$ 8 mm (mean $\pm$ SD). Jellyfish occurred with the highest abundances in the shallow-most water layer between 0 and 2 m and in only lower abundance in the water columns between 2 and 8 m. In the deepest water layer close to the bottom, the abundances of *Beroe* sp. were the significantly lowest over the entire water column (LRχ^2 = 105, df = 3, $p < 0.001$).

Another temporally dominant but more agile species compared to the jellyfish were the chaetognaths. This group also occurred with the highest abundances during the winter months (Fig. 9 – lower panel) and were also completely missing during the polar summer. Compared to the jellyfish, however, which were almost equally distributed over the water column except for the deepest stratum, Chaetognath occurred highly stratified in the water columns, with the high-

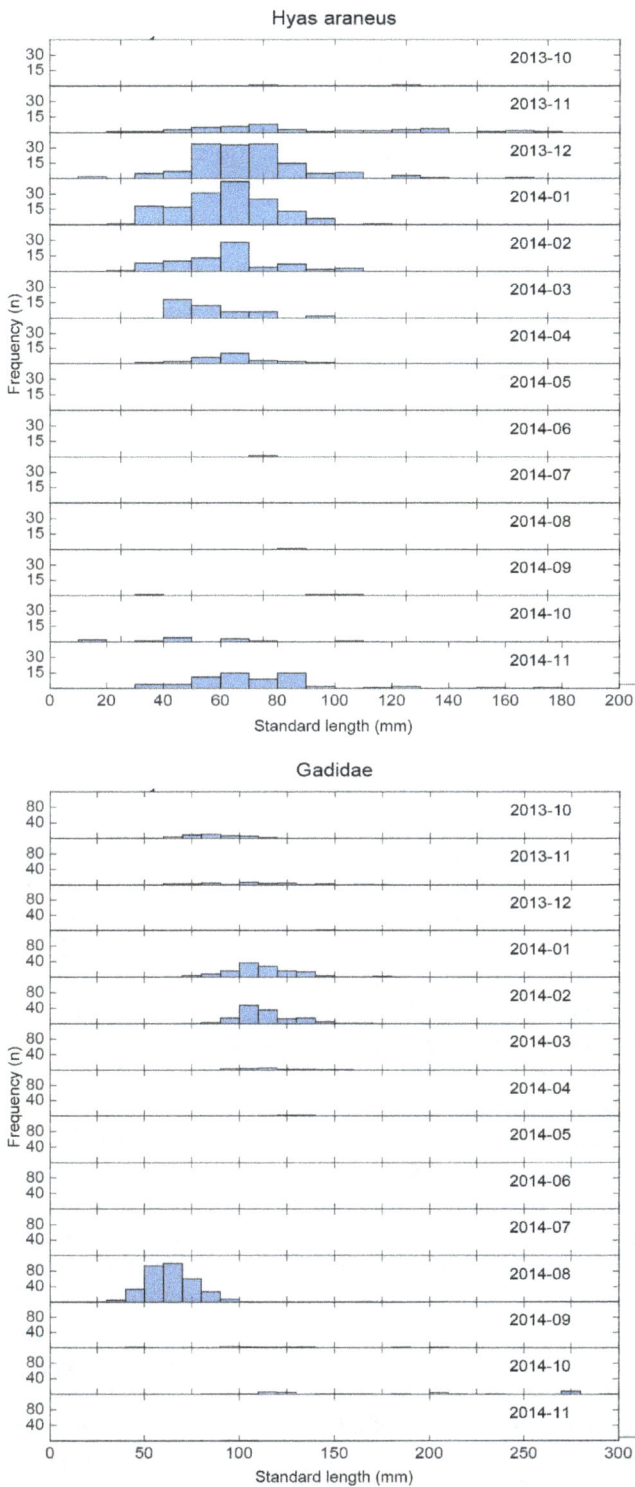

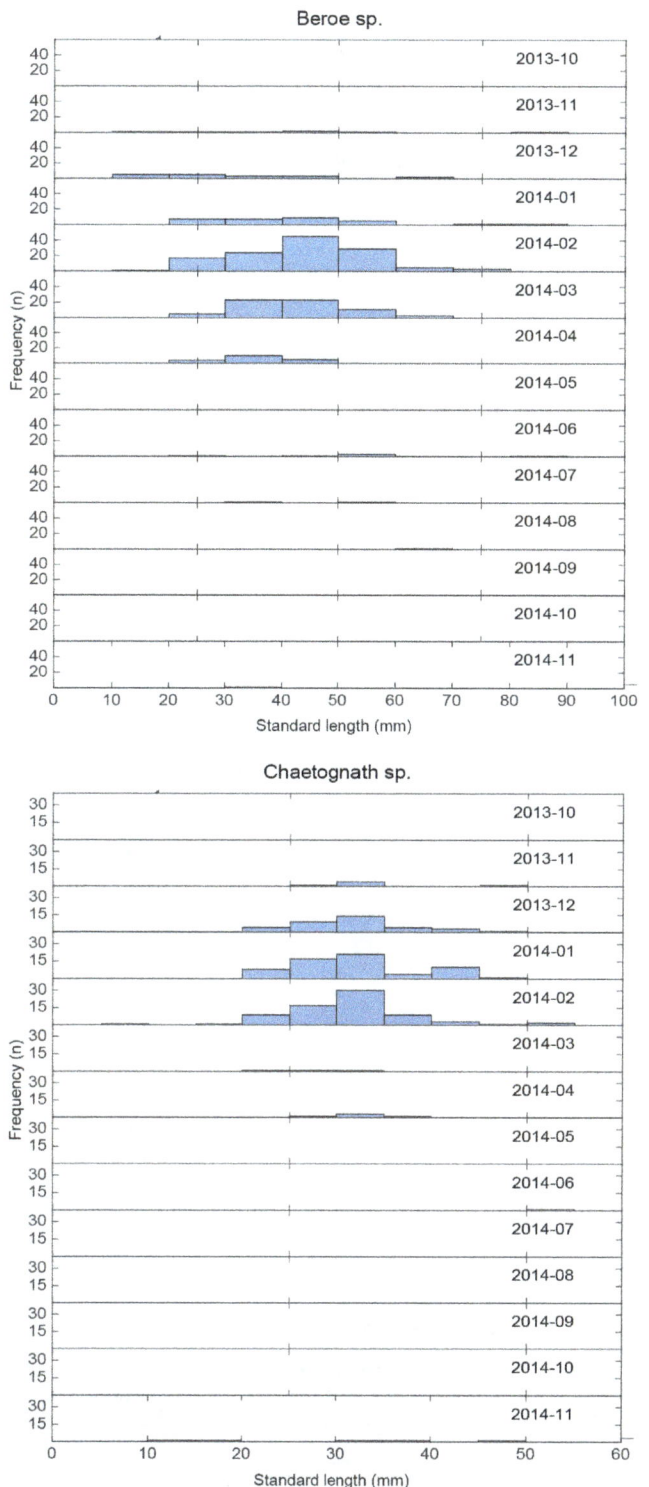

Figure 8. Length–frequency distributions of selected species or species groups (see panels) over the seasonal cycle.

Figure 9. Length–frequency distributions of selected species or groups (see panels) over the seasonal cycle.

est abundances in the 2–4 m depth layer; no specimen was found in the surface layer shallow than 2 m, and significantly lower abundances were also found in the deeper water layers ($\text{LR}\chi^2 = 490$, $\mathrm{d}f = 3$, $p < 0.001$). With lengths between 20 and 50 mm (mean $\pm$ SD $= 32 \pm 8$ mm), chaetognaths formed a major part of the pelagic winter community in the shallow areas. A detailed image based on species identification

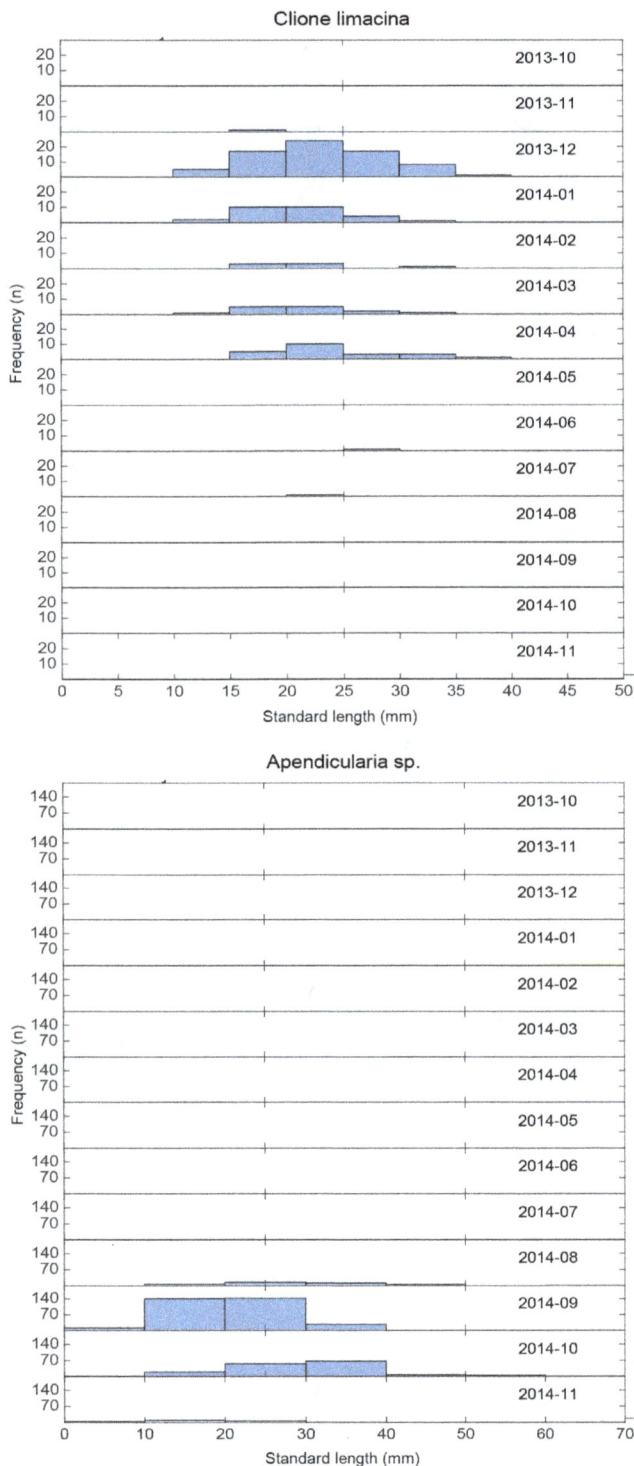

Figure 10. Length–frequency distributions of selected species (see panels) over the seasonal cycle.

as well as on the size distribution of the observed chaetognaths suggests that the majority of the observed specimens belong to the species *Parasagitta elegans* (Verrill, 1873).

Temporally, almost synchronized with the chaetognaths, pteropods (Fig. 10 – upper panel) also occurred in the water column and were observed in higher abundances until April. On the images, only *Clione limacina* was observed with body sizes from 10 to 40 mm and a mean size of 23.1 ± 5.5 mm (mean $\pm$ SD). Similar to the above-described chaetognaths and jellyfish, *Clione limacine* also occurred highly stratified in the water column, with a peak abundance in the 2–4 m depth layer and significantly lower abundances both in the surface layer and in deeper water strata (LR$\chi^2 = 143$, d$f = 4$, $p < 0.001$).

The only species that reached higher abundances not in winter but during the summer months were the appendicularia (Fig. 10 – lower panel). Especially during the months August to October a mass invasion of appendicularia in the upper water columns was observed. As for the other pelagic species, those higher abundances were mainly observed in the 2 to 4 m water layer, while no appendicularia were observed in the uppermost layer close to the surface and significantly lower abundances were observed below 4 m water depth (LR$\chi^2 = 1039$, d$f = 3$, $p < 0.001$).

4   Discussion

Shallow water areas are well known as important habitats for shallow water fish communities (Reyjol et al., 2005). Due to the often higher structural complexity of shallow coastal waters compared to the deeper parts of the ocean, coastal habitats are often observed as important spawning areas and nursery grounds that form the biological backbone of a diverse and stable benthic and fish community in the associated marine habitats. For the same reason, however, studying higher tropic biota in coastal environments is challenging with regard to a detailed assessment of their temporal and spatial dynamics, especially of mobile communities. The high structural complexity, especially of shallow water hard bottom or reef habitats, often prevents classical ship-supported and space-integrative sampling methods such as trawling or box coring (Brickhill et al., 2005; Fischer et al., 2007a; Wilding et al., 2007). Assessments in these structurally complex environments often require small-scaled and highly specialized "sampling" methodologies often based on optical mapping or imaging technologies operated by divers or ROVs, depending on the water depth. Brickhill et al. (2005), Fischer et al. (2007b), and Wehkamp and Fischer (2014) discussed the potential of such techniques specifically for the assessment of fish–habitat relationships in temperate and boreal habitats such as the southern North Sea. They concluded that in these waters, the comparatively restricted transparency of the water, the lower water temperatures, and the harsher weather conditions often result in only short operation times that result in low numbers of freeze-frame sub-samples taken in most studies, preventing a thorough analysis of the species–habitat relationships due to an insufficiently fine-scale sampling frequency. These limiting factors, especially of diver-operated in situ video technologies, often lead to extremely high variability in organism counts per frame, with too many zero counts, especially when the target organisms are mobile. This leads to a dramatic loss of statistical power in the subsequent data analysis (Brickhill et al., 2005).

These limitations are even more distinct in polar areas where the diver-supported access to the ecosystem is both temporally restricted and extremely expensive. Sampling structurally complex coastal habitats in polar areas is often only possible during a restricted period of time in the polar summer when light is available and the temperatures allow for in situ methods. Therefore, our knowledge of polar shallow water ecosystems and especially their role as nursery and juvenile habitat is extremely restricted. Most of the recent studies (e.g. Hop et al., 2002, 2012; Svendsen et al., 2002) in our addressed study area have been conducted during summer, when the fjord system is accessible by research vessels. Although the summer productive period is of great importance for Arctic ecosystems, several crucial processes (e.g. reproduction) take place during other seasons and especially during the polar winter. During these times, however, almost no information is available in most Arctic fjord systems (Kwasniewski, 2003). Understanding polar ecosystems in the context of global warming and expected or already observed ecosystem changes (Müller et al., 2011; Bartsch et al., 2016) is, however, crucial for thoroughly understanding the ecosystem behaviour in polar areas.

In this study, we do not provide results from experimental work in Kongsfjorden based on discrete studies with a clear short-term ecological hypothesis. In contrast, we provide data from a 1-year long quantitative assessment of hydrographic parameters together with quantitative data on the macrobiota community assessed by a remote controlled cable-connected underwater observatory installed in a typical shallow water habitat in the Kongsfjorden. Using a remote controlled vertical profiling system, we were able to continuously assess temperature, salinity, turbidity, and other hydrographic parameters together with the shallow water macrobiotic community over the entire water column from the benthic over the epi-benthic to the pelagic realm at a high temporal resolution. To our knowledge, this is the first dataset both from Kongsfjorden and from the entire Arctic that reveals such a year-round assessment of the shallow water macrobiotic community together with the quantitative data of the water temperature, salinity, and turbidity and, therefore, allows a deeper insight into the coupling of the seasonal dynamics of the biology and the hydrography compared to pure summer studies. The data reveal a distinct winter community in the fjords' shallow water ecosystem, which by far exceeds the summer community in both abundance and species diversity. Although we have not yet calculated biomass per m^3 for the assessed species, our data clearly show that the species abundance and species richness are highest during the polar winter that begins in December when no more light is available

under water. During this time, except for the appendicularia, most species, including fish (mainly gadids of the species *Gadus morhua* and *Boerogadus saida*), jellyfish (mainly *Beroe* sp.), chaetognaths (*Parasagitta elegans*), pteropods (*Clione limacina*), and smaller benthic and epi-benthic crustaceans (most possibly *Mysis oculata*, C. Buchholz, personal communication, 2016) invade the shallow water zone and build up highest abundances. During this study, an overall peak abundance was observed in February when the common sea spider *Hyas araneus* clearly dominated the community in numbers and biomass for a short time. Only 1 month later in March, however, *Hyas araneus* almost completely disappeared when fish, jellyfish, and pteropods formed the predominant community with respect to the overall abundances. The "winter" community persisted until April and then almost vanished. The time of the winter community "disappearance" highly corresponds to the increasing availability of light under water. Although sunlight is available at NyÅlesund again already during the middle of March (http://www.awipev.eu/awipev-observatories/current-weather/), the inclination angle of the light is still low until April, so that only a small fraction of the sunlight penetrates the water column (personal observation). However, to really correlate the presence of the "winter community" with the availability of light underwater, discrete measurements of the light intensity and light quality are necessary in the different depth strata to reveal whether light is an ultimate factor in the temporal occurrence of the fjords' shallow water winter community or only a proxy associated with another environmental factory. Our data suggest that especially water temperature may also have a significant influence on the spatio-temporal occurrence of the winter community. Our daily sampled temperature profiles clearly show that water temperature in the shallow water areas of Kongsfjorden can change within short times, even in winter, between < 0 and up to 4 °C. In particular, the peak abundance in the common sea spider *Hyas araneus* corresponds to the time of higher water temperature during February, and the collapse of the spider abundance occurred when the water temperatures decreased from 4 °C to only approximately 2 °C again. A similar temporal pattern could also be observed in the overall species abundance in April, when a short cold phase in the water temperature occurred. However, these seemingly corresponding changes in the biotic community and the changes in the abiotic environments may also be purely by chance, and we do not know yet whether there are functional relationships between these observations. The permanent installation of the cabled underwater observatory at NyÅlesund allows us to formulate and test such a hypothesis of a persisting shallow water "winter community" in the fjord system as well as the hypothesized controlling or at least affecting abiotic factors.

Our data additionally reveal another distinct community during the summer months when the temperatures increased up to 8 °C in the fjord. Then, appendicularia occurred in higher abundances for a restricted time, i.e. from August to

October, in the shallow water with a peak in abundances in September. In contrast to the winter community, which was mainly benthic or at least benthos-associated, this summer community was almost completely dominated by a single appendicularia species, most certainly belonging to the genus *Oikopleura* sp. (Dahms et al., 2015).

Besides appendicularia, juvenile cod fish were also found in September in the deeper littoral water layers closely associated with benthic habitats. The detailed length–frequency analysis of this cohort reveals that these fish were the YOY offspring of the same year (YOY cohort 2014) with an average standard length of 65 ± 16 mm. The data also reveal that these fish seem to stay in the littoral zone (even though the overall abundances strongly decreased over winter) and continuously grow and reach an average standard length of 100 to 125 mm in February–March at age class 1, when they seem to quantitatively leave the shallow water habitats. This outcome indicates a complex migration pattern of YOY cod in this area with a short winter phase in the littoral zone of the fjord system of Spitzbergen and a later migration towards deeper or offshore habitats as adults. Such temporally restricted shallow water phases have been observed already for several other cod species, especially during their juvenile phase (Pihl, 1982). This has been regarded as a juvenile behaviour to prevent predation by older conspecifics in the deeper adult habitats (Ruiz et al., 1993) as well as an improvement in the foraging efficiency of the juveniles during their non-piscivore microzoobenthic benthic feeding phase (Pihl, 1982).

In contrast to the clearly visible seasonal growth pattern in the cod species, no distinct growth could be observed in any of the other species, even in the highly abundant common sea spider, which showed a persisting size range between approximately 50 and 80 mm during all the winter months, except for the month of November in both years, when larger animals between 120 and 180 m were observed in the area, even though in much lower abundances.

As clearly stated before, this study does not provide a singular hypothesis-driven question; instead, it focuses on a basic assessment of the temporal (and with respect to the water column also spatial) pattern in the macrobiota community distribution and possible hydrographic factors that influence the shallow water biota. The results of this study are by far incomplete and only represent a 1-year study at a specific site in the Kongsfjorden ecosystem, which may or may not be representative of the shallow water community of this area. However, the study presents a continuous year-round dataset at a temporal resolution of 1 week, which is, to our knowledge, not available in any other fjord system, and especially not in the Arctic environment, where winter data are missing at almost every level. However, even though the data provide a unique year-round insight into a polar shallow water fjord community, we can assume that the technology used here has a certain bias with respect to species selectivity. Therefore, these data have to be taken with care. For instance, comparing our stereo-optically assessed fish data with data from classical sampling devices in Kongsfjord (Brand and Fischer, 2016; Hop et al., 2002; Renaud et al., 2011) or even with sporadic diver observations (Brand and Fischer, 2016; Hop et al., 2002), it becomes clear that our optical sensors are also species selective. Brand and Fischer (2016) for example reported for the summer month a distinct occurrence of the benthic sculpin *Myoxocephalus scorpius*, a typical temperate and highly camouflaged benthic fish species in fyke-net catches. Although we detected *Myoxocephalus scorpius* during summer also on the stereoscopic images, the overall abundance remained quite low. Unfortunately, the fyke-net catches of Brand and Fischer (2016), as with most other available marine studies of the fjord, are only available for the polar summer months, when our stereo-optical data revealed the lowest overall biota abundance at all. However, taking into account that fyke nets are highly time integrative and catch fish only directly at the bottom, the fyke-net and optical data may be complementary rather than contradictory. In the study of Brand and Fischer (2016), fyke nets with a mesh size of 12 mm and a steering net of 18 mm were used. This type of net gear is highly selective for strictly benthic fish species with a high potential of entanglement, such as sculpins. In contrast, a stereo-optical method is most probably less selective for benthic highly camouflaged fish species and may significantly underestimate fish with these characteristics.

Instead, our overall image assessment procedure was thoroughly performed by two different persons and showed similar results with respect to the quantitative detection of even small benthic mysids. Therefore, we assume that we would have also detected sculpins if available in higher abundances and thus conclude that the quantitative relation of the average abundance between the major fish species found on the images might be more precise, as found in the fyke net catches. This outcome seems to be supported also by the available diver observations in that area, at least during summer. Hop et al. (2002) and Renaud et al. (2011) both reported the cod species *Gadus morhua* as one of the most abundant species in the area, which would be in accordance with our findings. Nevertheless, the comparison of these two methods shows that there is a large uncertainty with respect to the methodological approach that should be used in future studies. Furthermore, our in situ optical methods allow for a low-invasive abundance estimate, for a precise length–frequency analysis of the mapped fish, and also for a continuous year-round assessment of the community. However, it does not allow for further investigations such as stomach content analysis and precise aging based on scale or otolith analysis. If we manage to combine such continuous hydrographic and community observations using cable-connected observatories with classical ground truthing fishing or sampling methods, we may reduce our scientific fishing effort to a limited number of specimens, which are needed for specific detailed analysis such as stomach content and otolith-based aging, and obtain the required more invasive stock abundance and growth data via non-invasive optical methods. These approaches may finally enable the reduction of our fishing effort without losing the required data density and therefore contribute to the increasing scientific demand of a resource conservative science also in fish and community ecology, especially in ecologically sensitive areas such as the polar fjords or marine protected areas.

Next steps and needs

In addition to the ecological and hydrographical results from the Kongsfjorden ecosystem presented here, the study demonstrates the advantages of permanently operated cabled observatory technology – especially when combined with other research methods in a multidisciplinary approach integrating biology with the understanding of the physical environment. Cabled observatories with continuous power supply and network access allow the use of state of the art IT technology and smart-monitoring approaches under water. These are often not applicable in mooring-based sensor technology because no feedback to the operator is possible and therefore the researcher himself cannot react to specific environmental situations during the measuring process. Furthermore, complex sensor systems like profiling videos or stereo-imaging systems often cannot be operated unsupervised for longer times because the controlling software is either too complex, the power consumption is too high, or the required test and development phases for unsupervised operation of such systems are too long and therefore too expensive. Cabled observatories with permanent access, power supply, and systems control allow even complex sensor systems to be operated for longer periods because in case of failures, the system can give an alert to an operator elsewhere to request remote control and if necessary sensor reset. Based on our experiences with the cabled observatory in Svalbard, we assume that such underwater research facilities, if operated within an international and well-focused research strategy, may significantly promote our knowledge, especially in remote and sensitive areas like the polar regions.

Competing interests. The authors declare that they have no conflict of interest.

Acknowledgements. We express our strong thanks to the AWIPEV staff, i.e. Rene Buergi and Verena Mohaupt, who made the continuous operation of the underwater observatory in this remote site possible. We furthermore want to thank the numerous divers from the AWI dive group, who did great work during our maintenance missions, as well as María Algueró Muñiz and Cornelia Bucholz for species identification of the jellyfish and the mysids. Special thanks also go to Christian Wiencke, who strongly supported the idea of a cabled underwater observatory at AWIPEV in the initialization phase. Furthermore, we want to explicitly express our thanks to the two reviewers, who gave us great support during the review process.

This work has been supported through the Coastal Observing System for Northern and Arctic Seas (COSYNA).

Edited by: P. Testor

References

Aguzzi, J., Mànuel, A., Condal, F., Guillén, J., Nogueras, M., Del Rio, J., Costa, C., Menesatti, P., Puig, P., Sardà, F., Toma, D., and Palanques, A.: The New Seafloor Observatory (OBSEA) for Remote and Long-Term Coastal Ecosystem Monitoring, Sensors, 11, 5850–5872, doi:10.3390/s110605850, 2011.

Bartsch, I., Paar, M., Fredriksen, S., Schwanitz, M., Daniel, C., Hop, H., and Wiencke, C.: Changes in kelp forest biomass and depth distribution in Kongsfjorden, Svalbard, between 1996–1998 and 2012–2014 reflect Arctic warming, Polar Biol., 39, 2021–2036, doi:10.1007/s00300-015-1870-1, 2016.

Brand, M. and Fischer, P.: Species composition and abundance of the shallow water fish community of Kongsfjorden, Svalbard, Polar Biol., 1–13, doi:10.1007/s00300-016-2022-y, 2016.

Brickhill, M. J., Lee, S. Y., and Connolly, R. M.: Fishes associated with artificial reefs: attributing changes to attraction or production using novel approaches, J. Fish Biol., 67, 53–71, 2005.

Buckland, S. T., Magurran, A. E., Green, R. E., and Fewster, R. M.: Monitoring change in biodiversity through composite indices, Philos. T. R. Soc. B, 360, 243–254, doi:10.1098/rstb.2004.1589, 2005.

Cottier, F., Tverberg, V., Inall, M., Svendsen, H., Nilsen, F., and Griffiths, C.: Water mass modification in an Arctic fjord through cross-shelf exchange: The seasonal hydrography of Kongsfjorden, Svalbard, J. Geophys. Res.-Oceans, 110, C12005, doi:10.1029/2004JC002757, 2005.

Dahms, H.-U., Joo, H.-M., Lee, J. H., Yun, M. S., Ahn, S. H., and Lee, S. H.: Demersally drifting invertebrates from Kongsfjorden, Svalbård (Arctic Ocean) – a comparison of catches from drift-pump and drift-nets, Ocean Science Journal, 50, 639–648, doi:10.1007/s12601-015-0058-5, 2015.

Fischer, P.: Fish, macroinvertebrate and hydrographic data including ctd profiling data from the shallow water area of Kongsfjord, Svalbard from 2013 to 2014, PANGAEA, doi:10.1594/PANGAEA.874141, 2017.

Fischer, P. and Eckmann, R.: Seasonal changes in fish abundance, biomass and species richness in the littoral zone of a large European lake, Lake Constance, Germany, Arch. Hydrobiol., 139, 433–448, 1997a.

Fischer, P. and Eckmann, R.: Spatial distribution of littoral fish species in a large European lake, Lake Constance, Germany, Arch. Hydrobiol., 140, 91–116, 1997b.

Fischer, P., Weber, A., Heine, G., and Weber, H.: Habitat structure and fish: assessing the role of habitat complexity for fish using a small, semiportable, 3-D underwater observatory, Limnol. Oceanogr.-Meth., 5, 250–262, 2007.

Hegseth, E. N. and Tverberg, V.: Effect of Atlantic water inflow on timing of the phytoplankton spring bloom in a high Arctic fjord (Kongsfjorden, Svalbard), J. Marine Syst., 113, 94–105, doi:10.1016/j.jmarsys.2013.01.003, 2013.

Hop, H., Pearson, T., Hegseth, E. N., Kovacs, K. M., Wiencke, C., Kwasniewski, S., Eiane, K., Mehlum, F., Gulliksen, B., Wlodarska-Kowalczuk, M., Lydersen, C., Weslawski, J. M., Cochrane, S., Gabrielsen, G. W., Leakey, R. J. G., Lønne, O. J., Zajaczkowski, M., Falk-Petersen, S., Kendall, M., Wängberg, S.-Å., Bischof, K., Voronkov, A. Y., Kovaltchouk, N. A., Wiktor, J., Poltermann, M., Prisco, G., Papucci, C., and Gerland, S.: The marine ecosystem of Kongsfjorden, Svalbard, Polar Res., 21, 167–208, doi:10.1111/j.1751-8369.2002.tb00073.x, 2002.

Hop, H., Wiencke, C., Vögele, B., and Kovaltchouk, N. A.: Species composition, zonation, and biomass of marine benthic macroalgae in Kongsfjorden, Svalbard, Bot. Mar., 55, , 399–414, doi:10.1515/bot-2012-0097, 2012.

Jørgensen, L. L. and Gulliksen, B.: Rocky bottom fauna in arctic Kongsfjord (Svalbard) studied by means of suction sampling and photography, Polar Biol., 24, 113–121, doi:10.1007/s003000000182, 2001.

Kwasniewski, S.: Distribution of *Calanus* species in Kongsfjorden, a glacial fjord in Svalbard, J. Plankton Res., 25, 1–20, doi:10.1093/plankt/25.1.1, 2003.

Müller, R., Bartsch, I., Laepple, T., and Wiencke, C.: Impact of oceanic warming on the distribution of seaweeds in polar and cold-temperate waters, in: Biology of Polar benthic algae, de Gruyter, edited by: Wiencke, C., 237–270, 2011.

Norwegian Polar Institute (2014): Kartdata Svalbard 1:100 000 (S100 Kartdata) / Map Data [Data set], Norwegian Polar Institute, https://doi.org/10.21334/npolar.2014.645336c7, last access: 31 March 2017.

Paar, M., Voronkov, A., Hop, H., Brey, T., Bartsch, I., Schwanitz, M., Wiencke, C., Lebreton, B., Asmus, R., and Asmus, H.: Temporal shift in biomass and production of macrozoobenthos in the macroalgal belt at Hansneset, Kongsfjorden, after 15 years, Polar Biol., 39, 2065–2076, doi:10.1007/s00300-015-1760-6, 2015.

Peterson, B. J., Holmes, R. M., McClelland, J. W., Vörösmarty, C. J., Lammers, R. B., Shiklomanov, A. I., Shiklomanov, I. A., and Rahmstorf, S.: Increasing river discharge to the Arctic Ocean, Science, 298, 2171–2173, doi:10.1126/science.1077445, 2002.

Pihl, L.: Food intake of young cod and flounder in a shallow bay on the Swedish west coast, Neth, J. Sea Res., 15, 419–432, doi:10.1016/0077-7579(82)90068-0, 1982.

Renaud, P., Tessmann, M., Evenset, A., and Christensen, G.: Benthic food-web structure of an Arctic fjord (Kongsfjorden, Svalbard), Mar. Biol. Res., 7, 13–26, doi:10.1080/17451001003671597, 2011.

Reyjol, Y., Fischer, P., Lek, S., Rösch, R., and Eckmann, R.: Studying the spatiotemporal variation of the littoral fish community in a large prealpine lake, using self-organizing mapping, Can. J. Fish. Aquat. Sci., 62, 2294–2302, doi:10.1139/f05-097, 2005.

Ruiz, G. M., Hines, A. H., and Posey, M. H.: Shallow water as a refuge habitat for fish and crustaceans in non-vegetated estuaries: an example from Chesapeake Bay, Mar. Ecol.-Prog. Ser., 99, 1–6, doi:10.3354/meps099001, 1993.

Stempniewicz, L., Błachowiak-Samołyk, K., and Węsławski, J. M.: Impact of climate change on zooplankton communities, seabird populations and Arctic terrestrial ecosystem – A scenario, Deep-Sea Res. Pt. II, 54, 2934–2945, doi:10.1016/j.dsr2.2007.08.012, 2007.

Svendsen, H., Beszczynska-Møller, A., Hagen, J. O., Lefauconnier, B., Tverberg, V., Gerland, S., Ørbæk, J. B., Bischof, K., Papucci, C., Zajaczkowski, M., Azzolini, R., Bruland, O., and Wiencke, C.: The physical environment of Kongsfjorden–Krossfjorden, an Arctic fjord system in Svalbard, Polar Res., 21, 133–166, doi:10.3402/polar.v21i1.6479, 2002.

Voronkov, A., Hop, H., and Gulliksen, B.: Diversity of hard-bottom fauna relative to environmental gradients in Kongsfjorden, Svalbard, Polar Res., 32, 11208, doi:10.3402/polar.v32i0.11208, 2013.

Walczowski, W., Piechura, J., Goszczko, I., and Wieczorek, P.: Changes in Atlantic water properties: an important factor in the European Arctic marine climate, ICES J. Mar. Sci., 69, 864–869, doi:10.1093/icesjms/fss068, 2012.

Wehkamp, M. and Fischer, P.: A practical guide to the use of consumer-level digital still cameras for precise stereogrammetric *in situ* assessments in aquatic environments, Underwater Technol., 32, 111–128, 2014.

Wehkamp, S. and Fischer, P.: Impact of coastal defence structures (tetrapods) on a demersal hard-bottom fish community in the southern North Sea, Mar. Environ. Res., 83, 82–92, doi:10.1016/j.marenvres.2012.10.013, 2013a.

Wehkamp, S. and Fischer, P.: Impact of hard-bottom substrata on the small-scale distribution of fish and decapods in shallow subtidal temperate waters, Helgoland Mar. Res., 67, 59–72, doi:10.1007/s10152-012-0304-5, 2013b.

Wehkamp, S. and Fischer, P.: The impact of coastal defence structures (tetrapods) on decapod crustaceans in the southern North Sea, Mar. Environ. Res., 92, 52–60, doi:10.1016/j.marenvres.2013.08.011, 2013c.

Werner, E. E.: Species Packing and Niche Complementarity in Three Sunfishes, Am. Nat., 111, 553–578, doi:10.1086/283184, 1977.

Wiencke, C.: The coastal ecosystem of Kongsfjorden, Svalbard. Synopsis of biological research performed at the Koldewey Station in the years 1991–2003, edited by: Wiencke, C., Ber. Polarforsch. Meeresforsch., 492, 1–244, 2004.

Wilding, T. A., Rose, C. A., and Downie, M. J.: A novel approach to measuring subtidal habitat complexity, J. Exp. Mar. Biol. Ecol., 353, 279–286, doi:10.1016/j.jembe.2007.10.001, 2007.

Willis, K., Cottier, F., Kwasniewski, S., Wold, A., and Falk-Petersen, S.: The influence of advection on zooplankton community composition in an Arctic fjord (Kongsfjorden, Svalbard), J. Marine Syst., 61, 39–54, doi:10.1016/j.jmarsys.2005.11.013, 2006.

4

A measurement system for vertical seawater profiles close to the air–sea interface

Richard P. Sims[1,2], **Ute Schuster**[2], **Andrew J. Watson**[2], **Ming Xi Yang**[1], **Frances E. Hopkins**[1], **John Stephens**[1], and **Thomas G. Bell**[1]

[1]Plymouth Marine Laboratory, Plymouth, UK
[2]University of Exeter, Exeter, UK

Correspondence to: Thomas G. Bell (tbe@pml.ac.uk)

Abstract. This paper describes a near-surface ocean profiler, which has been designed to precisely measure vertical gradients in the top 10 m of the ocean. Variations in the depth of seawater collection are minimized when using the profiler compared to conventional CTD/rosette deployments. The profiler consists of a remotely operated winch mounted on a tethered yet free-floating buoy, which is used to raise and lower a small frame housing sensors and inlet tubing. Seawater at the inlet depth is pumped back to the ship for analysis. The profiler can be used to make continuous vertical profiles or to target a series of discrete depths. The profiler has been successfully deployed during wind speeds up to $10\,\mathrm{m\,s^{-1}}$ and significant wave heights up to 2 m. We demonstrate the potential of the profiler by presenting measured vertical profiles of the trace gases carbon dioxide and dimethylsulfide. Trace gas measurements use an efficient microporous membrane equilibrator to minimize the system response time. The example profiles show vertical gradients in the upper 5 m for temperature, carbon dioxide and dimethylsulfide of $0.15\,°\mathrm{C}$, 4 μatm and 0.4 nM respectively.

1 Introduction

Exchange between the ocean and atmosphere is an important process for many gases. Important examples include carbon dioxide (CO_2), for which the oceans account for 25 % of the sink for anthropogenic emissions (Le Quéré et al., 2016), and dimethylsulfide (DMS), which has an oceanic source and influences cloud properties with implications for the global energy balance (Quinn and Bates, 2011). The magnitude and direction of air–sea gas transfer is typically represented by Flux $= K\,\Delta C$ (Liss and Slater, 1974), where ΔC is the concentration difference across the air–sea interface and K is the gas transfer velocity. Direct flux measurements (Bell et al., 2013; Yang et al., 2013; Miller et al., 2010) are only possible for a small number of gases and are not made routinely. Most flux estimates use a wind-speed-based parameterization of K (e.g. Wanninkhof, 2014) coupled with measurements of ΔC.

CO_2 is the most well-observed trace gas in the surface ocean, with 14.5 million measurements compiled into a global database, the Surface Ocean CO_2 Atlas (SOCAT), http://www.socat.info/ (Bakker et al., 2016). Global trace gas databases also exist for gases such as methane and nitrous oxide https://memento.geomar.de/ (Bange et al., 2009), dimethylsulfide http://saga.pmel.noaa.gov/dms/ (Lana et al., 2011) and halocarbons https://halocat.geomar.de/ (Ziska et al., 2013). Accurate estimation of air–sea flux requires concentration measurements that are representative of the interfacial concentration difference. Surface seawater samples are often collected from the underway seawater intake of research vessels, typically at 5–7 m depth. A source of potential error in air–sea flux calculations arises from the assumption of vertical homogeneity within the mixed layer (Robertson and Watson, 1992). If vertical concentration gradients exist in the mixed layer, then underway seawater is not representative of the interfacial layer, which could create a global sampling bias (McNeil and Merlivat, 1996).

Vertical gradients in trace gas concentrations have been observed under conditions that are favourable for near-surface stratification (Royer et al., 2016). At low wind speeds, high solar irradiance can suppress the depth of shear-induced mixing to create a near-surface layer several degrees warmer than the water below (Ward et al., 2004; Fairall et al., 1996). Near-surface stratification in the marine environment can also be induced by freshwater inputs such as rain (Turk et al., 2010) and riverine discharge. Changes in surface seawater temperature and salinity alter the solubility of dissolved gases and thus the amount available for air–sea exchange (Woolf et al., 2016). Dissolved gases isolated in the upper few metres of the ocean may additionally be modified by physical processes such as air–sea exchange and photochemistry. Marine biota confined within the stratified layer (Durham et al., 2009) can also alter trace gas concentrations. For the purposes of this paper, near-surface gradients are defined as physical and/or chemical gradients in the upper 10 m of the ocean.

Identifying and quantifying near-surface gradients in trace gas concentrations is challenging. Ship motion often inhibits near-surface measurements made with the standard oceanographic approach of sampling with Niskin bottles mounted on a CTD rosette. Substantial vertical movement of the rosette limits how close to the surface a sample can be taken. For example, a crane arm 4 m above the sea surface and 11 m from the centreline of a ship that is rolling by $\pm 4°$ will induce ~ 1.5 m sample depth variation every few seconds. CTD/Niskin bottle sampling requires that the rosette is kept below the sea surface. Sampling within 2 m of the sea surface is often impossible, even under relatively calm conditions.

We present a near-surface ocean profiling buoy (NSOP) designed for measuring near-surface profiles. The design principles for NSOP were as follows:

1. platform diameter less than the wavelength of most open ocean waves, allowing it to ride the swell;

2. short sampling arm close to the sea surface to reduce vertical movements induced by platform motion;

3. capable of deployment close to the ship (to retrieve water for trace gas analysis), but away from major turbulence and motion due to the ship itself.

Example profiles from a cruise on the European continental shelf (RRS *Discovery*, DY033, July 2015) and in the English Channel on board the RV *Plymouth Quest* (part of the Western Channel Observatory; Smyth et al. 2010, April 2014) are discussed.

2 Methods

2.1 NSOP description

NSOP is a repurposed ocean buoy (1.6 m diameter) with a central lifting eyelet (Fig. 1). The top of the buoy is 0.5 m above the sea surface. Mounted on top of the buoy are a line of sight, remotely operated winch (Warrior Winch, model C8000) and a gel battery (Haze, model HZY-S112-230). The

winch feeds Kevlar rope through a block and tackle with a 3 : 1 ratio to reduce rope pay-out speed to ~ 0.05 m s^{-1}. The block and tackle is attached to the end of an outstretched arm 0.25 m from the outer edge of the buoy. The winch line is attached to an open frame (0.35 m diameter, 0.8 m height) with the capacity to house multiple sensors. Desired sampling depth is targeted using knowledge of the winch pay-out speed. Rope pay-out is then timed with a stopwatch. This approach only approximately regulates the sampling depth because (i) winch pay-out varies slightly depending on the amount of rope on the spool and (ii) variable horizontal current strength affects the vertical versus horizontal position of the sampling frame. To minimize horizontal movement of the sampling frame we attached a 10 kg weight to the base of the frame.

The primary sensor on the sampling frame is a small CTD (Valeport miniCTD) set to sample at a high frequency (> 1 Hz). Under calm conditions it is possible to sample as close as 0.1 m from the air–sea interface when the miniCTD and tubing are mounted near the top of the frame. Rougher conditions demand that the frame be kept deeper (~ 0.5 m) as motion can momentarily bring the sensors and tubing out of the water. An emergency tag line was attached to the sampling frame in case the winch line failed. Seawater for trace gas analysis was pumped back to the ship at 3.5 L min^{-1} through a 50 m PVC hose (0.5 in inner diameter). A heavy-duty peristaltic pump (Watson Marlow, model 701IB/R) primed with water from the ship's underway supply was used to overcome the large hydraulic head (~ 4 m). The open end of the tubing was located at the same depth as the miniCTD. Water arriving to the ship's laboratory was divided, with ~ 3.0 L min^{-1} for flow-through analysis (e.g. equilibrator for trace gases) and ~ 0.5 L min^{-1} for discrete samples (e.g. total alkalinity).

We assessed the depth resolution capability of NSOP at a particular depth by looking at pressure variations under calm conditions with a fixed amount of winch rope paid out. In calm to moderate conditions (< 2.5 m significant wave height) the amount of vertical movement indicated by the standard deviation (SD) in the depth is ± 0.18 m (Fig. S1, Supplement). During four deployments in rough conditions (> 2.5 m significant wave height), the depth variability increased as the sampling frame was lowered (at 5 m, SD was ± 0.275 m).

2.2 NSOP deployment

On a large research vessel such as the RRS *Discovery*, the deployment and recovery of NSOP requires close coordination between the bridge and three personnel on deck. The NSOP was always deployed while the ship was on station and not at the same time as other overboard deployments. Ship orientation during deployments was typically with bow into the wind but also accounted for swell and current direction/speed. The NSOP was lifted by the aft crane (Fig. 1).

Once the NSOP had been lowered to the surface it was detached from the crane via a quick release. Two slack lines were looped through eyelets on the free-floating NSOP to maintain its position close to the ship. A third slack line was connected to the top of the buoy and passed through a block on a fully extended crane arm of 7 m to maintain this distance between NSOP and the ship. The slack lines successfully inhibited the tendency of NSOP to drift horizontally without disrupting its ability to ride the swell. The instrument frame acted like a sea anchor and minimized rotation of NSOP. A 4 m lifting strop used for recovery was connected to the lifting eyelet and loosely lashed to the aft slack line. During retrieval, the slack lines were hauled in and the crane and jib arms brought towards the ship to bring NSOP alongside. The lifting strop was then parted from the slack line and attached to the crane to lift NSOP back on deck. For additional photographs of a NSOP deployment and videos of NSOP during a deployment and in operation see Fig. S2 and videos.

Turbulence from the ship's propellers has the potential to mix the water column and destroy any near-surface gradients. The ship did not use the aft thrusters whenever conditions were suitable (mild sea state, weak currents and no local hazards). Keeping the NSOP away from the ship limited disruption of near-surface gradients by the thrusters and reduced the risk of line entanglement in the aft propellers. Our winch did not have a groove bar to feed the rope onto the winch drum, leading to an increased likelihood of snagging during spooling. To minimize snagging, the rope was manually fed onto the winch spool before deployments. Visual monitoring of the NSOP frame, slack lines and winch spool is important during deployment.

The NSOP has been successfully deployed in "moderate" sea states up to Beaufort force 5 ($\sim 10\,\mathrm{m\,s^{-1}}$ wind speed and wave heights of $\sim 2.0\,\mathrm{m}$). Deployment length typically varied from 1 to 3 h.

The NSOP can be used in two profiling modes: "continuous" and "discrete". Continuous profiling maximizes vertical coverage and involves the winch continuously paying rope in and out at $\sim 0.05\,\mathrm{m\,s^{-1}}$. A complete down/up profile to 10 m can be conducted in approximately 7 min. Depth resolution during continuous profiling is determined by the measurement response time. Instruments with rapid response times such as the miniCTD temperature and conductivity sensors (0.15 and 0.09 s) have theoretical depth resolutions of 0.75 and 0.45 cm respectively. Actual depth resolution will also be affected by the sampling depth variability of the NSOP instrument frame. A measurement setup with a longer response time (such as for seawater CO_2) requires a different approach (see Sect. 2.5).

During discrete profiling, the winch pays out a fixed amount of rope (typically 0.5 m) and the sampling frame is left at a fixed depth. After a fixed sampling period, more rope is paid out. The process is repeated down and then up such that a set of discrete depths are sampled in a "stepped" profile. The discrete profiling depth resolution is determined by

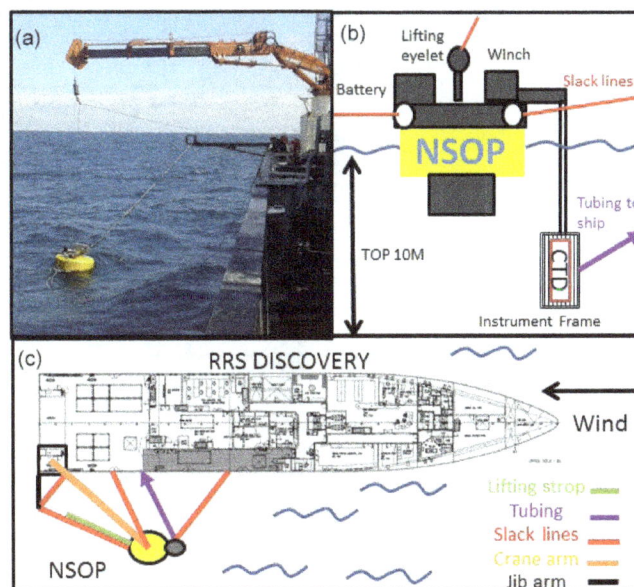

Figure 1. Different points of view of an NSOP deployment: **(a)** image from a deployment on RRS *Discovery* in May 2015 (Cruise DY030), **(b)** schematic cross section of NSOP including tubing back to ship (purple) and slack lines (red), and **(c)** top-down schematic from a research ship including ship orientation. Not to scale.

the depth fluctuations when sampling at a fixed depth (see Sect. 2.1). Discrete profiles are a more appropriate approach for measurement systems with a longer response time. A discrete profile with 0.5 m steps down to 5 m and back to the surface using a 2.5 min sampling period takes about an hour. The sampling period at each depth and frequency/distribution of depths within the profile can be adjusted to suit sampling priorities.

The maximum deployment time is limited by the capacity of the winch battery. When under no load, the battery allows for approximately 3 h of operation in the continuous mode. Discrete profiling requires substantially less winch usage such that battery drainage is even less of a concern.

2.3 CO_2 analysis

The CO_2 measurement system (Fig. 2) is a modified version of the system described by Hales et al. (2004). Seawater from the NSOP inlet was passed through the equilibrator (see Sect. 2.3.1) at $\sim 3\,\mathrm{L\,min^{-1}}$ and the flow rate monitored (Cynergy ultrasonic flow meter, model UF25B). A compressed nitrogen gas supply, maintained at a constant flow rate of $100\,\mathrm{mL\,min^{-1}}$ (Bronkhurst mass flow controller, model F-201-CV-100) flows through the equilibrator in the opposite direction to the seawater flow. The gas has high water vapour content after equilibration and is dried (Permapure nafion dryer, model MD-110-48S-4). The dried sample then

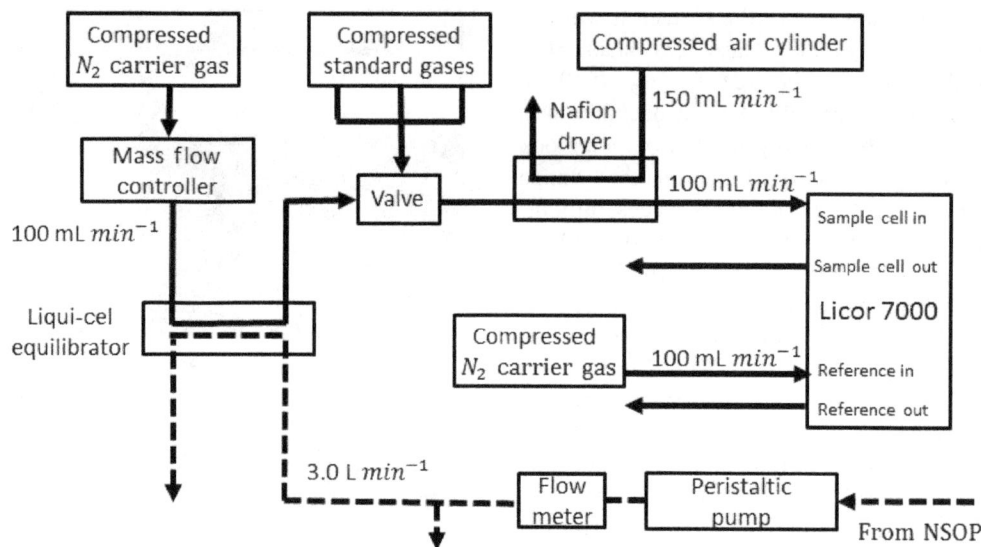

Figure 2. CO_2 system schematic. Solid and dashed arrows correspond to gas and water flows respectively. The LI-COR reference cell is flushed with equilibrated gas at 100 mL min^{-1}. A manual selection valve was used to switch between equilibrated gas and the CO_2 standards.

enters the analytical cell of a NDIR LI-COR 7000, which is protected with a 0.2 µm filter (Pall, Acro 50).

CO_2 measurements at atmospheric pressure as recommended by Dickson et al. (2007) were not possible due to the nature of the experimental setup. The continuous gas flow through the system caused a small 0.4 kPa pressure increase in the LI-COR measurement cell; this was in good agreement with a similar observation by B. Hales (0.5 kpa > ambient pressure; personal communication, 2014). The elevated pressure was taken to be representative of the equilibrator pressure and was used to obtain the partial pressure of CO_2 in the equilibrator ($p CO_{2(eq)}$).

The LI-COR was calibrated using three CO_2 standard gases before and after each NSOP deployment. The concentrations of the standard gases (BOC Ltd) were determined by referencing against US National Oceanic and Atmospheric Administration certified standards (244.91, 388.62, 444.40 ppm) in the laboratory. The seawater temperature at the entry and exit ports of the equilibrator was recorded at 1 Hz (Omega ultra-precise 1/10 DIN immersion RTD) using stackable microcontrollers (Tinkerforge master brick 2.1 and PTC bricklet). Equilibrator temperature probes and the miniCTD temperature sensor were calibrated before and after each cruise against an accurate reference sensor (Fluke, model 5616-12, ±0.011 °C) in a stable water bath (Fluke 7321).

Equilibrator

The showerhead equilibrator is the most commonly used equilibrator for CO_2 but takes ∼ 100 s to equilibrate (Dickson et al., 2007; Kitidis et al., 2012; Körtzinger et al., 2000; Webb et al., 2016). This equilibration time is too slow for ef-

fective use during NSOP deployments. We used a polypropylene membrane equilibrator (Liqui-Cel, model 2.5 × 8) with liquid and gas volumes of 0.4 and 0.15 L and a surface area of 1.4 m^2. Due to its large surface area to volume ratio and membrane porosity (50 %), the Liqui-Cel expedites gas transfer and efficiently achieves equilibration (Loose et al., 2009), with a 3 s response time for CO_2 (Hales et al., 2004). Membrane equilibrators have been used by others for trace gas analysis (Hales et al., 2004; Marandino et al., 2009).

Fugacity of seawater CO_2 is calculated from the LI-COR gas phase CO_2 measurement. This approach assumes that the gas phase sample has equilibrated fully with the seawater. We performed equilibration efficiency experiments in a seawater tank using a showerhead equilibrator as a reference. Liqui-Cel equilibration efficiency declined after prolonged exposure to seawater, likely due to biofouling of the membranes. In a fouled equilibrator, equilibration efficiency was a function of the flow rate on both the water and gas side of the membrane. An increased gas flow rate reduces the residence time inside the Liqui-Cel and allows less time to equilibrate (Fig. 3a). Increasing the waterside flow rate moves the gas phase closer to equilibrium because the transfer coefficient in the membrane increases (Fig. 3b).

Cleaning with an acid–base sequence restored the efficiency of a fouled equilibrator. It was necessary to actively pump chemicals through the Liqui-Cel to achieve a full recovery in efficiency. For more details on cleaning techniques, see Supplement. Efficiency reductions in membrane equilibrators like the Liqui-Cel have not been reported by previous studies. Some authors have used 5–50 µm filters to minimize biofouling (Hales et al., 2004) but this was not possible with the NSOP experimental design. If filtering seawater is not possible, we recommend flushing with freshwater after use,

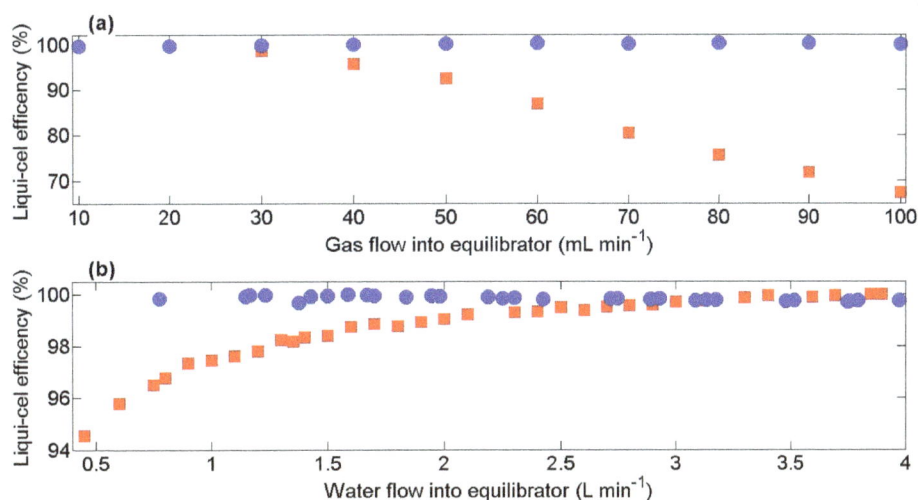

Figure 3. Liqui-Cel CO_2 equilibration efficiency (Liqui-Cel mixing ratio/showerhead mixing ratio) for **(a)** changing gas flow at a fixed water flow rate of $4\,L\,min^{-1}$ and **(b)** changing water flow at a fixed gas flow of $100\,mL\,min^{-1}$. Blue: unfouled equilibrator. Red: fouled equilibrator.

regular cleaning of the Liqui-Cel and daily tests to quantify equilibration efficiency. Trace gas measurement systems that use an internal liquid phase standard (e.g. dimethylsulfide, Sect. 2.4) account for any changes in equilibrator efficiency.

2.4 DMS analysis

DMS was measured with atmospheric pressure chemical ionization mass spectrometry (API-CIMS), using a system modified following Saltzman et al. (2009). Measurements were calibrated using an isotopic liquid standard of tri-deuterated DMS (see Bell et al., 2013 for details). Isotopic standard was injected at $120\,\mu L\,min^{-1}$ into the $3\,L\,min^{-1}$ seawater flow from NSOP before it entered the Liqui-Cel equilibrator. Compressed nitrogen gas was passed through the equilibrator in the counter direction to the seawater flow at $1\,L\,min^{-1}$. The use of an internal standard meant that any incomplete equilibration of the ambient non-isotopic DMS was also true for the isotope. The gas stream exited the equilibrator and was dried (Permapure nafion dryer, model MD-110-48S-4) before entering the mass spectrometer for analysis. DMS was detected at m/z (mass/charge) 63 and the isotopic standard detected at m/z 66. The concentration of DMS was calculated using the ion signals and relevant flow rates (Bell et al., 2015). This approach has been shown to compare well with other analytical techniques for DMS (Royer et al., 2014; Walker et al., 2016).

2.5 NSOP delay and response time

We used different approaches to assess the delay between instantaneous miniCTD measurements and water arriving to the ship for analysis. The delay between seawater entering the inlet and reaching the equilibrator was calculated as $114\,s$ using the internal volume of NSOP tubing (0.5 in inner diam-

eter, 54 m length) and a seawater flow rate of $4.15\,L\,min^{-1}$. Delay correlation analysis between the NSOP miniCTD temperature sensor and a second sensor positioned at the entrance to the equilibrator gives a similar delay of $112\,s$. Note that the total delay of the system is greater because it also includes the time that equilibrated gas takes to reach the LI-COR. We determined the total delay by transferring the seawater inlet quickly between two buckets with distinctly different CO_2 concentrations and timing how long it took for the signal to be detected by the LI-COR ($139\,s$; Fig. 4).

The response time of the NSOP setup was determined by simulating step changes in gas concentrations. A model fit to the exponential change in signal was used to estimate the response time (Fig. 4). We estimate the system response time (e-folding time) for CO_2 as $24\,s$, which is slightly faster than the $34\,s$ reported by Webb et al. (2016). The e-folding time in the DMS signal is estimated as $11\,s$, which is consistent with the rapid gas flow rate through the analytical system.

Continuous profiling with the CO_2 system and a $24\,s$ response time yields a depth resolution of $1.2\,m$, which is greater than the required resolution to assess near-surface gradients. DMS has a faster response time than CO_2, but in continuous profiling mode this only translates to a depth resolution of $0.6\,m$, slightly less than the 1.2–$2\,m$ reported by Royer et al. (2014). A depth resolution of $<0.5\,m$ was desired to capture upper ocean vertical gradients in CO_2 and DMS so NSOP was operated in discrete profiling mode.

2.6 Data processing

During discrete profiling, distinct sample depths were identified from the rapid changes in pressure during depth transitions. Data were binned into discrete depth bins using CTD pressure measurements. Trace gas data were assigned to depth bins after adjusting for the calculated transit time

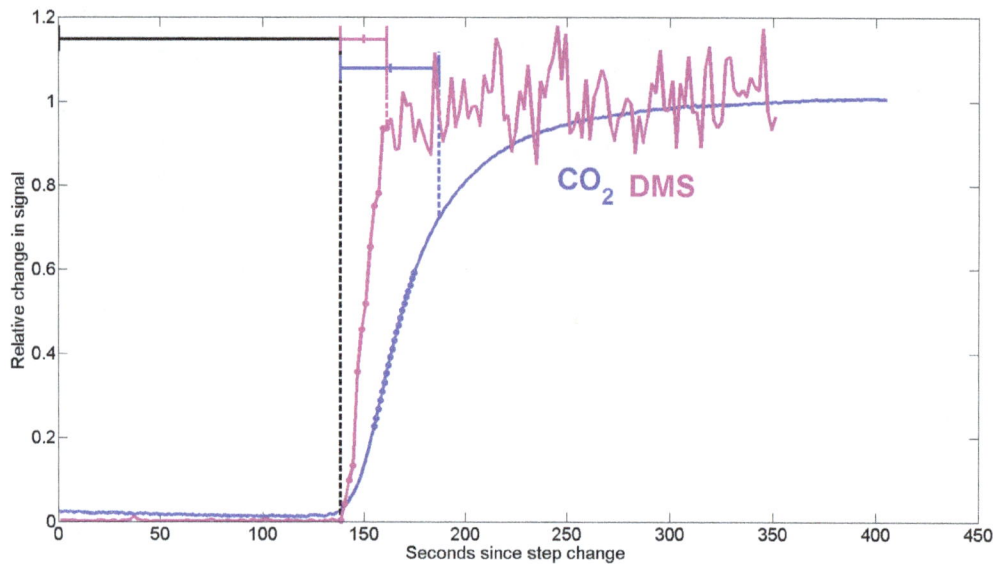

Figure 4. Instrument responses to step changes in seawater CO_2 (blue) and DMS (magenta). Step changes from 350 to 400 µatm for CO_2 and 0 to 2 nmol L^{-1} for DMS have been scaled down so that the initial and end concentrations are between 0 and 1. Time is referenced against the point when the step change was initiated. The response is seen in both instruments after a delay of 138 s (black dashed line). Two e-foldings are indicated by vertical dashed lines for CO_2 (blue) and DMS (magenta). The data points marked by circles were used to make an exponential fit to the data to determine the response time (Sect. 2.5).

through the NSOP tubing (Sect. 2.5). CO_2 data from the beginning (2 e-foldings + 15 s buffer = 63 s) and end (15 s buffer) of each depth bin was excluded from analysis to account for the response time of the system and the transition time between sample depths. The same approach was taken for DMS, where the faster response time resulted in a smaller portion of data excluded at the beginning of each depth bin (2 e-foldings + 15 s buffer = 37 s).

The CO_2 mixing ratio (xCO_2) measured in the LI-COR is converted to equilibrator fugacity ($fCO_{2(eq)}$) using calibration standards, in situ seawater salinity, and the pressure and temperature in the equilibrator (SOP 5# Underway pCO_2; Dickson et al., 2007). Vertical profiles of seawater CO_2 fugacity ($fCO_{2(sw)}$) are calculated using average equilibrator fugacity ($fCO_{2(eq)}$), equilibrator temperature ($T_{(eq)}$) and in situ seawater temperature ($T_{(sw)}$) at each depth (Takahashi et al., 1993).

2.7 Seawater sample collection using NSOP

The NSOP setup enables vertical profiles of discrete seawater samples to be collected from upstream of the equilibrator, with a split in the tubing diverting ~ 0.5 L min^{-1} into a sink. For example, discrete seawater samples (250 ml) have been successfully collected and analysed for Total Alkalinity (TA). Samples were collected and poisoned following best practice recommendations (SOP#1; Dickson et al., 2007). Bottle filling plus one overfill took ~ 60 s. Start and end times were recorded so that collection depth could be retrospectively de-

termined from the CTD pressure data. Analytical methods and an example depth profile are provided in the Supplement.

3 Field measurements/observations

Presented below are example profiles collected using NSOP. The first deployment was in the open ocean (30 July 2015, central Celtic Sea; 49.4213° N, −8.5783° E) from the RRS *Discovery* (100 m length, 6.5 m draught). The second deployment was in coastal waters (15 April 2014, Plymouth Sound; 50.348° N, −4.126° E) from the RV *Plymouth Quest* (20 m length, 3 m draught). A map of deployment sites is supplied in the Supplement.

3.1 Open-ocean deployment

NSOP was deployed at 14:05 (UTC) on 30 July 2015. During the 6 h preceding deployment, the ship was on station and encountered persistently strong solar radiance (>600 W m^{-2}), mild winds (<6 m s^{-1}) and calm sea state (significant wave height < 1.6 m). This combination of low wind speeds and high irradiance (Fig. S5) is favourable for near-surface stratification (Donlon et al., 2002).

Figure 5 presents the time series data collected by NSOP for depth, temperature, salinity and $fCO_{2(sw)}$. Discrete profiling began at 14:05 (UTC) at 0.7 m depth, which was as close to the surface as the frame could be located without the possibility of breaking the surface. Depth bins were identified based on rapid depth transitions (Fig. 5a). Bottles were filled for discrete samples during the downcast.

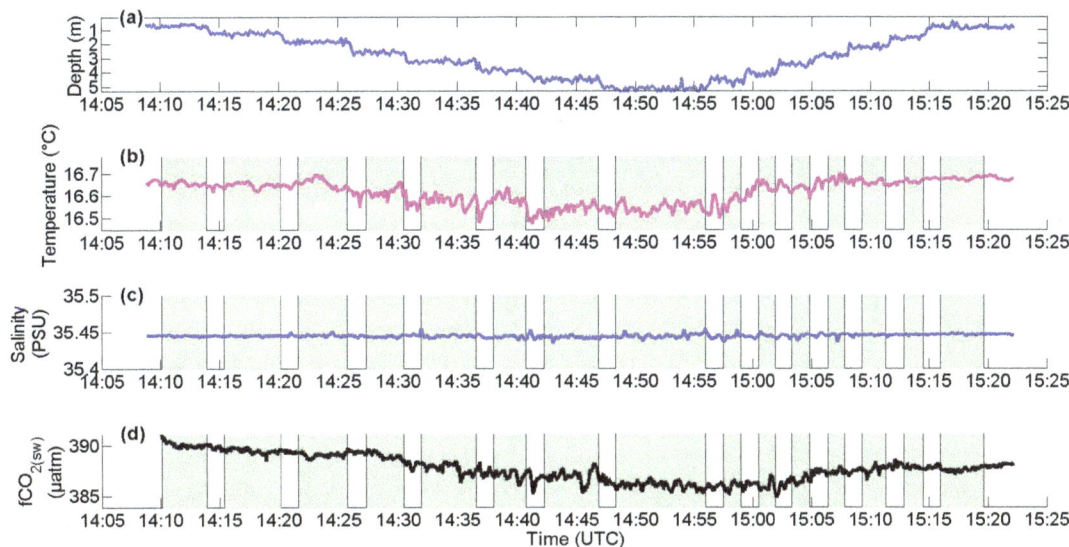

Figure 5. Time series measurements made during an NSOP deployment in the Celtic Sea on 30 July 2015. Data are 1 Hz depth **(a)**, seawater temperature **(b)**, salinity **(c)** and $f\mathrm{CO}_{2\mathrm{(sw)}}$ **(d)**. Data used for depth bin analysis (Sect. 2.6) are identified by a shaded background.

Profiling lasted 75 min and finished back at the surface at 15:20 (UTC). Seawater temperature was $16.61 \pm 0.06\,°\mathrm{C}$. At 14:20 (UTC), $f\mathrm{CO}_{2\mathrm{(atm)}}$ was 398 µatm and $f\mathrm{CO}_{2\mathrm{(sw)}}$ was 389 µatm at 0.67 m, meaning the ocean was undersaturated with respect to the atmosphere. The temperature and seawater CO_2 were the expected magnitude for summer in the Celtic Sea (Frankignoulle and Borges, 2001). Salinity was homogeneous throughout the NSOP deployment, only varying by ± 0.004.

Depth-binned salinity and temperature data did not show any significant variability (Fig. 6a). A slight temperature gradient was observed, with $0.15\,°\mathrm{C}$ difference between 5 m and the surface and a fairly constant reduction with depth ($0.03\,°\mathrm{C}$ per metre). The temperature profile was similar for down- and upcasts, although some continued warming of surface waters was evident in the upcast. The temperature measured by NSOP at 5.15 m depth agrees well with the coincident temperature measured by the bow thermistor at 5.5 m ($<0.02\,°\mathrm{C}$ difference) (Fig. 6c). There is no evidence that the ship's thrusters/propellers disrupted the near-surface gradients.

We compare the NSOP temperature profile with thermistor readings from a series of Sea-Bird Scientific (SBE 56) sensors (0.3, 0.6, 1.5, 3.5 and 7 m depth) mounted on a nearby temperature chain moored ~ 2.8 km away ($49.403°\,\mathrm{N}$, $-8.606°\,\mathrm{E}$) from the deployment site . The vertical profile implied by the NSOP deployment agrees with the mooring data (Fig. 6c), and corroborates the warming of the upper few metres of the ocean observed during the deployment. The agreement between these independent datasets suggests that it is unlikely that NSOP caused any significant localized warming of surface waters. The mean difference between NSOP temperature from discrete depths and the mooring sensors is $0.02\,°\mathrm{C}$. The surface data from the NSOP upcast show less agreement with the mooring, with NSOP temperatures $\sim 0.05\,°\mathrm{C}$ lower than the 0.3 m and 0.6 m mooring sensors. During the profile the ship drifted ~ 1 km from the start position of the profile and a further 0.2 km from the mooring. The small offset between the NSOP surface temperatures and the mooring may be driven by horizontal variability between the deployment and mooring locations. It is also possible that turbulence mixed warm surface waters down into cooler sub-surface layers. Turbulence could have been generated around the NSOP sampling frame or by an increase in wave-driven mixing when the significant wave height increased at $\sim 15\!:\!00$ UTC (Fig. S4a).

Seawater density (Fig. 7a) was calculated using the salinity and temperature profile data (Fig. 6a, b) and the 1983 Unesco equation of state (Millero and Poisson, 1981). As expected, with little variation in the salinity, changes in the density profile are dominated by temperature. The down- and upcasts for CO_2 show excellent agreement below 2.5 m. Surface water (<2 m) CO_2 is 2–4 µatm higher than at 5 m (Fig. 7b). Elevated surface CO_2 could be explained by a sustained flux from the atmosphere into a near-surface stratified layer with inhibited deepwater exchange. Under this assumption a vertical gradient in seawater CO_2 would need to be established shortly after the temperature gradient. A paired t test showed that the $f\mathrm{CO}_2$ measured in the surface bins on the downcast and upcast are were significantly different ($p = <0.001$). The deepening of the surface stratified layer could explain the more homogeneous CO_2 during the upcast. It is worth noting that in addition to physical processes, plankton trapped within the surface layer could also modify the surface CO_2. Trace gas concentrations may also be different in the sea surface microlayer but sampling that close to the surface is

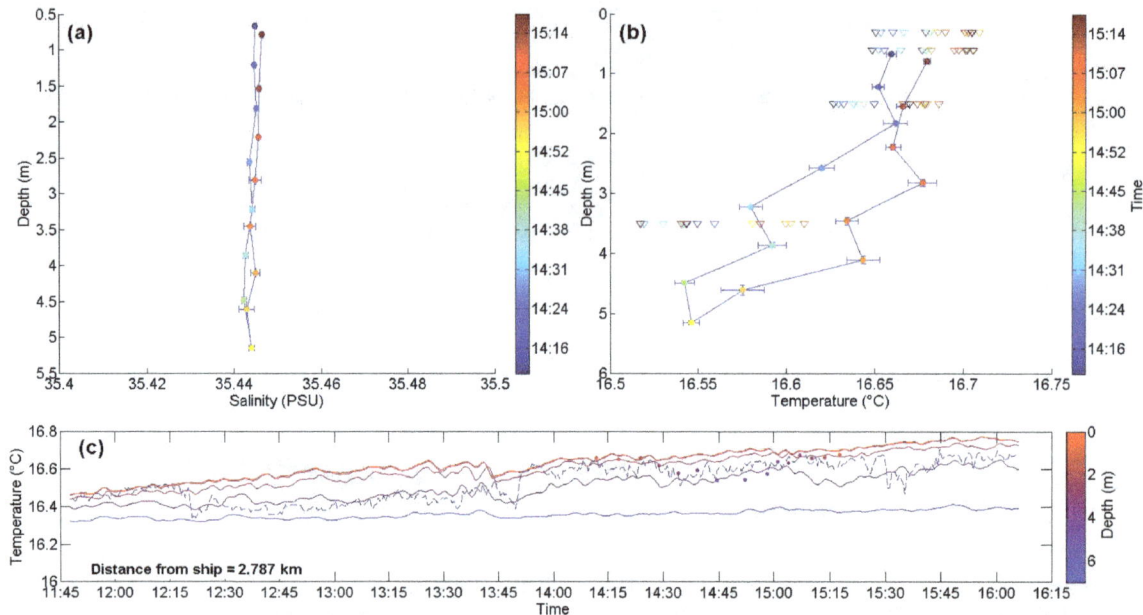

Figure 6. Salinity and temperature in the central Celtic Sea on 30 July 2015. NSOP profiles of salinity (**a**) and temperature (**b**) were derived using depth bins as described in Sect. 2.6. Data points are coloured by sampling time. Vertical and horizontal error bars show 2 standard errors of the mean in each depth bin. Coloured triangles in (**b**) are time-averaged temperature for four depths (0.3, 0.6, 1.5 and 3.5 m) at the nearby central Celtic Sea temperature mooring (49.403° N, −8.606° E). (**c**) Time series of temperature at the mooring. Time series of temperature at depths (0.3, 0.6, 1.5 and 3.5 m) are solid lines whereas the dashed line is the underway temperature at 5.5 m from RRS *Discovery* (located 2.8 km from the mooring). The mooring and underway temperatures are coloured according to their sample depth, where red is the air–sea interface. The circles are binned temperature data from NSOP, which have also been coloured to reflect the depth of collection.

beyond the capabilities of NSOP. Complementary measurements of the sea surface microlayer could be made using other state-of-the-art purpose-built sampling platforms such as the Sea Surface Scanner (Ribas-Ribas et al., 2017).

To assess measurement accuracy the NSOP Liqui-Cel CO_2 system was compared against an independent CO_2 system that had a showerhead equilibrator coupled to the ship's seawater supply pumped from 5.5 m below the sea surface (Hardman-Mountford et al., 2008; Kitidis et al., 2012). Technical issues meant that the underway CO_2 system installed on the RRS *Discovery* was not functioning during the deployment detailed above. However, during a deployment on 19 July 2015, the $f CO_{2(sw)}$ measured by NSOP at 5 m agreed well with independent measurements from the underway system, with difference of 1.7 ± 4.18 µatm. The agreement between the two systems is in line with previous intercomparisons (Ribas-Ribas et al., 2014; Körtzinger et al., 2000).

3.2 Coastal deployment

DMS profiles were collected on a small research vessel on 15 April 2014. The NSOP was deployed within the Plymouth Sound at 12:00 UTC and recovered 95 min later (Fig. 8). In the sheltered environment behind the breakwater the standard deviation in depth was ± 0.10 m, smaller than observed

during open ocean profiles. Seawater temperature and salinity demonstrate clear structure, with lower temperatures and higher salinities associated with sub-surface water. Two river estuaries (Plym and Tamar) converge and flow out to the open ocean through the Plymouth Sound. We likely observed a freshwater surface lens that was protected from wave-driven mixing and had been warmed over the course of the day. We used a different miniCTD during this deployment and were thus also able to collect fluorescence data (Fig. 8d).

Temperature profiles (Fig. 9a) show a sharp discontinuity in the downcast at ~ 5 m whereas in the upcast the thermocline had shoaled to ~ 3.5 m. The salinity profiles suggest similar mixing depths to the temperature profiles, with lower salinity water at the surface (Fig. 9b). The increase in fluorescence with depth (Fig. 9c) is either due to reductions in chlorophyll concentration close to the sea surface or because of quenching of the phytoplankton photosynthetic apparatus, which is often observed in surface waters that experience strong irradiance (Sackmann et al., 2008; Biermann et al., 2015). DMS concentrations reduce steadily with depth (Fig. 9d), which is likely explained by changes in DMS production and consumption rates by the biological community (Galí et al., 2013). The DMS profiles from the upcast and the downcast are very similar, with the largest difference at the very surface. A large difference in the surface-most data point can also be seen in the temperature data, and may re-

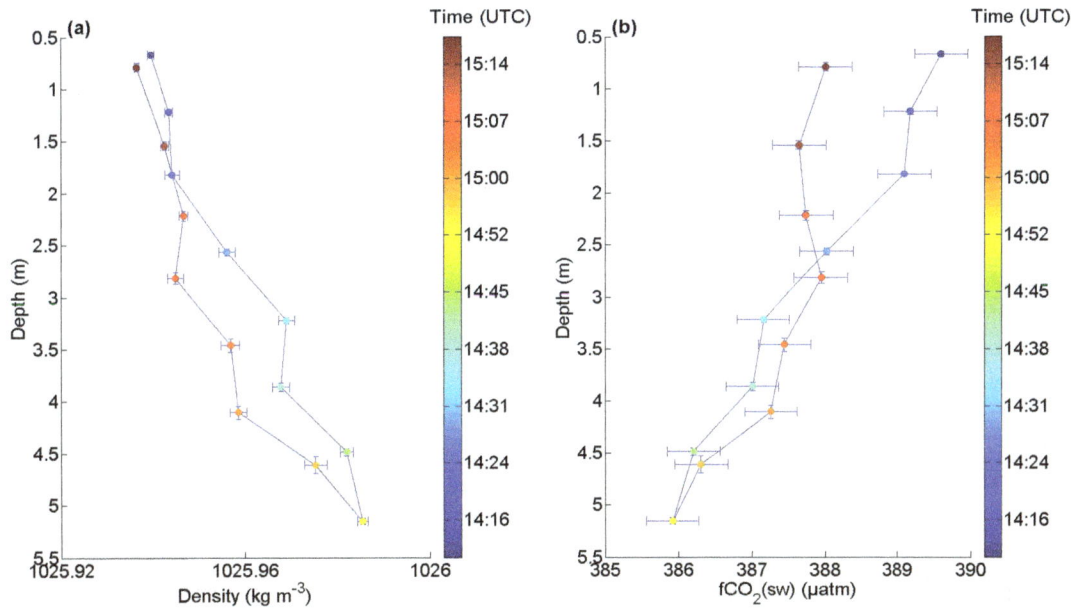

Figure 7. NSOP density (**a**) and $f\mathrm{CO}_{2(sw)}$, (**b**) profiles from the Celtic Sea on 30 July 2015. Data points are coloured by sample time. Vertical error bars correspond to 2 standard errors of the mean in each depth bin. The horizontal error bars in (**a**) are 2 standard errors of the mean, whereas in (**b**) they are the propagated error from the binned measurements used to calculate $f\mathrm{CO}_{2(sw)}$.

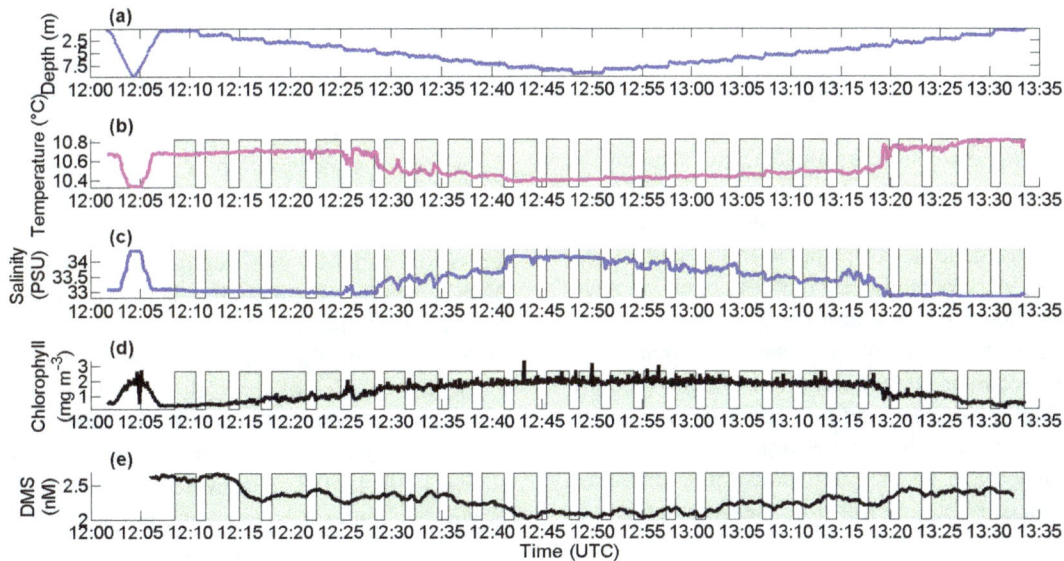

Figure 8. Time series measurements during an NSOP deployment in Plymouth Sound on 15 April 2014: depth (**a**), temperature (**b**), salinity (**c**), chlorophyll fluorescence (**d**) and $\mathrm{DMS}_{(sw)}$ (**e**). Data used for depth bin analysis (Sect. 2.6) are identified by a shaded background. The beginning of the time series is an example of a continuous profile (see Sect. 2.2).

flect mixing with sub-surface waters due to the motion of NSOP or short timescale variations in the physical environment.

4   Summary

This paper describes a near-surface ocean profiler (NSOP) designed to measure vertical trace gas profiles near the air–sea interface. NSOP is unique in approach as its sampling frame is lowered from a buoy that rides the ocean swell, reducing relative motion of the frame and hence fluctuations in sampling depth. The NSOP design facilitates near-surface

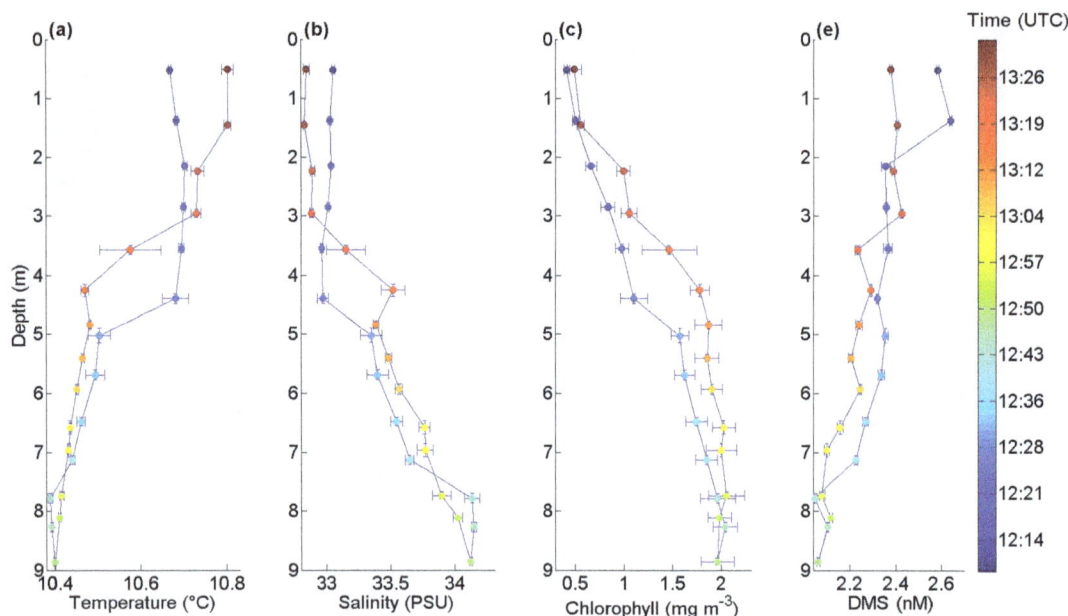

Figure 9. NSOP profiles collected in Plymouth Sound on 15 April 2014: temperature (**a**), salinity (**b**), chlorophyll fluorescence (**c**) and DMS$_{(sw)}$ (**d**). Data are coloured by sample time. Vertical and horizontal error bars are 2 standard errors of the mean (SEM) in each depth bin.

(< 0.5 m) sampling, significantly improving the capability to resolve vertical gradients. Other benefits include the ability to sample away from ship-driven turbulence and the flexibility to make a large range of near-surface measurements. The NSOP sampling frame houses the miniCTD and also has the capacity to incorporate additional sensors (e.g. turbulence, dissolved oxygen and other measures of phytoplankton abundance and photosynthetic health). The ability to collect water from discrete depths facilitates the collection of near-surface samples that require additional processing or take longer to analyse (e.g. TA, dissolved inorganic carbon, nutrients, the DMS-precursor DMSP, dissolved organic carbon). The NSOP is highly versatile and can be used for continuous or discrete profiling. Further development could adjust winch payout speed and enable continuous, high-resolution depth profiles for slower response time measurements (e.g. $f\mathrm{CO}_{2(sw)}$).

Near-surface stratification in the upper few metres of the ocean due to temperature and salinity gradients is a well-documented phenomenon. The presence or absence of chemical and biological gradients within near-surface stratified layers has been difficult to assess. NSOP is a platform with the capability to successfully resolve gradients in these near-surface layers. The data presented in this paper demonstrate that near-surface gradients in trace gases can lead to substantially different fluxes depending upon the seawater depth that is used to calculate the flux. Assuming that the effect

of temperature and salinity gradients on the flux can be accounted for using remote sensing methods (e.g Shutler et al., 2016), then the change in flux is directly proportional to the change in ΔC. In the case of the coastal DMS profile, a higher concentration (2.58 ± 0.02 nM) was observed 0.5 m below the sea surface compared to concentrations at 5 m (2.36 ± 0.03 nM). Assuming that the atmospheric concentration of DMS was negligible (a typical approach for DMS fluxes (see Lana et al., 2011), computing the flux with the 5 m waterside concentration instead of the 0.5 m waterside concentration means the flux is underestimated by 9.3 %. In the case of the Celtic Sea CO_2 profile, the concentration at 0.5 m (389.60 ± 0.36 µatm) was higher than at 5 m (385.92 ± 0.36 µatm). The atmospheric CO_2 concentration was 398.1 ± 0.3 µatm, which means that the surface water was less undersaturated than implied by the seawater concentration at 5 m. Using the 5 m waterside CO_2 concentration leads to an overestimation of the ΔC and flux by 43.5 % compared to using the 0.5 m waterside CO_2 concentration. The magnitudes of these concentration gradients are significant. However, such gradients (in magnitude and direction) do not persist for all hours of the day, under different environmental conditions and in all regions of the global ocean. A subsequent publication will discuss NSOP data collected during four cruises as well as the wider prevalence and implications of near-surface CO_2 gradients.

Competing interests. The authors declare that they have no conflict of interest.

Acknowledgements. We thank the captains and crews of the RV *Plymouth Quest* and RRS *Discovery* for their assistance with deploying NSOP, Christopher Balfour and Dave Sivyer for maintenance of the central Celtic Sea mooring near-surface temperature sensors, Vassilis Kitidis for supplying underway CO_2 data and Burke Hales for advice concerning Liqui-Cel CO_2 measurements. This research was made possible by PML internal funding, a NERC funded studentship (NE/L000075/1), temperature sensors on the central Celtic Sea mooring (NE/K002058/1) and the NERC Shelf Sea Biogeochemistry pelagic research programme (NE/K002007/1). The RRS *Discovery* underway data were supplied by the Natural Environment Research Council.

Edited by: Piers Chapman

References

Bakker, D. C. E., Pfeil, B., Landa, C. S., Metzl, N., O'Brien, K. M., Olsen, A., Smith, K., Cosca, C., Harasawa, S., Jones, S. D., Nakaoka, S.-I., Nojiri, Y., Schuster, U., Steinhoff, T., Sweeney, C., Takahashi, T., Tilbrook, B., Wada, C., Wanninkhof, R., Alin, S. R., Balestrini, C. F., Barbero, L., Bates, N. R., Bianchi, A. A., Bonou, F., Boutin, J., Bozec, Y., Burger, E. F., Cai, W.-J., Castle, R. D., Chen, L., Chierici, M., Currie, K., Evans, W., Featherstone, C., Feely, R. A., Fransson, A., Goyet, C., Greenwood, N., Gregor, L., Hankin, S., Hardman-Mountford, N. J., Harlay, J., Hauck, J., Hoppema, M., Humphreys, M. P., Hunt, C. W., Huss, B., Ibánhez, J. S. P., Johannessen, T., Keeling, R., Kitidis, V., Körtzinger, A., Kozyr, A., Krasakopoulou, E., Kuwata, A., Landschützer, P., Lauvset, S. K., Lefèvre, N., Lo Monaco, C., Manke, A., Mathis, J. T., Merlivat, L., Millero, F. J., Monteiro, P. M. S., Munro, D. R., Murata, A., Newberger, T., Omar, A. M., Ono, T., Paterson, K., Pearce, D., Pierrot, D., Robbins, L. L., Saito, S., Salisbury, J., Schlitzer, R., Schneider, B., Schweitzer, R., Sieger, R., Skjelvan, I., Sullivan, K. F., Sutherland, S. C., Sutton, A. J., Tadokoro, K., Telszewski, M., Tuma, M., van Heuven, S. M. A. C., Vandemark, D., Ward, B., Watson, A. J., and Xu, S.: A multi-decade record of high-quality fCO_2 data in version 3 of the Surface Ocean CO_2 Atlas (SOCAT), Earth Syst. Sci. Data, 8, 383–413, https://doi.org/10.5194/essd-8-383-2016, 2016.

Bange, H. W., Bell, T. G., Cornejo, M., Freing, A., Uher, G., Upstill-Goddard, R. C., and Zhang, G.: MEMENTO: a proposal to develop a database of marine nitrous oxide and methane measurements, Environ. Chem., 6, 195–197, 2009.

Bell, T. G., De Bruyn, W., Miller, S. D., Ward, B., Christensen, K. H., and Saltzman, E. S.: Air-sea dimethylsulfide (DMS) gas transfer in the North Atlantic: evidence for limited interfacial gas exchange at high wind speed, Atmos. Chem. Phys., 13, 11073–11087, https://doi.org/10.5194/acp-13-11073-2013, 2013.

Bell, T. G., De Bruyn, W., Marandino, C. A., Miller, S. D., Law, C. S., Smith, M. J., and Saltzman, E. S.: Dimethylsulfide gas transfer coefficients from algal blooms in the Southern Ocean, Atmos. Chem. Phys., 15, 1783–1794, https://doi.org/10.5194/acp-15-1783-2015, 2015.

Dickson, A. G., Sabine, C. L., and Christian, J. R.: Guide to best practices for ocean CO_2 measurements, Measurements, PICES Special Publication, 3, 91–102, 2007.

Donlon, C., Minnett, P., Gentemann, C., Nightingale, T., Barton, I., Ward, B., and Murray, M.: Toward improved validation of satellite sea surface skin temperature measurements for climate research, J. Clim., 15, 353–369, 2002.

Durham, W. M., Kessler, J. O., and Stocker, R.: Disruption of vertical motility by shear triggers formation of thin phytoplankton layers, Science, 323, 1067–1070, 2009.

Fairall, C., Bradley, E. F., Godfrey, J., Wick, G., Edson, J. B., and Young, G.: Cool-skin and warm-layer effects on sea surface temperature, J. Geophys. Res., 101, 1295–1308, 1996.

Frankignoulle, M. and Borges, A. V.: European continental shelf as a significant sink for atmospheric carbon dioxide, Global Biogeochem. Cy., 15, 569–576, 2001.

Galí, M., Simó, R., Vila-Costa, M., Ruiz-González, C., Gasol, J. M., and Matrai, P.: Diel patterns of oceanic dimethylsulfide (DMS) cycling: Microbial and physical drivers, Global Biogeochem. Cy., 27, 620–636, 2013.

Hales, B., Chipman, D., and Takahashi, T.: High-frequency measurement of partial pressure and total concentration of carbon dioxide in seawater using microporous hydrophobic membrane contractors, Limnol. Oceanogr.-Meth., 2, 356–364, 2004.

Hardman-Mountford, N. J., Moore, G., Bakker, D. C., Watson, A. J., Schuster, U., Barciela, R., Hines, A., Moncoiffé, G., Brown, J., and Dye, S.: An operational monitoring system to provide indicators of CO_2-related variables in the ocean, ICES J. Mar. Sci., 65, 1498–1503, 2008.

Kitidis, V., Hardman-Mountford, N. J., Litt, E., Brown, I., Cummings, D., Hartman, S., Hydes, D., Fishwick, J. R., Harris, C., and Martinez-Vicente, V.: Seasonal dynamics of the carbonate system in the Western English Channel, Cont. Shelf Res., 42, 30–40, 2012.

Körtzinger, A., Mintrop, L., Wallace, D. W., Johnson, K. M., Neill, C., Tilbrook, B., Towler, P., Inoue, H. Y., Ishii, M., and Shaffer, G.: The international at-sea intercomparison of fCO_2 systems during the R/V *Meteor* Cruise 36/1 in the North Atlantic Ocean, Mar. Chem., 72, 171–192, 2000.

Lana, A., Bell, T., Simó, R., Vallina, S. M., Ballabrera-Poy, J., Kettle, A., Dachs, J., Bopp, L., Saltzman, E., and Stefels, J.: An updated climatology of surface dimethylsulfide concentrations and emission fluxes in the global ocean, Global Biogeochem. Cy., 25, GB1004, https://doi.org/10.1029/2010GB003850, 2011.

Le Quéré, C., Andrew, R. M., Canadell, J. G., Sitch, S., Korsbakken, J. I., Peters, G. P., Manning, A. C., Boden, T. A., Tans, P. P., Houghton, R. A., Keeling, R. F., Alin, S., Andrews, O. D., Anthoni, P., Barbero, L., Bopp, L., Chevallier, F., Chini, L. P., Ciais, P., Currie, K., Delire, C., Doney, S. C., Friedlingstein, P., Gkritzalis, T., Harris, I., Hauck, J., Haverd, V., Hoppema, M., Klein Goldewijk, K., Jain, A. K., Kato, E., Körtzinger, A., Landschützer, P., Lefèvre, N., Lenton, A., Lienert, S., Lombardozzi, D., Melton, J. R., Metzl, N., Millero, F., Monteiro, P. M. S., Munro, D. R., Nabel, J. E. M. S., Nakaoka, S.-I., O'Brien, K., Olsen, A., Omar, A. M., Ono, T., Pierrot, D., Poulter, B., Rödenbeck, C., Salisbury, J., Schuster, U., Schwinger, J., Séférian, R., Skjelvan, I., Stocker, B. D., Sutton, A. J., Takahashi, T., Tian, H., Tilbrook, B., van der Laan-Luijkx, I. T., van der Werf, G. R., Viovy, N., Walker, A. P., Wiltshire, A. J., and Zaehle, S.:

Global Carbon Budget 2016, Earth Syst. Sci. Data, 8, 605–649, https://doi.org/10.5194/essd-8-605-2016, 2016.

Liss, P. S. and Slater, P. G.: Flux of Gases across the Air-Sea Interface, Nature, 247, 181–184, 1974.

Loose, B., Stute, M., Alexander, P., and Smethie, W.: Design and deployment of a portable membrane equilibrator for sampling aqueous dissolved gases, Water Resour. Res., 45, 2009.

Marandino, C. A., De Bruyn, W. J., Miller, S. D., and Saltzman, E. S.: Open ocean DMS air/sea fluxes over the eastern South Pacific Ocean, Atmos. Chem. Phys., 9, 345–356, https://doi.org/10.5194/acp-9-345-2009, 2009.

McNeil, C. L. and Merlivat, L.: The warm oceanic surface layer: Implications for CO_2 fluxes and surface gas measurements, Geophys. Res. Lett., 23, 3575–3578, 1996.

Miller, S. D., Marandino, C., and Saltzman, E. S.: Ship-based measurement of air-sea CO_2 exchange by eddy covariance, J. Geophys. Res.-Atmos., 115, 2010.

Millero, F. J. and Poisson, A.: International one-atmosphere equation of state of seawater, Deep-Sea Res. Pt. A, 28, 625–629, 1981.

Quinn, P. and Bates, T.: The case against climate regulation via oceanic phytoplankton sulphur emissions, Nature, 480, 51–56, 2011.

Ribas-Ribas, M., Rerolle, V., Bakker, D. C., Kitidis, V., Lee, G., Brown, I., Achterberg, E. P., Hardman-Mountford, N., and Tyrrell, T.: Intercomparison of carbonate chemistry measurements on a cruise in northwestern European shelf seas, Biogeosciences, 11, 4339–4355, https://doi.org/10.5194/bg-11-4339-2014, 2014.

Ribas-Ribas, M., Mustaffa, N. I. H., Rahlff, J., Stolle, C., and Wurl, O.: Sea Surface Scanner (S3): A Catamaran for High-resolution Measurements of Biogeochemical Properties of the Sea Surface Microlayer, J. Atmos. Ocean. Technol., 1433–1448, 2017.

Robertson, J. E. and Watson, A. J.: Thermal skin effect of the surface ocean and its implications for CO_2 uptake, Nature, 358, 738–740, 1992.

Royer, S.-J., Galí, M., Saltzman, E. S., McCormick, C. A., Bell, T. G., and Simó, R.: Development and validation of a shipboard system for measuring high-resolution vertical profiles of aqueous dimethylsulfide concentrations using chemical ionisation mass spectrometry, Environ. Chem., 11, 309–317, 2014.

Royer, S. J., Galí, M., Mahajan, A. S., Ross, O. N., Pérez, G. L., Saltzman, E. S., and Simó, R.: A high-resolution time-depth view of dimethylsulphide cycling in the surface sea, Sci. Rep., 6, 32325, https://doi.org/10.1038/srep32325, 2016.

Sackmann, B. S., Perry, M. J., and Eriksen, C. C.: Seaglider observations of variability in daytime fluorescence quenching of chlorophyll-a in Northeastern Pacific coastal waters, Biogeosciences Discuss., 5, 2839–2865, https://doi.org/10.5194/bgd-5-2839-2008, 2008.

Saltzman, E. S., De Bruyn, W. J., Lawler, M. J., Marandino, C. A., and McCormick, C. A.: A chemical ionization mass spectrometer for continuous underway shipboard analysis of dimethylsulfide in near-surface seawater, Ocean Sci., 5, 537–546, https://doi.org/10.5194/os-5-537-2009, 2009.

Shutler, J. D., Land, P. E., Piolle, J.-F., Woolf, D. K., Goddijn-Murphy, L., Paul, F., Girard-Ardhuin, F., Chapron, B., and Donlon, C. J.: FluxEngine: A Flexible Processing System for Calculating Atmosphere–Ocean Carbon Dioxide Gas Fluxes and Climatologies, J. Atmos. Ocean. Technol., 33, 741–756, 2016.

Smyth, T. J., Fishwick, J. R., Lisa, A.-M., Cummings, D. G., Harris, C., Kitidis, V., Rees, A., Martinez-Vicente, V., and Woodward, E. M.: A broad spatio-temporal view of the Western English Channel observatory, J. Plankt. Res., 32, 585–601, 2010.

Takahashi, T., Olafsson, J., Goddard, J. G., Chipman, D. W., and Sutherland, S.: Seasonal variation of CO_2 and nutrients in the high-latitude surface oceans: A comparative study, Global Biogeochem. Cy., 7, 843–878, 1993.

Turk, D., Zappa, C. J., Meinen, C. S., Christian, J. R., Ho, D. T., Dickson, A. G., and McGillis, W. R.: Rain impacts on CO_2 exchange in the western equatorial Pacific Ocean, Geophys. Res., Lett., 37, 2010.

Walker, C. F., Harvey, M. J., Smith, M. J., Bell, T. G., Saltzman, E. S., Marriner, A. S., McGregor, J. A., and Law, C. S.: Assessing the potential for dimethylsulfide enrichment at the sea surface and its influence on air-sea flux, Ocean Sci., 12, 1033–1048, https://doi.org/10.5194/os-12-1033-2016, 2016.

Wanninkhof, R.: Relationship between wind speed and gas exchange over the ocean revisited, Limnol. Oceanogr.-Meth., 12, 351–362, 2014.

Ward, B., Wanninkhof, R., McGillis, W. R., Jessup, A. T., DeGrandpre, M. D., Hare, J. E., and Edson, J. B.: Biases in the air-sea flux of CO_2 resulting from ocean surface temperature gradients, J. Geophys. Res.-Ocean. (1978–2012), 109, 2004.

Webb, J. R., Maher, D. T., and Santos, I. R.: Automated, in situ measurements of dissolved CO_2, CH_4, and δ13C values using cavity enhanced laser absorption spectrometry: Comparing response times of air-water equilibrators, Limnol. Oceanogr.-Meth., 14, 323–337, 2016.

Woolf, D., Land, P. E., Shutler, J. D., Goddijn-Murphy, L., and Donlon, C. J.: On the calculation of air-sea fluxes of CO_2 in the presence of temperature and salinity gradients, J. Geophys. Res.-Ocean., 121, 1229–1248, 2016.

Yang, M., Beale, R., Smyth, T., and Blomquist, B.: Measurements of OVOC fluxes by eddy covariance using a proton-transfer-reaction mass spectrometer – method development at a coastal site, Atmos. Chem. Phys., 13, 6165–6184, https://doi.org/10.5194/acp-13-6165-2013, 2013.

Ziska, F., Quack, B., Abrahamsson, K., Archer, S. D., Atlas, E., Bell, T., Butler, J. H., Carpenter, L. J., Jones, C. E., Harris, N. R. P., Hepach, H., Heumann, K. G., Hughes, C., Kuss, J., Krüger, K., Liss, P., Moore, R. M., Orlikowska, A., Raimund, S., Reeves, C. E., Reifenhäuser, W., Robinson, A. D., Schall, C., Tanhua, T., Tegtmeier, S., Turner, S., Wang, L., Wallace, D., Williams, J., Yamamoto, H., Yvon-Lewis, S., and Yokouchi, Y.: Global sea-to-air flux climatology for bromoform, dibromomethane and methyl iodide, Atmos. Chem. Phys., 13, 8915–8934, https://doi.org/10.5194/acp-13-8915-2013, 2013.

5

Lagrangian simulation and tracking of the mesoscale eddies contaminated by Fukushima-derived radionuclides

Sergey V. Prants, Maxim V. Budyansky, and Michael Y. Uleysky

Laboratory of Nonlinear Dynamical Systems, Pacific Oceanological Institute of the Russian Academy of Sciences, 43 Baltiyskaya st., 690041 Vladivostok, Russia

Correspondence to: Sergey V. Prants (prants@poi.dvo.ru)

Abstract. A Lagrangian methodology is developed to simulate, track, document and analyze the origin and history of water masses in ocean mesoscale features. It aims to distinguish whether water masses inside the mesoscale eddies originated from the main currents in the Kuroshio–Oyashio confluence zone. By computing trajectories for a large number of synthetic Lagrangian particles advected by the AVISO velocity field after the Fukushima accident, we identify and track the mesoscale eddies which were sampled in the cruises in 2011 and 2012 and estimate their risk of being contaminated by Fukushima-derived radionuclides. The simulated results are compared with in situ measurements, showing a good qualitative correspondence.

1 Introduction

High tsunami waves after the Tohoku earthquake on 11 March 2011 damaged the cooling system of the Fukushima Nuclear Power Plant (FNPP). Due to lack of electricity, it was not possible to cool nuclear reactors and the fuel storage pools that caused numerous explosions at the FNPP (for details see Povinec et al., 2013). The Fukushima accident was classified at the maximum level of 7, similar to the Chernobyl accident which happened in 1986 in the former Soviet Union. Radionuclides were released from the FNPP through two major pathways: direct discharges of radioactive water and atmospheric deposition onto the North Pacific Ocean. Indirect estimation of that deposition is in the range 6.4–35 PBq (Kumamoto et al., 2014). The total amount of

^{137}Cs isotope released into the ocean was estimated to be 3.6 ± 0.7 PBq by the end of May 2011 (Tsumune et al., 2013).

A few special research vessel (R/V) cruises were conducted, just after the accident and later, to measure radioactivity in sea water, zooplankton, fish and in other marine organisms. ^{137}Cs and ^{134}Cs isotopes with 30.17 and 2.06 years half-life, respectively, were detected over a broad area in the western North Pacific in 2011 and 2012 (Honda et al., 2012; Buesseler et al., 2012; Inoue et al., 2012a, b; Tsumune et al., 2012, 2013; Kaeriyama et al., 2013; Oikawa et al., 2013; Aoyama et al., 2013; Kameník et al., 2013; Kumamoto et al., 2014; Kaeriyama et al., 2014; Budyansky et al., 2015). ^{137}Cs concentration levels off Japan before the accident were estimated at the background level to be 1–3 mBq kg^{-1}, while ^{134}Cs was not detectable. Because of a comparatively short half-life time, any measured concentrations of ^{134}Cs could only be Fukushima derived.

The studied area is shown in Fig. 1a. It is known as the Kuroshio–Oyashio confluence zone or a subarctic frontal area (Kawai, 1972). The Kuroshio Extension prolongs the Kuroshio Current which turns to the east at about 35° N and flows as a strong meandering jet constituting a front separating the warm subtropical and cold subarctic waters. It is a region with one of the most intense air–sea heat exchange and the highest eddy kinetic-energy level. The Kuroshio–Oyashio confluence zone is populated with several mesoscale eddies that transfer heat, salt, nutrients, carbon, pollutants and other tracers across the ocean. They originate, besides from the Kuroshio Extension, from the Tsugaru Warm Current, flowing between the Honshu and Hokkaido islands, and from the cold Oyashio Current flowing out of the Arctic along the Kamchatka Peninsula and the Kuril Islands

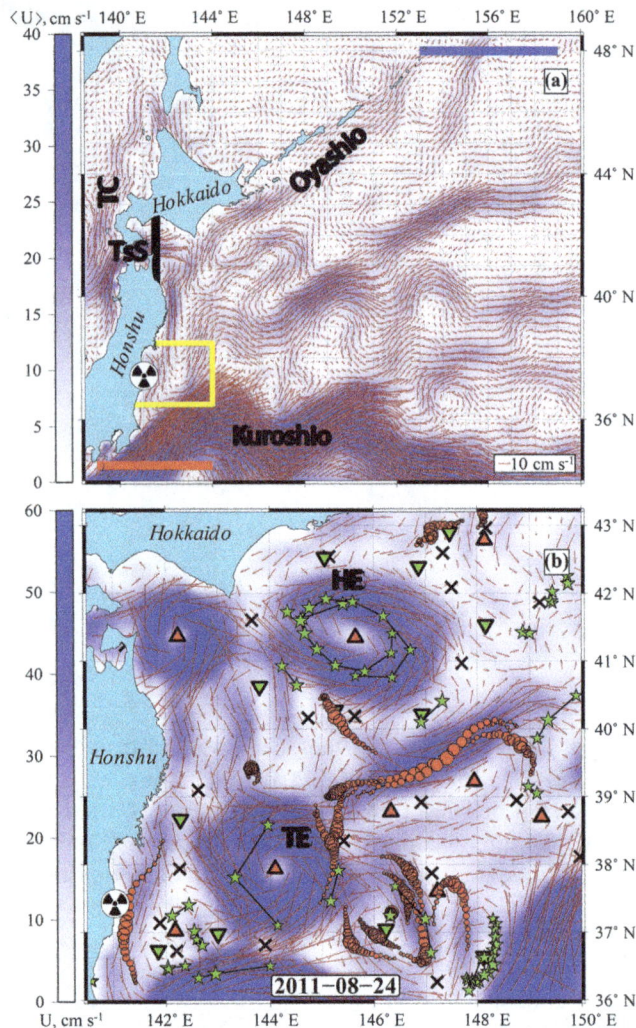

Figure 1. (a) The AVISO velocity field in the Kuroshio–Oyashio confluence zone, averaged from 1993 to 2016. TsS stands for the Tsugaru Strait. Location of the FNPP is shown by the radioactivity sign. The area just around the FNPP is shown by the yellow lines. **(b)** The velocity field on 24 August, 2011, with the Tohoku (TE) and Hokkaido (HE) eddies studied in the paper and with tracks of some available drifters (the red circles) and Argo floats (the green stars) present in the area at that time. Elliptic and hyperbolic stagnation points with zero mean velocity are indicated by triangles and crosses, respectively, with upward- and downward-oriented triangles denoting anticyclones and cyclones, respectively.

(Fig. 1a). The lifetime of those eddies ranges from a few weeks to a few years.

The standard approach in simulating transport phenomena, such as propagation of oil after the explosion at the Blue Horizon mobile drilling rig in the Gulf of Mexico in April 2010 and propagation of radioactive isotopes after the accident at the FNPP, is to run global or regional numerical models of circulation to simulate propagation of pollutants and try to forecast their trajectories. The outcomes pro-

vide "spaghetti-like" plots of individual trajectories which are hard to interpret. Moreover, as the majority of real trajectories in a chaotic environment are very sensitive to small and inevitable variations in initial conditions, they are practically unpredictable even over a comparatively short time.

A specific Lagrangian approach, based on dynamical systems theory, has been developed in recent decades with the aim of finding more or less robust material structures in chaotic flows governing mixing and transport of Lagrangian particles and creating transport barriers preventing propagation of a contaminant across them (for reviews see Samelson and Wiggins, 2006; Mancho et al., 2006; Koshel' and Prants, 2006; Haller, 2015). Identification of such structures in the ocean would help to predict, for short and medium-length periods of time, where a contaminant will move even without a precise solution of the Navier–Stokes equations. This approach has been successfully used in simulating propagation of oil in the Gulf of Mexico (Mezić et al., 2010; Huntley et al., 2011; Olascoaga and Haller, 2012) and propagation of Fukushima-derived radionuclides in the Pacific ocean (Prants et al., 2011b; Budyansky et al., 2015; Prants et al., 2014).

The present authors have developed a set of Lagrangian tools for tracking the origin, history and fate of water masses advected by analytic, altimetric and numerical velocity fields generated by eddy-resolved regional circulation models (Budyansky et al., 2009; Prants et al., 2011a, b; Prants, 2013; Prants et al., 2013; Prants, 2014; Budyansky et al., 2015). Each elementary volume of water can be attributed to physico-chemical properties (temperature, salinity, density, radioactivity, etc.) which characterize this volume as it moves. In addition, each water parcel can be attributed to other types of diagnostics which are exclusively a function of its trajectory. We call them "Lagrangian indicators". They are, for example, distance traveled by a fluid particle for some period of time; absolute, zonal and meridional displacements of particles from their original positions; the number of their cyclonic and anticyclonic rotations; time of residence of fluid particles inside a given area; exit time out off that area; and the number of times particles visited different places in a studied region.

The Lagrangian indicators contain information about the origin, history and fate of the corresponding water masses and allow the identification of water masses that move coherently, either by propagating together or by rotating together. Even if adjacent waters are indistinguishable, say, by temperature (e.g., the satellite SST images indicate no thermal front), the corresponding water masses could still be distinguishable by, for example, their origin, traveling history and other factors. The Lagrangian indicators are computed by integrating advection equations (Eq. 1) for a large number of synthetic particles forward and backward in time. When integrating Eq. (1) forward in time, one computes particle trajectories to know the fate of the corresponding particles, and when integrating Eq. (1) backward in time, one could know where the particles came from and the history of their travel.

The purpose of this paper is threefold. Firstly, we develop a Lagrangian methodology in order to track and document the origin and history of water masses constituting prominent mesoscale features. It allows the distinction of water masses inside mesoscale eddies originating from the main currents in the Kuroshio–Oyashio confluence zone. Secondly, we apply that methodology in order to identify and track the mesoscale eddies, advected by the altimetric AVISO velocity field, with a risk of being contaminated by Fukushima-derived radionuclides. Finally, the simulation results are compared qualitatively with in situ sampling of those eddies in the R/V cruises. The location and form of the simulated eddies are verified, when possible, by tracks of surface drifters and diving Argo floats available at the sites aoml.noaa.gov/phod/dac and www.argo.net, respectively.

2   Data and methodology

All the simulation results are based on integrating equations of motion for a large number of synthetic particles (tracers) advected by the AVISO velocity field.

$$\frac{d\lambda}{dt} = u(\lambda, \varphi, t), \quad \frac{d\varphi}{dt} = v(\lambda, \varphi, t), \tag{1}$$

where u and v are angular zonal and meridional velocities, and φ and λ are latitude and longitude, respectively. The altimetry-based velocities were obtained from the AVISO database (aviso.altimetry.fr) archived daily on a $1/4° \times 1/4°$ grid. The velocity field was interpolated using a bicubical spatial interpolation and third-order Lagrangian polynomials in time. In integrating Eq. (1) we used a fourth-order Runge–Kutta scheme with an integration step of 0.001 days.

The velocity field is from altimetry data, which provide the geostrophical component of the real near-surface velocities valid at the mesoscale. In order to display the enormous amount of information, we plot maps of specific Lagrangian indicators versus particle's initial positions. The region under study is seeded with a large number of Lagrangian particles whose trajectories are computed for a given period of time. The results obtained are processed to get a data file with the field of a specific Lagrangian indicator in this area. Finally, its values are coded by color and represented as a map in geographic coordinates.

It is informative also to identify "instantaneous" stagnation elliptic and hyperbolic points on the Lagrangian maps. We mark them by triangles and crosses, respectively. They are points with zero velocity which are computed daily with the AVISO velocity field. The elliptic points are called stable and the hyperbolic ones are unstable. Their local stability properties are characterized by a standard method calculating eigenvalues of the Jacobian matrix of the velocity field. The elliptic points, situated mainly in the centers of eddies, are those points around which the motion is stable and circular. Upward (downward) orientation of one of the triangle's top

on the maps means anticyclonic (cyclonic) rotations of water around them. The hyperbolic points, situated mainly between and around eddies, have stable manifolds along which water parcels converge to such a point and unstable manifolds along which they diverge. The stagnation points are moving Eulerian features and may undergo bifurcations in the course of time. In spite of nonstationarity of the velocity field, some of them may exist for weeks and much more. The hyperbolic points and their attracting and repelling manifolds were recently identified with the help of drifter's tracks in the Gulf of La Spezia in the northwestern Mediterranean Sea (Haza et al., 2010), in the Gulf of Lion (Nencioli et al., 2011), in the Gulf of Mexico (Olascoaga et al., 2013) and in the northwestern Pacific (Prants et al., 2016).

The altimetry-based Lagrangian maps allow accurate identification and tracking of mesoscale eddies and document their transformation due to interactions with currents and other eddies. Inspecting daily-computed Lagrangian maps for a long period of time (up to 2 years in this paper) and computing stagnation elliptic points daily, one can track the origin and fate of water masses within a given eddy if it is sufficiently large and long lived (i.e., more than a week). For this purpose Lagrangian diagnostics are more appropriate than commonly used Eulerian techniques, because Lagrangian maps are imprints of the history of water masses involved in the vortex motion, whereas vorticity, Okubo–Weiss parameter and similar indicators are only instantaneous snapshots (see Olascoaga et al., 2013, and Prants et al., 2016, for comparison).

Being motivated by the problem of identification of Fukushima-contaminated waters in the core and at the periphery of persistent mesoscale eddies in the area, we develop in this paper a specific Lagrangian technique designed to distinguish water masses of a different origin inside the eddies with a risk of being contaminated. With this aim we specify, besides Fukushima-derived waters, water masses originated from the main currents in the Kuroshio–Oyashio confluence zone. The integration was performed backward in time. We removed from consideration all the particles entered into any AVISO grid cell with two or more corners touching the land in order to avoid artifacts due to the inaccuracy of the altimetry-based velocity field near the coast.

In what follows, we define the "yellow" waters on the maps as those which have a large risk of being contaminated because they came after the accident from the area just around the FNPP, enclosed by the yellow straight lines in Fig. 1a, for the period from the day of the accident, March 11, 2011, to May 18, 2011, when direct releases of radioactive isotopes to the ocean and atmosphere stopped. The "red" waters are salty and warm Kuroshio waters. To be more exact, they came from the red zonal line (34.5° N, 139–144° E) in Fig. 1a, crossing the Kuroshio main jet. The "black" waters came from the warm Tsushima Current flowing via the Tsugaru Strait out off the Japan Sea and across that strait (the black line with 40–43° N, 141.55° E). The "blue" waters

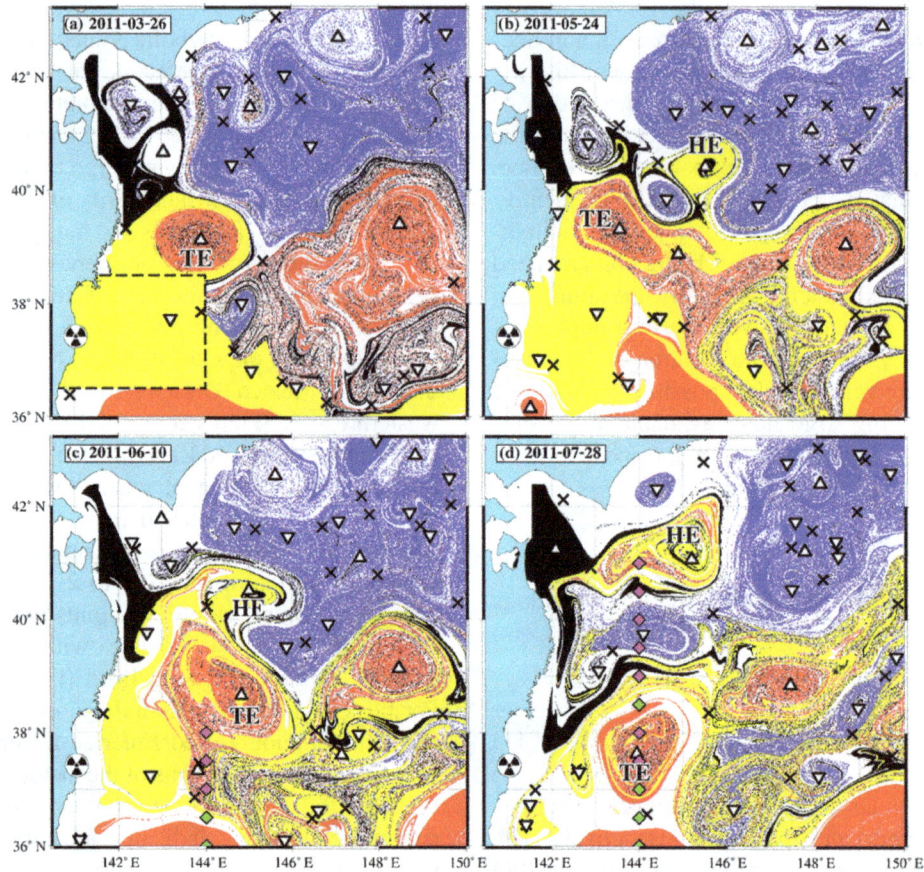

Figure 2. The Lagrangian maps show evolution of the Tohoku eddy (TE) from after the accident to the days of its sampling and the origin of waters in its core and at the periphery. The red, black and blue colors specify the tracers which came for 2 years in the past to their places on the maps from the Kuroshio, Oyashio and Tsushima currents, respectively, more exactly, from the corresponding line segments shown in Fig. 1a. The yellow color marks the Lagrangian particles coming from the area around the FNPP in Fig. 1a (shown in Fig. 2a by the dashed line), after the day of the accident on 11 March 2011. The TE was sampled on 10 and 11 June 2011 by Buesseler et al. (2012) along the transect 35.5–38° N, 144° E shown in (c) and at the end of July 2011 by Kaeriyama et al. (2013) along the transect 35° N–41° N, 144° E shown in (d). The locations of stations with surface seawater samples (collected by Buesseler et al., 2012 and Kaeriyama et al., 2013) with measured radiocesium concentrations at the background level are indicated by the green diamonds. Stations where the concentrations were measured to be much higher are marked by the magenta diamonds.

are fresher and colder waters originating from the Oyashio Current and crossing the blue zonal line (48° N, 153–159° E) shown in Fig. 1a. The "white" waters on the Lagrangian maps have not been specified as originating from one of the segments mentioned above. They could reach their places on the maps from anywhere besides those segments.

We are interested in advective transport for a comparatively long period of time, up to 2 years. It is hardly possible to adequately simulate motion of a specified passive particle in a chaotic flow, but it is possible to reproduce transport of a statistically significant number of particles. Our results are based not on simulation of individual trajectories but on statistics for 490 000 Lagrangian particles. We cannot, of course, guarantee that we compute "true" trajectories for individual particles. The description of the general pattern of transport for half a million particles is much more robust.

However, we do not try to quantitatively simulate the concentration of radionuclides or estimate the content of water masses of different origin inside the studied eddies.

3 Results

A few mesoscale eddies were present in the studied area on the day of the accident. The cyclonic eddies with the centers, marked by the downward-oriented triangles on the Lagrangian maps, prevailed in the area to the north of the Subarctic Front, the boundary between the subarctic (blue) and subtropical (red) waters in Fig. 2. The anticyclonic eddies with the centers, marked by the upward-oriented triangles, prevailed to the south of the front.

The large anticyclonic Tohoku eddy (TE,) with the center at around 39° N, 144° E in March 2011, was sampled after the accident in the two R/V cruises in June (Buesseler et al., 2012) and July 2011 (Kaeriyama et al., 2013), showing large concentrations of ^{137}Cs and ^{134}Cs. The anticyclonic Hokkaido eddy (HE), genetically connected with the TE, originated in the middle of May 2011 with the center at around 40° N, 145° E. After that it captured some contaminated water from the TE. It was sampled at the end of July 2011 (Kaeriyama et al., 2013).

The anticyclonic Tsugaru eddy (TsE) was genetically connected with the HE. It originated in the beginning of February 2012 with the center at around 41.9° N, 148° E and captured some contaminated water from the HE. The TsE was sampled in the R/V *Professor Gagarinskiy* cruise on 5 July, 2012, and found to have concentrations of ^{137}Cs and ^{134}Cs over the background level at the surface and at intermediate depths (Budyansky et al., 2015). All these eddies will be studied in this section from the Lagrangian point of view in order to simulate and track by which transport pathways they could have gained water masses from the Fukushima area or from other origins and to compare qualitatively the simulation results with in situ measurements.

3.1   The Tohoku eddy

We tracked with daily-computed Lagrangian maps the birth, metamorphoses and decay of the mesoscale anticyclonic TE. It originated in the middle of May 2010 with the elliptic point at around 38° N, 144° E at that time as the result of interaction of a warm anticyclonic Kuroshio ring with a cyclone with mixed Kuroshio and Oyashio core waters. It has interacted with other eddies almost for a year, with multiple splitting and merging in the area to the east off the Honshu Island. Just after the accident, it began to gain yellow water from the area around the FNPP with a high risk of contamination. That eddy is clearly seen in an earlier simulation just after the accident in Fig. 3b by Prants et al. (2011b) and on the Lagrangian map in Fig. 2a as a red patch labeled as TE with the center at 39° N, 144° E on 26 March 2011.

The maps in Fig. 2 and in the subsequent figures were computed, as was explained in Sect. 2. The red color in the core of the TE means that its core water was of subtropical origin. More precisely, the red tracers were advected for 2 years from the red line segment in Fig. 1a to the current place on the map. In March 2011 yellow water, coming from the area around the FNPP with a comparatively high risk of being contaminated, wrapped round the TE. A thin streamer of Tsugaru black water, coming from the black line segment in Fig. 1a, wrapped a periphery of the TE at the end of March. Yellow waters propagated gradually to the east and south due to a system of currents wrapping around the eddies present in the area. The straight zonal boundary along 36.5° N and meridional boundary along 144° E, separating water masses of different origin in Fig. 2a on 26 March 2011,

are just fragments of the boundary in Fig. 1a restricting the area around the FNPP. These boundaries separate the yellow tracers which were present within the area from those which have not yet managed to penetrate inside the area for 15 days after the accident.

In April and May 2011 the TE had a sandwich-like structure, with the red subtropical core belted with a narrow streamer of Fukushima yellow waters which, in turn, was encircled by a red streamer of Kuroshio subtropical water (Fig. 2b). A new eddy configuration appeared at the end of May in Fig. 2b, with the TE interacting with a blue cyclone with the center at 39.9° N, 144.7° E and a newborn yellow anticyclone which we call the Hokkaido eddy with the center at 40.4° N, 145.5° E. The core of that cyclone consisted of a blue subarctic Oyashio water with low risk of being contaminated, but the HE core water came from the area around the FNPP with a high risk of being contaminated.

In the course of time the TE moved gradually to the south. Its periphery was sampled at the beginning of June by Buesseler et al. (2012), and the whole eddy was crossed at the end of July 2011 by Kaeriyama et al. (2013). Fukushima-derived cesium isotopes were measured on 10 and 11 June during the R/V *Ka'imikai-o-Kanaloa* cruise (Buesseler et al., 2012) along the 144° E meridional transect where the cesium concentrations were found to be in the range from the background level, $C_{137} = 1.4$–3.6 mBq kg^{-1} (stations 13 and 14), to a high level up to $C_{137} = 173.6 \pm 9.9$ mBq kg^{-1} (station 10). The ratio ^{134}Cs/^{137}Cs was close to 1.

For ease of comparison, we mark, with the green diamonds in Fig. 2c, the locations of stations 13 and 14 with collected surface seawater samples by Buesseler et al. (2012) in which the cesium concentrations were measured to be at the background level ($\lesssim 3.6$ mBq kg^{-1}). The stations 10, 11 and 12, where the concentrations were found to be much larger, are indicated by the magenta diamonds. Our simulation in Fig. 2c shows that stations 13 and 14 on the days of sampling were located in red and white waters with a low risk of containing Fukushima-derived radionuclides.

Transport and mixing at and around stations 10, 11 and 12 with high measured values of the cesium concentrations (Buesseler et al., 2012) were governed mainly by the interaction of the TE with the yellow mesoscale cyclone with the center at 37.2° N, 142.8° E. This cyclone formed in the area in April and captured yellow waters with a high risk of contamination. Unfortunately, it has not been sampled in the R/V *Ka'imikai-o-Kanaloa* cruise. The surface seawater samples at stations 10, 11 and 12 were collected on the days of sampling at the eastern periphery of that cyclone and at the southern periphery of the TE with the yellow streamer there. Station 10, with the highest measured level of the ^{137}Cs concentration of $C_{137} = 173.6 \pm 9.9$ mBq kg^{-1}, was located at 38° N, 144° E inside the wide streamer of yellow water around the TE. Stations 11 and 12, with $C_{137} = 103.7 \pm 5.9$ mBq kg^{-1} and $C_{137} = 93.6 \pm 4.9$ mBq kg^{-1}, respectively, were located within the narrow streamers with yellow simulated water in Fig. 2c intermitted with narrow streamers of red water. So, we estimate the likelihood of finding Fukushima-derived radionuclides there (the magenta diamonds) to be much higher than at stations 13 and 14 (the green diamonds), and it is confirmed by a qualitative comparison with measured data.

A specific configuration of mesoscale eddies occurred in the area to the northeast of the FNPP at the end of July 2011, the days of sampling by Kaeriyama et al. (2013) along the 144° E meridian from 35 to 41° N during the R/V *Kaiun maru* cruise. That transect is shown in Fig. 2d. It crosses the TE and the cyclone with blue Oyashio water, which is genetically linked to the blue cyclone at 39.9° N, 144.7° E in Fig. 2b. The transect also partly crosses the periphery of the anticyclonic HE. The measured ^{137}Cs concentrations in surface seawater samples at the stations C43–C55 were found to be in the range from the background level, 1.9 ± 0.4 mBq kg^{-1} (station C52), to a much higher level of 153 ± 6.8 mBq kg^{-1} (station C47). The colored tracking maps in Fig. 5 by Prants et al. (2014) show where the simulated tracers of that transect were moving from 11 March to 10 April, 2011, being advected by the AVISO velocity field.

The risk of radioactive contamination of the markers placed at 36–36.5° N was estimated by Prants et al. (2014), to be small, because they were advected mainly by the Kuroshio Current from the southwest to the east (the corresponding concentrations were measured by Kaeriyama et al., 2013, to be 2–5 mBq kg^{-1}). The present simulation in Fig. 2d also shows that stations C51, 52 and 53 (the green diamonds), with the measured cesium concentrations at the background level on the days of sampling by Kaeriyama et al. (2013) were located in the red waters (stations C51 and C53) advected by the main Kuroshio jet from the southwest and in the white waters (station C52) between the TE and the jet. Therefore, we estimate the likelihood of finding Fukushima-derived radionuclides there to be comparatively low.

The transect 36.5–38° N in Fig. 2d (the red one in Fig. 5 by Prants et al., 2014) crossed the TE. The ^{137}Cs concentrations at the stations C49 and C50 of that transect were measured to be 36 ± 3.3 and 50 ± 3.6 mBq kg^{-1} (Kaeriyama et al., 2013). Comparing those results with simulated ones, we note the presence of yellow water in the TE core at the locations of those stations. Surface samples at station C48 (38.5° N) were measured to contain the ^{137}Cs concentration to be at the background level 2.7 ± 0.6 mBq kg^{-1} (Kaeriyama et al., 2013). The corresponding green diamond is located in our simulation in the area with red and white waters.

Inspecting the Lagrangian maps on the days between 6 June and 28 July (not shown), we have found that the yellow cyclone with the center at 37.2° N, 142.8° E in Fig. 2c collapsed at the end of June. Its yellow core water with a high risk of being contaminated was wrapped around the neighbor anticyclone TE in the form of a wide yellow streamer visible in Fig. 2d. The highest concentration, $C_{137} = 153 \pm 6.8$ mBq kg^{-1}, was measured by Kaeriyama et al. (2013) at

station C47 (39° N), situated in the area of that streamer. Stations C46 (39.5° N) with $C_{137} = 83 \pm 5.0$ mBq kg^{-1} is situated in the close proximity to a yellow streamer sandwiched between white and black waters.

A comparatively high concentration, $C_{137} = 65 \pm 4.3$ mBq kg^{-1}, was measured by Kaeriyama et al. (2013) at station C45 (40° N) during the days of sampling in the core of the blue cyclone with the center at 39.7° N, 144.2° E (Fig. 2d). Our simulation shows that it was formed mainly by Oyashio blue waters (with a low risk of being contaminated by Fukushima-derived radionuclides) and partly by white waters.

When comparing simulation results in Fig. 2d with the measurements by Kaeriyama et al. (2013), we have found that the simulation is consistent with samplings at stations C48, 51, 52 and 53 in the sense that the cesium concentrations were measured to be at the background level in those places on the maps where there is no signs of yellow water with a high risk of containing Fukushima-derived radionuclides. Our simulation is also consistent, at least quantitatively, with samplings at stations C47, 49 and 50 with high measured levels of the cesium concentrations because the yellow water is present there in our simulation.

However, there is an inconsistency of simulation with samplings at stations C45 and C46, where there are practically no yellow tracers but rather only blue and white ones. The reasons for this inconsistency might be different. In this paper we track only those tracers which originated from the blue, red and black segments as well as the yellow rectangular around the FNPP shown in Fig. 1a. So we did not specify the origin of white waters. They could reach their places on the maps from anywhere besides those segments and the area around the FNPP. They could in principle contain Fukushima-derived radionuclides that were deposited at the sea surface from the atmosphere after the accident and then advected by eddies and currents in the area. Moreover, they could be those tracers which were located inside AVISO grid cells near the coast around the FNPP just after the accident and were then advected outside. We removed from consideration all the tracers entered into any AVISO grid cell with two or more corners touching the land because of inaccuracy of the altimetry-based velocity field there and in order to avoid artifacts.

Thus, the white streamers inside the core and at the periphery of the blue cyclone with the center at 39.7° N, 144.2° E (nearby stations C45 and C46 with high measured concentrations of cesium by Kaeriyama et al., 2013) could, in principle, contain contaminated water. However, it has not been proved in our simulation due to the above-mentioned reasons.

3.2 The Hokkaido eddy

Now we consider the anticyclonic HE. It originated in the middle of May (see the yellow patch in Fig. 2b with the cen-

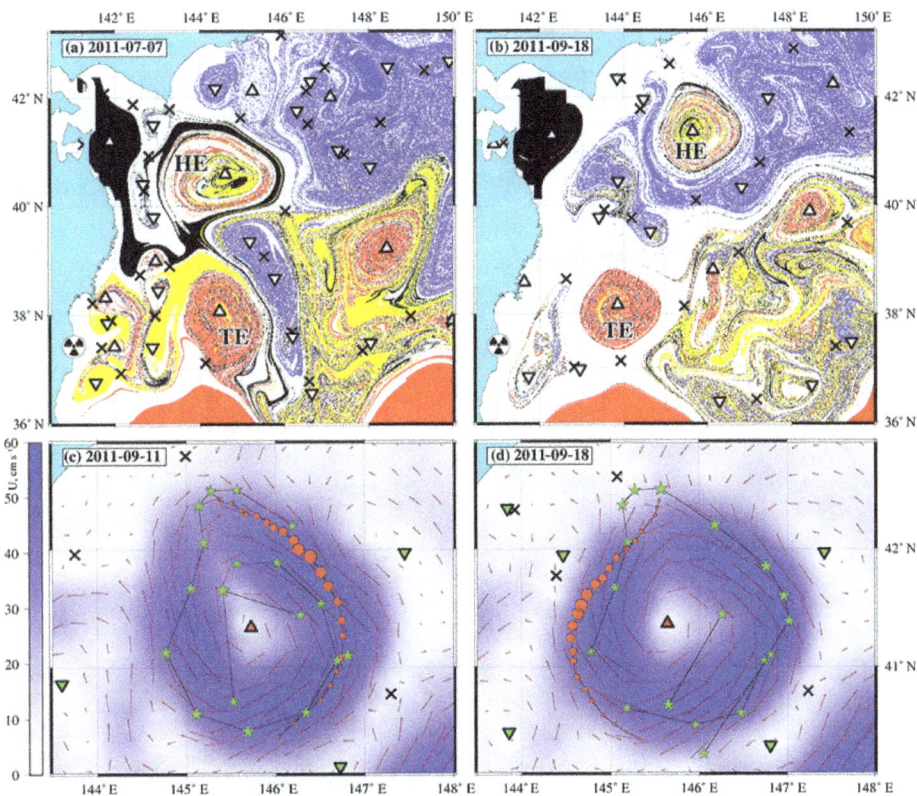

Figure 3. **(a)–(b)** The Lagrangian maps show evolution of the Hokkaido eddy (HE) after the FNPP accident to the days of its sampling and the origin of waters in its core and at the periphery. **(c)–(d)** A fragment of the track of the drifter no. 39123 is indicated by the full circles for 3 days before the day indicated with the size of circles increasing in time. Tracks of three Argo floats are shown by the stars. The largest star corresponds to the day indicated and the other ones each show float positions 7 days before and after that date.

ter at 40.3° N, 145.5° E), being genetically linked to the TE. During May, the TE gradually lost a Fukushima yellow water from its periphery to form the core of the HE. Fig. 3a shows the HE with a yellow core surrounded by modified subtropical red water which, in turn, is surrounded by Tsugaru black water.

The sampling of that eddy and its periphery by Kaeriyama et al. (2013) along the 144° E meridian at the end of July showed comparatively high concentrations of $C_{137} = 60 \pm 4.0$ and 71 ± 4.6 mBq kg^{-1} at stations C44 (40.5° N) and C43 (41° N), respectively. Station C43 was located inside the anticyclone HE filled mainly by yellow waters, and we estimate the likelihood of finding Fukushima-derived radionuclides there to be large. Station C44 was located at the southern periphery of the anticyclone HE at the boundary between white and blue waters but in close proximity to a yellow streamer.

The location of the HE on 24 August 2011 is shown in the AVISO velocity field in Fig. 1b. To verify the simulated locations of the HE and its form, we plot in Fig. 3c and d fragments of the tracks of a drifter and three Argo floats captured by that eddy in September 2011. A fragment of the track of the drifter no. 39123 is shown by the red circles with the size increasing in time for 3 days be-

fore the dates indicated in Figs. 3c and d and decreasing for 3 days after those dates, i.e., the largest circle corresponds to the drifter position at the indicated date. It was launched after the accident on 18 July 2011 at the point 45.588° N, 151.583° E in the Oyashio Current, advected by the current to the south and eventually captured by the HE moving around clockwise. Fragments of the clockwise tracks of the three Argo floats are shown by stars in Fig. 3c and d for 7 days before and 7 days after the indicated dates. The float no. 5902092 was released long before the accident on September 9, 2008 at the point 32.699° N, 145.668° E to the south of the Kuroshio Extension jet and was able to cross the jet and go far north. The float no. 2901019 was released before the accident on 19 April 2010 at the point 41.723° N, 146.606° E. The float no. 2901048 was released just after the accident on 10 April 2011 at the point 37.469° N, 141.403° E nearby the FNPP.

Our simulation shows that the HE contained, after its formation in the middle of May 2011, a large amount of yellow water probably contaminated by the Fukushima-derived radionuclides. This conclusion is supported by an increased concentration of radiocesium measured in its core at station C43 by Kaeriyama et al. (2013) at the end of July 2011. The

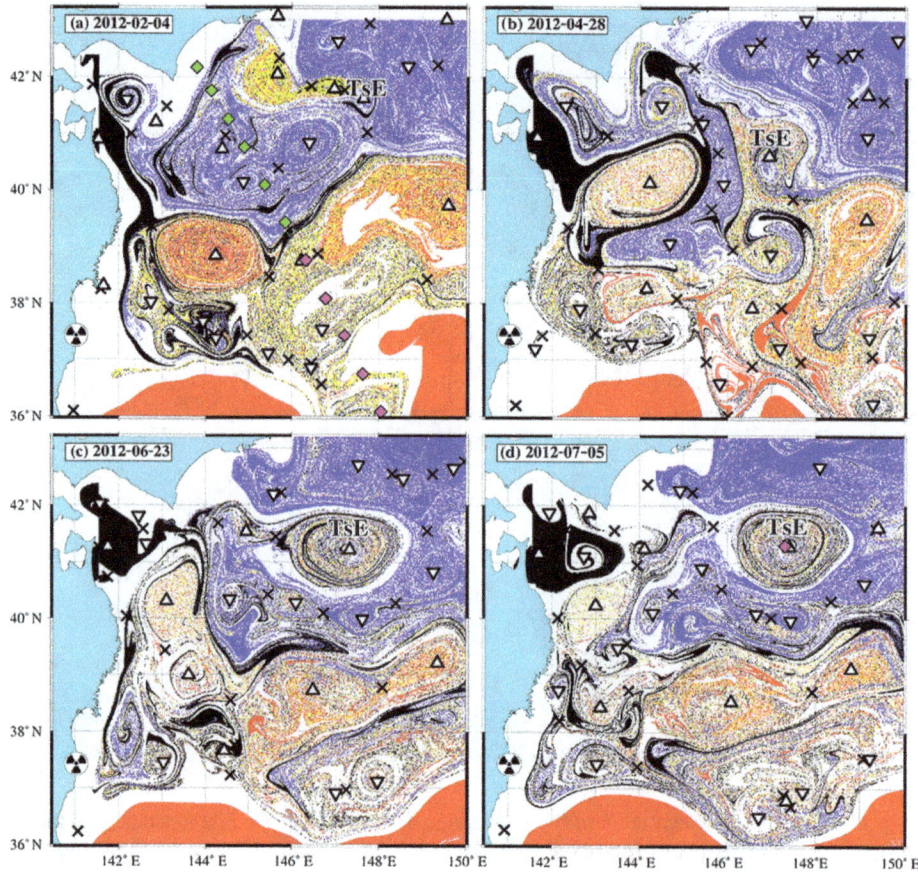

Figure 4. The Lagrangian maps in the study area in the first half of 2012. **(a)** The locations of stations in the beginning of February with surface seawater samples (collected by Kumamoto et al., 2014) with measured radiocesium concentrations at the background level (the green diamonds) and with higher concentration levels (the magenta diamonds). **(b–d)** The Lagrangian maps show evolution of the Tsugaru eddy (TsE), which originated on 4 February 2012 **(a)** after splitting of the HE and was sampled by Budyansky et al. (2015) at station 84 on 5 July 2012 and shown to have increased radiocesium concentrations (the magenta diamond in **d**).

HE persisted in the area around 42° N, 148° E up to the end of January of the next year. It eventually split on 31 January 2012 into two anticyclones.

3.3   The Tsugaru eddy

The anticyclonic TsE originated on 4 February 2012 after decay of the HE (the yellow patch with the elliptic point at 42° N, 145.6° E in Fig. 4a). The elliptic point at the center of the TsE appeared at 41.8° N, 146.9° E. Just after its birth, the HE begun to transport its yellow water around the TsE with the core consisted of an Oyashio blue water (Fig. 4b). The strong Subarctic Front is visible in Fig. 4 as a contrast boundary between Oyashio blue water and Fukushima-derived yellow water, with the Tsugaru black water in between.

Seawater samples for radiocesium measurements in the frontal area were collected during the R/V *Mirai* cruise from 31 January to 5 February 2012 along one of the observation lines of the World Ocean Circulation Experiment (WOCE) in the western Pacific, specifically the

WOCE-P10–P10N line (Kumamoto et al., 2014). We impose on the simulated Lagrangian map in Fig. 4a locations of stations to the north of the Kuroshio Extension (> 36° N) with measured levels of the cesium concentrations. As before, the green diamonds mark locations of those stations, P10–114 (42.17° N, 143.8° E), P10–112 (41.75° N, 144.13° E), P10–110 (41.25° N, 144.51° E), P10–108 (40.76° N, 144.88° E), P10–106 (40.08° N, 145.37° E) and P10–104 (39.42° N, 145.85° E), where the cesium concentrations in surface seawater samples were measured by Kumamoto et al. (2014) to be at the background level.

The stations, P10–102 (38.75° N, 146.32° E), P10–100 (38.08° N, 146.77° E), P10–98 (37.42° N, 147.2° E), P10-96 (36.74° N, 147.63° E) and P10–94 (36.08° N, 148.05° E), where the concentrations were found to be larger (but not exceeding 25.19 ± 1.24 mBq kg^{-1} for ^{137}Cs), are indicated by the magenta diamonds. It is worth stressing a good qualitative correspondence with our simulation results 10 months after the accident in the sense that stations with measured background level are in the area of Oyashio blue waters with low risk of being contaminated, whereas stations with comparatively high levels of radiocesium concentrations are in the area of the Fukushima-derived yellow waters with increased risk of contamination.

As to the TsE, it was sampled later, in 5 July 2012, during the cruise of the R/V *Professor Gagarinskiy* (Budyansky et al., 2015) when it was a comparatively large mesoscale eddy around 150 km in diameter with the elliptic point at $41.3°$ N, $147.3°$ E consisting of intermittent strips of blue and yellow waters (Fig. 4d), which were wrapped around during its growth from February to July 2012. Station 84 in that cruise was located near the elliptic point of that eddy (called "G" by Budyansky et al., 2015). The concentrations of ^{137}Cs at the surface and at 100 m depth were measured as 11 ± 0.6 and 18 ± 1.3 mBq kg^{-1}, respectively, an order of magnitude larger than the background level. As to the ^{134}Cs concentration, it was measured to be smaller, 6.1 ± 0.4 and 10.4 ± 0.7 mBq kg^{-1}, due to a shorter half-lifetime of that isotope. In fact, it was one of the highest cesium concentrations measured inside all the eddy features sampled in the cruise 15 months after the accident.

The maximal concentration of radionuclides was observed, as expected, not at the surface but within subsurface and intermediate water layers (100–500 m) in the potential density range of 26.5–26.7 due to a convergence and subduction of surface water inside anticyclonic eddies. The corresponding tracking map in Fig. 10c by Budyansky et al. (2015) confirms its genetic link with the TE, and, therefore, a probability of detecting increased cesium concentrations was expected to be comparatively large. We were able to track all the modification of the TsE up to its death on 16 April 2013 in the area around $40°$ N, $147.5°$ E.

4 Conclusions

We elaborated a specific Lagrangian methodology for simulating, tracking and documenting the origin and history of water masses in ocean mesoscale features. Integrating advection equations for passive particles in the AVISO velocity field backward in time, we have computed Lagrangian maps clearly demonstrating which waters the mesoscale eddies in the Kuroshio–Oyashio confluence zone were composed of. It allowed the simulation of the ways in which they gained and lost water with a risk of being contaminated by Fukushima-derived radionuclides. We have studied three genetically linked persistent mesoscale anticyclonic eddies in the area, TE, HE and TsE, which were sampled in the R/V cruises in 2011 and 2012 and shown to contain higher concentrations of radiocesium isotopes. The simulated Lagrangian maps allowed the documentation and analysis of how they interact and pass radioactive water to each other. The simulated results have been shown to be in a good qualitative correspondence compared with in situ measurements.

We hope that the proposed methodology could be applied to simulate propagation of pollutants after future possible accidents and identify and track contaminated persistent features in the ocean. The Lagrangian methodology seems to be useful, as well, for planning courses of the R/V cruises. It allows not only tracking of mesoscale eddies in the studied area but also identification of the origin of water masses and to estimate a priori concentrations of radionuclides, pollutants or other Lagrangian tracers inside the eddies planned to be sampled.

Competing interests. The authors declare that they have no conflict of interest.

Acknowledgements. The methodological part of the work was supported by the Russian Foundation for Basic Research (project no. 16-05-00213) and the simulations were supported by the Russian Science Foundation (project no. 16-17-10025). The altimeter products were distributed by AVISO with support from CNES.

Edited by: M. Hecht

References

Aoyama, M., Uematsu, M., Tsumune, D., and Hamajima, Y.: Surface pathway of radioactive plume of TEPCO Fukushima NPP1 released ^{134}Cs and ^{137}Cs, Biogeosciences, 10, 3067–3078, https://doi.org/10.5194/bg-10-3067-2013, 2013.

Budyansky, M. V., Uleysky, M. Y., and Prants, S. V.: Detection of barriers to cross-jet Lagrangian transport and its destruction in a meandering flow, Phys. Rev. E, 79, 056215, https://doi.org/10.1103/physreve.79.056215, 2009.

Budyansky, M. V., Goryachev, V. A., Kaplunenko, D. D., Lobanov, V. B., Prants, S. V., Sergeev, A. F., Shlyk, N. V., and Uleysky, M. Y.: Role of mesoscale eddies in transport of Fukushima-derived cesium isotopes in the ocean, Deep-Sea Res. Pt. I, 96, 15–27, https://doi.org/10.1016/j.dsr.2014.09.007, 2015.

Buesseler, K. O., Jayne, S. R., Fisher, N. S., Rypina, I. I., Baumann, H., Baumann, Z., Breier, C. F., Douglass, E. M., George, J., Macdonald, A. M., Miyamoto, H., Nishikawa, J., Pike, S. M., and Yoshida, S.: Fukushima-derived radionuclides in the ocean and biota off Japan, P. Natl. Acad. Sci. USA, 109, 5984–5988, https://doi.org/10.1073/pnas.1120794109, 2012.

Haller, G.: Lagrangian Coherent Structures, Annual Rev. Fluid Mech., 47, 137–162, https://doi.org/10.1146/annurev-fluid-010313-141322, 2015.

Haza, A. C., Özgökmen, T. M., Griffa, A., Molcard, A., Poulain, P.-M., and Peggion, G.: Transport properties in small-scale coastal flows: relative dispersion from VHF radar measurements in the Gulf of La Spezia, Ocean Dynam., 60, 861–882, https://doi.org/10.1007/s10236-010-0301-7, 2010.

Honda, M. C., Aono, T., Aoyama, M., Hamajima, Y., Kawakami, H., Kitamura, M., Masumoto, Y., Miyazawa, Y., Takigawa, M., and Saino, T.: Dispersion of artificial caesium-134 and -137 in the western North Pacific one month after the Fukushima accident, Geochem. J., 46, e1–e9, 2012.

Huntley, H. S., Lipphardt, B. L., and Kirwan, A. D.: Monitoring and Modeling the Deepwater Horizon Oil Spill: A Record-Breaking Enterprise, chap. Surface Drift Predictions of the Deepwater Horizon Spill: The Lagrangian Perspective, 179–195, American Geophysical Union, Washington, D. C., https://doi.org/10.1029/2011GM001097, 2011.

Inoue, M., Kofuji, H., Hamajima, Y., Nagao, S., Yoshida, K., and Yamamoto, M.: ^{134}Cs and ^{137}Cs activities in coastal seawater along Northern Sanriku and Tsugaru Strait, northeastern Japan, after Fukushima Dai-ichi Nuclear Power Plant accident, J. Environ. Radioactiv., 111, 116–119, https://doi.org/10.1016/j.jenvrad.2011.09.012, 2012a.

Inoue, M., Kofuji, H., Nagao, S., Yamamoto, M., Hamajima, Y., Yoshida, K., Fujimoto, K., Takada, T., and Isoda, Y.: Lateral variation of ^{134}Cs and ^{137}Cs concentrations in surface seawater in and around the Japan Sea after the Fukushima Dai-ichi Nuclear Power Plant accident, J. Environ. Radioactiv., 109, 45–51, https://doi.org/10.1016/j.jenvrad.2012.01.004, 2012b.

Kaeriyama, H., Ambe, D., Shimizu, Y., Fujimoto, K., Ono, T., Yonezaki, S., Kato, Y., Matsunaga, H., Minami, H., Nakatsuka, S., and Watanabe, T.: Direct observation of ^{134}Cs and ^{137}Cs in surface seawater in the western and central North Pacific after the Fukushima Dai-ichi nuclear power plant accident, Biogeosciences, 10, 4287–4295, https://doi.org/10.5194/bg-10-4287-2013, 2013.

Kaeriyama, H., Shimizu, Y., Ambe, D., Masujima, M., Shigenobu, Y., Fujimoto, K., Ono, T., Nishiuchi, K., Taneda, T., Kurogi, H., Setou, T., Sugisaki, H., Ichikawa, T., Hidaka, K., Hiroe, Y., Kusaka, A., Kodama, T., Kuriyama, M., Morita, H., Nakata, K., Morinaga, K., Morita, T., and Watanabe, T.: Southwest Intrusion of ^{134}Cs and ^{137}Cs Derived from the Fukushima Dai-ichi Nuclear Power Plant Accident in the Western North Pacific, Environ. Sci. Technol., 48, 3120–3127, https://doi.org/10.1021/es403686v, 2014.

Kameník, J., Dulaiova, H., Buesseler, K. O., Pike, S. M., and Šťastná, K.: Cesium-134 and 137 activities in the central North Pacific Ocean after the Fukushima Dai-ichi Nuclear Power Plant accident, Biogeosciences, 10, 6045–6052, https://doi.org/10.5194/bg-10-6045-2013, 2013.

Kawai, H.: Hydrography of the Kuroshio Extension, in: Kuroshio: Physical Aspects of the Japan Current, edited by: Stommel, H. M. and Yoshida, K., University of Washington Press, Seattle, 235–352, 1972.

Koshel', K. V. and Prants, S. V.: Chaotic advection in the ocean, Physics-Uspekhi, 49, 1151–1178, https://doi.org/10.1070/PU2006v049n11ABEH006066, 2006.

Kumamoto, Y., Aoyama, M., Hamajima, Y., Aono, T., Kouketsu, S., Murata, A., and Kawano, T.: Southward spreading of the Fukushima-derived radiocesium across the Kuroshio Extension in the North Pacific, Scientific Reports, 4, 1–9, https://doi.org/10.1038/srep04276, 2014.

Mancho, A. M., Small, D., and Wiggins, S.: A tutorial on dynamical systems concepts applied to Lagrangian transport in oceanic flows defined as finite time data sets: Theoretical and computational issues, Phys. Rep., 437, 55–124, https://doi.org/10.1016/j.physrep.2006.09.005, 2006.

Mezić, I., Loire, S., Fonoberov, V. A., and Hogan, P.: A New Mixing Diagnostic and Gulf Oil Spill Movement, Science, 330, 486–489, https://doi.org/10.1126/science.1194607, 2010.

Nencioli, F., d'Ovidio, F., Doglioli, A. M., and Petrenko, A. A.: Surface coastal circulation patterns by in-situ detection of Lagrangian coherent structures, Geophys. Res. Lett., 38, L17604, https://doi.org/10.1029/2011gl048815, 2011.

Oikawa, S., Takata, H., Watabe, T., Misonoo, J., and Kusakabe, M.: Distribution of the Fukushima-derived radionuclides in seawater in the Pacific off the coast of Miyagi, Fukushima, and Ibaraki Prefectures, Japan, Biogeosciences, 10, 5031–5047, https://doi.org/10.5194/bg-10-5031-2013, 2013.

Olascoaga, M. J. and Haller, G.: Forecasting sudden changes in environmental pollution patterns, P. Natl. Acad. Sci. USA, 109, 4738–4743, https://doi.org/10.1073/pnas.1118574109, 2012.

Olascoaga, M. J., Beron-Vera, F. J., Haller, G., Triñanes, J., Iskandarani, M., Coelho, E. F., Haus, B. K., Huntley, H. S., Jacobs, G., Kirwan, A. D., Lipphardt, B. L., Özgökmen, T. M., Reniers, A. J. H. M., and Valle-Levinson, A.: Drifter motion in the Gulf of Mexico constrained by altimetric Lagrangian coherent structures, Geophys. Res. Lett., 40, 6171–6175, https://doi.org/10.1002/2013gl058624, 2013.

Povinec, P. P., Hirose, K., and Aoyama, M.: Fukushima Accident: Radioactivity Impact on the Environment, Elsevier, Amsterdam, https://doi.org/10.1016/B978-0-12-408132-1.01001-9, 2013.

Prants, S. V.: Dynamical systems theory methods to study mixing and transport in the ocean, Phys. Scripta, 87, 038115, https://doi.org/10.1088/0031-8949/87/03/038115, 2013.

Prants, S. V.: Chaotic Lagrangian transport and mixing in the ocean, The European Physical Journal Special Topics, 223, 2723–2743, https://doi.org/10.1140/epjst/e2014-02288-5, 2014.

Prants, S. V., Budyansky, M. V., Ponomarev, V. I., and Uleysky, M. Y.: Lagrangian study of transport and mixing in a mesoscale eddy street, Ocean Model., 38, 114–125, https://doi.org/10.1016/j.ocemod.2011.02.008, 2011a.

Prants, S. V., Uleysky, M. Y., and Budyansky, M. V.: Numerical simulation of propagation of radioactive pollution in the ocean from the Fukushima Dai-ichi nuclear power plant, Dokl. Earth Sci., 439, 1179–1182, https://doi.org/10.1134/S1028334X11080277, 2011b.

Prants, S. V., Ponomarev, V. I., Budyansky, M. V., Uleysky, M. Y., and Fayman, P. A.: Lagrangian analysis of mixing and transport of water masses in the marine bays, Izvestiya, Atmos. Ocean. Phys., 49, 82–96, https://doi.org/10.1134/S0001433813010088, 2013.

Prants, S. V., Budyansky, M. V., and Uleysky, M. Yu.: Lagrangian study of surface transport in the Kuroshio Extension area based on simulation of propagation of Fukushima-derived radionuclides, Nonlin. Processes Geophys., 21, 279–289, https://doi.org/10.5194/npg-21-279-2014, 2014.

Prants, S. V., Lobanov, V. B., Budyansky, M. V., and Uleysky, M. Y.: Lagrangian analysis of formation, structure, evolution and splitting of anticyclonic Kuril eddies, Deep-Sea Res. Pt. I, 109, 61–75, https://doi.org/10.1016/j.dsr.2016.01.003, 2016.

Samelson, R. M. and Wiggins, S.: Lagrangian Transport in Geophysical Jets and Waves: The Dynamical Systems Approach, vol. 31 of Interdisciplinary Applied Mathematics, Springer Science+Business Media, LLC, https://doi.org/10.1007/978-0-387-46213-4, 2006.

Tsumune, D., Tsubono, T., Aoyama, M., and Hirose, K.: Distribution of oceanic ^{137}Cs from the Fukushima Dai-ichi Nuclear Power Plant simulated numerically by a regional ocean model, J. Environ. Radioactiv., 111, 100–108, https://doi.org/10.1016/j.jenvrad.2011.10.007, 2012.

Tsumune, D., Tsubono, T., Aoyama, M., Uematsu, M., Misumi, K., Maeda, Y., Yoshida, Y., and Hayami, H.: One-year, regional-scale simulation of ^{137}Cs radioactivity in the ocean following the Fukushima Dai-ichi Nuclear Power Plant accident, Biogeosciences, 10, 5601–5617, https://doi.org/10.5194/bg-10-5601-2013, 2013.

6

Quality assessment of the TOPAZ4 reanalysis in the Arctic over the period 1991–2013

Jiping Xie[1], **Laurent Bertino**[1], **François Counillon**[1], **Knut A. Lisæter**[1], **and Pavel Sakov**[2]

[1]Nansen Environmental and Remote Sensing Center, Bergen 5006, Norway
[2]Bureau of Meteorology, Melbourne VIC3001, Australia

Correspondence to: Jiping Xie (jiping.xie@nersc.no)

Abstract. Long dynamical atmospheric reanalyses are widely used for climate studies, but data-assimilative reanalyses of ocean and sea ice in the Arctic are less common. TOPAZ4 is a coupled ocean and sea ice data assimilation system for the North Atlantic and the Arctic that is based on the HYCOM ocean model and the ensemble Kalman filter data assimilation method using 100 dynamical members. A 23-year reanalysis has been completed for the period 1991–2013 and is the multi-year physical product in the Copernicus Marine Environment Monitoring Service (CMEMS) Arctic Marine Forecasting Center (ARC MFC). This study presents its quantitative quality assessment, compared to both assimilated and unassimilated observations available in the whole Arctic region, in order to document the strengths and weaknesses of the system for potential users. It is found that TOPAZ4 performs well with respect to near-surface ocean variables, but some limitations appear in the interior of the ocean and for ice thickness, where observations are sparse. In the course of the reanalysis, the skills of the system are improving as the observation network becomes denser, in particular during the International Polar Year. The online bias estimation successfully maintains a low bias in our system. In addition, statistics of the reduced centered random variables (RCRVs) confirm the reliability of the ensemble for most of the assimilated variables. Occasional discontinuities of these statistics are caused by the changes of the input data sets or the data assimilation settings, but the statistics remain otherwise stable throughout the reanalysis, regardless of the density of observations. Furthermore, no data type is severely less dispersed than the others, even though the lack of consistently reprocessed observation time series at the beginning of the reanalysis has proven challenging.

1 Introduction

The Arctic Ocean plays an important role in the global climate system, where the sea ice at the interface between atmosphere and ocean regulates the fluxes of heat, moisture and momentum. The recent warming of the Arctic and the change of its water cycle has been linked to the following manifestations: a significant reduction and thinning of the sea ice cover (Johannessen et al., 2004; Shimada et al., 2006; Rothrock et al., 2008; Kwok and Rothrock, 2009), more freshwater in the Arctic in the 2000s (Haine et al., 2015) and more mobility and faster deformations of the Arctic sea ice (Rampal et al., 2009; Spreen et al., 2011). The interpretation of such changes is severely hampered by the sparseness of the concerned observations, which should not be improved dramatically in the near future. It can be assisted by free-running model simulations, but those are usually hampered by mislocations of ice edge and certain water masses. One possibility is to study surrogate locations where similar processes are assumed to take place. Another solution is to correct the dynamical model by assimilating observations available over relevant timescales.

The latter activities thus necessitate a state-of-the-art reanalysis system able to accurately honor the observations in a physically consistent manner. Recent efforts in Arctic Ocean state estimation have delivered either long-window optimizations (Nguyen et al., 2009, 2011) or, more often, short-window estimations (Schweiger et al., 2011; Mathiot et al., 2012; Sakov et al., 2012; Chevallier et al., 2013). Long-window optimizations deliver continuous model trajectories, which are physically more consistent than those using short windows. On the other hand, slicing the optimization problem into short windows makes the estimation problem more linear or better conditioned (fewer unknowns and observations) and delivers more accurate products. Besides the window length, the choice of a background error covariance matrix is also a critical aspect in a data-scarce area such as the Arctic. The background error covariance used in an ocean data assimilation system can be – by increasing order of complexity – based on fixed multivariate spatial statistics (Cummings et al., 2009), an empirical estimation by a time-invariant ensemble (Oke et al., 2008) or a seasonally variable ensemble (Brasseur et al., 2005; Xie et al., 2011). In the case of ice–ocean systems, sea ice data assimilation often relies on rudimentary ice-only nudging methods (Schweiger et al., 2011; Tietsche et al., 2013); however, the possibility to account for flow-dependent coupled ice–ocean data assimilation updates has already been demonstrated in Lisæter et al. (2003). The pilot TOPAZ4 reanalysis of Sakov et al. (2012) has shown that the forecast error covariance from a dynamical ensemble mitigates the physical inconsistencies that could be expected from a short assimilation window.

The TOPAZ4 system is a coupled ocean–sea ice data assimilation system of the physical environment in the North Atlantic and Arctic oceans (see Fig. 1), which was initially used for short-term forecasting (Bertino and Lisæter, 2008) and later on for reanalysis (Sakov et al., 2012). TOPAZ4 represents the Arctic component of the CMEMS system (marine.copernicus.eu) where it is also used with coupling to an ecosystem model (Samuelsen et al., 2015; Simon et al., 2015). The present paper follows the pilot TOPAZ4 reanalysis by Sakov et al. (2012) in which the performance of the same system has been demonstrated for the period of 2003–2008. They proposed an implementation of the EnKF data assimilation method that avoids ensemble collapse, provides reliable state-dependent error estimates and improves the match to independent observations compared to a free-running simulation.

Forced by the European Centre for Medium-Range Weather Forecasts (ECMWF) ERA-Interim reanalysis (Dee et al., 2011), TOPAZ4 assimilates most available measurements including along-track sea level anomalies (SLAs) from satellite altimeters, sea surface temperatures (SSTs), sea ice concentrations (SICs) and sea ice drift (SID) from satellites as well as in situ temperature and salinity profiles. The proposed reanalysis is 4 times longer (1991–2013) than the pilot reanalysis, and includes data-scarce periods with poor observational coverage and more intense observing efforts, such as during the International Polar Year (IPY, 2007–2009). The focus of this study is to provide a quantitative assessment of the reanalysis performance in the pan-Arctic region (defined as north of 63° N) in order to guide the user through its skills and limitations. In particular, we investigate the stability of the ensemble reliability through changes of the Arctic observational network, the variability of the system accuracy

in different sub-areas, its seasonal cycle and its trend in the course of the reanalysis.

The outline of this paper is as follows: in Sect. 2, the reanalysis system is described including the model, the data assimilation scheme and their implementation. Section 3 evaluates the reliability of the reanalysis ensemble. In Sect. 4, we compare the ensemble mean against available observations: altimetry, SST, T–S profiles, ice concentration, ice drift and ice thickness. For each of these quantities, we assess the variability of the system performance in space or in time. Section 5 summarizes and discusses the potential improvements of our system for the next version of the reanalysis.

2 The reanalysis system

2.1 The HYCOM ice–ocean model

The TOPAZ4 system uses version 2.2 of the Hybrid Coordinate Ocean Model (HYCOM) developed at the University of Miami (Bleck, 2002; Chassignet et al., 2003). It uses 28 hybrid z-isopycnal layers, and the top layer has a minimum thickness of 3 m. The model grid has a horizontal resolution of 12–16 km, which is eddy permitting from the Equator to the Nordic Seas but is still far from being eddy resolving in the Arctic. The lateral boundaries of temperature and salinity are relaxed to a combination of the World Atlas of 2005 (WOA05; Locarnini et al., 2006) and version 3.0 of the Polar Science Center Hydrographic Climatology (PHC; Steele et al., 2001). HYCOM is coupled to a sea ice model in which the ice thermodynamics are described in Drange and Simonsen (1996) and the elastic–viscous–plastic rheology in Hunke and Dukowicz (1997). The surface momentum fluxes use a bulk formula parameterization (Kara et al., 2000), and the related thermodynamic fluxes are computed as described in Drange and Simonsen (1996).

The model has been initialized from the same climatology data as used at the boundaries. The Pacific water inflow is imposed by a barotropic inflow through the Bering Strait at the model boundary and balanced by an outflow at the southern boundary of the domain. Unlike in Sakov et al. (2012), the inflow varies seasonally as found in observations (Woodgate et al., 2005): with a maximum in June (1.3 Sv), a minimum in January (0.4 Sv), and the mean transport is 0.8 Sv.

2.2 Data assimilation with the EnKF

Given observations, a model forecast and assumptions on their respective uncertainties and at time t_i, the analyzed model states can be estimated by data assimilation using the least squares minimization (Evensen, 1994, 2003):

$$\mathbf{X}_i^a = \mathbf{X}_i^f + \mathbf{K}_i(\mathbf{Y}_i - \mathbf{H}\mathbf{X}_i^f), \tag{1}$$

where $\mathbf{Y}_i$ is the matrix of perturbed observations, $\mathbf{X}_i$ is the ensemble of model state vectors and $\mathbf{H}$ is the observation

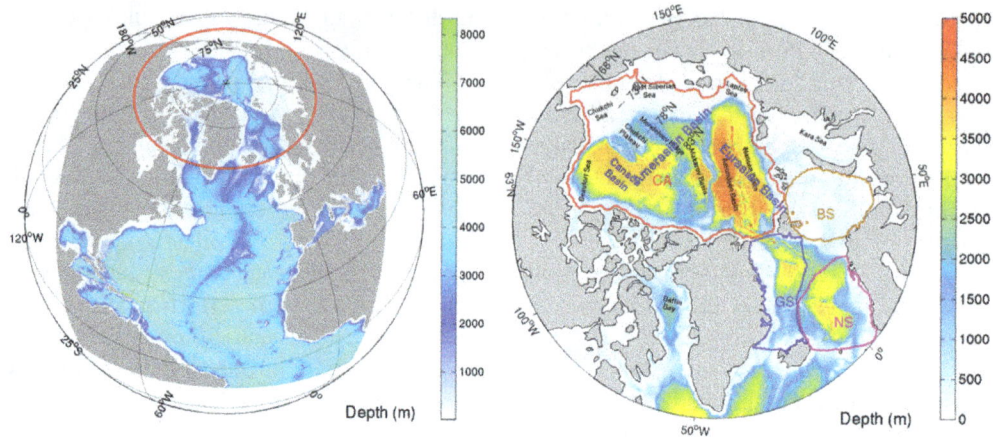

Figure 1. Left: bottom topography in the whole TOPAZ4 domain. The red line delimits the pan-Arctic region north of 63° N. Right: definition of sub-basins and marginal seas. The domain is divided into the four subregions delimited by the colored lines: the central Arctic in red (CA), the Greenland Sea in blue (GS), the Barents Sea in orange (BS) and the Norwegian Sea in magenta (NS).

operator denoting the projection from the model state variables to the measurements. The superscripts "a" and "f" refer to the analyzed and the forecast states, respectively. We use the deterministic form of the EnKF (DEnKF; Sakov and Oke, 2008), which solves the analysis without the requisite to perturb the observations. The term in the parentheses in Eq. (1) is the departure from the model simulations to the observations (named innovations). As opposed to Sakov et al. (2012), the 1 % multiplicative inflation, which becomes problematic when used with spatially varying observational network (Anderson et al., 2001), has been removed near to the end of the reanalysis (January 2010). Multiplicative inflation leads to an exponential increase of the spread in absence of observation (such as in the interior of the Arctic Ocean). When combined with a multivariate update, it will amplify the biases of the observed variables. For instance, the passive microwave satellite images of sea ice confuse melt ponds (not considered in TOPAZ4) with open water (Ivanova et al., 2015). This results in a bias that in turn leads to a degradation of the stratification in the Arctic due to the multiplicative inflation. The bias estimation procedure has also been modified as explained below (see Sect. 2.4).

2.3 Assimilated observations

The observations assimilated into the reanalysis are same types as used in Sakov et al. (2012) except for some updates in the data sources. They are the satellite SST, SLA, in situ temperature and salinity profiles, SIC and low-resolution SID data from satellites. An overview of the observations used in the reanalysis is given in Table 1. The preprocessing, temporal averaging and observation errors are mostly following the procedure described in Sakov et al. (2012).

At the beginning of the time period, the assimilated SST data are the 1° resolution Reynolds SST from NOAA (Reynolds and Smith, 1994). In June 1998, they are re-

placed by the high-resolution OSTIA data (Stark et al., 2007) from the UK Met Office. The assimilated SLA data are the delayed-time product (version vxxc) from Collecte Localisation Satellites (CLS) which is validated, unfiltered and not subsampled by CLS. The SIC from the Ocean & Sea Ice Satellite Application Facility (OSISAF) are assimilated into the TOPAZ4 system. Before 19 June 2002, this assimilated product is derived from the Special Sensor Microwave/Imager (SSM/I) at 25 km resolution, and later is derived from the Advanced Microwave Scanning Radiometer for EOS (AMSR-E) 89 GHz brightness temperature at 12.5 km resolution. In the last 3 years, this product has been upgraded to a 10 km resolution. The temperature and salinity profiles include Argo floats, ice-tethered profiles (ITPs) from the Damocles project and a large collection of hydrographic cruise data. With the exception of the Reynolds SST, all assimilated data are available through the CMEMS portal.

2.4 Bias estimation in the TOPAZ4 reanalysis

Two bias fields (for SST and mean sea surface height (MSSH)) are estimated online by model state augmentation, thus the analysis state of Eq. (1) is modified as

$$\begin{pmatrix} \overline{\mathbf{x}}_i^a \\ \mathbf{c}_i^a \end{pmatrix} = \begin{pmatrix} \overline{\mathbf{x}}_i^f \\ \mathbf{c}_i^f \end{pmatrix} + \mathbf{K}_i (\mathbf{y}_i - \mathbf{H}\overline{\mathbf{x}}_i^f + \mathbf{H}\mathbf{c}_i^f), \qquad (2)$$

where $\overline{\mathbf{x}}_i$ is the ensemble mean of the model state vector at the analysis time i, $\mathbf{y}_i$ is the vector of observations and $\mathbf{c}_i^f$ represents the estimated bias correction inherited from the analyzed bias correction at time $i - 1$. In order to avoid inconsistencies between assimilation of SST and temperature profile, the SST bias is propagated downwards into the model mixed layer and decays exponentially (into the $\mathbf{H}$ operator).

The initial biases for each ensemble member are random values, homogeneous in space and uniformly distributed. The

Table 1. Overview of assimilated observations per cycle, with average numbers for the cycles during which the observations are present.

Type	Number	After SO	Spacing	Resolution	Period	Provider
SLA	9×10^4	5×10^4	Track	7 km	1992–2013	CLS
SST	6×10^3	6×10^3	Gridded	100 km	1990–1998	Reynolds SST from NCDC (http://www.nhc.noaa.gov/aboutsst.shtml)
SST	2×10^6	2.4×10^5	Gridded	5 km	1998–2013	OSTIA from UK Met Office
In situ T/S	3×10^4	5×10^3	Point	–	1990–2013	Ifremer + other
SIC (SSM/I)	9×10^4	5×10^4	Gridded	25 km	1990–2002	OSISAF
SIC (AMSR-E)	1.6×10^5	5×10^4	Gridded	12.5 km*	2002–2013	OSISAF
SIC (AMSR-E)	1.6×10^5	5×10^4	Gridded	12.5 km	2008–2009	AMSR-E (http://nsidc.org/data/amsre/)
Ice drift (CERSAT)	6×10^3	10^3	Gridded	35 km	2002–2010	Ifremer
Ice drift (OSISAF)	4×10^3	10^3	Gridded	62.5 km	2011–2013	OSISAF
Total	2.3×10^6	4×10^5				

* The resolution of ice concentration product increased to 10 km. Unless specified, all observations are from http://marine.copernicus.eu. NCDC is the National Climatic Data Center.

initial SST biases are sampled in the interval $[-4, 4]\,°C$, and within $[-0.6, 0.6]\,m$ for the MSSH.

The bias fields are updated according to the sample covariance from the forecast ensemble, but are not integrated forward. To avoid a collapse of the bias ensembles, a multiplicative inflation is used (2 % for SLA and 6 % for SST). The multiplicative inflation of bias did not handle well the changes of observation coverage: it has been re-initialized and capped at 5 °C for SST bias in April 2001 (hereafter called event E1). Later on, in May 2006, it was re-initialized again and replaced by an additive inflation of identical amplitude (event E2), using an auto-regressive temporal process with one order, which definitively prevented further divergence. After several assimilation steps, the bias fields converge to temporally stable and spatially variable fields. Figure 2 shows the bias estimates at the end of the reanalysis for the SSH and the SST. The bias patterns compare well with those obtained in Sakov et al. (2012)[1]. There are small discrepancies because the bias is estimated at a different time – December 2009 in Sakov et al. (2012) instead of December 2013 here – and the bias estimation is the result of a longer estimation period for which the signal-to-noise ratio is reduced. The misfits using the online-bias-corrected values are slightly lower than the bias estimate of the last analysis step (not shown). Although the static part of the bias would theoretically be better estimated on the last assimilation of the reanalysis, the online bias approach can follow decadal trends in the errors, as well as seasonal biases and changes of the observational network. The online bias estimate is provided together with the model output. In the following validation sections, the online bias estimates c_i^a are used to offset the reanalysis state.

---

[1] Sakov et al. (2012) present the mean SSH bias of the opposite sign.

3 Probabilistic reliability analysis

The main selling point of an ensemble data assimilation system is the probabilistic evaluation of the uncertainties, which follows the model dynamics and thus varies both in time and space. This ability comes at a risk of divergence of the Kalman filter: if the ensemble collapses, the Kalman gain tends to zero and the assimilation system behaves as one – expensive – free run. The EnKF is designed to support a very heterogeneous observational network: when observations become denser, the ensemble spread is supposed to shrink, but the forecast accuracy should be improved accordingly. However, in practice, maintaining the reliability through the course of the reanalysis requires careful analysis and handling of ill-specified model or observation error terms, and verifies that one observational data set is not "over-assimilated" at the expense of the others. Here, a simple method is used to assess the system reliability and whether the uncertainty predicted by the EnKF is commensurate with actual deviations from observations. The ensemble resolution, as well as more oceanographic interpretation of the bias, will be presented in Sect. 4.

The ensemble statistics of the assimilated variables have been stored at each assimilation time (every week) and in observational space. This allows the evaluation using the modified reduced centered random variable (RCRV; Talagrand et al., 1999; Candille et al., 2007) to measure the reliability of the TOPAZ4 system. Considering one observation y and the ensemble mean of model state $\overline{\mathbf{x}}^f$, the scalar variable q can be defined as the innovation normalized by the observation and model uncertainties:

$$q = \frac{y - \mathbf{H}\overline{\mathbf{x}}^f}{\sqrt{\sigma_o^2 + \sigma_{en}^2}}, \qquad (3)$$

where σ_o is the observation error and σ_{en} is the standard deviation of the corresponding forecast ensemble, including the

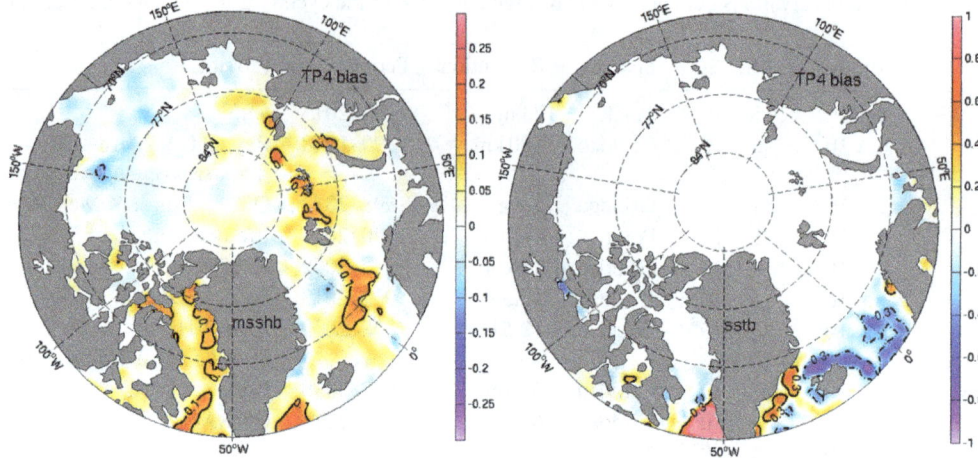

Figure 2. Estimates of the mean SSH bias (left) and the SST bias (right) obtained at the last analyzed date by online parameter estimation. In the left panel, the solid (dashed) line indicates the 10 (−10) cm isolines. In the right panel, the solid (dashed) line indicates the 0.3 °C (−0.3 °C) isolines. There is no bias estimation for SST in the white area north of 70° N.

uncertainty of bias estimation for SLA and SST. In the framework of the Kalman filter, q is assumed to be a reduced centered Gaussian variable.

In the following, we will assess the time evolution of the averaged bias:

$$b = E[q] = \frac{1}{M} \sum_j^M \frac{y_j - \mathbf{H}\overline{\mathbf{x}}^f}{\sqrt{\sigma_{oj}^2 + \sigma_{enj}^2}}, \qquad (4)$$

where M is the total number of observations at the assimilation time. Furthermore, the standard deviation of q,

$$d = \sqrt{\frac{M}{M-1} E\left[(q-b)^2\right]}, \qquad (5)$$

measures the ensemble dispersion with respect to the normalized misfits.

The first two moments of the RCRV, b and d, provide simple diagnostics of whether the forecast ensemble obtained from TOPAZ4 provides a reliable estimate of the uncertainty of the ensemble mean, which is trusted in view of the observations with the assumed uncertainties. Assuming that we can neglect all cross-covariances between innovations, a perfectly reliable system would have no bias (i.e., $b = 0$) and a dispersion equal to 1 (Candille et al., 2007). A d smaller than 1 is a sign that the assimilation system could be too optimistic about its uncertainties and vice versa. Both cases indicate that the EnKF system is not well calibrated, which in turn leads to suboptimal performance of the reanalysis system.

The two first moments of the reanalysis RCRV are presented for the different observational types. The time series of b and d in the 23 years are shown in Figs. 3 and 4.

The dispersion and seasonal bias of SLA increased after the launch of ENVISAT in 2002, when previously unobserved areas at high latitude got to be included in the calculation of the statistics. We can notice that the bias stabilizes

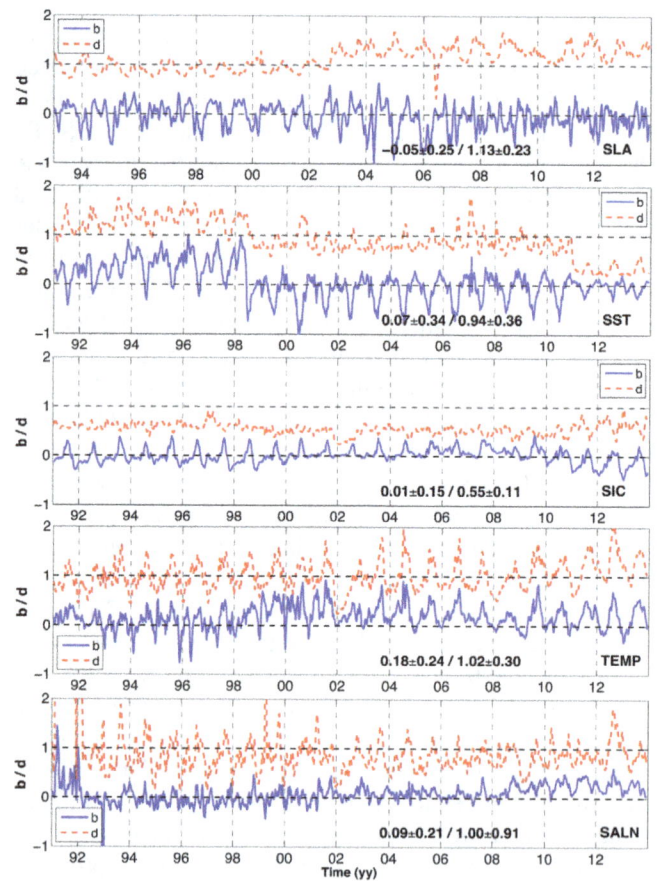

Figure 3. Time series of b (blue line) and d (dashed red line) of SLA, SST, SIC, in situ temperature and salinity observations, respectively, in the Arctic region. They are filtered by a smoothing average within 28 days. The average (standard deviation) of b and d is shown in the panels.

later on when the multiplicative inflation is replaced by the auto-regressive bias correction (event E2 in 2006).

The SST panel of Fig. 3 exhibits a cold winter bias and a slight overdispersion during the time when Reynolds SST is assimilated (until 1998). The transition to OSTIA initially improves the reliability statistics with a dispersion close to 1 and a reduced bias fluctuating around 0, which relate to the changes of observation errors and land mask. The warm bias is dominant in summer. During the last 3 years of the reanalysis, the summer warm bias b is reduced but the dispersion shrinks dramatically. This coincided with the time when the observation error was increased and the quality control of the observations (based on observation uncertainty) was softened, which resulted in assimilating more observations in the Gulf Stream and near the ice edge. Although it is somewhat counterintuitive that increasing the observation error leads to a degradation of the reliability, this can happen if the misfits to the observations increase more than the model uncertainty. Furthermore, the new observation coverage includes regions close to the ice edge where the spatiotemporal interpolation of SST may have degraded the reliability (this will be further discussed in Sect. 4.2).

In the SIC panel of Fig. 3, the dispersion is underestimated throughout the reanalysis, with d on average at 0.55. The bias fluctuates around 0 with a standard deviation of 0.15 mostly related to a summer bias (Lisæter et al., 2003). A bias degradation and a dispersion improvement are jointed with clear seasonality during the last 3 years, which relates to the aforementioned change of SST assimilation settings.

The RCRVs for in situ temperatures reveal a cold bias in the reanalysis, especially salient after 1998 following developments of the observational network. A seasonal cycle in both b and d is detected during the IPY period, which may have been present before but insufficiently observed. The RCRVs for in situ salinities are initially noisy due to lack of observations. The IPY data also reveal a fresh bias as they sample regions of the central Arctic that were previously unobserved. The ensemble dispersion of salinity is good, with a tendency to be on the low side, and especially after 2002 the observation samples increase remarkably due to Argo floats.

The RCRVs for SID show initially too little dispersion ($d = 0.56$) from 2002 to 2010, shown in Fig. 4 (consistent with Sakov et al., 2012). Afterward, the dispersions increase when the drag coefficient is reduced in 2011, leaving more freedom for the ice to drift following the ocean currents, but the system becomes overdispersive ($\sim d = 1.36$) when the SID data source is switched from 3-day drifts on 35 km resolution to 2-day drifts on a 62.5 km resolution grid. The system shows no clear bias but the bias variability increases with the new observation product; its features will be discussed in Sect. 4.

Overall, the statistics presented are relatively stable throughout the reanalysis. There is a good balance between the different data types assimilated: none of the data types are severely less dispersed than the others. For most of the

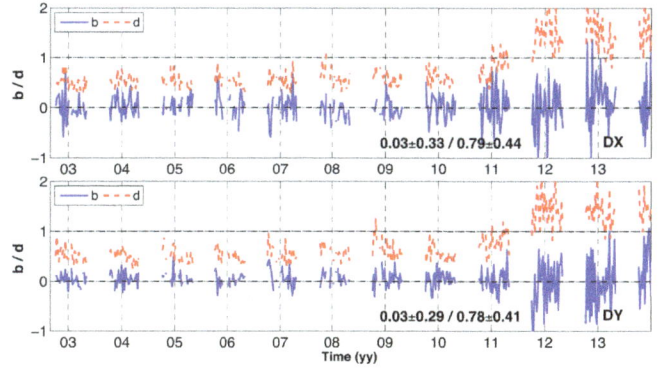

Figure 4. Time series of b (blue line) and d (dashed red line) about the zonal (DX) and meridional (DY) drifts of sea ice in the Arctic. The average (standard deviation) of b and d is shown in the panels.

assimilated observation data sets, the biases fluctuate around 0 with amplitudes no larger than 0.1 (except for the in situ temperatures); the dispersions mostly fluctuate around 1 and the departures from 1 are smaller than 0.15 (except for the assimilated SIC and SID) without any sign of general ensemble collapse. However, there are some clear discontinuities caused by the introduction of new data sets with different spatial coverage (polar orbit, land mask, sea ice mask) or the related error variance adjustments. Providing a consistent reanalysis is thus challenging in the absence of continuous reprocessed observations marked with the time period.

4 Quantitative deterministic accuracy

In this section, we investigate whether the accuracy of the reanalysis ensemble mean (also called resolution in Candille et al., 2007) varies spatially, seasonally or interannually. Such information is necessary for potential users of the reanalysis product. It also pinpoints the model limitations that motivate further developments of modeling and assimilation approach. The misfits of the reanalysis are calculated by the daily averages of the ensemble mean and the observations. The bias and the root mean square differences (RMSDs) of the misfits are calculated as described in Eqs. (6) and (7):

$$\text{Bias} = \frac{1}{N}\sum_{i=1}^{N}(\mathbf{H}_i\bar{\mathbf{x}}_i^{\text{f}} - \boldsymbol{y}_i - \mathbf{H}\mathbf{c}_i^{\text{f}}) \tag{6}$$

$$\text{RMSD} = \sqrt{\frac{1}{N}\sum_{i=1}^{N}(\mathbf{H}_i\bar{\mathbf{x}}_i^{\text{f}} - \boldsymbol{y}_i - \mathbf{H}\mathbf{c}_i^{\text{f}})^2}, \tag{7}$$

where $\bar{\mathbf{x}}_i^{\text{f}}$ is the forecasted daily average from the ensemble mean, which is compared to the observations $\boldsymbol{y}_i$ on the same day. N is the number of times sampling was conducted over the diagnostic period (either 365 or 366 yearly). For SST and SLA, the bias term of $\mathbf{c}_i^{\text{f}}$ is the online estimated correction ($\mathbf{c}_i^{\text{f}} = \mathbf{c}_{i-1}^{\text{a}}$, as in Eq. 2). Error bars are used to represent the standard deviations of these quantities – i.e., the variability

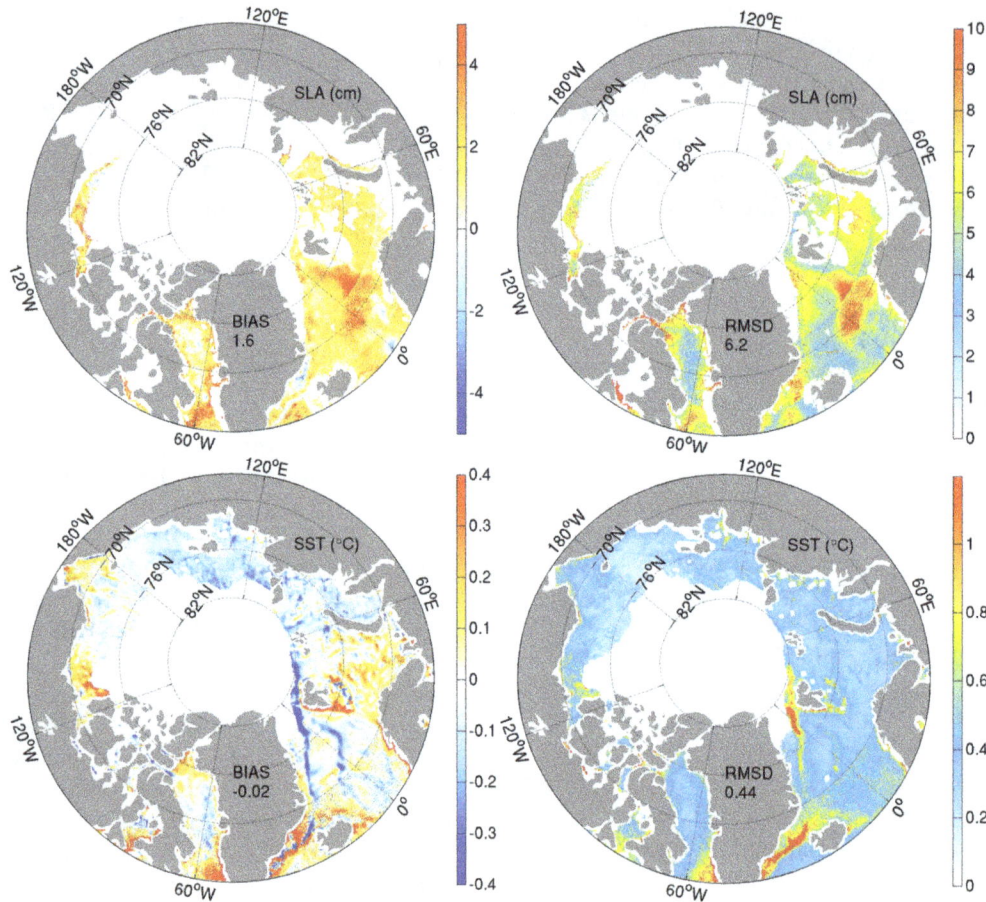

Figure 5. Top: residual bias (left) and RMSD (right) between the daily average SLA from the reanalysis and the assimilated along-track SLA data averaged over the period 1993–2013 (unit: cm). Bottom: the corresponding residual bias (left) and RMSD (right) between the daily average SST from the reanalysis and the assimilated observations averaged over the period 1999–2013 (unit: °C). Areas with less than 30 observations have been masked in white.

of the RMSD or bias estimate through the calculation period. For assimilated observations, the bias is the same as the *b* term in the RCRV.

4.1 Sea level anomalies

The SLA accuracy in the reanalysis is evaluated in the pan-Arctic region (defined to the north of 63° N; see Fig. 1). The spatial variability of the bias and RMSD, calculated over the whole reanalysis period (1993–2013), is shown at the top of Fig. 5. The residual bias is mainly positive, with much smaller amplitude than the estimated bias (see Fig. 2). Some positive biases reach over 4 cm around the Lofoten Basin and south of the Baffin Bay. Except for the sea ice edge in the Greenland Sea, the high RMSDs (over 9 cm) match the areas of large bias shown in Fig. 5. The spatially averaged bias is 1.6 cm, and the RMSD is about 6.2 cm.

The yearly time series of the SLA misfits and the observation number are shown on the left side of Fig. 6. The number of assimilated observations evolves with the launch or com-

pletion of satellite missions. The number of observation increases in 2000 with the launch of the GEOSAT Follow On (GFO) mission. The missions of Topex, Jason 1 and Jason 2 do not contribute directly to the pan-Arctic region as their inclination is 66°, unlike 70° for GFO. A low observation period is in 2009–2010 with the end of GFO mission (Le Traon et al., 2015), followed by an increase in 2011 with Cryosat-2, a decrease in 2012 with the end of Envisat and a last increase with the Saral/AltiKa mission in 2013. From 1993 to 2013, the RMSD decreases gradually from over 9 cm to less than 6 cm. After 2000, the residual bias stabilizes around 1 cm but remains positive. The RMSD gradually reduces with the introduction of new and more accurate observations. The reduced altimeter constellation in 2009–2010 does not cause an increase of the misfits. This demonstrates the advantage of assimilating multiple types of observations, as improved SSH may also be the result of improved SST or temperature and salinity profiles. Meanwhile, the temporal standard deviation of the RMSD during the year (shown as the half-error

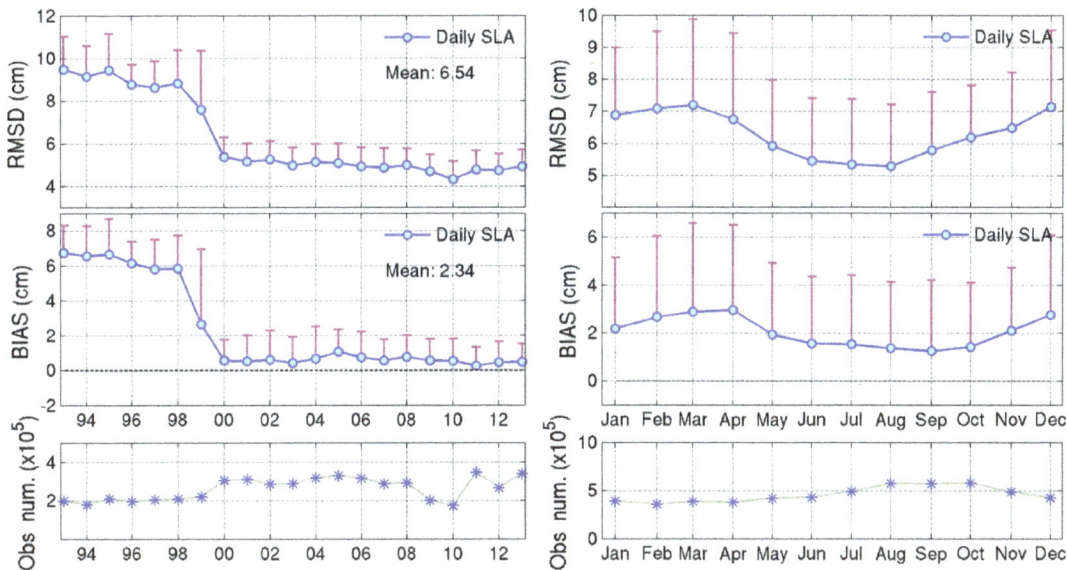

Figure 6. Left: yearly averaged estimates of daily SLA RMSD (upper) and the residual bias (middle) of the TOPAZ reanalysis calculated against the along-track SLA available in the pan-Arctic region (unit: cm). The error bars denote the standard deviations of the daily statistics within each year. The bottom panel is the number of available observations in each year. Right: similar plot for monthly averaged estimate of daily SLA RMSD (upper), and the residual bias (middle). The error bars denote the standard deviations of the daily statistic within each month. The bottom panel shows the number of observations available for each month in the pan-Arctic during 1993–2013.

bar) also reduces from 1–2 cm to less than 1 cm, indicating the system is getting more stable with time.

The seasonal cycle of the accuracy is shown on the right side of Fig. 6. The SLA being masked by sea ice, the number of observations varies seasonally in opposition to the sea ice cover. The RMSD ranges from 5 to 7 cm as a consequence of the seasonal spatial coverage. The residual bias is positive throughout 1 year but reaches a maximum in April. This may be explained as well by the seasonal sea ice coverage, but also by a possible underestimation of the thermal expansion. The standard deviations of the residual bias and RMSD have no visible seasonality.

4.2 Sea surface temperatures

The spatial variability of the SST misfits during 1999–2013 is shown at the bottom of Fig. 5. Note that SST is masked under sea ice, as done during assimilation. There are stripes of cold residual bias and high RMSD along the ice edge from north of the Svalbard Island until south of the Greenland Sea. These are contradictory to the sea ice concentration biases in the same areas in Sect. 4.4, where a cold bias corresponds with too little ice. The accuracy of SST observations near ice edge is poor and relies on strong ad hoc assumptions. Another salient feature is the warm bias (>0.3 °C) north of the Denmark Strait where the recirculation of Atlantic water inflow is excessive in TOPAZ4 (Lien et al., 2016). This pattern was also visible in the estimated bias shown in Fig. 2, suggesting that the estimated bias accounts for most of the bias but that it still underestimates the true bias. An addi-

tional stripe of the cold residual bias and higher RMSD is clear along Mohns Ridge, also pointing to topographic steering issues. In the Barents Sea, a relative weak bias is noticeable. Besides these areas, most of the SST RMSD is lower than 0.6 °C. On average, in the whole Arctic region, the SST RMSD is about 0.44 °C during the period 1999–2013.

The evolution of SST accuracy of the TOPAZ4 reanalysis is shown on the left side of Fig. 7, together with the number of observations. In June 1998, the coarse-resolution Reynolds SST is swapped to the higher-resolution OSTIA SST and the number of observations increases drastically. On average, over the period 1991–2013, the SST RMSD is about 0.63 °C, and the bias −0.08 °C. In the first years, the SST RMSDs are initially about 1 °C but decrease gradually down to 0.8 °C before 1998. During this period, the model has a cold SST bias around −0.3 °C with 0.1 °C standard deviation. After the introduction of OSTIA, the SST bias settles down closer to zero, but a slight positive in summer is still noticeable before 2011. Meanwhile, the RMSD decreases rapidly below 0.6 °C as a direct consequence of the bias reduction and the more abundant observations. In 2010, the RMSD reaches the minimum below 0.4 °C. At that time, the ensemble spread was getting too small, and the system performance was too constrained by SST, as can be seen in the standard deviation of RMSD. It was thus decided to artificially increase the SST observation errors, which resulted in a small increase of the misfit up to 0.5 °C. It is clear from the above that the transition to high-resolution SST in our system has led to a higher SST accuracy.

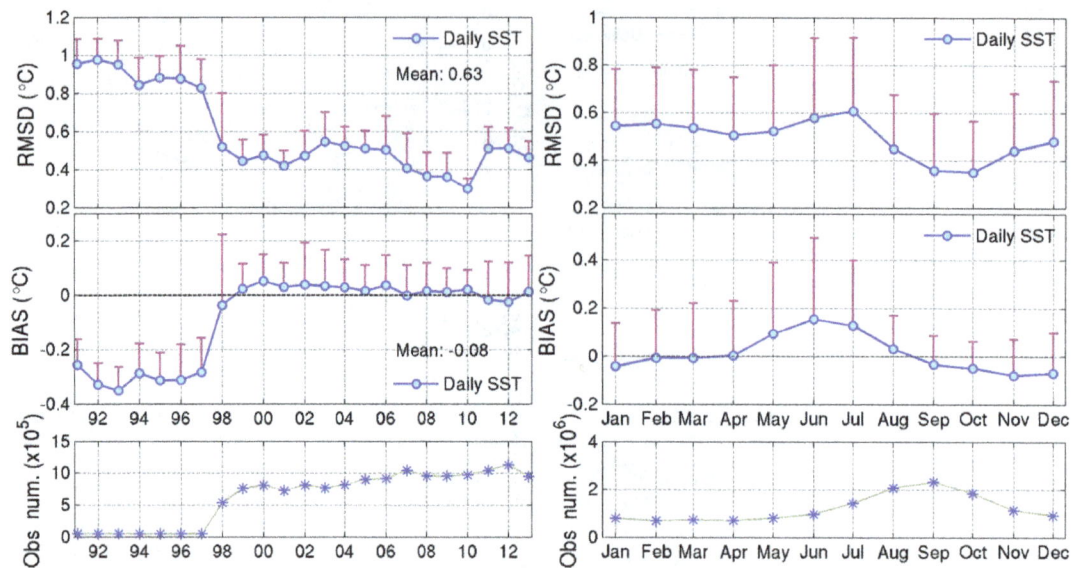

Figure 7. Same as the previous figure but for SST over the period 1991–2013 (unit: °C).

Furthermore, the seasonal performance of SST is shown in Fig. 7. As for SLA, the number of observations varies seasonally with the sea ice mask and causes the changes of the bias and RMSD. The RMSD is minimum in September and October with less than 0.4 °C owing to more observations, and is maximum at 0.6 °C in June and July when the bias is maximum as well. The reason for the larger bias in summer months is indeterminate but should relate to the inaccuracies of the mixed layer depths and the atmospheric radiative forcing.

4.3 In situ temperature and salinity profiles

There are 1.1×10^5 temperature and salinity profiles assimilated in the pan-Arctic region during the period 1991–2013, but their distributions and the respective uncertainties are very uneven both in time and space, with more observations in ice-free areas and during the IPY. In order to limit variability of the uncertainty, the bias normalized by the uncertainties of the observation and model error (i.e., b as defined in Eq. 4), is shown in Fig. 8. For temperature, there is a cold (warm) bias along the west (east) coast of the Svalbard archipelago, which indicates a northward Atlantic water flow that is too weak across the Fram Strait and a southward flow of Arctic water east of Svalbard that is too weak. There are biases that are too saline on both coasts of the Svalbard archipelago and along the Norwegian coast. They likely result from an underestimation of river discharges.

To investigate the vertical structures of the biases, the averaged temperature and salinity profiles from the reanalysis and the climatology WOA13 (Locarnini et al., 2013), and their misfits are shown in Fig. 9. The analysis is separated

into four subregions: the central Arctic, the Barents Sea, the Greenland Sea and the Norwegian Sea (see Fig. 1).

In the central Arctic, the average profiles depict well the cold halocline water near the surface and warm saline water around 400 m associated with Atlantic water (AW). Near the surface (deeper than 200 m), the salinity misfits of TOPAZ4 are slightly smaller than the climatology. The core Atlantic water is clearly too diffuse in TOPAZ4 (not pronounced enough and vertically too broad), leading to a cold bias (−0.3 °C) and 0.5 °C RMSD around that depth. Another large RMSD is noticeable around 1000 m (0.6 °C and 0.3 psu). Since the bias at that depth is low and since the climatology has lower RMSD, it suggests that TOPAZ4 has too much variability at depths. That variability is likely due to the data assimilation setup with the combined effect of multiplicative inflation and spurious correlations (see Sect. 2.2).

In the Greenland Sea, the temperature RMSDs and biases are again slightly smaller than the climatology near the surface (upper 200 m), but degrade very near below, reaching the maxima of RMSD (> 1 °C and 0.1 psu) and bias around 800 m.

In the Norwegian Sea, the features are similar: the model has some skills near the surface but deteriorates at depths where the AW is present but is too diffuse. It is too broad and does not capture the maximum at the same depth as in the observation. It is a well-known limitation of ocean models nowadays (Ilıcak et al., 2016).

In the Barents Sea, the RMSD for temperature and salinity can be reduced near the surface, even compared to that of the climatology. But the AW (temperature > 3 °C and salinity > 35 psu, Blindheim and Østerhus, 2003) of the TOPAZ4 is too warm and saline, which suggests there is too much AW inflow or too weak a vertical mixing.

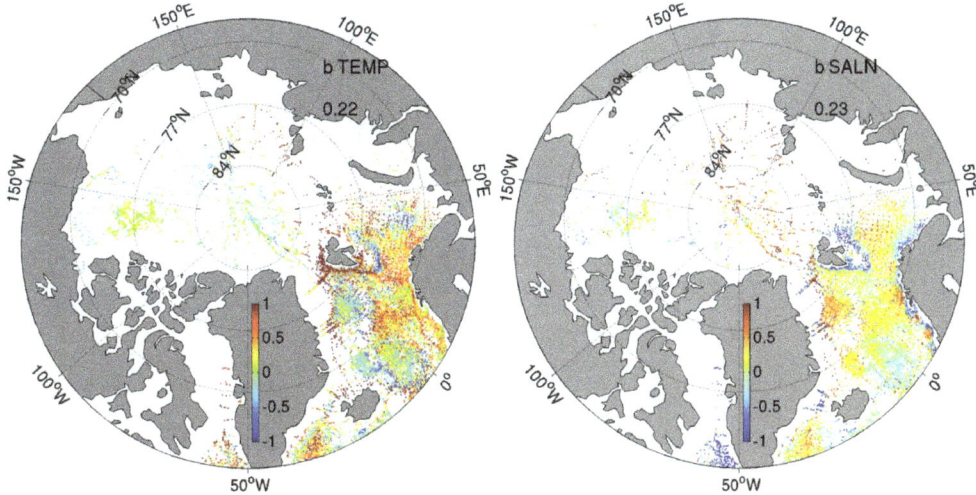

Figure 8. Spatial distribution of b for in situ temperature (left) and salinity (right) during the period from 1991 to 2013. The observation number in a grid is required to be more than 30. Note that profiles may end at different depths and cause spottiness.

Furthermore, we investigate the time evolution of the misfits throughout the reanalysis. Figure 10 shows the time series of the root mean square innovations (RMSIs) of temperatures and salinities in the whole Arctic at depths of 300–800 m, indicative of the Atlantic water layers. As in Sakov et al. (2012), the total uncertainty is added to assess the time reliability of the system. However, in this study, we use the formulation of σ_{tot} from Rodwell et al. (2016), which assumes that for a perfect reliable system RMSI is equal to σ_{tot}, with bias included:

$$\sigma_{tot}^2 = \mathrm{BIAS}^2 + \sigma_{en}^2 + \sigma_o^2. \qquad (8)$$

Here, the term "BIAS" refers to the innovation mean equivalent to the misfit at assimilation time.

For temperature profiles, the BIAS is negative, especially during the period of 1994–2005, indicating a warm bias at 300–800 m depths. This bias is persistent in the whole period, but reduces during the international Polar Year (IPY) period. Concurrently, the RMSI (red line in Fig. 10) also decreases after 2006. Since the reliability remains constant during the IPY (see Sect. 3), the enhanced accuracy can be considered a performance improvement, directly caused by the intensive observation efforts. The diagnosed uncertainty σ_{tot} (blue dashed line) and the RMSI are evolving in phase, which indicates a good potential for probabilistic forecasting. After the E2 event, the diagnosed σ_{tot} slightly underestimates the RMSI, which may result from the removal of the multiplicative inflation.

For salinity, the model seems too saline until the start of the IPY. The bias does not reemerge post-IPY when the number of salinity observations is very much reduced but still covers the same regions. The RMSI is also reduced during the IPY. Although there is some similarity in the evolution of the two curves, the diagnosed σ_{tot} is overestimating the

RMSI. This result seems to contradict the underdispersion in Fig. 3, but the difference relates to the depths at which the metrics are calculated (300–800 m here against full observation depth in Fig. 3). The cause of the overestimation stems from too large an observation error (not shown) and suggests a revision of the observation error settings for salinity profiles.

4.4 Sea ice concentration

Relative to the daily sea ice concentration product from OSISAF (CMEMS OSI TAC product), the spatial variability of the SIC misfits is shown in Fig. 11. As a large seasonal variability in the sea ice extent, this is carried out at two characteristic times of one year: the maximum (March) and minimum ice extent (September).

In March, there is a dipole anomaly on either side of the ice edge in the Greenland Sea. The ice edge in TOPAZ4 is transiting too sharply from pack ice to open water because the heat capacity of the ice is neglected. This leads to a dipole bias (positive inside the ice and negative outside) during the melting season. There is also a weak bias over regions that are usually ice-free. Indeed, OSISAF does not employ weather filtering and places a thick band of low concentration ($< 10\%$) in ice-free regions (Ivanova et al., 2015).

In September, TOPAZ4 shows a negative bias in the Greenland Sea. At that time of the year, the sea ice flows southwards and TOPAZ4 tends to underestimate the southern extension of the sea ice tongue along Greenland. This indicates that the dynamical forcing is biased or that the drag coefficients are incorrect as the ice is in free drift there.

The RMSD is approximately 5 % in most of Arctic region except close to the sea ice edge where the RMSD exceeds 25 %, which coincides with regions where the bias is high. Data assimilation does constrain the sea ice concentrations

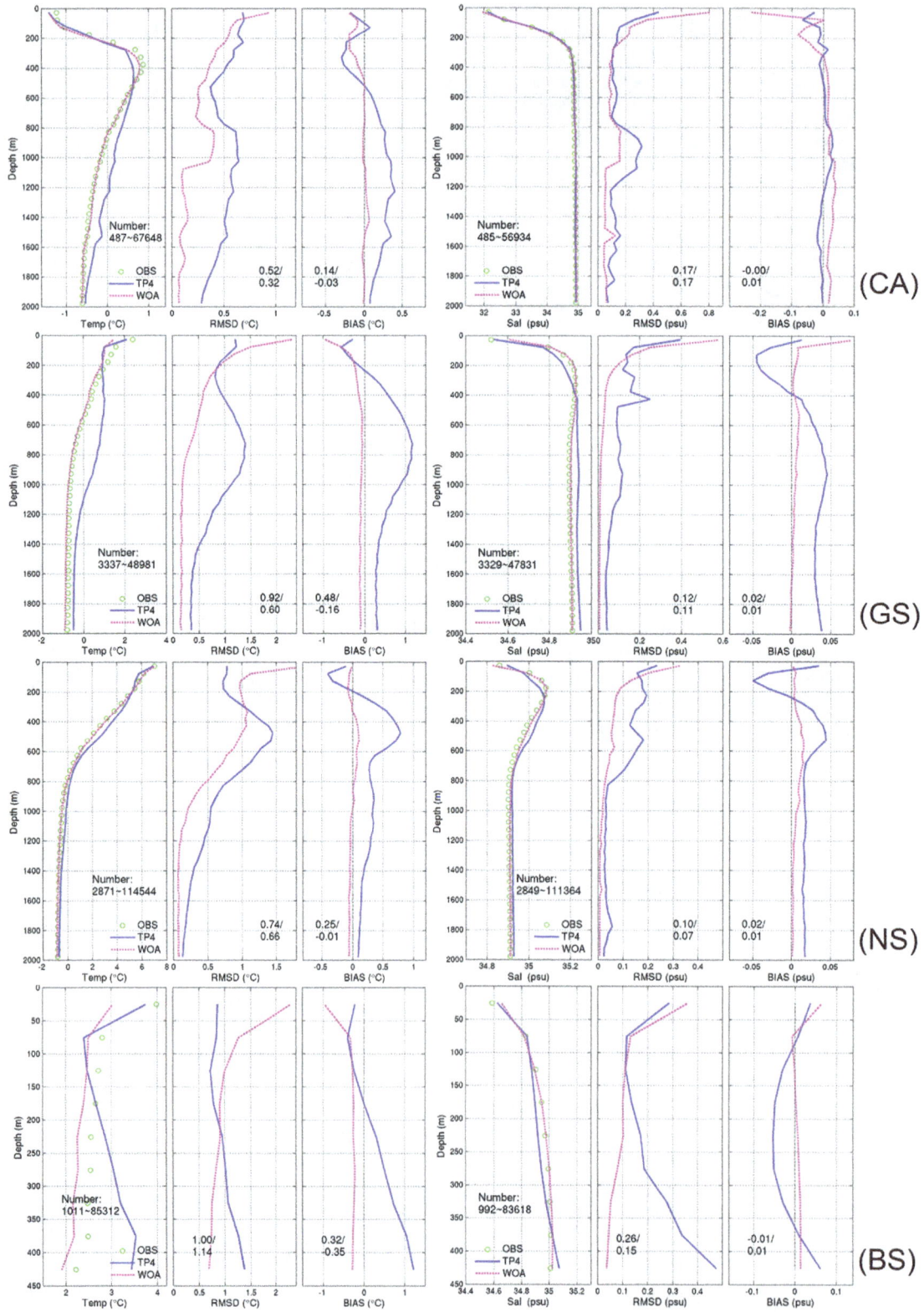

Figure 9. The mean profiles of temperature (left) and salinity (right) and the corresponding bias and RMSD in each of the marginal seas of the pan-Arctic region. The green circles indicate the observations, the blue lines indicate the TOPAZ reanalysis and the pink lines are from the WOA13 climatology. The numbers in the first-column sub-panels are the minimal and maximal number of observations available in each of the 50 m depths; the upper numbers in the other-column sub-panels are the mean estimates in vertical for TOPAZ reanalysis, and the lower numbers are for WOA13.

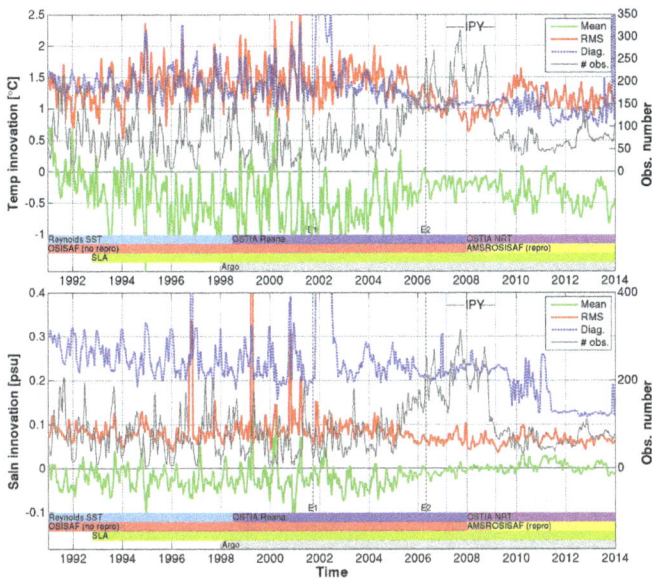

Figure 10. Time series of innovation statistics for temperature (top) and salinity (bottom) observed at the depth between 300 and 800 m depths. The bias is plotted with a green line, the RMSD is in red and the number of assimilated observations is plotted with a grey line. The blue dashed line indicates σ_{tot} as defined in Eq. (8). The time series are filtered with a 28-day moving window. The vertical dashed lines indicate the change events tuning the bias correction in the course of the TOPAZ reanalysis.

but the model biases (lack of resolution of ocean currents, biases of ice drift or ice thickness) still cause locally high residual errors of ice concentrations.

In order to assess the interannual variability of the performance of TOPAZ4, we have decided to use the standard sea ice extent (SIE) metric. SIE is calculated as the surface area in which the ice concentration is larger than 15 %.

As the variability in the decadal trend of SIE in the Arctic is large, we present the interannual evolution in the whole Arctic and in two subregions: the Greenland Sea and Barents Sea (Fig. 12). TOPAZ4 shows good agreement with the OS-ISAF observations in the pan-Arctic region and the mean SIE in the 23 years is 8.03×10^6 instead of 7.96×10^6 km^2 in the observations. The decreasing trend of SIE during the period 1991–2013 is -6.16×10^4 km^2 yr^{-1}, which compares well to the trend of the observations (-6.34×10^4 km^2 yr^{-1}).

In the Greenland Sea, the SIE in TOPAZ4 is underestimated, which clearly relates to the bias in the southern extent of the sea ice tongue along the coast of Greenland. The bias in TOPAZ4 is on average -3.6×10^4 km^2 and the decreasing trend in TOPAZ4 is -3.1×10^3 km^2 yr^{-1}, which is larger than observed (-2.3×10^3 km^2 yr^{-1}). In the Barents Sea, the variability agrees well, although TOPAZ4 underestimates the SIE slightly. The decreasing trend is comparable.

The seasonality of the SIE in OSISAF and TOPAZ4 is investigated in Fig. 13. It is clear that the seasonal cycle of the ice extent is generally well simulated by the reanalysis in the pan-Arctic area. In the summer months from June to August, a slight underestimation of the ice extent is apparent, and the minimal ice extent comes a little too early compared to the observations. In the Greenland Sea, the underestimation of sea ice extent is larger. The underestimation of sea ice extent starts in February and increases during the sea ice melt, reaching a maximum (of about 1×10^5 km^2) in July. In the Barents Sea, the seasonal cycle is well simulated but some differences are noticeable there in the beginning of the year, reaching a maximum in April, and returning to zero in August and September when there is no ice.

4.5 Sea ice drift

The sea ice drifts from the buoy data of the International Arctic Buoy Program (IABP) are available at 12 h frequency from 1991 to 2011. It is an independent data set and is used here for validation. To avoid the "survival bias" caused by the retreat of sea ice from the marginal seas and unresolved coastal effects, the buoy drift vectors are limited to the central Arctic, as shown with the red line in the right panel of Fig. 1. The waters shallower than 30 m and closer than 50 km to the coastline are excluded. This data set has been gridded to be compared with the model. Each grid cell is filled (i.e., considered reliable) if the calculation involves at least 30 buoys within a day. A coarser grid than the model resolution is used (four grid cells which correspond to approximately 60×60 km^2) to avoid having too many empty cells. The daily average from the measurement is the mean of the 12 h drifting speed. For comparison, the model drifting speed is calculated from the daily average of eastward and northward velocity. Several approximations are made during this comparison; we compare Eulerian to Lagrangian drift which is expected to be faster; the model ice drift is calculated from daily averages of u and v instead of daily ice drift, which is faster by approximately 0.5 km per day (not shown).

On average, over the period 1991–2011, the mean drift fields of sea ice are presented in Fig. 14. As the resulting drift estimate appeared noisy, a smoothing with the neighboring grid cells has been applied. Both observations and TOPAZ4 show a similar pattern with a pronounced Beaufort Gyre, although the center of the gyre is slightly shifted. We can also notice that TOPAZ4 globally overestimates the ice drift with a bias of 1.7 km d^{-1}. In the Chukchi Sea, TOPAZ4 underestimates the drift by approximately -2 km d^{-1}.

Over the period 1991–2011, the monthly time series of the ice drift speeds are compared in Fig. 15. They are averaged in the central Arctic from the reanalysis and the buoy data, respectively. On average, the drift speed is about 7 km d^{-1} in buoy data, and about 9.4 km d^{-1} in the TOPAZ4 reanalysis. The fast bias is clear until the end of 2010. From that time onward, the drag coefficient of the atmosphere on sea ice has been reduced from 2.14×10^{-3} to 1.6×10^{-3}. We can see that the bias is much reduced during the last year.

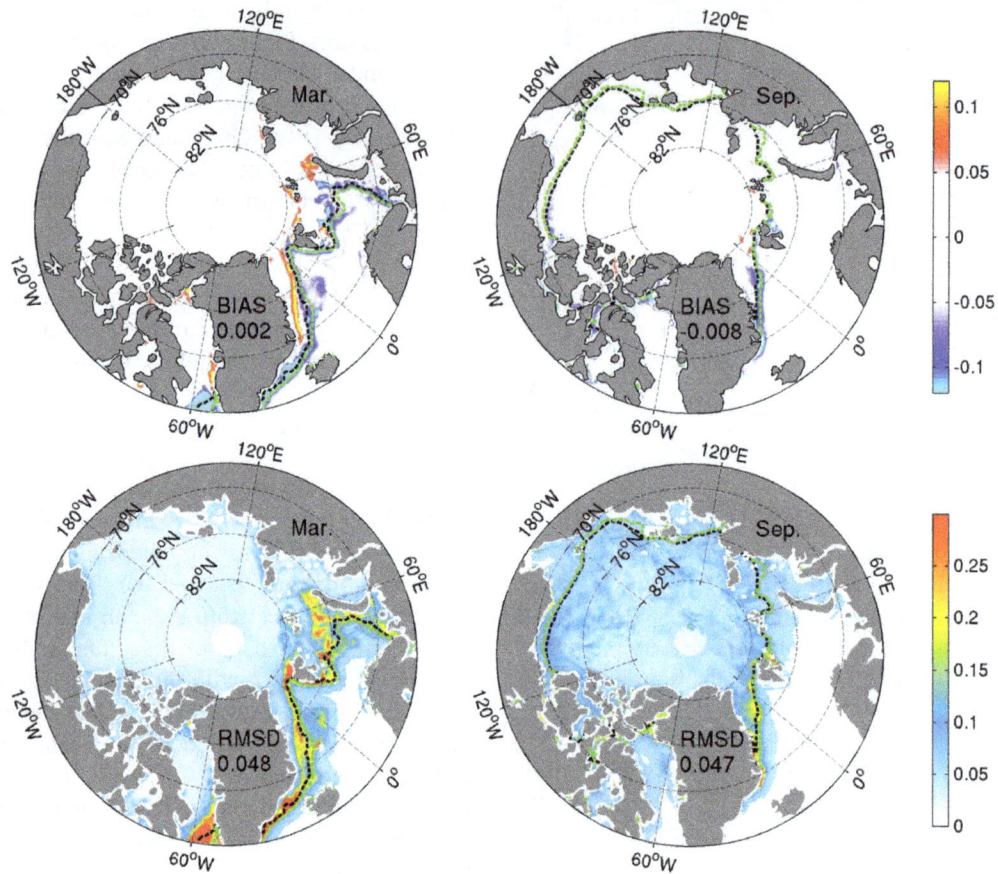

Figure 11. Spatial bias (upper) and RMSD (lower) of sea ice concentration in the TOPAZ reanalysis for March (left) and September (right), calculated from the daily averages for the period 1991–2013. The dashed black (green) lines delimit the monthly mean sea ice edges (at 15 %) in the TOPAZ reanalysis (OSISAF).

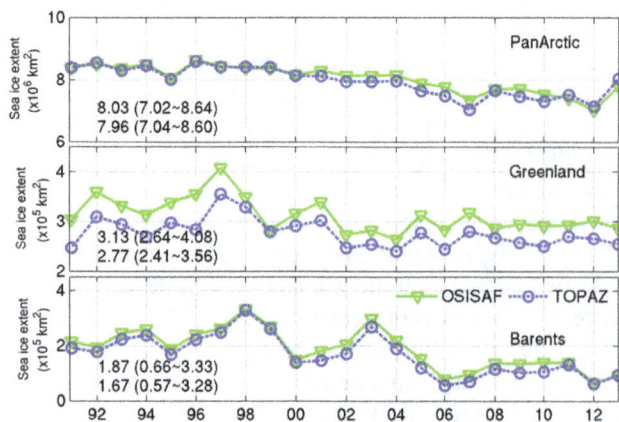

Figure 12. Yearly time series of the sea ice extent in the pan-Arctic region, the Greenland Sea and the Barents Sea from TOPAZ reanalysis (dashed) and OSISAF (solid).

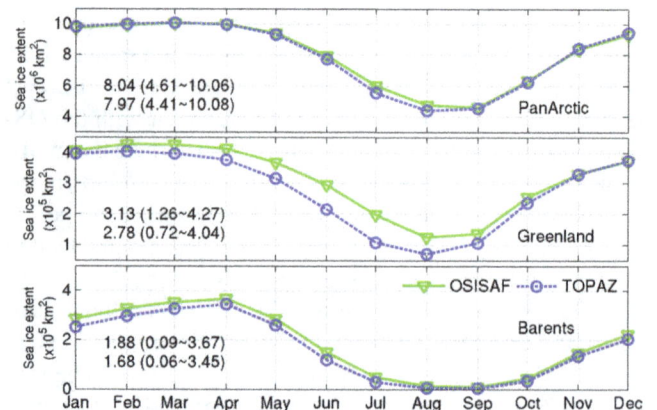

Figure 13. Seasonality of the sea ice extents in the TOPAZ reanalysis (blue line) and OSISAF (green line) in the pan-Arctic Ocean, Greenland Sea and Barents Sea regions.

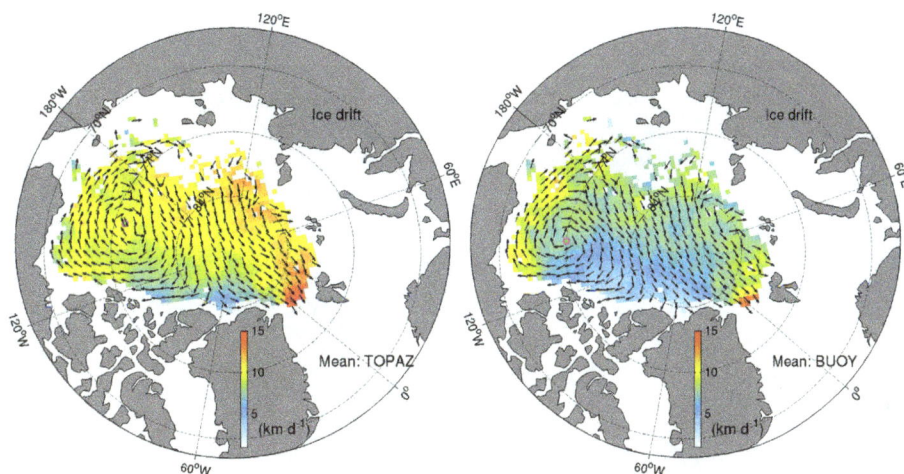

Figure 14. Sea ice drift vectors (arrows) and speeds (color shading) averaged over the period 1991–2011 for (left) TOPAZ reanalysis and (right) IABP buoys. The center of the anticyclonic Beaufort Gyre is marked with a magenta circle in the TOPAZ reanalysis (155° W, 78.1° N) and in the observations (145° W, 77° N), respectively.

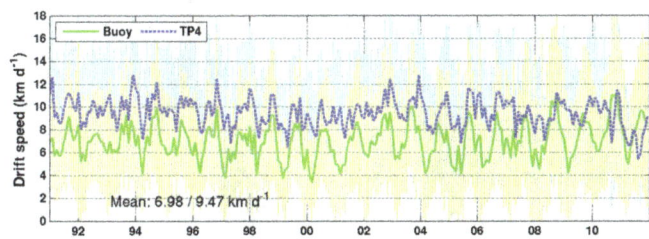

Figure 15. Monthly time series of the daily averaged sea ice drift speeds in the central Arctic from the TOPAZ reanalysis (blue line) and the IABP buoys (green line) during 1991–2011. The error bars represent the standard deviations of the daily estimates for each month.

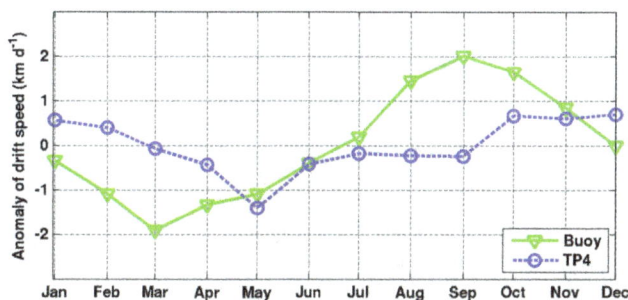

Figure 16. Seasonality of the sea ice drift velocities from the reanalysis and the buoy during 1991–2011.

The RMSD is on average $5.1\,\mathrm{km\,d^{-1}}$, of which $2.5\,\mathrm{km\,d^{-1}}$ can be attributed to the bias. The correlation between the two curves is about 0.6.

In addition, the monthly seasonality cycle of the ice drift over the period 1991–2011 is plotted in Fig. 16. While the buoys show a clear seasonality in the ice drift, being slowest in March and fastest in September, the seasonality in the TOPAZ4 reanalysis is weaker and reaches a minimum in May (delayed by 2 months).

4.6 Sea ice thickness

The sea ice thickness in Arctic has attracted much attention in recent years because it has been found to be sensitive to global warming (Kwok et al., 2009; Zygmuntowska et al., 2014). In this study, sea ice thickness is an independent data set, as it has not been assimilated. The observations of ice thickness with basin scale are still very few. A satellite-derived product for the Arctic Ocean ice provides the estimations of sea ice thickness for February–March and

October–November between 2003 and 2008 (ICESat, Kwok et al., 2009). Figure 17 shows the spatial distributions of the mean sea ice thicknesses and their differences. The spatial correlations are 0.74 and 0.87 for spring and fall, respectively. On average, TOPAZ4 is too thin compared to ICESat with a bias of -0.79 and $-0.64\,\mathrm{m}$ in spring and in fall. In spring, TOPAZ4 is too thin, in particular north of Ellesmere Island by approximately 2 m. There is a positive bias centered in the Beaufort Gyre in spring. In fall, this bias is wider and displaced slightly to the east.

Another source of validation is the Unified Sea Ice Thickness Climate Data Record (Lindsay, 2013) resulting from a concerted effort to collect as many observations as possible of Arctic sea ice draft, freeboard and thickness. The sea ice draft is measured by the sonar of US Navy Submarines from National Snow and Ice Data Center (USSUB-DG and USSUB-AN; Wadhams and Horne, 1980; Wensnahan and Rothrock, 2005; Rothrock and Wensnahan, 2007), and the sea ice thickness by flight campaigns from NASA Operation IceBridge (IceBridge; Kurtz et al., 2013), as shown

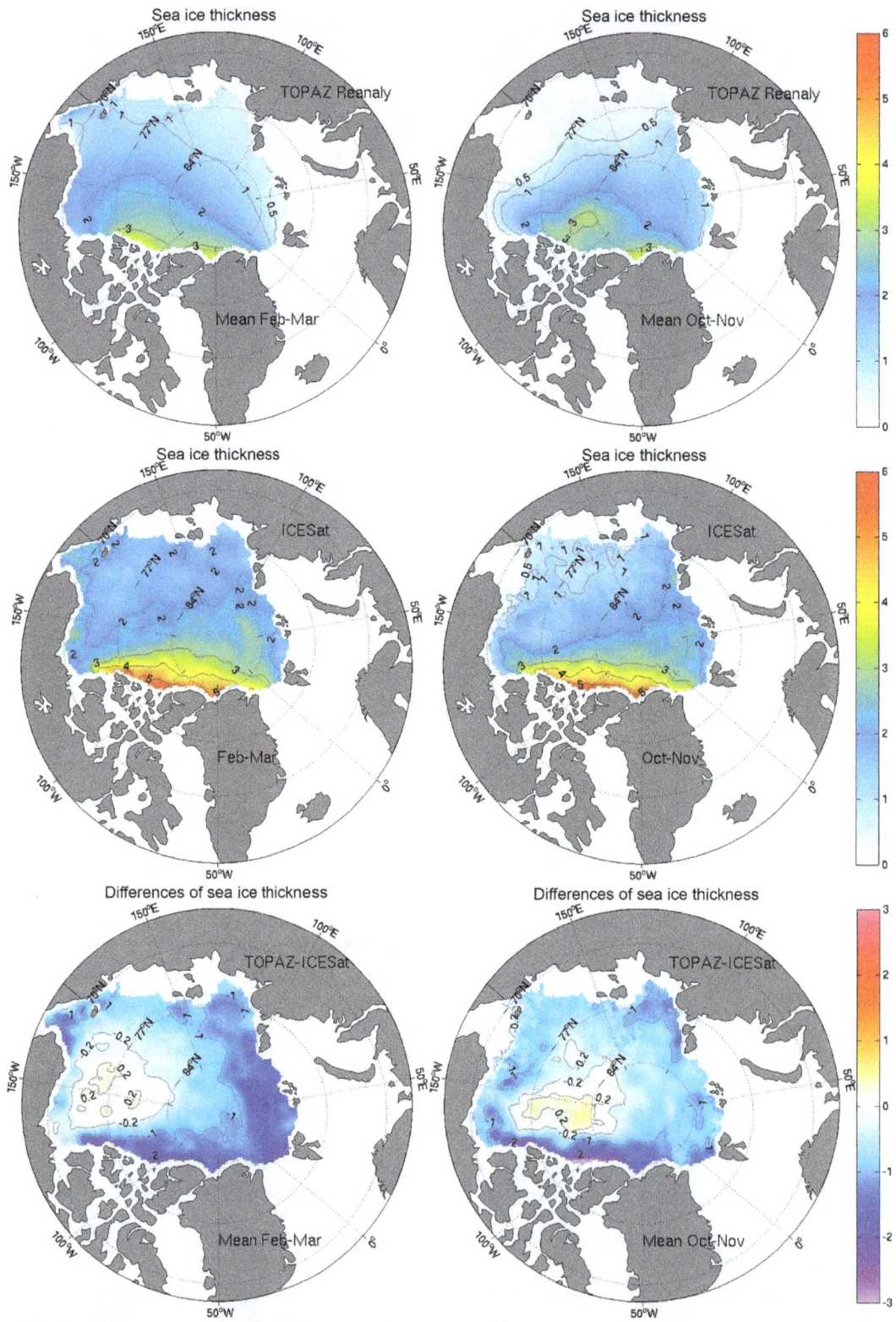

Figure 17. Mean sea ice thicknesses from TOPAZ (upper) and ICESat (middle), and their difference (bottom) for February–March (left column) and October–November (right column) averaged over the period 2003–2008.

in Fig. 18a. The sea ice draft data have been diagnosed in TOPAZ4 as proposed by Eq. (4) of Alexandrov et al. (2010):

$$D_i = H_i \cdot \frac{\rho_i}{\rho_w} + H_{sn} \cdot \frac{\rho_{sn}}{\rho_w}, \tag{9}$$

where D_i is ice draft, H_i is ice thickness and H_{sn} is the snow thickness. The ρ_i, ρ_w and ρ_{sn} are the densities for sea ice, water and snow (respectively, 900, 1000 and $300\,\mathrm{kg\,m^{-3}}$).

The IceBridge ice thickness covers the period of 2009–2011. TOPAZ4 reanalysis is too thin with a bias of 1.1 m, a RMSD of 1.4 m and a correlation of 0.5. The bias against the sea ice draft is smaller with 0.3–0.4 m, and a RMSD about 0.6–0.7 m. The correlation coefficients are relatively good with 0.86 and 0.69, which is higher than for the IceBridge data. These discrepancies are likely to be related to the spatial distribution of the different data sets. Hence, IceBridge data are concentrated around the northern coast of Greenland where TOPAZ4 showed largest bias in the comparison with ICESat.

As another diagnostic of interest, the daily time series of sea ice volume from TOPAZ4 in the Arctic in 1991–2013 is shown by the blue curve in the left panel of Fig. 19. Before 2001, the sea ice volume varies stably around $1.4 \times 10^4\,\mathrm{km^3}$, with a significant seasonal variability between $8 \times 10^3\,\mathrm{km^3}$ and $1.9 \times 10^4\,\mathrm{km^3}$. Afterwards, in the period 2001–2010, the sea ice volume decreases dramatically. This reduction of sea ice volume is qualitatively consistent with the limited satellite records. First, the estimate from Kwok et al. (2009), derived from the ICESat record from 2003 to 2008, shows a similar trend. After revising the uncertainties of input data (snow depth, sea ice density and ice concentrations), Zygmuntowska et al. (2014) corrected the estimates of the mean sea ice volume, shown as the starred line in Fig. 18. With respect to these sea ice volume estimates, TOPAZ4 still has too little ice. In the right panel of Fig. 19, the seasonal cycles of sea ice volume from TOPAZ4 and the standard deviation in the 23 years are shown by the blue curve and the cyan error bars, respectively. In May, the maximum sea ice volume is about $1.5 \times 10^4\,\mathrm{km^3}$, and in September is less than $5 \times 10^3\,\mathrm{km^3}$. The sea ice volumes from Zygmuntowska et al. (2014) are plotted on top of the averaged TOPAZ4 seasonal cycle in the period 1991–2013. These correspond well to the model climatology, but still betray an underestimation because the measurements are representative of a period of lower ice volume.

The TOPAZ4 seasonal cycle of ice volume seems to change in amplitude during different time eras, although the reasons lie in two successive changes of the settings of the EnKF. In December 2001, the variance of precipitation errors is increased from 1.10^{-17} to $1.10^{-12}\,\mathrm{m^2\,s^{-2}}$, as an adjustment for a slow decrease of ensemble spread. These perturbations being truncated Gaussian, the truncation resulted in excessive snow precipitations. The excessive snow depths have then isolated the ice from the atmosphere and reduced the amplitude of the yearly cycle from 1.08 to 0.74 m (see

Fig. 20); this also delayed the phase of the cycle. In January 2011, an unbiased log-normal law replaced the truncated Gaussian perturbations with an amplitude of 30 %. The amplitude and phase of the seasonal cycle returned to more correct values. The sensitivity experiments in Finck et al. (2013) verified the above-mentioned issue.

5 Summary and discussions

This study is conducted to present and validate the official physical multi-year CMEMS product for the Arctic region. The proposed reanalysis is unique compared to other reanalysis products (see Table 1 of Chevallier et al., 2016). It proposes a long high-resolution dynamical reconstruction of the ocean and sea ice, and assimilates a complete set of observations available in the Arctic region with an advanced ensemble data assimilation method and with strongly coupled data assimilation between ocean and sea ice. The above results present a concise account of the strengths and weaknesses of the resulting data set. The above findings can be summarized variable by variable:

SLA In the period 1993–2013, the RMSD of daily SLA in the reanalysis is gradually decreased from over 9 cm to less than 6 cm in the pan-Arctic region. The introduction of a bias estimation scheme proves very efficient in constraining the bias. The largest RMSDs over 9 cm are found around the Lofoten Basin. There is also a patch of larger misfit near the ice edge, but observations are also less accurate there. There is a weak seasonality in the performance of the system with the best results in the summer. The system is slightly overdispersive mostly due to bias estimation.

SST The SST RMSD is about 0.63 °C over the period 1991–2013, and after 1999 it is reduced to about 0.44 °C with a smaller bias around −0.02 °C. The transition to high-resolution OSTIA SST is highly beneficial for constraining the bias and the RMSD, but an overestimation of the observation error from the provider was needed to avoid a collapse of the ensemble spread. The performance of the system varies seasonally following the observational amounts and a larger bias during summer months. The system dispersion is close to 1 in most of the years but can be over- or underdispersive depending on the settings of observation errors and bias estimation.

Temperature and salinity profiles The misfits of the reanalysis are small near the surface (in the top 100 to 200 m), even compared to those of the WOA13 climatology. Below this depth, the model shows large biases and performs poorer (RMSD > 1 °C and about 0.1 psu). Some of the biases relate to the limitations of the model to maintain the Atlantic water (as expected from Ilıcak et al., 2016) and others relate to a degradation intro-

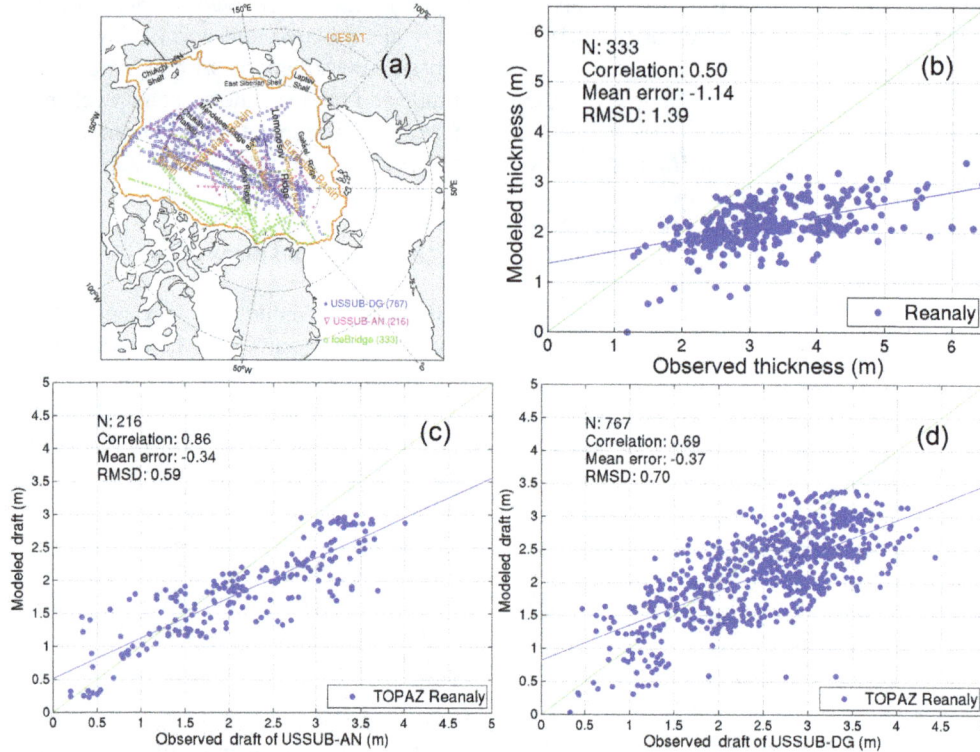

Figure 18. Validation the sea ice thickness in the TOPAZ reanalysis versus available in situ observations. **(a)** Locations of in situ observations available from IceBridge, USSUB-AN and USSUB-DG in the central Arctic. Regression analysis of TOPAZ reanalysis **(b)** vs. IceBridge; **(c)** vs. USSUB-AN; **(d)** vs. USSUB-DG.

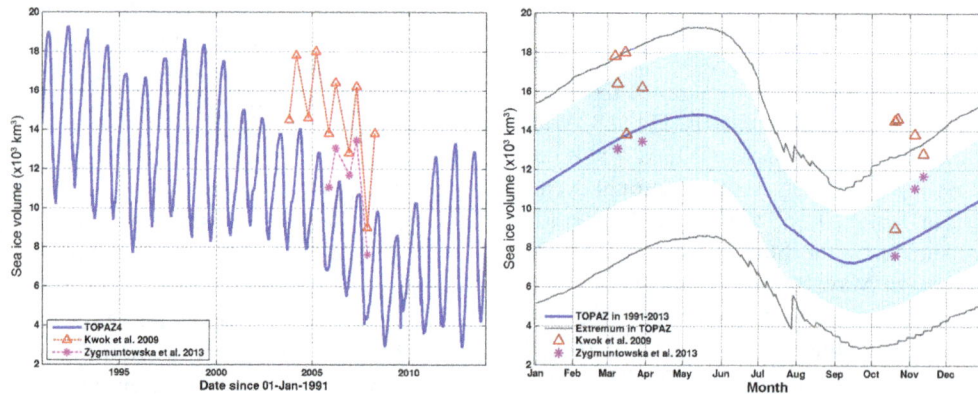

Figure 19. Left: time series of the daily averaged sea ice volume in the Arctic from the TOPAZ4 (blue line) and the observations from Kwok et al. (2009) and from Zygmuntowska et al. (2014). Right: daily time series of the averaged sea ice volume in the Arctic from the TOPAZ4 for the period 1991–2013 (blue line) and the standard deviation shown as the cyan error bar. The grey lines represent the extreme volumes in the 23 years. The triangle and start markers are the observations estimated by Kwok et al. (2009) and Zygmuntowska et al. (2014), respectively.

duced by data assimilation (a flat multiplicative infla- tion). A large improvement occurred at the times when the inflation method was upgraded and when there were more available observations during the IPY. The system reliability is overall stable in time, in spite of the very inhomogeneous data sampling over the past 23 years.

Sea ice concentration and extent TOPAZ4 agrees well with the OSI-SAF sea ice concentrations. On average, the RMSDs are lower than 5 % and the biases close to zero. The misfits are larger close to the ice edge, and poorest in the Greenland Sea. The errors are related to biases in the thermodynamics and dynamics of the sea ice model. The bias is largest during the

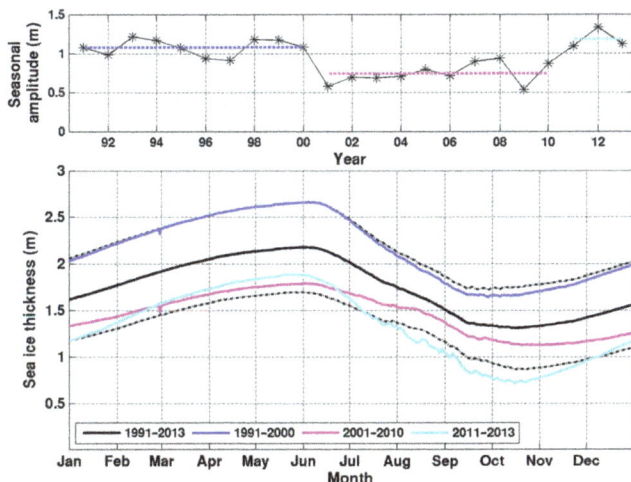

Figure 20. Top: yearly time series of the seasonal amplitudes of the mean sea ice thickness in the central Arctic, shown with the solid black line. The dashed lines represent the averaged estimate for 1991–2000, 2001–2010 and 2011–2013 (1.08, 0.74 and 1.18 m, respectively). Bottom: daily time series of the mean sea ice thickness in the central Arctic for three different time periods. The black dashed lines denote the standard deviation for the 23 yearly estimates.

summer season. The performance is stable throughout the reanalysis but the dispersion is consistently too low ($d = 0.55$), probably due to a too rudimentary thermodynamical sea ice model.

Sea ice drift The averaged drift in TOPAZ4 shows comparable patterns to independent observation from IABP buoys with the classical Beaufort Sea gyre and transpolar drift. However, the center of the gyre is slightly misplaced. The RMSD of drift speed in the reanalysis is about 5.1 km d^{-1}, and has a fast bias by about 2.5 km d^{-1}. The monthly time variability compares well, but TOPAZ4 has too weak a seasonal cycle and shifted by 2 months. From 2011 onwards, the atmospheric drag coefficient was adjusted and the ice drift speed agrees better with observations after the change. Still, with RMSDs of 5 km d^{-1} close to the signal itself, improving the performance of ice drift appears a priority for future operational use. The dispersion is also low but becomes too large after switching to a different observational product.

Sea ice thickness TOPAZ4 shows some large biases (approximately −1.1 m) compared to ice thickness from ICESat and IceBridge as well as compared to ice draft data, although the thick ICESat ice draft may have been overestimated (Khvorostovsky and Rampal, 2016). The thickness bias is largest north of Ellesmere Island with bias up to 2 m. The spatial pattern and regression compare reasonably well. The ice is too thin in the period

2001–2010 due to excessive snow depths and the seasonal cycle is too small during that time.

RCRV diagnostics have shown a good balance between the different data types assimilated: none of the data types are severely less dispersed than the others. The results from the 23-year reanalysis show overall a reasonable stability over time and good agreements with observations. However, some clear discontinuities are caused by transitions from one data set to other new observations in areas that were completely unobserved, and also by changes in the data assimilation settings. Assessing the system for such a long period also reveals some limitations that are either inherent to the data assimilation implementation or due to model flaws. In the following, we list the possible reasons and the means to tackle these in the future version of the ARC MFC system.

– The Atlantic waters have a signature that is too diffused. In order to improve their advection, we will double the horizontal and vertical resolution (50 hybrid layers and 5 km horizontal resolution). The parameterization of diapycnal mixing will be reduced under sea ice as proposed in Morison et al. (1985). We also foresee that increasing the resolution will be well useful for resolving the circulation in the Nordic Seas and reduce the seasonal biases of SST and SSH.

– The system has too sharp an ice edge. The current thermodynamic model does not account for the heat capacity of the sea ice. TOPAZ will be upgraded to the community sea ice mode CICE (Hunke et al., 2010), which uses a complex thermodynamic parameterization.

– Observations detect melt ponds as open water, whereas melt ponds are not simulated in the current TOPAZ4. This creates bias in sea ice during summer months that is transferred to the interior of the ocean via coupled data assimilation. In the future, we will choose the best alternative between using an existing melt pond model or detect and remove the signature of the melt ponds from the observations.

– Comparisons against sea ice drift and ice thickness highlighted more severe limitations: ice that is too thin , a thickness gradient that is too smooth from Greenland into the Beaufort Gyre; the center of the Beaufort Gyre being slightly misplaced, the sea ice drift being too fast. These biases can be reduced by optimizing the sea ice strength (P^*) and the drag parameters both in ocean and atmospheric (Massonnet et al., 2014). However, optimal values of these parameters are moving targets in view of their limited physical realism. The methodology proposed by Barth et al. (2015), to estimate biases in atmospheric wind from ice drift will also be considered. But the RMSDs of ice drift are relatively high (5 km d^{-1} for an ice drift generally inferior to 10 km d^{-1}) although comparable to short-term forecasts in Schweiger and Zhang (2015). These fluctuating misfits are less likely to be reduced by model tuning.

- There are further indications that the viscous-plastic and the related elastic–viscous–plastic rheologies have inherent limitations for simulating long-term properties of the ice drift – e.g., the acceleration of sea ice drift, the phase of its seasonal cycle (Rampal et al., 2011). A high-priority objective is therefore to couple TOPAZ to the neXtSIM sea ice model that is based on an elasto-brittle rheology. Recent studies with a forced version of neXtSIM (Bouillon and Rampal, 2015; Rampal et al., 2016) suggest that the model is capable of reproducing the sea ice deformations over a wide range of spatial and temporal scales and reduces the error of the sea ice drift. It is of interest to understand to which extent the coupling feedback will respond to this improved dynamical model.

- The online bias estimation appeared quite successful to limit bias in our model, but its implementation in the EnKF was very sensitive to the choice of inflation method used. The latest configuration that combined r factor inflation and autoregressive additive inflation for parameters is our recommendation in a realistic system with a strongly variable observation network.

- The EnKF has proven capable to assimilate a large variety of observations, but more observations should be assimilated, like the sea ice thickness of thin ice from the European Space Agency's (ESA) Soil Moisture and Ocean Salinity (SMOS) in Kaleschke et al. (2012) and Tian-Kunze et al. (2014). Also the complementary thickness of thick ice from ICESat (Kwok et al., 2009; Khvorostovsky and Rampal, 2016) and CryoSat-2 (Wingham et al., 2006; Laxon et al., 2013), and SMOS sea surface salinity (Reul et al., 2012) will be tested in order to determine how to better assimilate into the system in the near future.

- Although efforts were made to freeze as much of the assimilation setting as possible, some change have been necessary: e.g., replacing the multiplicative inflation by additive inflation or changes of observation product. These have caused discontinuities in the accuracy and in the reliability of the system. These discontinuities may become problematic for the interpretation of mechanisms of variability in the Arctic. For optimizing its consistency, a reanalysis should limit its observation network to that available through the whole reanalysis period, as done in Counillon et al. (2016) with assimilation of SST only. However, such a type of reanalysis prioritizes consistency at the expense of accuracy, which is not the purpose of the TOPAZ system. In a future reanalysis production, consistently reprocessed data sets

from the ESA climate change initiatives (ESA CCIs) will be assimilated over the whole period (these were not available at the start of this reanalysis). The monitoring of reliability metrics can be automated and the results presented here indicate that the reliability should then remain stable.

- The next physical ARC MFC reanalysis will provide a stochastic product, in order to provide a natural framework for estimating the system accuracy in space and time, and to provide input data for probabilistic weather or stand-alone sea ice models.

Competing interests. The authors declare that they have no conflict of interest.

Acknowledgements. Thanks to P. Rampal for processing the buoy data set for sea ice drift and for the useful discussions. We thank to the US National Snow and Ice Data Center (NSIDC) for providing the IceBridge data. This study was supported by successive MyOcean projects from the European Commission (grant number 218812), the Arctic element of the Copernicus Marine Services and a grant of CPU time from the Norwegian Supercomputing Project (NOTUR II grant number nn2993k). We thank two anonymous reviewers for constructive suggestions that have improved this manuscript.

Edited by: E. J. M. Delhez

References

Alexandrov, V., Sandven, S., Wahlin, J., and Johannessen, O. M.: The relation between sea ice thickness and freeboard in the Arctic, The Cryosphere, 4, 373–380, doi:10.5194/tc-4-373-2010, 2010.

Anderson, J. L.: An ensemble adjustment Kalman filter for data assimilation, Mon. Weather Rev., 129, 2884–2903, doi:10.1175/1520-0493(2001)129<2884:AEAKFF>2.0.CO;2, 2001.

Barth, A., Canter, M., Schaeybroeck, B. V., Vannitsem, S., Massonnet, F., Zunz, V., Mathiot, P., Alvera-Azcárate, A., and Beckers, J.–M.: Assimilation of sea surface temperature, sea ice concentration and sea ice drift in a model of the Southern Ocean, Ocean Model., 93, 22–39, doi:10.1016/j.ocemod.2015.07.011, 2015.

Bertino, L. and Lisæter, K. A.: The TOPAZ monitoring and prediction system for the Atlantic and Arctic Oceans, Journal of Operational Oceanography, 1, 15–19, doi:10.1080/1755876X.2008.11020098, 2008.

Bleck, R.: An oceanic general circulation model framed in hybrid isopycnic-Cartesian coordinates, Ocean Model., 4, 55–88, doi:10.1016/S1463-5003(01)00012-9, 2002.

Blindheim, J. and Østerhus, S.: The Nordic Seas, Main Oceanographic Features, in: The Nordic Seas: An Integrated Perspective, edited by: Drange, H., Dokken, T., Furevik, T., Gerdes, R., and Berger, W., Geoph. Monog. Series, 158, 11–37, 2003.

Bouillon, S. and Rampal, P.: Presentation of the dynamical core of neXtSIM, a new sea ice model, Ocean Model., 91, 23–37, doi://10.1016/j.ocemod.2015.04.005, 2015.

Brasseur, P., Bahurel, P., Bertino, L., Birol, F., Brankart, J.-M., Ferry, N., Losa, S., Remy, E., Schröter, J., Skachko, S., Testut, C.-E., Tranchant, B., Van Leeuwen, P. J., and Verron, J.: Data assimilation for marine monitoring and prediction: The MERCATOR operational assimilation systems and the MERSEA developments, Q. J. Roy. Meteor. Soc., 131, 3561–3582, doi:10.1256/qj.05.142, 2005.

Candille, G., Côté, C., Houtekamer, P. L., and Pellerin, G.: Verification of an Ensemble Prediction system against observations, Mon. Weather Rev., 135, 2688–2699, doi:10.1175/MWR3414.1, 2007.

Chassignet, E. P., Smith, L. T., and Halliwell, G. R.: North Atlantic Simulations with the Hybrid Coordinate Ocean Model (HYCOM): Impact of the vertical coordinate choice, reference pressure, and thermobaricity, J. Phys. Oceanogr., 33, 2504–2526, doi:10.1175/1520-0485(2003)033>2504:NASWTH<2.0.CO:2, 2003.

Chevallier, M., Salas-Mélia, D., Voldoire, A., and Déqué, M.: Seasonal forecasts of the Pan-Arctic sea ice extent using a GCM-based seasonal prediction system, J. Climate, 26, 6092–6104, doi:10.1175/JCLI-D-12-00612.1, 2013.

Chevallier, M., Smith, G., Lemieux, J.-F., Dupont, F., Forget, G., Fujii, Y., Hernandez, F., Msadek, R., Peterson, K. A., Storto, A., Toyoda, T., Valdivieso, M., Vernieres, G., Zuo, H., Balmaseda, M., Chang, Y.-S., Ferry, N., Garric, G., Haines, K., Keeley, S., Kovach, R. M., Kuragano, T., Masina, S., Tang, Y., Tsujino, H., and Wang, X: Intercomparison of the Arctic sea ice cover in global ocean-sea ice reanalyses from the ORA-IP project, Clim. Dynam., 1–30, doi:10.1007/s00382-016-2985-y, 2016.

Counillon, F., Keenlyside, N., Bethke, I., Wang, Y., Billeau, S., Shen, M. L., and Bentsen, M.: Flow-dependent assimilation of sea surface temperature in isopycnal coordinates with the Norwegian Climate Prediction Model, Tellus A, 68, 32437, doi:10.3402/tellusa.v68.32437, 2016.

Cummings, J., Bertino, L., Brasseur, P., Fukumori, I., Kamachi, M., Martin, M., Mogensen, K., Oke, P., Testut, C. E., Verron, J., and Weaver, A.: Ocean data assimilation systems for GODAE, Oceanography, 22, 96–109, doi:10.5670/oceanog.2009.69, 2009.

Dee, D. P., Uppala, S. M., Simmons, A. J., Berrisford, P., Poli, P., Kobayashi, S., Andrae, U., Balmaseda, M. A., Balsamo, G., Bauer, P., Bechtold, P., Beljaars, A. C. M., van de Berg, L., Bidlot, J., Bormann, N., Delsol, C., Dragani, R., Fuentes, M., Geer, A. J., Haimberger, L., Healy, S. B., Hersbach, H., Hólm, E. V., Isaksen, L., Kållberg, P., Köhler, M., Matricardi, M., McNally, A. P., Monge-Sanz, B. M., Morcrette, J.-J., Park, B.-K., Peubey, C., de Rosnay, P., Tavolato, C., Thépaut, J.-N., and Vitart, F.: The ERA-Interim reanalysis:configuration and performance of the data assimilation system, Q. J. Roy. Meteor. Soc., 137, 553–597, doi:10.1002/qj.828, 2011.

Drange, H. and Simonsen, K.: Formulation of air-sea fluxes in the ESOP2 version of MICOM, Technical Report No. 125, Nansen Environmental and Remote Sensing Center, 23 pp., 1996.

Evensen, G.: Sequential data assimilation with a nonlinear quasi-geostrophic model using Monte Carlo methods to forecast error statistics, J. Geophys. Res., 99, 10143–10162, doi:10.1029/94JC00572, 1994.

Evensen, G.: The ensemble Kalman filter: theoretical formulation and practical implementation, Ocean Dynam., 53, 343–367, doi:10.1007/s10236-003-0036-9, 2003.

Finck, N., Counillon, F., Bertino, L., Bouillon, S., and Rampal, P.: Validation of sea ice quantities of TOPAZ for the period 1990–2010, Technical Report No. 332, Nansen Environmental and Remote Sensing Center, 30 pp., 2013.

Haine, T., Curry, B., Gerdes, R., Hansen, E., Karcher, M., Lee, C., Rudels, B., Spreen, G., Steur, L., Stewart, K. D., and Woodgate R.: Arctic freshwater export: Status, mechanisms, and prospects, Global Planet. Change, 125, 13–35, doi:10.1016/j.gloplacha.2014.11.013, 2015.

Hunke, E. C. and Dukowicz, J. K.: An elastic-viscous-plastic model for sea ice dynamics, J. Phys. Oceanogr., 27, 1849–1867, doi:10.1175/1520-0485(1997)027<1849:AEVPMF>2.0.CO;2, 1997.

Hunke, E. C., Lipscomb, W. H., and Turner, A. K.: CICE: the Los Alamos Sea Ice Model Documentation and Software User's Manual Version 4.1 LA-CC-06-012, T-3 Fluid Dynamics Group, Los Alamos National Laboratory, Los Alamos NM 87545, 76 pp., 2010.

Ilıcak, M., Drange, H., Wang, Q., Gerdes, R., Aksenov, Y., Bailey, D., Bentsen, M., Biastoch, A., Bozec, A., Böning, C., Cassou, C., Chassignet, E., Coward, A. C., Curry, B., Danabasoglu, G., Danilov, S., Fernandez, E., Fogli, P. G., Fujii, Y., Griffies, S. M., Iovino, D., Jahn, A., Jung, T., Large, W. G., Lee, C., Lique, C., Lu, J., Masina, S., George Nurser, A., Roth, C., Salas y Mélia, D., Samuels, B. L., Spence, P., Tsujino, H., Valcke, S., Voldoire, A., Wang, X., and Yeager, S. G.: An assessment of the Arctic Ocean in a suite of interannual CORE-II simulations. Part III: Hydrography and fluxes, Ocean Model., 100, 141–161, doi:10.1016/j.ocemod.2016.02.004, 2016.

Ivanova, N., Pedersen, L. T., Tonboe, R. T., Kern, S., Heygster, G., Lavergne, T., Sørensen, A., Saldo, R., Dybkjær, G., Brucker, L., and Shokr, M.: Inter-comparison and evaluation of sea ice algorithms: towards further identification of challenges and optimal approach using passive microwave observations, The Cryosphere, 9, 1797–1817, doi:10.5194/tc-9-1797-2015, 2015.

Johannessen, O. M., Bengtsson, L., Miles, M. W., Kuzmina, S. I., Semenov, V. A., Alekseev, G. V., Nagurny, A. P., Zakharov, V. F., Bobylev, L. P., Pettersson, L. H., Hasselmann, K., and Cattle, H. P.: Arctic climate change – observed and modelled temperature and sea-ice variability, Tellus A, 56, 328–341, doi:10.1111/j.1600-0870.2004.00060.x, 2004.

Kaleschke, L., Tian-Kunze, X., Maaß, N., Mäkynen, M., and Drusch, M.: Sea ice thickness retrieval from SMOS brightness temperatures during the Arctic freeze-up period, J. Geophys. Lett., 39, L05501, doi:10.1029/2012GL050916, 2012.

Kara, A., Rochford, P. A., and Hurlburt, H. E.: Efficient and accurate bulk parameterizations of air-sea fluxes for use in general circulation models, J. Atmos. Ocean. Tech., 17, 1421–1438, doi:10.1175/1520-0426(2000)017<1421:EAABPO>2.0.CO;2, 2000.

Khvorostovsky, K. and Rampal, P.: On retrieving sea ice freeboard from ICESat laser altimeter, The Cryosphere, 10, 2329–2346, doi:10.5194/tc-10-2329-2016, 2016.

Kurtz, N. T., Farrell, S. L., Studinger, M., Galin, N., Harbeck, J. P., Lindsay, R., Onana, V. D., Panzer, B., and Sonntag, J. G.: Sea ice thickness, freeboard, and snow depth products from Operation IceBridge airborne data, The Cryosphere, 7, 1035–1056, doi:10.5194/tc-7-1035-2013, 2013.

Kwok, R. and Rothrock, D.: Decline in Arctic sea ice thickness from submarine and ICESat records: 1958–2008, Geophys. Res. Lett., 36, L15501, doi:10.1029/2009GL039035, 2009.

Kwok, R., Cunningham, G. F., Wensnahan, M., Rigor, I., Zwally, H. J., and Yi, D.: Thinning and volume loss of the Arctic Ocean sea ice cover: 2003–2008, J. Geophys. Res., 114, C07005, doi:10.1029/2009JC005312, 2009.

Laxon, S. W., Giles, K. A., Ridout, A. L., Wingham, D. J., Willatt, R., Cullen, R., Kwok, R., Schweiger, A., Zhang, J., Haas, C., Hendricks, S., Krishfield, R., Kurtz, N., Farrell, S., and Davidson, M.: CryoSat-2 estimates of Arctic sea ice thickness and volume, Geophys. Res. Lett., 40, 732–737, doi:10.1002/grl.50193, 2013.

Le Traon, P.-Y., Antoine, D., Bentamy, A., Bonekamp, H., Breivik, L. A., Chapron, B., Corlett, G., Dibarboure, G., DiGiacomo, P., Donlon, C., Faugère, Y., Font, J., Girard-Ardhuin, F. , Gohin, F., Johannessen, J. A., Kamachi, M., Lagerloef, G., Lambin, J., Larnicol, G., Le Borgne, P., Leuliette, E., Lindstrom, E., Martin, M. J., Maturi, E., Miller, L., Mingsen, L., Morrow, R., Reul, N., Rio, M. H., Roquet, H., Santoleri, R., and Wilkin, J.: Use of satellite observations for operational oceanography: recent achievements and future prospects, Journal of Operational Oceanography, 8, s12–s27, doi:10.1080/1755876X.2015.1022050, 2015.

Lien, V. S., Hjøllo, S. S., Skogen, M. D., Svendsen, E., Wehde, H., Bertino L., Counillon, F., Chevallier, M., and Garric, G.: An assessment of the added value from data assimilation on modelled Nordic Seas hydrography and ocean transports, Ocean Model., 99, 43–59, doi:10.1016/j.ocemod.2015.12.010, 2016.

Lindsay, R. W.: Unified sea ice thickness climate data record collection spanning 1947–2012, Boulder, Colorado USA: National Snow and Ice Data Center, doi:10.7265/N5D50JXV, 2013.

Lisæter, K., Rosanova, J., and Evensen, G.: Assimilation of ice concentration in a coupled ice-ocean model, using the Ensemble Kalman filter, Ocean Dynam., 53, 368–388, doi:10.1007/s10236-003-0049-4, 2003.

Locarnini, R., Antonov, J., and Garcia, H.: World Ocean Atlas 2005, Volume 1: Temperature, vol. 61, US Dept. of Commerce, National Oceanic and Atmospheric Administration, 2006.

Locarnini, R. A., Mishonov, A. V., Antonov, J. I., Boyer, T. P., Garcia, H. E., Baranova, O. K., Zweng, M. M., Paver, C. R., Reagan, J. R., Johnson, D. R., Hamilton, M., and Seidov, D.: World Ocean Atlas 2013, Volume 1, Temperature, edited by: Levitus, S. and Mishonov, A., NOAA Atlas NESDIS, 40 pp., 2013.

Massonnet, F., Goosse, H., Fichefet, T., and Counillon, F.: Calibration of sea ice dynamic parameters in an ocean-sea ice model

using an ensemble Kalman filter, J. Geophys. Res., 119, 4168–4184, doi:10.1002/2013JC009705, 2014.

Mathiot, P., König Beatty, C., Fichefet, T., Goosse, H., Massonnet, F., and Vancoppenolle, M.: Better constraints on the sea-ice state using global sea-ice data assimilation, Geosci. Model Dev., 5, 1501–1515, doi:10.5194/gmd-5-1501-2012, 2012.

Morison, J. H., Long, C. E., and Levine, M. D.: Internal wave dissipation under sea ice, J. Geophys. Res., 90, 11959–11966, doi:10.1029/JC090iC06p11959, 1985.

Nguyen, A., Menemenlis, D., and Kwok, R.: Improved modeling of the Arctic halocline with a subgrid-scale brine rejection parameterization, J. Geophys. Res., 114, C11014, doi:10.1029/2008JC005121, 2009.

Nguyen, A., Menemenlis, D., and Kwok, R.: Arctic ice-ocean simulation with optimized model parameters: Approach and assessment, J. Geophys. Res., 116, C04025, doi:10.1029/2010JC006573, 2011.

Oke, P. R., Brassington, G. B., Griffin, D. A., and Schiller, A.: The Bluelink ocean data assimilation system (BODAS), Ocean Model., 21, 46–70, doi:10.1016/j.ocemod.2007.11.002, 2008.

Rampal, P., Weiss, J., and Marsan, D.: Positive trend in the mean speed and deformation rate of Arctic sea ice, 1979–2007, J. Geophys. Res., 114, C05013, doi:10.1029/2008JC005066, 2009.

Rampal, P., Weiss, J., Dubois, C., and Campin, J. M.: IPCC climate models do not capture Arctic sea ice drift acceleration: Consequences in terms of projected sea ice thinning and decline, J. Geophys. Res., 116, C00D07, doi:10.1029/2011JC007110, 2011.

Rampal, P., Bouillon, S., Ólason, E., and Morlighem, M.: neXtSIM: a new Lagrangian sea ice model, The Cryosphere, 10, 1055–1073, doi:10.5194/tc-10-1055-2016, 2016.

Reul, N., Tenerelli, J., Boutin, J., Chapron, B., Paul, F., Brion, E., Gaillard, F., and Archer, O.: Overview of the first SMOS sea surface salinity products. Part I: Quality assessment for the second half of 2010, IEEE T. Geosci. Remote, 50, 1636–1647, doi:10.1109/TGRS.2012.2188408, 2012.

Reynolds, R. and Smith, T.: Improved global sea surface temperature analyses using optimum interpolation, J. Climate, 7, 929–948, doi:10.1175/1520-0442(1994)007<0929:IGSSTA>2.0.CO;2, 1994.

Rodwell, M. J., Lang, S. T. K., Ingleby, N. B., Bormann, N., Hólm, E., Rabier, F., Richardson, D. S., and Yamaguchi, M.: Reliability in ensemble data assimilation, Q. J. Roy. Meteor. Soc., 142, 443–454, doi:10.1002/qj.2663, 2016.

Rothrock, D. A. and Wensnahan, M.: The accuracy of sea-ice drafts measured from U. S. Navy submarines, J. Atmos. Ocean. Tech., 24, 1936–1949, doi:10.1175/JTECH2097.1, 2007.

Rothrock, D. A., Percival, D. B., and Wensnahan, M.: The decline in arctic sea-ice thickness: Separating the spatial, annual, and interannual variability in a quarter century of submarine data, J. Geophys. Res., 113, C05003, doi:10.1029/2007JC004252, 2008.

Sakov, P. and Oke, P. R.: A deterministic formulation of the ensemble Kalmanfilter: an alternative to ensemble square root filters, Tellus A, 60, 361–371, doi:10.1111/j.1600-0870.2007.00299.x, 2008.

Sakov, P., Counillon, F., Bertino, L., Lisæther, K. A., Oke, P. R., and Korablev, A.: TOPAZ4: an ocean-sea ice data assimilation system for the North Atlantic and Arctic, Ocean Science, 8, 633–656, doi:10.5194/os-8-633-2012, 2012.

Samuelsen, A., Hansen, C., and Wehde, H.: Tuning and assessment of the HYCOM-NORWECOM V2.1 biogeochemical modeling system for the North Atlantic and Arctic oceans, Geosci. Model Dev., 8, 2187–2202, doi:10.5194/gmd-8-2187-2015, 2015.

Schweiger, A., Lindsay, R., Zhang, J. L., Steele, M., Stern, H., and Kwok, R.: Uncertainty in modeled Arctic Sea Ice volume, J. Geophys. Res., 116, C00D06, doi:10.1029/2011JC007084, 2011.

Schweiger, A. J. and Zhang, J.: Accuracy of short-term sea ice drift forecasts using a coupled ice-ocean model, J. Geophys. Res., 120, 7827–7841, doi:10.1002/2015jc011273, 2015.

Shimada, K., Kamoshida, T., Itoh, M., Nishino, S., Carmack, E., McLaughlin, F., Zimmermann, S., and Proshutinsky, A.: Pacific Ocean inflow: Influence on catastrophic reduction of sea ice cover in the Arctic Ocean, Geophys. Res. Lett., 33, L08605, doi:10.1029/2005GL025624, 2006.

Simon, E., Samuelsen, A., Bertino, L., and Mouysset, S.: Experiences in multiyear combined state-parameter estimation with an ecosystem model of the North Atlantic and Arctic Oceans using the Ensemble Kalman Filter, J. Marine Syst., 152, 1–17, doi:10.1016/j.jmarsys.2015.07.004, 2015.

Spreen, G., Kwok, R., and Menemenlis, D.: Trends in Arctic sea ice drift and role of wind forcing: 1992–2009, Geophys. Res. Lett., 38, L19501, doi:10.1029/2011GL048970, 2011.

Stark, J., Donlon, C., Martin, M., and McCulloch, M.: OSTIA: An operational, high resolution, real time, global sea surface temperature analysis system, OCEAN 2007-Eurrope, IEEE, 1–4, doi:10.1109/OCEANSE.2007.4302251, 2007.

Steele, M., Morley, R., and Ermold, W.: PHC: A global ocean hydrography with a high-quality Arctic Ocean, J. Climate, 14, 2079–2087, doi:10.1175/1520-0442(2001)014<2079:PAGOHW>2.0.CO;2, 2001.

Talagrand, O., Vautard, R., and Strauss, B.: Evaluation of probabilistic prediction system. Proc. Workshop on Predictability, Reading, United Kingdom, ECMWF, 1–25, 1999.

Tian-Kunze, X., Kaleschke, L., Maaß, N., Mäkynen, M., Serra, N., Drusch, M., and Krumpen, T.: SMOS-derived thin sea ice thickness: algorithm baseline, product specifications and initial verification, The Cryosphere, 8, 997–1018, doi:10.5194/tc-8-997-2014, 2014.

Tietsche, S., Notz, D., Jungclaus, J. H., and Marotzke, J.: Assimilation of sea-ice concentration in a global climate model – physical and statistical aspects, Ocean Sci., 9, 19–36, doi:10.5194/os-9-19-2013, 2013.

Wadhams, P. and Horne, R. J.: An analysis of ice profiles obtained by submarine in the Beaufort Sea, J. Glaciol., 25, 401–424, 1980.

Wensnahan, M. and Rothrock, D. A.: Sea-ice draft from submarine-based sonar: Establishing a consistent record from analog and digitally recorded data, Geophys. Res. Lett., 32, L11502, doi:10.1029/2005GL022507, 2005.

Wingham, D. J., Francis, C. R., Baker, S., Bouzinac, C., Brockley, D., Cullen, R., Chateau-Thierry, P., Laxon, S. W., Mallow, U., Mavrocordatos, C., Phalippou, L., Ratier, G., Rey, L., Rostan, F., Viau, P., and Wallis, D. W.: CryoSat: A mission to determine the fluctuations in Earth's land and marine ice fields, Adv. Space Res., 37, 841–871, doi:10.1016/j.asr.2005.07.027, 2006.

Woodgate, R., Aagaard, K., and Weingartner, T.: Monthly temperature, salinity, and transport variability of the Bering Strait through flow, Geophys. Res. Lett., 32, L04601, doi:10.1029/2004GL021880, 2005.

Xie, J., Counillon, F., Zhu, J., and Bertino, L.: An eddy resolving tidal-driven model of the South China Sea assimilating along-track SLA data using the EnOI, Ocean Sci., 7, 609–627, doi:10.5194/os-7-609-2011, 2011.

Zygmuntowska, M., Rampal, P., Ivanova, N., and Smedsrud, L. H.: Uncertainties in Arctic sea ice thickness and volume: new estimates and implications for trends, The Cryosphere, 8, 705–720, doi:10.5194/tc-8-705-2014, 2014.

7

Freshening of Antarctic Intermediate Water in the South Atlantic Ocean in 2005–2014

Wenjun Yao[1], Jiuxin Shi[1], and Xiaolong Zhao[2]

[1]Physical Oceanography Laboratory/CIMST, Ocean University of China and Qingdao National Laboratory for Marine Science and Technology, Qingdao, China
[2]North China Sea Marine Forecasting Center, State Oceanic Administration, Qingdao, 266061, Shandong, China

Correspondence to: Wenjun Yao (wjimyao@gmail.com)

Abstract. Basin-scale freshening of Antarctic Intermediate Water (AAIW) is reported to have occurred in the South Atlantic Ocean during the period from 2005 to 2014, as shown by the gridded monthly means of the Array for Real-time Geostrophic Oceanography (Argo) data. This phenomenon was also revealed by two repeated transects along a section at 30° S, performed during the World Ocean Circulation Experiment Hydrographic Program. Freshening of the AAIW was compensated for by a salinity increase of thermocline water, indicating a hydrological cycle intensification. This was supported by the precipitation-minus-evaporation change in the Southern Hemisphere from 2000 to 2014. Freshwater input from atmosphere to ocean surface increased in the subpolar high-precipitation region and vice versa in the subtropical high-evaporation region. Against the background of hydrological cycle changes, a decrease in the transport of Agulhas Leakage (AL), which was revealed by the simulated velocity field, was proposed to be a contributor to the associated freshening of AAIW. Further calculation showed that such a decrease could account for approximately 53 % of the observed freshening (mean salinity reduction of about 0.012 over the AAIW layer). The estimated variability of AL was inferred from a weakening of wind stress over the South Indian Ocean since the beginning of the 2000s, which would facilitate freshwater input from the source region. The mechanical analysis of wind data here was qualitative, but it is contended that this study would be helpful to validate and test predictably coupled sea–air model simulations.

1 Introduction

Thermocline and intermediate waters play an important part in global overturning circulation by ventilating the subtropical gyres in different parts of the world oceans (Sloyan and Rintoul, 2001). They also constitute the northern limb of the Southern Hemisphere supergyre (Ridgway and Dunn, 2007; Speich et al., 2002).

Previous studies have addressed the variability of intermediate water. Wong et al. (2001) found that the intermediate water had freshened between the 1960s and the period 1985–1994 in the Pacific Ocean. Bindoff and McDougall (2000) reported that there had been freshening of water between 500 and 1500 db from 1962 to 1987 along 32° S in the Indian Ocean. Curry et al. (2003) showed a salinity reduction on the isopycnal surface of intermediate water for the period from the 1950s to the 1990s in the western Atlantic. The freshening variability can be traced back to the signature of water in the formation regions (Church et al., 1991). The freshening examples given above are in agreement with the enhancement of the hydrological cycle, in which the wet (precipitation (P) > evaporation (E), P dominance) subpolar regions have been getting wetter and vice versa for the dry (E dominance) subtropical regions over the last 50 years (Held and Soden, 2006; Skliris et al., 2014).

Antarctic Intermediate Water (AAIW) is characterized by a salinity minimum (core of AAIW) centered at the depths of 600 and 1000 m (Fig. 1), which lies within the potential density (with reference to sea surface) range of $\sigma_0 = 27.1$–$27.3\,\mathrm{kg\,m^{-3}}$ (Piola and Georgi, 1982). The AAIW is found from just north of the Subantarctic Front (SAF; Orsi et al., 1995) in the Southern Ocean and can be traced as far as 20° N

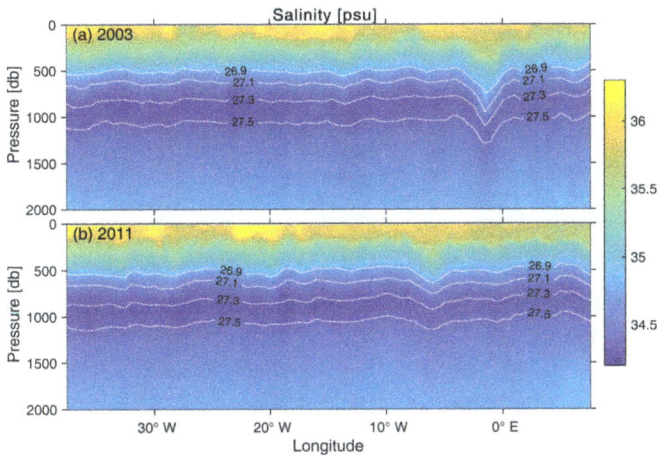

Figure 1. WOCE salinity sections along 30° S in the South Atlantic Ocean (positions shown in Fig. 2) observed in (a) 2003 and (b) 2011. Overlaid white solid–dotted lines are γ^n surfaces ranging from 26.9 to 27.5 kg m^{-3}, with a 0.2 kg m^{-3} interval.

(Talley, 1996). It is generally accepted that the variability of AAIW is largely controlled by air–sea–ice interaction (Close et al., 2013; Naveira Garabato et al., 2009; Santoso and England, 2004), but the argument about its origin and formation process continues. For example, there is the circumpolar formation theory of AAIW along the SAF, through mixing with Antarctic Surface Water (AASW) along isopycnals (Fetter et al., 2010; Sverdrup et al., 1942). Alternatively, it has been proposed that there is a local formation of AAIW in specific regions, as a by-product of Subantarctic Mode Water (SAMW) relating to deep convection (McCartney, 1982; Piola and Georgi, 1982). The first standpoint states that the AAIW is primarily derived from entirely subpolar sources; meanwhile the second one emphasizes the role that air–sea interaction plays in the oceans south of South America.

In the South Atlantic, AAIW constitutes the return branch of the Meridional Overturning Circulation (MOC) (Donners and Drijfhout, 2004; Speich et al., 2007; Talley, 2013). As an open-ocean basin, the South Atlantic is fed by two different sources of AAIW (Sun and Watts, 2002). The first is younger, fresher and has a lower apparent oxygen utilization (AOU) and originates from the Southeast Pacific (McCartney, 1977; Talley, 1996) and the winter waters west of the Antarctic Peninsula (Naveira Garabato et al., 2009; Santoso and England, 2004). These source regions of AAIW are mostly dominated by the net surface freshwater flux from atmosphere to ocean ($P > E$), which facilitates the freshening of AAIW with time. The second is the older, saltier and higher AOU AAIW which comes from the Indian Ocean, transported by the Agulhas Leakage (AL) as Agulhas rings (Fig. 2). The mixture of the above two types of AAIW can lead to a transition of hydrographic properties across the subtropical South Atlantic (Boebel et al., 1997).

The influence of AL on the variability of AAIW in the South Atlantic has been demonstrated to be substantial (Hummels et al., 2015; Schmidtko and Johnson, 2012), as 50–60 % of the Atlantic AAIW originates from the Indian Ocean (Gordon et al., 1992; McCarthy et al., 2012), with increased (decreased) transport of AL relating to salinification (freshening) of AAIW. AL has apparently increased during the period from the 1950s to the early 2000s (Durgadoo et al., 2013; Lübbecke et al., 2015), but there have been no studies addressing the influence of AL on the AAIW in South Atlantic since 2000.

With the instigation of the Array for Real-time Geostrophic Oceanography (Argo) program, in situ hydrographic observation has tremendously expanded since 2003 (Roemmich et al., 2015), particularly in the Southern Ocean (SO) where historical data are sparse and intermittent. This decreases the uncertainty of estimates for the research on both seasonal and decadal variations of subsurface and intermediate waters.

The present work reports the freshening of AAIW in the South Atlantic over the preceding decade (2005–2014) using gridded monthly data based on Argo data. Against the background of an enhanced hydrological cycle, decreased transport of AL contributed to such freshening and may be driven by a weakening of wind stress in the South Indian Ocean during the same period.

2   Data and methods

Based on individual temperature (T) and salinity (S) profiles from Argo, International Pacific Research Centre (IPRC) gridded monthly-mean data for the period 2005–2014 have been produced using variational interpolation. The IPRC data have 27 levels from 0 to 2000 m depth vertically, on a nominal 1° × 1° grid globally and at monthly temporal resolution (http://apdrc.soest.hawaii.edu/projects/Argo/data/gridded/On_standard_levels/index-1.html). To reduce the error from low vertical resolution of data when computing the hydrographic values on isopycnal surfaces, T and S profiles were first interpolated onto 1 m vertical depth intervals using a spline method in the intermediate water depth, and a linear method in the thermocline depth. Because the IPRC data were interpolated from randomly distributed Argo profiles, it is necessary to demonstrate the robust nature of their signals by comparing them with the other Argo gridded products. As a result, the Japan Agency of Marine-Earth Science and Technology (JAMSTEC, Hosoda et al., 2008) T and S data from 2005 to 2014, with 1° longitude and 1° latitude resolution, were also collected for comparison and verification. The number of Argo profiles is rapidly increasing year by year, and part of their distribution has been outlined in previous studies, inter alia Hosoda et al. (2008) and Roemmich et al. (2015).

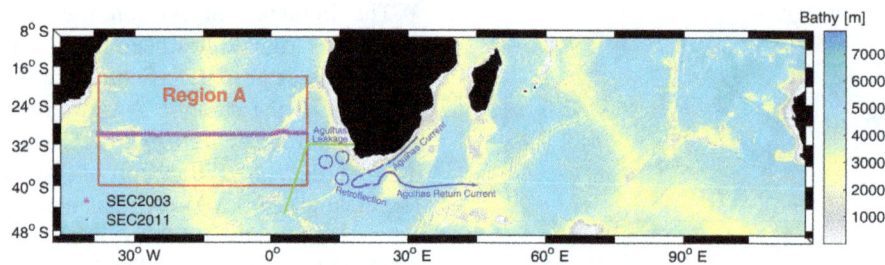

Figure 2. Bathymetry of the South Indian–Atlantic oceans. Color shading is ocean depth. Red box delineates the area for the basin-wide average of gridded data (hereafter referred to as Region A). The green line shows the Good Hope section, which is used to calculate the leakage transport to the South Atlantic. Magenta stars represent transatlantic CTD stations measured in 2003, with blue dots showing the 2011 measurements. The Agulhas Current, Retroflection, Agulhas Return Current and Agulhas Leakage (as eddies) are also shown.

Two hydrographic cruises of repeated transects along 30° S were conducted during the World Ocean Circulation Experiment (WOCE) Hydrographic Program (http://www.nodc.noaa.gov/woce/wdiu/diu_summaries/whp/index.htm). Their locations are presented in Fig. 2. The first transect consisted of 72 stations in 2003 by the R/V *Mirai* (Japan, Kawano et al., 2004); the second was in 2011 with 81 stations sampled from the *Ronald H. Brown* (United States, Feely et al., 2011). These two transects not only occupied almost identical station positions in the subtropical South Atlantic, but were also conducted in the same season (November and October respectively). Furthermore, the time interval between the two sections from November 2003 to October 2011 is very similar to the period covered by the IPRC data (January 2005–December 2014) and can therefore be used to validate those results.

To smooth out some of the higher frequency variability (i.e. mesoscale eddies and internal waves), the investigation of halocline variation should be along neutral density surfaces (McCarthy et al., 2011; McDougall, 1987). The layer of AAIW is defined using neutral density (γ^n, unit: $kg\,m^{-3}$; Jackett and McDougall, 1997) instead of potential density, with the upper and lower boundaries being $27.1\,\gamma^n$ and $27.6\,\gamma^n$ (Goes et al., 2014), respectively.

Monthly 10 m wind fields between years 1980 and 2014 from the ERA-Interim archive at the European Centre for Medium Range Weather Forecasts (ECMWF) (http://apps.ecmwf.int/datasets/data/interim-full-daily/levtype=sfc/) were used to investigate the decadal variability of wind stress (WS) over the South Indian Ocean. Another reanalysis wind product of National Centers for Environmental Prediction Department of Energy Atmospheric Model Intercomparison Project reanalysis 2 (NCEP-2, http://www.esrl.noaa.gov/psd/data/gridded/data.ncep.reanalysis2.html) was also used for the period 1980–2014. Additionally, the satellite-derived wind products of the Quick Scatterometer (QuikSCAT) for 2000–2007 and the Advanced Scatterometer (ASCAT) for 2008–2014 (both in ftp://ftp.ifremer.fr/ifremer/cersat/products/gridded/MWF/L3/) were used to compare and verify the decadal variability

of WS revealed by the ERA-Interim wind product. In this work, the WS over open ocean was calculated from 10 m wind field data using the equation adopted in Trenberth et al. (1989).

Reanalysis data including precipitation (P) and evaporation (E) from the ERA-Interim were used to reveal the freshwater input from the atmosphere to the ocean surface in the preceding decade.

The Simple Ocean Data Assimilation version 3.3.1 (SODA3.3.1, http://www.atmos.umd.edu/~ocean/), which is forced by the Modern-Era Retrospective Analysis for Research and Applications Version 2 (MERRA-2), spans the 36-year period 1980–2015 (Carton et al., 2017). The global simulated velocity field at specified depths provided by SODA makes it possible to evaluate the transport of AL.

3 Freshening of Antarctic Intermediate Water

3.1 Freshening observed from Argo gridded products

The Argo gridded products provide a globally distributed and continuous time series of T and S profiles down to 2000 m ocean depth. The present work focused on the AAIW in the South Atlantic Basin (Fig. 2, Region A), which encompasses most of the subtropical gyre and a part of the tropical regimes (Boebel et al., 1997; Talley, 1996). Computed from the Argo gridded data of IPRC, the biennial mean $\theta - S$ diagram (Fig. 3a) clearly shows that the AAIW has experienced a process of progressive basin-scale freshening during the period from January 2005 to December 2014. The linear trend of salinity (Fig. 3b) further reveals that the freshening takes up most of the AAIW layer but with a little salinification in the deeper part. Except around the $27.42\,\gamma^n$ neutral density surface, the AAIW variation is significant at the 95 % confidence level, using the F-test criteria. In comparison with Fig. 3a, it was found that the cutoff point of transformation from salinity decrease to increase is near the salinity minimum. Above the salinity minimum, the shift of $\theta - S$ trends towards cooler and fresher values along density surfaces and seems to be a response to the warming and fresh-

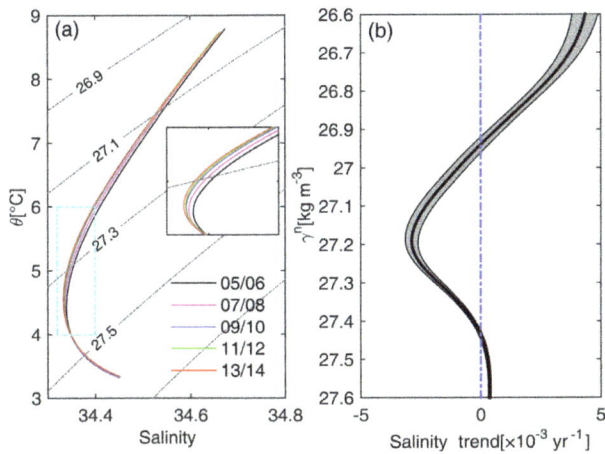

Figure 3. (a) Biennial mean θ–S diagram averaged over Region A for IPRC data with γ^n surfaces superimposed (grey solid–dotted lines). The inserted figure is the magnification of the area delineated by cyan solid–dotted box. The corresponding time for each θ–S curve is listed in their bottom-right corner (i.e. 05/06 for 2005–2006). **(b)** Salinity trend along γ^n surfaces for period January 2005–December 2014 is displayed by the thick black line, and the 95 % confidence intervals (F-test) are represented by the light grey shadings, calculated from IPRC data.

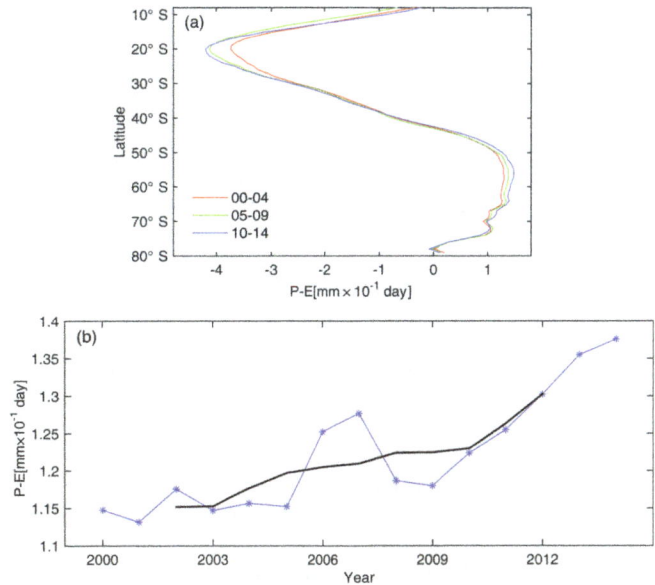

Figure 4. Calculated from ERA-Interim precipitation and evaporation data: **(a)** zonal mean (ocean areas only) of annual $P - E$ (freshwater input, mm day^{-1}); each line represents a 5-year averaged result. The corresponding time period (i.e. 00–04 for 2000–2004) is listed in the bottom-left corner. **(b)** Time series of annually $P - E$, averaged over the oceans in 45–65° S, 0–360° E band from 2000 to 2014 (blue star), and its 5-year running mean (black).

ening of surface waters where AAIW ventilates. Such thermohaline change has also been found in the Pacific and Indian oceans over a different time period (Wong et al., 1999). Church et al. (1991) and Bindoff and McDougall (1994) have researched the counterintuitive cooling of AAIW temperature induced by warming of surface water. They showed that a warming parcel in the mixed layer would subduct further equatorward, which would lead the $\theta - S$ curve to become cooler and fresher at a given density. The salinity decrease of the AAIW core indicates that such a change can only be induced by freshwater input from the source region, as mixing with more saline surrounding waters cannot give rise to a salt loss in the salinity minimum layer.

To demonstrate the robustness of AAIW variations revealed by the IPRC data, re-plots of Fig. 3a–b using another Argo gridded product from JAMSTEC are also shown for comparison (see Fig. S1 in the Supplement; only the AAIW layer is shown). Not only was the same variation along density surfaces in the AAIW layer found, but so too was a freshening of the salinity minimum. The isoneutral salinity increases in both IPRC and JAMSTEC data below the salinity minimum are quite small. The main discrepancy between them is that the salinity reduction in the JAMSTEC data is somewhat less than IPRC and at a higher 95 % confidence level (a mean of 0.006 between 27.1 γ^n and 27.6 γ^n).

The freshwater gain for the basin-scale salinity decrease of AAIW (mean salinity difference of 0.012 between 27.1 γ^n and 27.6 γ^n over a mean water mass thickness of 500 m) is estimated at 17 mm yr^{-1} in its source region. This as-

sumes that the South Atlantic only experienced freshwater input and nothing changed, thus the relationship between the salinity in 2005 and 2014 per unit area was roughly $S_{2005} \cdot 500 = S_{2014} \cdot (500 + \Delta d)$. Here $S_{2005} = S_{2014} + 0.012$ and Δd is the freshwater gain during the covered period. However, the depth-integrated salinity change over the water column (between 26.6 γ^n and 27.6 γ^n) was 0.0014, since a salinity increase of thermocline water balances the observed freshening of AAIW. This salinity budget implies contemporary hydrological cycle intensification in the Southern Hemisphere, which is illustrated by the P minus E change from 2000 to 2014, with $P - E$ increasing in the subpolar region and vice versa in the subtropical region (Fig. 4a). In these cases, the thermocline (intermediate) water that ventilates in the high-evaporation (precipitation) subtropical (subpolar) regions gets more saline (freshened), as shown by the hydrographic observations (Fig. 3b).

Against the background of hydrological cycle augmentation, the annual freshwater input in the AAIW ventilation region during the freshening period increased by 0.02 mm day^{-1}, about 17 % of the $P - E$ in 2005 (Fig. 4b). It is considered that the significant $P - E$ increase began around 2003 (Fig. 4b, 5-year running mean line), which means the observed freshened AAIW could be traced back to 2003. Though it was not possible to compute the direct freshwater input to the South Atlantic Basin in this study, the

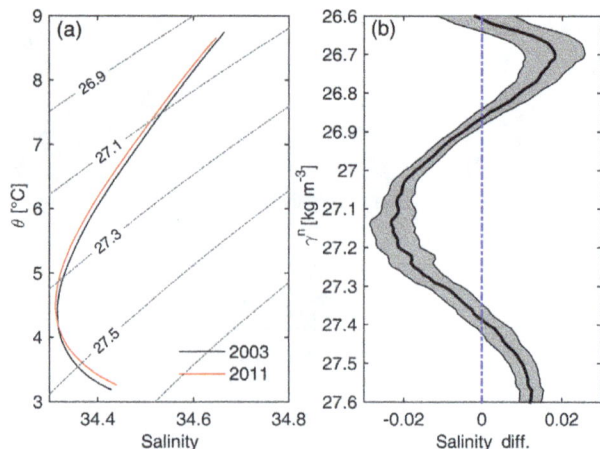

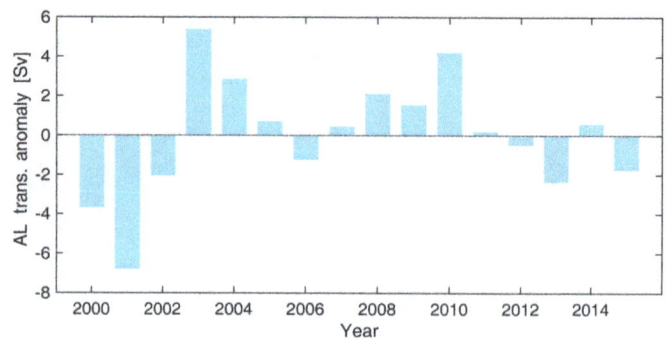

Figure 6. Computation of Agulhas Leakage transport anomaly from the SODA velocity field along the Good Hope line. Note that the depth integration is only for the AAIW layer.

Figure 5. (a) The same as Fig. 3a but for sectional mean of WOCE hydrographic casts. The corresponding year for each θ–S curve is listed in the bottom-right corner. **(b)** Sectional mean differences (thick black line) of WOCE hydrographic data along γ^n and their 95 % confidence intervals (grey shadings, t test).

Argo-era freshening of AAIW is qualitatively consistent with the freshwater gain in its source region.

3.2 Freshening in the quasi-synchronous WOCE CTD observations

Here, two synoptic transatlantic sections from WOCE hydrographic program were used to explore the decadal freshening signal identified in the above subsection. Similar to Fig. 3a, the sectional mean $\theta - S$ diagram (Fig. 5a) displays the same shift of thermohaline values, including freshening of the salinity minimum, salinity reduction in the upper AAIW layer and vice versa in the lower layer. Compared to the $\theta - S$ curves of IPRC data (Fig. 3a), the curves of WOCE (Fig. 5a) seem to be, in general, cooler θ and fresher S. It is suggested that this is because the IPRC mean is weighted towards the warmer and saltier waters in the north.

Unlike the Argo gridded product which has a continuous time series of T and S data, there are only two sections in the WOCE observations. Instead of calculating the linear trend of salinity (as was done with the IPRC data), the difference in salinity observed in 2003 and 2011 was estimated (Fig. 5b). The light grey shading denotes the 95 % confidence interval using simple t test criteria and considering the number of degrees of freedom. Above the salinity minimum, the freshening of AAIW revealed by the IPRC and the WOCE data are quite similar, with the maximum appearing near 27.2 γ^n. Because the last WOCE observation terminated in 2011 and the salinity reduction would continue at least up to 2014, as displayed in Fig. 3a, the magnitude of the freshening in WOCE (Fig. 5b) is smaller than IPRC (Fig. 3b). In the water layer below the salinity minimum (around 27.41 γ^n), the salinity increase shown in the WOCE data is relatively large (Fig. 5b). This is thought to be because the salinity rise reached its

maximum around 2011, which is shown in the time series of basin-wide averaged salinity on 27.45 γ^n- and 27.55 γ^n-density surfaces (see Fig. S2).

For the salinification of thermocline water, there is a large discrepancy between IPRC and WOCE data on neutral density surfaces 26.6–26.7 γ^n (Fig. 5b). It is considered that this would not affect the salinity budget over the water column (Fig. 5b), given that the salt gain of thermocline water would balance the observed freshened AAIW. In conclusion, the general trend and consistency of the detail therein of the salinity change over the last 10-year time period, revealed by the IPRC and the WOCE data, leads us to state that the freshening of AAIW is a robust and valid finding.

4 Decrease of Agulhas Leakage transport

AAIW in the South Atlantic is largely influenced by the AL through the intermittent pinching off of Agulhas rings (Fig. 2; Beal et al., 2011), transferring salty thermocline and intermediate water from the Indian Ocean to the South Atlantic (De Ruijter et al., 1999). The above discussion suggests that the freshening of AAIW was induced by the input of freshwater from the source regions, which consist of the southeast Pacific Ocean and the circumpolar subpolar oceans (see Sect. 1). As a result, if the transport of more saline water from the Indian Ocean decreased, it would promote the effect of this freshwater increase. In this section, the decrease of AL transport was evaluated by depth integration of the velocity field and further demonstrated by using an indirect indicator.

4.1 Evaluation from SODA velocity

In modeling studies, it is widely accepted to use a Lagrangian approach to quantify the leakage (Biastoch et al., 2009; van Sebille et al., 2009). Here, a simplified strategy was employed to compute the leakage by integrating the velocity within AAIW layer (approximately between 610 and 1150 m, according to Fig. 1), which was shown to result in a similar quantification to the Lagrangian one (Le Bars et al.,

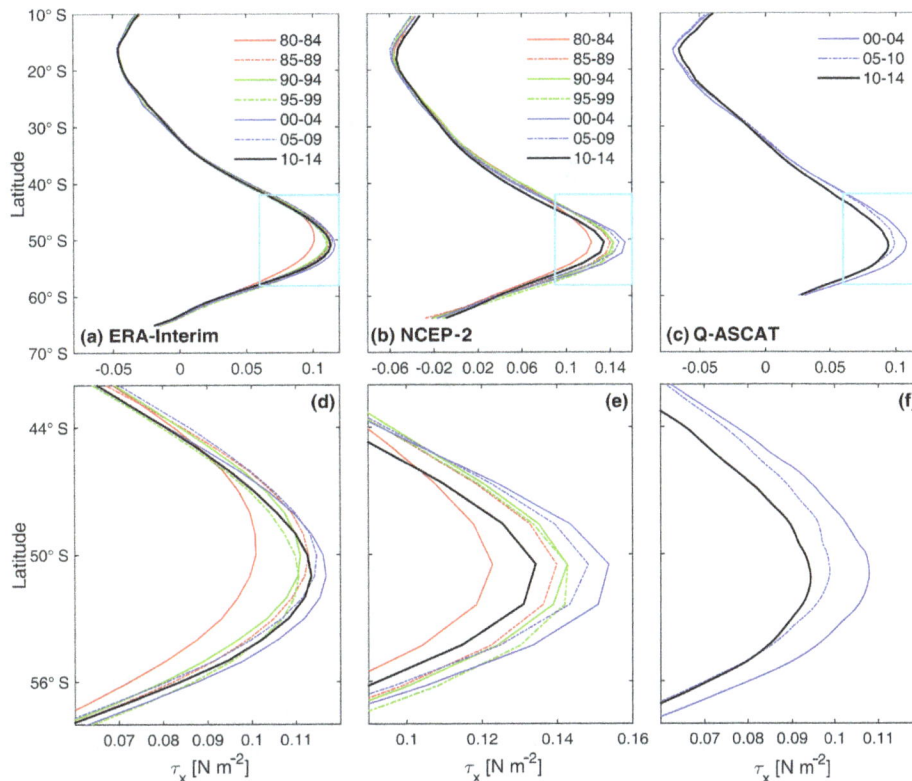

Figure 7. Zonally averaged wind stress calculated from the wind product of (**a**) ERA-Interim, (**b**) NCEP-2 and (**c**) QuikSCAT–ASCAT over the Indian Ocean (20–110° E) for different periods (i.e. 80–84 for January 1980–December 1984; 00–04 for January 2000–December 2004) listed in the top-right corners. Panels (**d, e, f**) are the magnification of cyan boxes in (**a, b, c**), respectively.

2014). The depth integration is along the Good Hope section (green line in Fig. 2), using the cross-component velocity. Note that the leakage calculation is from the continent to the zero line of the barotropic streamfunction, which is the separation of the Agulhas regime and the Antarctic Circumpolar Current (Biastoch et al., 2015).

Before showing the transport computed from the SODA velocity data, it is necessary to verify that the SODA hydrographic data show the same freshening of AAIW as other datasets. AAIW in the South Atlantic was also found to have freshened during period 2005–2014, though with relatively small magnitude (Fig. S3). Yearly leakage computation within the AAIW layer was carried out for the period 2000–2015 (Fig. 6). It shows that the leakage in the early 2010s is smaller than that in the middle and late 2000s, forming a decreasing trend in a nearly 10-year period. This estimation of leakage seems to be consistent with the indirect estimate of AL transport given below.

The following calculation is to simply estimate the contribution of the AL transport change to our observed freshening. As shown by Fig. 6, the decreased rate of AL transport could be taken to be 2 Sv in a 10-year time period, assuming that this rate increased year by year in the study period (i.e., 0.2 Sv in the first year, 0.4 Sv in the second year, and so on). Following Sun and Watts (2002), here we take

the salinity difference of $\Delta S = 0.1$ between the South Indian and the South Atlantic in the AAIW layers. The other parameters, including total seconds in a year, water thickness of the AAIW layer and the area of Region A, are taken to be $\Delta t = 365 \times 24 \times 3600\,\mathrm{s}$, $\Delta d = 500\,\mathrm{m}$ and $\Delta s_\mathrm{A} = 1.09 \times 10^{13}\,\mathrm{m}^2$, respectively. Therefore, the salinity decrease from 2005 to 2014, induced by the change of AL transport, should be $(0.2 + 0.4 + \ldots + 2) \times 10^6 \times \Delta t \times \Delta S/(\Delta s_\mathrm{A} \times \Delta d)$. This results in a salinity reduction of 0.0064, which could account for approximately 53.0 % of the observed freshening revealed by the IPRC data. Though our estimate here is quite rough, we can state that, during 2005–2014, the AL significantly influenced the salinity change in the South Atlantic Ocean within the AAIW layers.

4.2 Weakening of the westerlies in the South Indian Ocean

Continuous measurements of the AL transport have never been realized before. An earlier study suggested that an increased AL transport correlates well with a poleward shift of the westerly winds (Beal et al., 2011). However, after using reanalysis and climate models, Swart and Fyfe (2012) argued that the strengthening of Southern Hemisphere surface westerlies has occurred without major transgressions in its

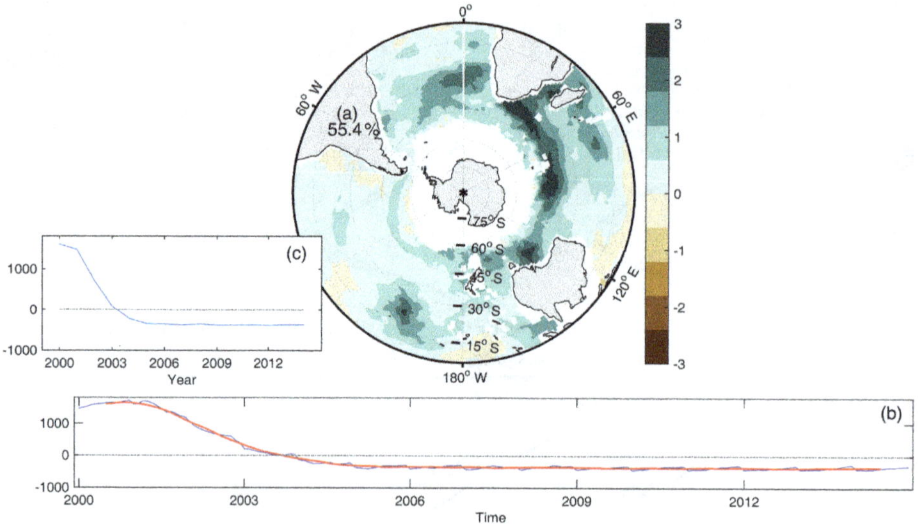

Figure 8. (a) Pattern and **(b)** time series (blue: monthly; red: 13-month smoothed) of EOF1 of salinity on 27.36 γ^n surface. **(c)** Yearly mean time series of EOF1. Calculated from SODA data.

latitudinal position over the period 1979–2010, during which period the AL has largely increased (Biastoch et al., 2009). A more recent study from Durgadoo et al. (2013) showed that the increase of AL is concomitant with an equatorward rather than a poleward shift of westerlies in their simulation cases. They also concluded that the intensity of westerlies is predominantly responsible for controlling this Indian–Atlantic transport. Many relevant studies agreed on this relationship, that the enhancement of westerly wind intensity is related to the increase of AL (Goes et al., 2014; Lee et al., 2011; Loveday et al., 2015).

The AL corresponds most significantly to westerly wind strength averaged over the Indian Ocean in contrast to that averaged circumpolarly or locally (Durgadoo et al., 2013). According to the work of Durgadoo et al. (2013), zonally averaged WS was calculated from the ERA-Interim wind product over the Indian Ocean (20–110° E) for every 5-year period since 1980 (Fig. 7a and d). Previous studies (Lee et al., 2011; Loveday et al., 2015) have found that the WS has increased considerably from the 1980s to the beginning of the 2000s (Fig. 7d), consistent with the contemporary increase of AL transport. Though there are oscillations during 1990s, the WS reached its peak around the years 2000–2004 (Fig. 7d), then began to decline. It can be concluded that the WS has weakened for period 2000–2014 (Fig. 7d), which implies a concurrent decrease of AL transport.

In addition to the ERA-Interim wind data, we have further checked the zonally averaged WS over the Indian Ocean (20–110° E), using another reanalysis product of NCEP-2 (Fig. 7b and e) and the combined QuikSCAT–ASCAT (Fig. 7c and f) satellite-derived wind products. The three zonally averaged WS agree that during the period 2000–2014, the westerlies reached a peak in the years 2000–2004, and then progressively subsided through 2005–2009 to 2010–2014. The pro-

cess of gradual decline of WS is most pronounced in the NCEP-2 data. It is noteworthy that none of the three products show a significant meridional shift of the latitude of maximum WS from 2000 to 2014, in corroboration with the conclusion of Swart and Fyfe (2012).

4.3 Evidence from other works

Many efforts have been made to estimate AL transport, especially using model simulations (Lübbecke et al., 2015; Loveday et al., 2015). In recent years, Le Bars et al. (2014) provided the time series of AL transport over the satellite altimeter era, computed from absolute dynamic topography data, which can show the decadal variation of AL present. In their result (Fig. 8 in Le Bars et al., 2014), the anomalies of AL from satellite altimetry reached a peak around 2003 (annual average), and then began to subside, apart from a mid-2011 increase. In addition, their negative trend of AL (Fig. 9 in Le Bars et al., 2014) over the period from October 1992 to December 2012 indicates that the transport was reduced during the 2000s in contrast to the 1990s. Another study by Biastoch et al. (2015) should be of help in the present discussion. Though the time series of AL obtained from models did not show a distinct decline of AL transport in the last decade, which seems partly due to the data filter applied and the end of the time series (Fig. 4 in Biastoch et al., 2015), it displays a maximum of salt transport around 2000 (Fig. 5 in Biastoch et al., 2015). This peak and the subsequent decline of salt transport are consistent with the freshening of AAIW over the similar time period considered here.

Thus, in addition to the freshwater input that gave rise to the salt loss of the AAIW in the South Atlantic Ocean, reduced transport of AL or salt would further enhance this signal. Unfortunately, the analyses of the contributions from both the source region and the AL were only quantitative. Future work should be focused on the quantification of each factor based on model simulations.

5 Conclusions and discussions

The analysis of IPRC gridded data shows that the AAIW in the South Atlantic has experienced basin-scale freshening for the period from January 2005 to December 2014 (Fig. 3a and b), with freshwater input estimated at $17\,\mathrm{mm\,yr^{-1}}$ in its source region. Two transects of the WOCE hydrographic program observed in 2003 and 2011 also reveal the above variation of AAIW in the last decade (Fig. 5a and b).

This freshening in the intermediate water layer is thought to be compensated for by increased salinity in shallower thermocline water, indicating a contemporary intensification of hydrological cycle (Figs. 3b and 5b). In this case the freshwater input from atmosphere to ocean surface increased in the subpolar high-precipitation region and vice versa in the subtropical high-evaporation region (Fig. 4a). Over the last 10-year time period, significant freshwater gain began around 2003 (Fig. 4b), suggesting that the observed freshened AAIW could be traced back to this time.

Against the background of hydrological cycle intensification, the decrease of AL transport is proposed to contribute to the freshening of AAIW in the South Atlantic, associated with a weakening of westerlies over the South Indian Ocean. This decrease was revealed by the leakage evaluation along the Good Hope section. The mechanical analysis shows that the WS over the South Indian Ocean reached its peak around 2000–2004 and began to subside through 2005–2009 to 2010–2014 (Fig. 7), reversing its increasing phase from the 1950s to the beginning of the 2000s, during which period the AL had increased (Durgadoo et al., 2013; Lübbecke et al., 2015). This indirectly estimated variability of AL is consistent with other studies covering a similar period (Biastoch et al., 2015; Le Bars et al., 2014). As the AAIW carried by the AL is more saline relative to its counterpart in the South Atlantic Ocean, its decrease would promote the effect of freshwater input from the source region. Our estimate further suggests that such an induced freshwater input by AL could account for approximately 53 % of the observed freshening.

One might ask if there are any other sources that could significantly affect the AAIW in the South Atlantic Ocean, for example the Southeast Pacific (see Sect. 1). To clarify this question, we displayed the first pattern of empirical orthogonal function (EOF1) and its time series (called the principal components) of salinity on the $27.36\,\gamma^{n}$ (around $27.2\,\sigma_{0}$) surface (Fig. 8) in the Southern Hemisphere, which explains 55.4 % of the variance. It shows that in 2000–2014, the most significant salinity reduction appeared in the South Indian Ocean, especially in the region of the Agulhas Current system. It also shows that compared to the west Atlantic, the east Atlantic (whose intermediate water is largely fed by its counterpart in the South Indian Ocean) experienced a major salinity reduction. In addition to these salinity changes, we also note that the salinity decrease in the southeast Pacific was considerably less than that in the South Indian and the South Atlantic. Therefore, it implies that the Southeast Pacific did not play an important role in our observed AAIW freshening.

The purpose of this work is to reveal the decadal freshening of AAIW in the South Atlantic Ocean over the last 10-year time period, and suggest the related contributing mechanism. Future work should be focused on the quantification of these two contributors, and the influence they have on the world ocean circulation, through modeling studies.

Competing interests. The authors declare that they have no conflict of interest.

Acknowledgements. This study is supported by the Chinese Polar Environment Comprehensive Investigation and Assessment Programs (grant nos. CHINARE-04-04, CHINARE-04-01).

Edited by: Piers Chapman

References

Beal, L. M., De Ruijter, W. P., Biastoch, A., and Zahn, R.: On the role of the Agulhas system in ocean circulation and climate, Nature, 472, 429–436, https://doi.org/10.1038/nature09983, 2011.

Biastoch, A., Böning, C. W., Schwarzkopf, F. U., and Lutjeharms, J.: Increase in Agulhas leakage due to poleward shift of Southern Hemisphere westerlies, Nature, 462, 495–498, https://doi.org/10.1038/nature08519, 2009.

Biastoch, A., Durgadoo, J. V., Morrison, A. K., van Sebille, E., Weijer, W., and Griffies, S. M.: Atlantic multi-decadal oscillation covaries with Agulhas leakage, Nat. Commun., 6, 10082, https://doi.org/10.1038/ncomms10082, 2015.

Bindoff, N. L. and McDougall, T. J.: Diagnosing climate change and ocean ventilation using hydrographic data, J. Phys. Oceanogr., 24, 1137–1152, https://doi.org/10.1175/1520-0485(1994)024<1137:DCCAOV>2.0.CO;2, 1994.

Bindoff, N. L. and McDougall, T. J.: Decadal changes along an Indian Ocean section at 32 S and their interpretation, J. Phys. Oceanogr., 30, 1207–1222, https://doi.org/10.1175/1520-0485(2000)030<1207:DCAAIO>2.0.CO;2, 2000.

Boebel, O., Schmid, C., and Zenk, W.: Flow and recirculation of Antarctic intermediate water across the Rio Grande rise, J. Geophys. Res.-Oceans, 102, 20967–20986, https://doi.org/10.1029/97JC00977, 1997.

Carton, J. A., Chepurin, G. A., and Chen, L.: An updated reanalysis of ocean climate using the Simple Ocean Data Assimilation version 3 (SODA3), in preparation, 2017.

Church, J. A., Godfrey, J. S., Jackett, D. R., and McDougall, T. J.: A model of sea level rise caused by ocean thermal expansion, J. Climate, 4, 438–456, https://doi.org/10.1175/1520-0442(1991)004<0438:AMOSLR>2.0.CO;2, 1991.

Close, S. E., Naveira Garabato, A. C., McDonagh, E. L., King, B. A., Biuw, M., and Boehme, L.: Control of mode and intermediate water mass properties in Drake Passage by the Amundsen Sea Low, J. Climate, 26, 5102–5123, https://doi.org/10.1175/JCLI-D-12-00346.1, 2013.

Curry, R., Dickson, B., and Yashayaev, I.: A change in the freshwater balance of the Atlantic Ocean over the past four decades, Nature, 426, 826–829, https://doi.org/10.1038/nature02206, 2003.

De Ruijter, W., Biastoch, A., Drijfhout, S., Lutjeharms, J., Matano, R., Pichevin, T., Van Leeuwen, P., and Weiger, W.: Indian-Atlantic interocean exchange: Dynamics, estimation and impact, J. Geophys. Res., 104, 20885-20910, 1999.

Donners, J. and Drijfhout, S. S.: The Lagrangian view of South Atlantic interocean exchange in a global ocean model compared with inverse model results, J. Phys. Oceanogr., 34, 1019–1035, https://doi.org/10.1175/1520-0485(2004)034<1019:TLVOSA>2.0.CO;2, 2004.

Durgadoo, J. V., Loveday, B. R., Reason, C. J. C., Penven, P., and Biastoch, A.: Agulhas Leakage Predominantly Responds to the Southern Hemisphere Westerlies, J. Phys. Oceanogr., 43, 2113–2131, https://doi.org/10.1175/JPO-D-13-047.1, 2013.

Feely, R. A., Wanninkhof, R., Alin, S., Baringer, M., and Bullister, J.: Global Repeat Hydrographic/CO_2/Tracer surveys in Support of CLIVAR and Global Cycle objectives: Carbon Inventories and Fluxes, 2011.

Fetter, A., Schodlok, M., and Zlotnicki, V.: Antarctic Intermediate Water Formation in a High-Resolution OGCM, Geophys. Res. Abstr., Vol. 12, , EGU General Assembly 2010, Vienna, Austria, 2010.

Goes, M., Wainer, I., and Signorelli, N.: Investigation of the causes of historical changes in the subsurface salinity minimum of the South Atlantic, J. Geophys. Res.-Oceans, 119, 5654–5675, https://doi.org/10.1002/2014JC009812, 2014.

Gordon, A. L., Weiss, R., Smethie Jr., W. M., and Warner, M. J.: Thermociine and Intermediate Water Communication, J. Geophys. Res., 97, 7223–7240, https://doi.org/10.1029/92JC00485, 1992.

Held, I. M. and Soden, B. J.: Robust responses of the hydrological cycle to global warming, J. Climate, 19, 5686–5699, https://doi.org/10.1175/JCLI3990.1, 2006.

Hosoda, S., Ohira, T., and Nakamura, T.: A monthly mean dataset of global oceanic temperature and salinity derived from Argo float observations, JAMSTEC Rep. Res. Dev., 8, 47–59, https://doi.org/10.5918/jamstecr.8.47, 2008.

Hummels, R., Brandt, P., Dengler, M., Fischer, J., Araujo, M., Veleda, D., and Durgadoo, J. V.: Interannual to decadal changes in the western boundary circulation in the Atlantic at 11° S, Geophys. Res. Lett., 42, 7615–7622, https://doi.org/10.1002/2015GL065254, 2015.

Jackett, D. R. and McDougall, T. J.: A neutral density variable for the world's oceans, J. Phys. Oceanogr., 27, 237–263, https://doi.org/10.1175/1520-0485(1997)027<0237:ANDVFT>2.0.CO;2, 1997.

Kanamitsu, M., Ebisuzaki, W., Woollen, J., Yang, S.-K., Hnilo, J. J., Fiorino, M., and Potter, G. L.: NCEP-DOE AMIP-II Reanalysis (R-2), B. Am. Meteor. Soc., 1631–1643, http://www.cpc.ncep.noaa.gov/products/wesley/reanalysis2/kana/reanl2-1.htm, 2002.

Kawano, T., Uchida, H., Schneider, W., Kumamoto, Y., Nishina, A., Aoyama, M., Murata, A., Sasaki, K., Yoshikawa, Y., and Watanabe, S.: Cruise Summary of WHP P6, A10, I3 and I4 Revisits in 2003, AGU Fall Meeting Abstracts, 2004.

Le Bars, D., Durgadoo, J. V., Dijkstra, H. A., Biastoch, A., and De Ruijter, W. P. M.: An observed 20-year time series of Agulhas leakage, Ocean Sci., 10, 601–609, https://doi.org/10.5194/os-10-601-2014, 2014.

Lee, S. K., Park, W., van Sebille, E., Baringer, M. O., Wang, C., Enfield, D. B., Yeager, S. G., and Kirtman, B. P.: What caused the significant increase in Atlantic Ocean heat content since the mid-20th century?, Geophys. Res. Lett., 38, L17607, https://doi.org/10.1029/2011GL048856, 2011.

Loveday, B., Penven, P., and Reason, C.: Southern Annular Mode and westerly-wind-driven changes in Indian-Atlantic exchange mechanisms, Geophys. Res. Lett., 42, 4912–4921, https://doi.org/10.1002/2015GL064256, 2015.

Lübbecke, J. F., Durgadoo, J. V., and Biastoch, A.: Contribution of increased Agulhas leakage to tropical Atlantic warming, J. Climate, 28, 9697–9706, https://doi.org/10.1175/JCLI-D-15-0258.1, 2015.

McCarthy, G., McDonagh, E., and King, B.: Decadal variability of thermocline and intermediate waters at 24° S in the South Atlantic, J. Phys. Oceanogr., 41, 157–165, https://doi.org/10.1175/2010JPO4467.1, 2011.

McCarthy, G. D., King, B. A., Cipollini, P., McDonagh, E. L., Blundell, J. R., and Biastoch, A.: On the sub-decadal variability of South Atlantic Antarctic Intermediate Water, Geophys. Res. Lett., 39, L10605, https://doi.org/10.1029/2012GL051270, 2012.

McCartney, M. S.: Subantarctic Mode Water, in: A Voyage of Discovery: George Deacon 70th Anniversary Volume, Pergamon, edited by: Angel, M. V., 103–119, Woods Hole Oceanographic Institution, 1977.

McCartney, M. S.: The subtropical recirculation of mode waters, J. Mar. Res., 40, 427–464, 1982.

McDougall, T. J.: Neutral surfaces, J. Phys. Oceanogr., 17, 1950–1964, https://doi.org/10.1175/1520-0485(1987)017<1950:NS>2.0.CO;2, 1987.

Naveira Garabato, A. C., Jullion, L., Stevens, D. P., Heywood, K. J., and King, B. A.: Variability of Subantarctic Mode Water and Antarctic Intermediate Water in the Drake Passage during the Late-Twentieth and Early-Twenty-First Centuries, J. Climate, 22, 3661–3688, https://doi.org/10.1175/2009jcli2621.1, 2009.

Orsi, A. H., Whitworth III, T., and Nowlin Jr., W. D.: On the meridional extent and fronts of the Antarctic Circumpolar Current, Deep-Sea Res. Pt. I, 42, 641–673, https://doi.org/10.1016/0967-0637(95)00021-W, 1995.

Piola, A. R. and Georgi, D. T.: Circumpolar properties of Antarctic intermediate water and Subantarctic Mode Water, Deep-Sea Res. Pt. A, 29, 687–711, https://doi.org/10.1016/0198-0149(82)90002-4, 1982.

Ridgway, K. R. and Dunn, J. R.: Observational evidence for a Southern Hemisphere oceanic supergyre, Geophys. Res. Lett., 34, L13612, https://doi.org/10.1029/2007gl030392, 2007.

Roemmich, D., Church, J., Gilson, J., Monselesan, D., Sutton, P., and Wijffels, S.: Unabated planetary warming and its ocean structure since 2006, Nature Climate Change, 5, 240–245, https://doi.org/10.1038/nclimate2513, 2015.

Santoso, A. and England, M. H.: Antarctic Intermediate Water circulation and variability in a coupled climate model, J. Phys. Oceanogr., 34, 2160–2179, https://doi.org/10.1175/1520-0485(2004)034<2160:AIWCAV>2.0.CO;2, 2004.

Schmidtko, S. and Johnson, G. C.: Multidecadal Warming and Shoaling of Antarctic Intermediate Water*, J. Climate, 25, 207–221, https://doi.org/10.1175/jcli-d-11-00021.1, 2012.

Skliris, N., Marsh, R., Josey, S. A., Good, S. A., Liu, C., and Allan, R. P.: Salinity changes in the World Ocean since 1950 in relation to changing surface freshwater fluxes, Clim. Dynam., 43, 709–736, https://doi.org/10.1007/s00382-014-2131-7, 2014.

Sloyan, B. M. and Rintoul, S. R.: Circulation, Renewal, and Modification of Antarctic Mode and Intermediate Water*, J. Phys. Oceanogr., 31, 1005–1030, https://doi.org/10.1175/1520-0485(2001)031<1005:CRAMOA>2.0.CO;2, 2001.

Speich, S., Blanke, B., de Vries, P., Drijfhout, S., Döös, K., Ganachaud, A., and Marsh, R.: Tasman leakage: A new route in the global ocean conveyor belt, Geophys. Res. Lett., 29, 1416, https://doi.org/10.1029/2001gl014586, 2002.

Speich, S., Blanke, B., and Cai, W.: Atlantic meridional overturning circulation and the Southern Hemisphere supergyre, Geophys. Res. Lett., 34, L23614, https://doi.org/10.1029/2007GL031583, 2007.

Sun, C. and Watts, D. R.: A view of ACC fronts in streamfunction space, Deep-Sea Res. Pt. I, 49, 1141–1164, https://doi.org/10.1016/S0967-0637(02)00027-4, 2002.

Sverdrup, H. U., Johnson, M. W., and Fleming, R. H.: The Oceans: Their physics, chemistry, and general biology, Prentice-Hall, New York, 1942.

Swart, N. and Fyfe, J.: Observed and simulated changes in the Southern Hemisphere surface westerly wind-stress, Geophys. Res. Lett., 39, L16711, https://doi.org/10.1029/2012GL052810, 2012.

Talley, L. D.: Antarctic intermediate water in the South Atlantic, in: The South Atlantic: Present and Past Circulation, 219–238, Springer, 1996.

Talley, L. D.: Closure of the global overturning circulation through the Indian, Pacific, and Southern Oceans: Schematics and transports, Oceanography, 26, 80–97, https://doi.org/10.5670/oceanog.2013.07, 2013.

Trenberth, K. E., Large, W. G., and Olson, J. G.: The effective drag coefficient for evaluating wind stress over the oceans, J. Climate, 2, 1507–1516, https://doi.org/10.1175/1520-0442(1989)002<1507:TEDCFE>2.0.CO;2, 1989.

van Sebille, E., Biastoch, A., Van Leeuwen, P., and De Ruijter, W.: A weaker Agulhas Current leads to more Agulhas leakage, Geophys. Res. Lett., 36, L03601, https://doi.org/10.1029/2008GL036614, 2009.

Wong, A. P., Bindoff, N. L., and Church, J. A.: Large-scale freshening of intermediate waters in the Pacific and Indian Oceans, Nature, 400, 440–443, https://doi.org/10.1038/22733, 1999.

Wong, A. P., Bindoff, N. L., and Church, J. A.: Freshwater and heat changes in the North and South Pacific Oceans between the 1960s and 1985–94, J. Climate, 14, 1613–1633, https://doi.org/10.1175/1520-0442(2001)014<1613:FAHCIT>2.0.CO;2, 2001.

8

The Coastal Observing System for Northern and Arctic Seas (COSYNA)

Burkard Baschek[1], Friedhelm Schroeder[1], Holger Brix[1], Rolf Riethmüller[1], Thomas H. Badewien[2], Gisbert Breitbach[1], Bernd Brügge[3], Franciscus Colijn[1], Roland Doerffer[1], Christiane Eschenbach[1], Jana Friedrich[1], Philipp Fischer[4], Stefan Garthe[5], Jochen Horstmann[1], Hajo Krasemann[1], Katja Metfies[4], Lucas Merckelbach[1], Nino Ohle[6], Wilhelm Petersen[1], Daniel Pröfrock[1], Rüdiger Röttgers[1], Michael Schlüter[4], Jan Schulz[2], Johannes Schulz-Stellenfleth[1], Emil Stanev[1], Joanna Staneva[1], Christian Winter[7], Kai Wirtz[1], Jochen Wollschläger[1], Oliver Zielinski[2], and Friedwart Ziemer[1]

[1]Institute of Coastal Research, Helmholtz-Zentrum Geesthacht, Geesthacht, Germany
[2]Institute for Chemistry and Biology of the Marine Environment, University of Oldenburg, Oldenburg, Germany
[3]Federal Maritime and Hydrographic Agency, Hamburg, Germany
[4]Alfred Wegener Institute, Helmholtz Center for Polar and Marine Research, Center for Polar and Marine Research, Bremerhaven, Germany
[5]Research and Technology Centre (FTZ), University of Kiel, Büsum, Germany
[6]Hamburg Port Authority, Hamburg, Germany
[7]MARUM, Center for Marine Environmental Sciences, Bremen University, Bremen, Germany

Correspondence to: Burkard Baschek (burkard.baschek@hzg.de)

Abstract. The Coastal Observing System for Northern and Arctic Seas (COSYNA) was established in order to better understand the complex interdisciplinary processes of northern seas and the Arctic coasts in a changing environment. Particular focus is given to the German Bight in the North Sea as a prime example of a heavily used coastal area, and Svalbard as an example of an Arctic coast that is under strong pressure due to global change.

The COSYNA automated observing and modelling system is designed to monitor real-time conditions and provide short-term forecasts, data, and data products to help assess the impact of anthropogenically induced change. Observations are carried out by combining satellite and radar remote sensing with various in situ platforms. Novel sensors, instruments, and algorithms are developed to further improve the understanding of the interdisciplinary interactions between physics, biogeochemistry, and the ecology of coastal seas. New modelling and data assimilation techniques are used to integrate observations and models in a quasi-operational system providing descriptions and forecasts of key hydrographic variables. Data and data products are publicly available free of charge and in real time. They are used by multiple interest groups in science, agencies, politics, industry, and the public.

1 Introduction

A large part of humanity lives near the coasts and depends on the coastal oceans. At the same time, global problems such as climate change, sea level rise, or ocean acidification influence the ecosystems and communities along the coasts in particular. Shelf seas host unique ecosystems and provide essential sources for life in the ocean and the bordering land, while regions like the North Sea are heavily used for a multitude of human activities, from tourism and ship traffic to the exploitation and exploration of food resources, energy, and raw materials. Shelf seas are also heavily influenced by terrestrial processes due to continuous influx of natural and anthropogenic material from river systems and the atmosphere. They therefore act as important interfaces for global material cycles, for example through the uptake, emission, and transport of carbon compounds.

Understanding coastal systems is therefore of a high value, not only from a scientific point of view, but also due to its societal value. Coastal research has, however, long been hampered by the effort involved in investigating the highly complex coastal systems, the diversity of disciplines and institutions involved, and the difficulties in obtaining long-term and high-resolution, consistent measurements.

Current observations in the North Sea reveal substantial changes in biogeochemistry and food webs accompanied by the occurrence of new and the disappearance of established species (Gollasch et al., 2009; Buschbaum et al., 2012). The causes of these shifts are only partially known. Changes in physical quantities (e.g. temperature, wind) as well as anthropogenic influences (e.g. pollution, over-fishing, invasive species) most probably act as major drivers (Emeis et al., 2015). In the Arctic, the thawing of permafrost has started to cause coastal erosion and an increase in greenhouse gas emissions (IPCC, 2014). These examples highlight the sensitivity and dynamic behaviour of such complex systems that are still barely understood and insufficiently documented and monitored.

Recent advances in technology enable the use of remotely controlled automated measurements and the development of "intelligent" integrated systems that combine measurements and numerical modelling to create a synoptic view of coastal systems. The Coastal Observing System for Northern and Arctic Seas (COSYNA) has been established to demonstrate the feasibility of this idea for shallow, coastal areas. COSYNA focuses on the complex interdisciplinary processes of the German Bight in the North Sea and the Arctic coast near Svalbard, to assess the impact of anthropogenic changes, and to provide a scientific infrastructure. The focus regions have been chosen because they are ideal test beds in terms of natural variability and processes, human use and change, as well as accessibility.

The principal objective of observations, instrument development, and modelling is to improve our understanding of the interdisciplinary interactions between physical, biogeochemical, and ecological processes in coastal seas, to investigate how they can be best described at present, and how they will evolve in the future. To this end, COSYNA combines its measurement capabilities in the German Bight in a network that is designed to expand beyond individual platforms, areas, campaigns, and quantities to generate a holistic view of the entire coastal system by analysing the multitude of measurements taking into consideration the combination of different data sources as well as integrating them into model analyses.

In COSYNA, data and knowledge tools are developed and provided to be of use for multiple interest groups in industry, agencies, politics, environmental protection, or the public. These data and products are publicly available free of charge and can be used to support national monitoring authorities to comply, for example, with the requirements of the European Water Framework Directive and the Marine Strategy Framework Directive. The coastal observatory involves national and international contributions to international programmes, such as the coastal module of the Global Ocean Observing System (coastal GOOS), the European Ocean Observing System (EOOS as supported by EuroGOOS), the Global Earth Observations System of Systems (GEOSS), Marine Geological and Biological Habitat Mapping (GEOHAB), and the COPERNICUS Marine Environment Monitoring Service (CMEMS).

COSYNA is coordinated by the Helmholtz-Zentrum Geesthacht (HZG), Germany, and has been jointly developed, implemented, and operated with 10 other German partner institutions (see Table 1).

The present Ocean Science and Biogeochemistry interjournal special issue, "COSYNA: integrating observations and modeling to understand coastal systems", collects contributions highlighting various aspects of the complex observing system. This article provides an overview of COSYNA, its observational and modelling approach, as well as the diverse associated scientific studies and activities. It aims at connecting the articles in the special issue to previously published results from COSYNA. To this end, we will first describe the focus regions (Sect. 2), objectives (Sect. 3), and the international context of COSYNA (Sect. 4), before giving an overview of the observations (Sect. 5), sensor and instrument development (Sect. 6), as well as modelling and data assimilation activities (Sect. 7). Data, data products, and outreach activities are then described (Sects. 8 and 9) before a brief outlook on future activities is given (Sect. 10).

2 Coastal focus regions

The focus regions of COSYNA, the German Bight of the North Sea and the Arctic coast at Svalbard, are representative of two extremes in the broad spectrum of northern and Arctic coasts. The German Bight is one of the most intensely used coastal seas worldwide, with often opposing interests of economy, nature conservation, and recreation. Arctic seas and coasts are among the areas most affected by and vulnerable to global warming. For a recent assessment of impacts of climate change on the North Sea region, see NOSCCA (2016).

2.1 The German Bight

The German Bight (Fig. 1) is located in the south-eastern corner of the North Sea, a temperate, semi-enclosed shelf sea. Sündermann et al. (1999) define its seaward boundaries at $6°30'$ E and $55°00'$ N. The German Bight is relatively shallow, with water depths of generally less than 40 m. The main topographical features are the glacially formed Elbe River valley that spreads out to the north-west and a chain of barrier islands along the Dutch, German, and Danish North Sea coast. The islands protect the major part of the Wadden Sea,

Table 1. COSYNA partners.

Helmholtz-Zentrum Geesthacht (co-ordination)	HZG
Alfred Wegener Institute, Helmholtz Centre for Polar and Marine Research	AWI
Center for Marine Environmental Sciences at Bremen University	MARUM
Institute for Chemistry and Biology of the Marine Environment at the University of Oldenburg	ICBM
Research and Technology Centre at the University of Kiel	FTZ
German Federal Maritime and Hydrographic Agency	BSH
Center for Earth System Research and Sustainability	CEN
Hamburg Port Authority	HPA
Lower Saxony State Department for Waterway, Coastal and Nature Conservation	NLWKN
Schleswig-Holstein's Agency for Coastal Defence, National Parks, and Marine Conservation	LKN
German Federal Waterways Engineering and Research Institute	BAW

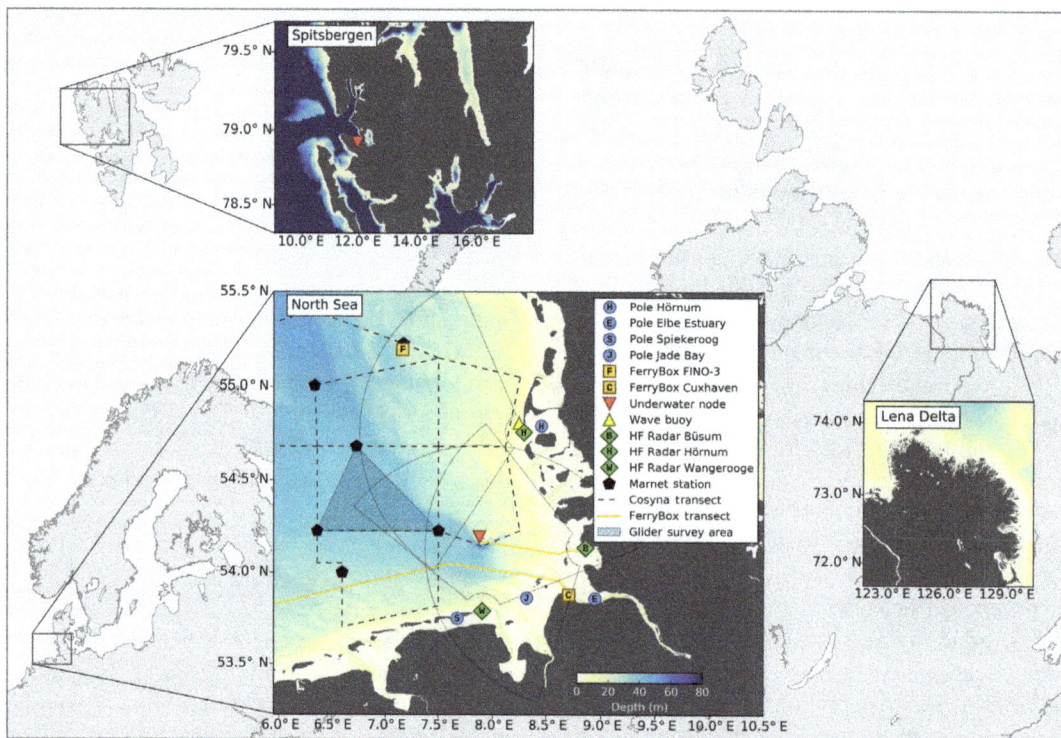

Figure 1. Map showing the pre-operational components of the COSYNA coastal observing system.

the largest unbroken system of intertidal sand and mud flats in the world.

The North Sea is characterised by the transition from oceanic to brackish water with variable freshwater input at the coasts. Physical drivers such as wind, sea surface temperature (SST), or tides control the natural variability in circulation and exchange processes with the open sea and the coastal fringe boundaries over a broad range of temporal and spatial scales (Schulz et al., 1999; Sündermann et al., 1999; Emeis et al., 2015; NOSCCA, 2016).

Strong tidal currents and intermittent strong wind events form a regime of high kinetic and turbulent energy with significant bed–water column exchange in the North Sea. Westerly winds typically prevail in the North Sea, but variations

exist and southerlies and easterlies may produce secondary circulation patterns (Otto et al., 1990). The currents are dominated by the M2 lunar tidal component that enters the North Sea from the north and moves as a Kelvin wave cyclonically through the North Sea (Otto et al., 1990; Howarth, 2001). Strong tidal currents in the channels connecting the Wadden Sea with the German Bight drive an intense exchange and a net import of suspended particulate matter and nutrients into the Wadden Sea (Burchard et al., 2008; Staneva et al., 2009; van Beusekom et al., 2012) and sustain its muddy component and the high productivity of the intertidal mud flats (Postma, 1984; van Beusekom et al., 1999; van Beusekom and de Jonge, 2002; Colijn and de Jonge, 1984). The tides thus cause a complex pattern of mixing conditions just off the

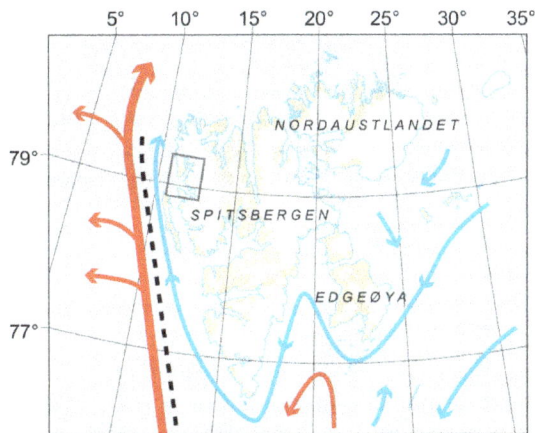

Figure 2. Spitsbergen with Kongsfjord (small rectangle) at the western coast of Svalbard. Arrows indicate the warmer Atlantic water masses (red) from the West Spitsbergen Current and colder less saline Arctic water (blue) from the East Spitsbergen Current (Cottier et al., 2005).

Figure 3. Research village NyÅlesund. The Spitsbergen Underwater-Node is located about 30 m in front of the Old Pier (**a**). The control station is located at the base of the Old Pier on land (**b**).

barrier islands and the mouths of the estuaries of the rivers Elbe, Weser, and Ems.

Global and local anthropogenic impacts overlay and interfere with these natural forcings. The global increase in CO_2 concentrations led to a long-term increase in SST that accelerated to $0.08\,°C\,yr^{-1}$ in the last decade (Loewe, 2009), while the average annual sea level rise reached $1.6\,mm\,yr^{-1}$ for the last 110 years (Wahl et al., 2013), and the average pH decreased from 8.08 to 8.01 in the years 1970 to 2006 (Lorkowski et al., 2012).

The North Sea is surrounded by densely populated, highly industrialised countries and is directly affected by multiple, often conflicting uses. One of the densest ship traffic lines worldwide crosses the German Bight and demands regular dredging of shipping channels and harbour basins. The Wadden Sea region, a UNESCO World Natural Heritage Site since 2009, is exposed to an import of pollutants and nutrients from land. The high biomass production caused by the latter resulted in the identification of the entire German Bight as a problem area by the OSPAR commission (OSPAR, 2008). Overfishing with bottom trawls impacts benthic invertebrate communities and leads to a decrease in biomass and species richness of fish communities (Emeis et al., 2015). As the latest development, the massive construction of offshore wind farms – underway or planned – is likely to have a significant impact on marine mammals (Koschinky et al., 2003) and seabirds (Garthe and Hüppop, 2004; M. Busch et al., 2013), but possibly also mixing (Lass et al., 2008; Ludewig, 2015; Carpenter et al., 2016) and nutrient transport.

2.2 The Arctic coast

While Spitsbergen (79° N) is geographically classified as fully Arctic, it is significantly influenced by Arctic and At-

lantic water masses from the Fram Strait (Fig. 2; Hop et al., 2002). Due to an increased advection rate of warmer Atlantic water masses in the fjord systems over the last decade, the first signs of an overall warming in the fjords have been observed, with a decrease in seasonal ice coverage (Stroeve et al., 2007) and significant changes throughout the food web (Hegseth and Tverberg, 2013; Van de Poll et al., 2016; Willis et al., 2006; Brand and Fischer, 2016).

The 20 km long Kongsfjord is located at the western coast of Svalbard and opens to a shelf system in a westerly direction. It has no sill and shares the outlet to the Atlantic with the more northern Krossfjord (Cottier et al., 2005). From this outlet, an underwater canyon runs through the shelf to the continental edge, establishing a connection to the deeper water masses of the West Spitsbergen Current off the shelf. Complex mixing processes between the Arctic shelf water masses, the Atlantic deep water masses, and the highly seasonal freshwater runoff from the inner part of the fjord result in strong environmental gradients from the inner parts of the fjords to its mouth (Svendsen et al., 2002). These gradients and their short- and long-term variability directly influence the pelagic and benthic realms of the fjord and the local food web (Stempniewicz et al., 2007). Due to the condensed temporal and spatial patterns of Atlantic and polar realms in a single fjord system, as well as the observed increase in mean water temperatures, the retreat of glaciers, and decrease in sea ice coverage over the last decades, the Kongsfjord ecosystem (Fig. 2) became an international focal point of climate change research.

The first research station addressing the Kongsfjord ecosystems was built by the Norsk Polar Institute in NyÅlesund (Fig. 3) at 78°55′ N, 11°56′ E in 1970. Since then, more than 15 nations have operated their own research stations in this northernmost year-round inhabited research village of the world, including German–French research station AW-IPEV (www.awipev.eu).

Even in Kongsfjord with its ideal and year-round available research infrastructure, most field research has been done in summer (Fischer et al., 2017) and only very little is known about the several month long polar winter with its prevailing darkness. The winter months are, however, essential for life cycles, the reproduction of many species (Fischer et al., 2017), and hence for the entire ecosystem (Hop et al., 2012). It is COSYNA's aim to help close this observational gap, providing year-round observations in this polar fjord system.

COSYNA activities also comprise remote sensing techniques that have been proved and tested in the North Sea to coastal waters in the Lena Delta, Siberia, for the quantification of suspended matter and chlorophyll as well as in situ measurements of inherent optical properties (Örek et al., 2013). The Lena Delta covers $32\,000\,\text{km}^2$ and discharges freshwater from a catchment area of $2\,400\,000\,\text{km}^2$ into the Arctic Ocean.

3 Objectives and benefits

Complex, highly interdisciplinary natural processes characterise the North Sea across several timescales and lengthscales. It is COSYNA's goal to help disentangle natural processes and anthropogenic impact in this region by combining consistent long-term time series at representative locations with process-oriented high-resolution observations. Numerical models of various resolutions are used to provide a context for observations ranging from the turbulent to basin-wide spatial scales. Observations are integrated into models using data assimilation techniques for resolutions, timescales, and quantities where such integration is possible and useful. It has therefore been COSYNA's approach to build an integrated observing system that is geared towards high flexibility and can be used on a variety of scales and problems that are of scientific or societal interest.

Routine observations of key variables and data assimilation techniques are employed to improve model performance for hindcasts, nowcasts, and short-term forecasts. The implementation of such a system achieves several objectives: it bridges spatial and temporal scales, while it establishes a backdrop against which key processes, such as exchange processes between the North Sea and Wadden Sea, the impact of extreme events, biological productivity variations, and the influences of e.g. offshore wind farm construction, can be investigated. The extensive development of offshore wind farms, for instance, requires sound environmental statistics and improved forecasts for planning and operation, while their influence on hydrodynamics, let alone biogeochemistry or biology, of the North Sea is still poorly understood.

The benefits of the COSYNA system are expected to be manifold. It contributes to technology development of key sensors and infrastructure, data interpretation algorithms such as for satellites and HF radar, as well as to modelling and data assimilation techniques suitable for operational use and monitoring. These developments and the creation of products of interest for various user groups contribute to the sciences while also benefitting society, e.g. by supplying coastal and seafloor observations of the North Sea in support of the European framework strategies and directives towards the goal of achieving a "good environmental status" of the marine environment.

As for the dissemination of data and products, COSYNA's objective is to make them available free of charge to the broadest possible audience in near-real time, while ensuring high quality standards and rigorous monitoring of data quality. Additional quality controls taking long-term perspectives into account are to be performed on an ongoing basis, ultimately resulting in data publications.

4 International context

With the initiation of the permanent Global Ocean Observing System – GOOS (Intergovernmental Oceanographic Commission, 1993) – and stepwise implementation of its many separate observing systems, new concepts regarding the worldwide systematic and sustained observation of the oceans have been put in place. Considering the role of coastal areas in ecological communities and their exposure to massive human utilisation, a GOOS coastal module was proposed to provide a basis for extended predictability of the coastal environment in both model and observations (Intergovernmental Oceanographic Commission, 1997). Awareness of the multitude of societal benefits (ABARE, 2006; https://ioos.noaa.gov/about/societal-benefits/) stimulated considerable investment in the worldwide implementation of integrated coastal ocean observatories (ICOOS).

In Europe, EuroGOOS (http://eurogoos.eu) is the pan-European GRA that co-ordinates six regional operational systems (ROOSes), such as the North West Shelf Operational Oceanographic System (NOOS, http://eurogoos.eu/roos/north-west-european-shelf-operational). In addition to providing operational oceanographic services and carrying out marine research, EuroGOOS puts considerable effort into unlocking fragmented and hidden marine data and making them openly available. Its data play a key role in the development of the European Marine Observation and Data Network (EMODnet) data portals (http://www.emodnet.eu). EMODnet is designed to cover all European coastal waters. The European ROOSes feed data into EMODnet either directly or through SeaDataNet (Schaap and Lowry, 2010; http://www.seadatanet.org/) and the Copernicus Marine Environment Monitoring Service (CMEMS, http://marine.copernicus.eu).

COSYNA contributes through the Helmholtz-Zentrum Geesthacht (HZG), as a EuroGOOS member, to the defi-

Table 2. Standard COSYNA observables.

Platform	Parameter
Meteorology	pressure, temperature, global radiation, and wind vector
Physical oceanography	pressure, temperature, salinity, current, wave height, and direction
Biogeochemistry	optical turbidity, total suspended matter concentration, chlorophyll *a* concentration, and dissolved oxygen

nition and implementation of operational services for near-coast, shallow ocean waters. Based on the FerryBox project funded by the EU in 2002–2005, HZG is co-chairing the FerryBox EuroGOOS Task Team (http://www.ferrybox.org). Via NOOS, the FerryBox data are fed into the EMODnet portals, while COSYNA's high-frequency radar data are delivered directly to the EMODnet Physics data portal and the glider data to the CMEMS data server.

5　Observations

The COSYNA observation network was designed to cover spatial scales ranging from a tidal catchment area in the Wadden Sea to the southern North Sea (Fig. 1). An additional observing station was installed at the western coast of Svalbard. Nearly all platforms deliver a set of COSYNA standard observables comprising key meteorological, oceanographic, and biogeochemical bulk parameters (Table 2). Tables 3 and 4 provide a comprehensive overview of the COSYNA platforms.

Four stationary systems were installed on poles placed in three tidal basins of East Frisia and one in the North Frisian Wadden Sea. They provide highly resolved COSYNA standard parameters (see Table 2) and allow the integration of energy and matter budgets over the sampled catchment areas. An additional pole and a stationary FerryBox monitor the exchange between the German Bight and the Elbe River as its main tributary.

To estimate transports across the northern cross section of the German Bight, a FerryBox was installed on the FINO3 (Forschungsplattformen in Nord- und Ostsee) wind-turbine research platform. Upstream of it, along the mean transport pathway in the German Bight, the FINO1 platform is located at the site of a station belonging to the Marine Environmental Monitoring Network in the North Sea and Baltic Sea (MARNET) operated by the German Federal Maritime and Hydrographic Agency (BSH). In general, MARNET complements the fixed COSYNA platforms (Table 3) towards the offshore regions of the German exclusive economic zone (EEZ). FerryBox systems operating on several ships of opportunity extend the COSYNA network to the North Sea scale, with several regular routes.

To provide a good spatial coverage, remote sensing with high-frequency (HF) radar and satellites is used. Two HF radar arrays are installed at the North Frisian coast and one

at the East Frisian coast with a nearly rectangular viewing angle to the other two systems. This configuration allows the determination of horizontal surface current vectors over most of the German Bight. The surface concentrations of total suspended matter, chlorophyll *a*, and yellow substances, "Gelbstoff", were obtained from 2003 to 2012 with MERIS (Medium Resolution Imaging Spectrometer) onboard ENVISAT, followed by MODIS (Moderate Resolution Imaging Spectroradiometer).

To go beyond the limitations in power and data transmission rates that most COSYNA platforms face, two COSYNA Underwater-Node Systems were developed and installed. They are pilots towards long-term observations of parameters beyond the COSYNA standard observables, such as optical systems for non-invasive determination of plankton or fish populations and their behaviour. The underwater node off the island of Helgoland is the first installation in a shallow water environment worldwide subject to strong wave forces. At Svalbard, the underwater node allows year-round observations under the sea ice under harsh environmental conditions. To explore physical and biogeochemical processes at the sediment–water interface over longer periods of time in high detail, three lander systems were developed that can be connected to the Underwater-Node Systems for longer operations.

Observations of the vertical distribution of variables over most of the water column were achieved with two alternating gliders operating for several weeks north-west off the island of Helgoland. Ship cruises with an undulating towed fish were carried out two to four times per year along a repeated grid covering the German Bight with the MARNET stations at its crossing points. For details on the moving platforms used in COSYNA, see Table 3.

All data are transferred in near-real time to the COSYNA data server and are publicly available in the COSYNA data portal (http://codm.hzg.de/codm/). Quality control processes are applied and data are flagged accordingly following SeaDataNet definitions[1].

5.1　Stationary measurements

Six fixed stations are the central element of COSYNA and serve as platforms to record point-like time series of meteoro-

---

[1] http://seadatanet.maris2.nl/v_bodc_vocab/browse.asp?order=entrykey&=L201

Table 3. Fixed platforms used in COSYNA. Abbreviations: M: meteorology, P: physical oceanography, B: biogeochemistry. For abbreviations of the partner institutions, see Table 1.

Platform	Years	Position	Mean tidal range (m)	Parameters	Partners
Pole Hörnum Basin	2002–2013 (Mar–Nov)	54°47.6′ N 008°27.1′ E	2.3	M, P, B	HZG
Pole Elbe Estuary	2012–2013 (Mar–Nov)	53°51.5′ N 008°56.6′ E	2.8	M, P, B	HPA, HZG
Pole Spiekeroog	2002–now (year round)	53°45.0′ N 007°40.3′ E	2.8	M, P, B	ICBM
FerryBox FINO-3	2011–2016 (year round)	55°11.7′ N 007°9.5′ E	0.9	P, B	HZG
FerryBox Cuxhaven	2010–now (year round)	53°52.6′ N 008°42.3′ E	2.9	P, B	HZG
Lander		n.a.	n.a.	P, B	MARUM, AWI, HZG
Underwater Node Helgoland	2012–now (year-round)	59°11′ N 008°52.8′ E		P, B	AWI, HZG
Underwater Node Spitsbergen	2012–now (year-round)	78°92′ N, 011°9′ E		P, B	AWI, HZG
Marine Radar Fino	2011–now (year-round)	55°11.7′ N 007°9.5′ E		M, P	HZG
Marine-Radar Sylt	2012–now (year-round)	54°49.2′ N 8°16.8′ E		M, P	HZG
HF-Radar Sylt	2009–now (year-round)	54°49.2′ N 8°16.8′ E		P	HZG
HF-Radar Büsum	2009–now (year-round)	54°7.2′ N 8°51.6′ E		P	HZG
HF-Radar Wangerooge	2009–now (year-round)	53°47.4′ N 7°55.2′ E		P	HZG

Table 4. Moving platforms used in COSYNA. Time resolution is given between repeated measurements at the same location. Abbreviations: M: meteorology, P: physical oceanography, B: biogeochemistry; S: water surface, U: upper water column, FC: full water column. The abbreviations of the partner institutions are explained in Table 1.

Platform	Vertical range	Time resolution	Parameters	Partner
FerryBox	U	1/2 day to a week	P, B	HZG
Glider	FC	days to months	P, B	HZG
Seabird	U	–	P	FTZ
Satellites	S	2 times in 3 days	B	HZG
Ship surveys	FC	months	M, P, B	HZG

logical and marine parameters. They provide high-frequency observations to resolve variability well below tidal periods in order to estimate statistically significant tidal fluxes as well as long-term records or trends over several years at the same location. Measuring poles were implemented at three tidal inlets, the inner Hörnum Basin, Jade Bay, and the Otzumer Balje close to the island of Spiekeroog, to capture the hydrodynamics and suspended particulate matter concentrations (SPMCs) typical of the East Frisian and North Frisian Wadden Sea. An additional pole was placed in the outer Elbe Estuary (Fig. 1).

While the inner Hörnum Basin represents the zero usage zone of the National Park of the North Frisian Wadden Sea,

Jade Bay is exposed to intense activity of building a new deep water port. The Otzumer Balje discharges a catchment area that is typical of the East Frisian Wadden Sea and was intensely investigated during the ELAWAT ecosystem research project (Dittmann, 1999). The Elbe pole was operated to contribute to the sediment management plan of the Elbe Estuary and to complement the data of the stationary Cuxhaven FerryBox on the southern side of the Elbe mouth. The FerryBox on FINO3 captures offshore conditions in the German Bight. All these stations are described in the following in more detail (Table 3).

Figure 4. The measuring poles at Spiekeroog (**a**) and in the inner Hörnum tidal basin (**b**). For details, see Sect. 5.1.

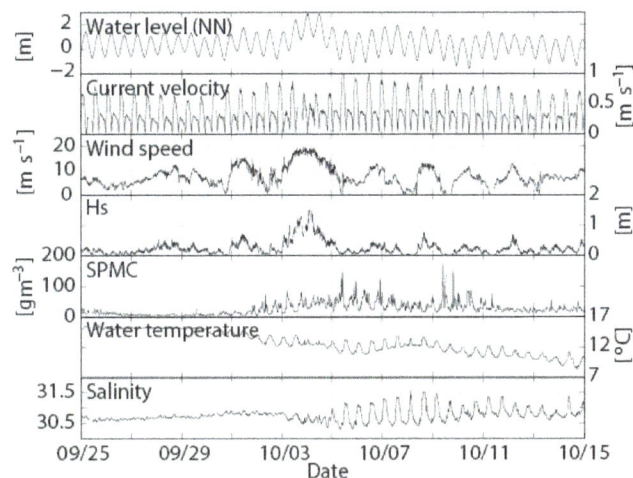

Figure 5. Time series of the measuring pole in the Hörnum Basin showing 3 weeks of data with a sampling frequency of 10 min.

5.1.1 Poles Hörnum Basin, Jade Bay, and Elbe Estuary

The poles at the inner Hörnum Basin, Jade Bay, and in the Elbe Estuary were mounted from March to November to prevent ice damage in the winter months. They consisted of a 15 m long steel tube, 5 m of which were jetted into the sea bed. A platform accessible via a ladder was mounted on top of the 40 cm diameter tube, resulting in an overall length of 18 m (Fig. 4). The platform carried meteorological sensors and a radiometer, solar panels for energy supply, an automated yet remotely controllable water sampler, and logger boxes for temporary data storage and wireless communication. A manual winch was used to retrieve the underwater instrument unit for maintenance. This unit was mounted with its lower end 1 m above the seafloor. It was equipped with sensors for all COSYNA standard observables of physical oceanography and biogeochemistry (Onken et al., 2007; Table 2).

In order to reduce sensor fouling, the underwater unit was cleaned at least twice a month. Possible sensor drift and cleansing effects were monitored by direct comparison with a well-calibrated reference system before, during, and after maintenance. Water samples were taken during maintenance to relate optical signals to SPMC.

To observe heat fluxes between the tidal flats and the water body, a vertical temperature sediment profiler was developed and deployed in the intertidal sediments close to the pole (Onken et al., 2010). It was operated for more than a year. At a distance of 5 nautical miles, an additional mooring with an upward-looking ADCP (acoustic Doppler current profiler) and a Datawell wave rider buoy was deployed.

In order to compute along-channel fluxes in the Hörnum Basin, occasional ship surveys were carried out over full tidal cycles relating across- and along-channel transects to the pole data. They were complemented by water samples and turbidity measurements. As an example, measurements over

3 weeks are shown (Fig. 5) comprising a significant wind event with peak velocities up to $20\,\mathrm{m\,s^{-1}}$ resulting in a sea level rise of more than 1.5 m and significant wave heights of up to 1.7 m. Water temperature and salinity after the storm exhibit the characteristic tidal (mainly M2) variability. Current velocities are predominantly at frequency M4, with a clear ebb–flood asymmetry. SPMC shows a complex variability reflecting the M4 tidal current dependencies as well as horizontal along-channel gradients. Interestingly, the onset of the rise in SPMC and its peak value lag behind the significant wave height by nearly one tidal period, indicating that the source of the additionally suspended material is located remotely from the pole.

The observations at the pole also indicate that the steady import of particulate matter is closely connected to the specific thermodynamic processes of the amphibic Wadden Sea area (Burchard et al., 2008; Onken et al., 2007; Onken and Riethmüller, 2010; Flöser et al., 2011).

5.1.2 Pole Spiekeroog

Time series of oceanographic, meteorological, and biogeochemical data have been continuously recorded since 2002 at a measuring pole of the Institute for Chemistry and Biology of the Marine Environment in the tidal channel of the Otzumer Balje close to the island of Spiekeroog (Figs. 1 and 4; Reuter et al., 2009; Badewien et al., 2009). The Spiekeroog time-series station (position 53°45′0.10″ N, 007°40′16.3″ E, mean sea level 13 m) consists of a 35.5 m long pole with a diameter of 1.6 m that is driven 10 m into the sediment. The temperature, conductivity, and pressure sensors are deployed within five horizontal tubes (1.5, 3.5, 5.5, 7.5, and 9 m above the seafloor) that are aligned in the main current direction. A platform is mounted on top of the pole, about 7 m above sea level. It consists of two laboratory containers hosting a second platform at 12 m above sea level that is equipped with solar panels, a wind turbine, and meteorological sensor systems. Oceanographic sensors are installed in special tubes within the pole that are oriented in the main direction of the tidal flow. An acoustic Doppler current profiler is mounted 1 m above the seafloor on a horizontal arm of 12 m length. The Spiekeroog time-series station is capable of withstanding storm events and ice conditions. It has been part of COSYNA since 2012.

The acquired data sets are fundamental for the improvement and validation of model results (Burchard and Badewien, 2015; Grashorn et al., 2015; Lettman et al., 2009; Staneva et al., 2009; Burchard et al., 2008) as well as to answer various research questions (Rullkötter, 2009; Badewien et al., 2009; Hodapp et al., 2015; Meier et al., 2015; Holinde et al., 2015) such as concerning the impact of storm surges, algal blooms on sediment dynamics, and exchange processes. The data sets are also valuable for assessing the long-term variability of oceanographic and biological parameters and determining anthropogenic impacts. The experience gained at the pole also helped to improve fouling-prone sensing methods and quality assurance (Garaba et al., 2014b; Schulz et al., 2015; Oehmcke et al., 2015).

5.1.3 Stationary FerryBoxes

As part of the COSYNA network, a stationary FerryBox was installed inside the pole of research platform FINO3. Water is pumped from approximately 5 and 16 m below mean sea level height for the continuous analysis of near-surface and seafloor waters. The FerryBox is equipped with sensors for standard oceanographic parameters (Table 1). Temporarily, nutrient analysers and a pCO$_2$ sensor were added.

Despite harsh operating conditions, the FerryBox has been operational since July 2011, with short interruptions during storm periods that were caused by sea spray and condensation that occurred notwithstanding the use of a heated steel cabinet for the protection of its electronics. Due to its remote position in the North Sea, personnel and spare parts had to be transported by helicopter to the platform for maintenance. Weather conditions therefore constrained the accessibility of the platform and sensors requiring regular maintenance could only be used temporarily. The software was operated remotely.

Since August 2010, a stationary FerryBox has also been installed in a container directly at the waterfront of Cuxhaven Harbour. It samples the tidally influenced, highly turbid lower Elbe River, the main freshwater discharge into the COSYNA observation area. The FerryBox was complemented by the Elbe Estuary measurement pole located 18 km upstream on the northern side of the river (Sect. 5.1.1) to contribute to a better understanding of the SPM dynamics and transport through the Elbe estuarine turbidity zone into the German Bight.

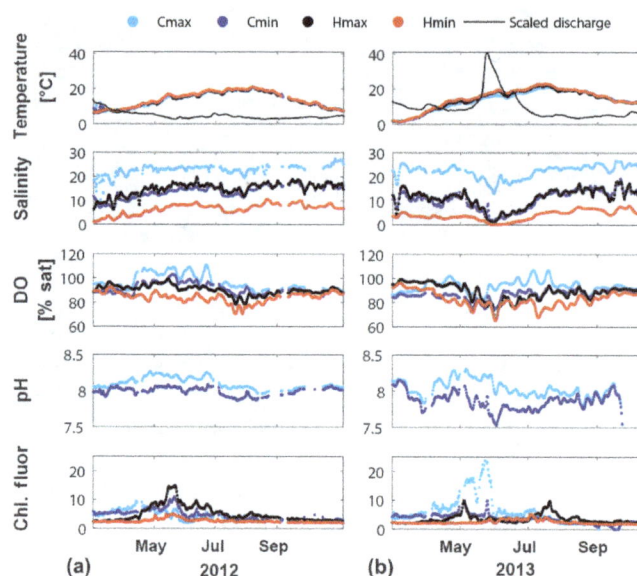

Figure 6. Time series of the stationary FerryBox located at Cuxhaven at the Elbe River mouth for 2012 **(a)** and 2013 **(b)**. Top to bottom: water temperature and Elbe River discharge (m^3 s^{-1}) at Neu Darchau station scaled by dividing it by 100 (thin black line), salinity, dissolved oxygen saturation (DO), pH, and chlorophyll a fluorescence. Shown are the Cuxhaven values at low tide (dark blue, Cmin), high tide (light blue, Cmax) and from the Elbe Estuary measurement pole at low tide (red, Hmin) and high tide (black, Hmax).

The water intake is located at a mean depth of 4 m. The oceanographic sensors are described in Sect. 5.4. The FerryBox is also equipped with a nitrate, phosphate, and silicate analyser as well as a fluorescence-based instrument for phytoplankton group determination. A meteorological station mounted on the top of the container provides wind speed and global radiation values.

Due to its easy and constant accessibility, the Cuxhaven FerryBox is an ideal platform for the testing of the long-term performance of new sensors under environmental conditions.

As an example, a time series of several parameters is shown for 2012 and 2013 (Fig. 6). A strong discharge period in the summer of 2013 led to a substantial decrease in salinity, with nearly freshwater conditions at low water for a 2-week period (Voynova et al., 2017).

5.2 Ocean gliders

Ocean gliders are autonomous underwater vehicles, propelled by a buoyancy engine. In the last decade they have become an established oceanographic platform in the open ocean, autonomously collecting data with a high temporal resolution along (re)programmable transects. Due to their operational flexibility and a long endurance of the order of months, gliders sample the oceans at low cost in a way no other platforms currently do (Testor et al., 2010).

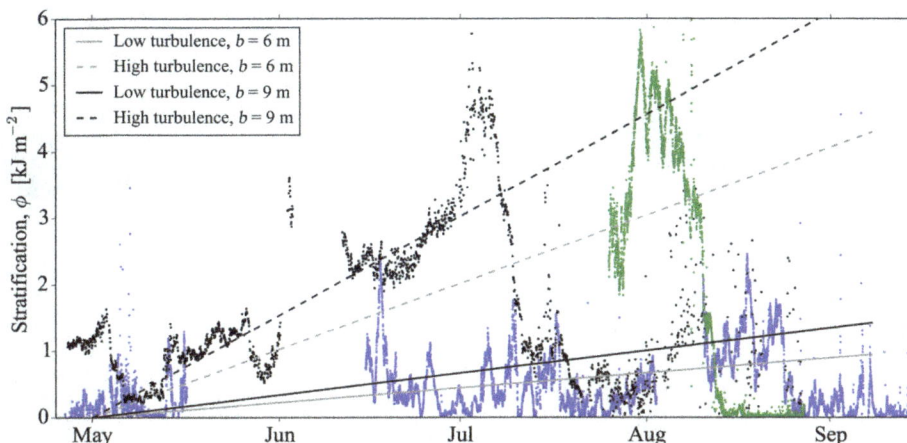

Figure 7. Measurements showing the observed buildup of stratification ϕ over the summer months unaffected by offshore construction (dots; Carpenter et al., 2016) and the estimated rate of stratification removal by the turbine foundation structures in offshore wind farms (straight lines). The stratification is computed as $\phi(t) = \int_0^H \left[\rho_{\mathrm{mix}} - \rho(z,t) \right] gz\,\mathrm{d}z$, with water depth H, density ρ, gravitational acceleration g, vertical coordinate z, and time t. Measurements are from a thermistor mooring at Marnet station NSB3 in 2009 (black dots); glider data are collected in the vicinity (54°40.8′ N, 6°43.9′ E) in 2014 (green dots) and from larger-scale transects passing through NSB3 in 2012 (blue points). The rate of stratification removal for thermocline thicknesses $b = 6, 9$ m is based on a simple 1-D analytical model (Carpenter et al., 2016).

The use of ocean gliders in shallow coastal waters is, however, challenging. COSYNA and a few other observatories have pioneered this particular use. Due to bathymetric constraints, currents can reach magnitudes in excess of the nominal glider speed, making it difficult to follow a prescribed transect. Intense commercial and recreational shipping traffic significantly increases the likelihood of a glider–ship collision (Merckelbach, 2013). This will almost certainly result in the loss of the glider and possibly in a hull rupture, if a fast lightweight craft is involved (Drücker et al., 2015). Therefore, COSYNA collaborates closely with the authority responsible for safety regulations in the German sector of the North Sea (Wasser- und Schifffahrtsamt) to develop prediction methodologies to mitigate the risk at sea involving gliders (Merckelbach, 2016).

COSYNA maintains three Slocum Littoral Electric gliders (Jones et al., 2005). These gliders have been used in the German sector of the North Sea in different operational modes. Gliders are particularly well suited for surveying repeated transects over long periods of time (months). Their long endurance makes it viable to run two gliders in an alternating service. While one glider is operational, the second one is refurbished. The gliders have also been deployed for shorter, targeted experiments. The use of multiple gliders provides additional spatial information. In order to fly gliders in formation, operational techniques have been developed so that they act as a single entity facilitating the interpretation of the spatial variability. The measurements taken with COSYNA gliders are available on CODM. With the help of a Java applet, glider data can be visualised in three dimensions (Breitbach et al., 2016).

The evolution of stratification during 2012 and part of 2014 is shown in Fig. 7 to illustrate glider measurements. The data were collected by two gliders in alternating service in 2012, and within a single experiment in 2014. From May to August, the potential energy and stratification of the water column increase due to solar heat flux. During that time, the water column is partially mixed by wind and waves at several instances. After September, mixing dominates and the heat fluxes are too low to create a stable stratification. Data from 2014 show interannual variability with a strong stratification in August and a subsequent complete mixing of the water column caused by a storm. After this event, the stratification was not restored.

5.3 High-frequency radar system

In order to detect surface currents, a high-frequency (HF) radar network was established in the German Bight of the North Sea. It consists of three "Wellen Radar" (WERA) systems (Gurgel et al., 1999) located on the islands of Sylt and Wangerooge and in Büsum (Fig. 8).

The radar signal propagates along the ocean surface beyond the horizon and is backscattered by surface waves with wavelengths between 5 and 50 m (half the electromagnetic wavelength of the radar). The WERA systems typically cover a range distance of 100 km with a resolution of 1.5 km. All systems transmit via a rectangular array of four antennas with a total power of 32 W. The systems on Sylt and in Büsum operate at 10.8 MHz with a linear receiver array consisting of 12 antennas, while the radar on Wangerooge operates at 12.1 MHz with a 16-antenna array.

Figure 8. HF radar system in the German Bight with its three stations in Büsum and on the islands of Sylt and Wangerooge. The right panel shows an example of the 2-D current field derived from overlapping radar signals.

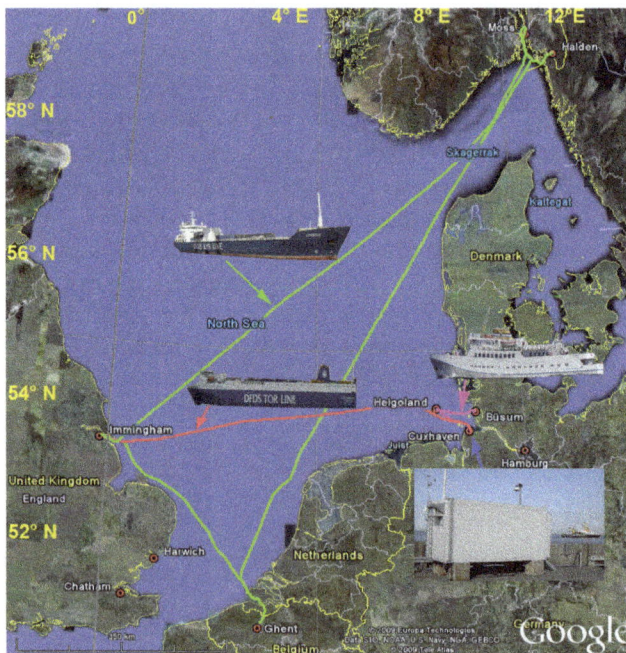

Figure 9. Map of FerryBox routes and stationary platforms equipped with FerryBoxes.

The acquired data are subject to quality control and are publicly available within 30 min of acquisition. In an additional processing step, the radial components of each radar site are assimilated into a numerical simulation model (Stanev et al., 2015) that is also used for short-term forecasts.

Since 2013, the HF radar network has also been used for ship detection, tracking, and fusing information of the radars with other sources of ship information such as from the Automated Identification System. Although the HF radar net-

work was set up for the retrieval of oceanographic parameters, leading to a limited resolution and detection performance, ship detection can be performed at each HF radar station every 33 s (Dzvonkovskaya et al., 2008). Tracking and fusion are performed as a post-processing task utilising state-of-the-art algorithms (Bruno et al., 2013; Maresca et al., 2014; Vivone et al., 2015).

5.4 FerryBox

In order to obtain oceanographic near-surface variables in a cost-effective way on a routinely basis, FerryBox systems have been developed within COSYNA and were installed on several ships-of-opportunity such as ferries or cargo ships, research vessels, or as stationary units (Fig. 9). They deliver key physical state variables of the North Sea and the Arctic coast off Svalbard and fill gaps concerning robust biogeochemical observations of the oceans. In particular, observations of the coastal carbon cycle with high temporal and spatial resolution along the ship tracks help to understand impacts of climate change or eutrophication on productivity, as well as the influence of single events such as storms or floods on the system. The recorded variables include temperature, conductivity, salinity (derived from temperature and conductivity), chlorophyll a fluorescence, turbidity, dissolved oxygen (DO), the partial pressure of CO_2 (pCO_2), pH, alkalinity, nutrients, and algal groups (derived from patterns of algal fluorescence by excitation at different wavelengths). The data are used for model validation (Petersen et al., 2011; Haller et al., 2015) and assimilation studies (Stanev et al., 2011; Grayek et al., 2011; Fig. 10).

The FerryBox is a modular system that can be easily extended with additional sensors. Compared to other platforms, such as buoys, the FerryBox systems have fewer limitations

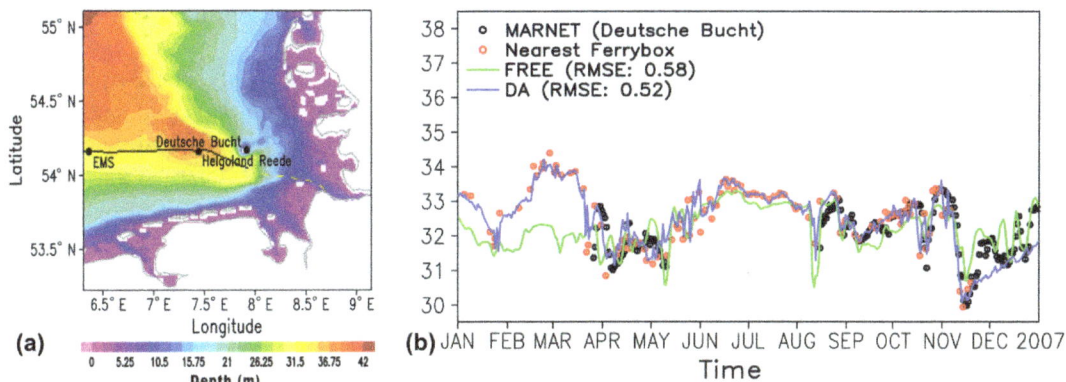

Figure 10. (a) Topography of the German Bight and FerryBox track. **(b)** Comparison of simulated sea surface temperature from a free model run and a run with data assimilation (DA) against MARNET and the nearest FerryBox observations (Grayek et al., 2011).

Figure 11. Set-up of the COSYNA Underwater-Node System with (1) land-based server and power supply, (2) cable connection (max. 10 km) to the first primary underwater node, (3) breakout box to connect the primary node to the underwater cable, (4) primary node system, (5, 6) cable connection (max. 70 m) to sensor units, and (7) cable connection to a second node. A third node can be connected to the second node.

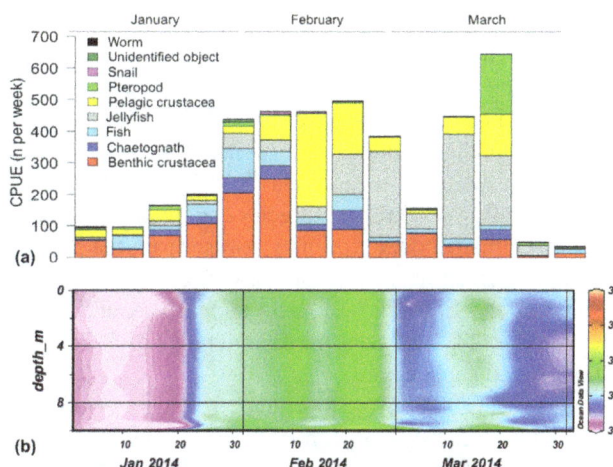

Figure 12. (a) The temporal abundances of the main biota groups assessed with a stereo-optic sensor attached to the Underwater-Node System in Spitsbergen from January 2014 to March 2014. CPUE (catch per unit effort) refers to the total number of organisms per group counted per week. **(b)** The temporal and spatial patterns of salinity in the depth range between 0 and 10 m assessed with one remote controlled vertical CTD profile per day during the same time period when the biota measurements (upper panel) were done.

due to space, power consumption, or harsh environmental conditions, allowing the operation of experimental and less robust sensors (Petersen, 2014). Due to a self-cleaning mechanism, the system maintenance intervals can be extended up to several months. All data are stored in the FerryBox system and are transferred to the COSYNA server when the vessel has a stable Internet connection.

5.5　Underwater-Node System

While cabled underwater observatory technology has been developed for deep sea research applications over the last decades, cabled underwater observatories for shallow water were only recently initiated due to the predicted dramatic effects of climate change especially in the world's coastal re-

gions. They are needed as core research infrastructures when either a continuous high-frequency or real-time monitoring of hydrographical or biological data is required or when scientific instrumentation requires more power than batteries can provide. Cabled underwater observatories enable new research approaches in marine science by providing long-term time series. Similar to atmospheric or terrestrial research, they are suitable to form the backbone of international coastal and climate change research.

The harsh environments of shallow waters with extreme wave impact, storms, sea ice, strong currents, as well as biofouling and the direct impact of fishing vessels require the development of very robust cabled systems. COSYNA started with this development in 2010, with the goal of observing multidisciplinary processes in the harsh environmental conditions in the North Sea and in the Arctic areas – in particular during storms and in winter when access with vessels is difficult or impossible.

The COSYNA Underwater-Node System is designed for water depth between 10 m (in high-energy environments like the North Sea) to a maximum of 300 m. It comprises a land based power unit and server providing 1000 VDC, a GBit-network connection, and virtual computer technology for up to 20 different users. This land-based control system is connected to the underwater node unit via a fibre-optic and power hybrid cable that can be up to 10 km long (Fig. 11).

The underwater unit is built as a basic lander system. Up to 10 underwater plugs provide power and network connection. The underwater unit can be outfitted with an uninterrupted low-power battery supply for 6–8 h operating time to enable temporary disconnection from the high-voltage electricity. From this central underwater node unit (Fig. 11 4), sensors or sensor units with a power consumption of up to 200 W (Fig. 11 5–6) can be connected via a cable up to 70 m long. Communication and data transfer with the attached sensors or sensor units are realised via TCP/IP. Completely separated ports allow scientists to directly communicate with the instruments independently of other users. From the primary node system, an uplink power and network connection allows the serial connection of a secondary and tertiary underwater node unit (Fig. 11 7) to reach a maximal range of 30 km from the land-based support unit.

Since 2012, COSYNA has operated two Underwater-Node Systems. One node system with 10 separated ports is located off the island of Helgoland at 59°11′, N/8°52.79 E at 10 m water depth close to the "Helgoland Roads" long-term time-series station and the MarGate AWI underwater experimental area (Wehkamp and Fischer, 2012, 2013a, b). It is operated as a permanent monitoring facility for the main hydrographical parameters in the southern North Sea (temperature, conductivity, O_2, pH, turbidity, currents), as a docking and support system for complex sensor systems with high power and data transfer demands, such as stereo-optical cameras (Wehkamp and Fischer, 2014), and as a test facility for the development and operation of the Underwater-Node Systems in the shallow environment of the North Sea. Since 2012, the Helgoland node system has endured two severe storms with wind speeds of up to 12 Bft (190 km h^{-1}), providing evidence that the operation of cabled observatories is possible under extreme conditions.

The southern North Sea is well known as a high-energy environment with wind speeds above 10 m s^{-1} (>6 Bft) during considerable phases of the year. Research cruises with intense sampling programmes are therefore often problematic and cabled observatories provide an invaluable extension for continuous and long-term monitoring programmes. They may therefore help fill a significant gap in our understanding of ecosystem behaviour in coastal environments beyond 6–8 Bft.

The second continuously operated COSYNA underwater observatory has been deployed since 2012 off Svalbard at 78°92′ N, 11°9′ E. It is located at the western coast of Spitsbergen close to the international research village of NyÅlesund. It comprises a FerryBox system and a COSYNA Underwater-Node System at the "Old Pier" (Fig. 3) close to the research village of NyÅlesund. It provides a continuous year-round monitoring system as well as an access point for international project partners. Since 2015, the COSYNA underwater observatory has been part of EU project Jerico-Next, the long-term research strategy of the NyÅlesund research council, and the Kongsfjord Flagship Program.

Also, the Svalbard observatory is operated as a permanent monitoring facility for the main hydrographical parameters in the fjord system (temperature, conductivity, O_2, pH, turbidity, currents) and as a docking and support system for complex sensor systems. It is fully remotely controlled and all sensors and sensor units can be accessed via the Internet from Germany. The Svalbard observatory is equipped with four access points and is specifically designed for national and international cooperation in the Kongsfjorden ecosystem. A main feature of the Svalbard observatory is a vertical profiling sensor unit, which allows one to remotely position attached sensors at a specific depth on a daily or even hourly basis. Thus, the entire water column can be sampled year-round, even under sea ice.

With the remotely controlled sensor set-up of the COSYNA Underwater-Node System, it was possible for the first time to gain data with a temporal resolution of up to 1 Hz with both CTD and ADCP sensors, and with highly complex sensors like a stereo-optical camera system that is able to measure abundance, species composition, and length frequency distributions of macroscopic organisms (Wehkamp and Fischer, 2014). No data set of this kind has previously been available from any Arctic ecosystem worldwide, thus providing unique insights into the dynamics of a polar ecosystem with a very high temporal and spatial resolution (Fig. 12).

5.6 Landers

Under the COSYNA framework, different autonomous seafloor observatories (landers) have been developed and are applied in various past and ongoing research programmes. These landers bridge the observational gap between long-term monitoring stations, remote sensing applications, and ship-based field campaigns. They are mobile, and can be used to spatially interpolate between monitoring stations and provide data with very high temporal resolution (Kwoll et al., 2013, 2014; Oehler et al., 2015a; Ahmerkamp et al., 2017). Lander operations aim at measuring various processes close to the seafloor or in the sediment and are designed to have minimal impact on the environment and quantities that are measured. The landers can be either operated autonomously for days or weeks at a time, or may be connected to the COSYNA Underwater-Node System that provides power and data connection for the landers.

The landers developed and used in COSYNA are (i) the SedObs (Sediment Dynamics Observatory) lander measuring seafloor dynamics, (ii) the NuSObs (Nutrient and Suspension Observatory) lander, and (iii) the FLUXSO (Fluxes on Sand Observatory) lander.

5.6.1 Lander SedObs

The Sediment Dynamics Observatory (SedObs) lander is used to investigate seafloor dynamics and to improve the fundamental knowledge of multi-phase flows and the interaction of physical and biological processes. The seafloor and lower water column are characterised by morphodynamic processes acting on a large range of spatial and temporal scales. Observations with SedObs focus on short-term dynamics from turbulence to tides or storm events. Particular focus is given to the interaction of water motion by currents and waves as well as the transport of sediments and other substances with the sea bed evolution under the influence of (micro-)biological stabilising and destabilising organisms (Ahmerkamp et al., 2015).

SedObs consists of a 2×2 m steel frame with a platform providing space for battery power supply and the installation of sensors (Fig. 13). The platform rests on four adjustable and inclined legs. Foot plates provide a stable stand, prohibit subsidence, and reduce scouring around the legs. Sensors can be attached to the legs for measurements close to the sea bed. The lander is deployed with a launching frame from a research vessel orienting it in the direction of the main currents. After release of the lander, the frame is recovered in order to minimise flow disturbances. For recovery, a floating buoy with a recovery line is released acoustically. Typical deployment times exceed 25 h to account for the diurnal inequality in tidal variations. Deployments can be extended to longer periods of several weeks depending on measuring frequency, battery, and storage limitations, and the increasing risk of damage by trawlers.

Flow velocities and turbulence above and below the lander are measured with two acoustic Doppler current profilers. The upward-looking ADCP also captures the directional surface wave spectrum. Two acoustic Doppler velocimeters record velocity at two levels with high frequency. Turbulence characteristics are computed from highly frequent velocity fluctuations (Amirshahi et al., 2016).

The small-scale bathymetry below the lander is measured with a 3-D acoustic ripple profiler (Bell and Thorne, 1997). The sensor is installed about 1.8 m above the seafloor covering a circular area of 6.2 m diameter. Sediment transport characteristics are measured with Sequoia Lisst 100X instruments providing in situ particle size distributions of suspended sediments. Characteristics of suspended matter con-

centration are provided by optical backscatter sensors and the backscattered signal strengths of the hydroacoustic instruments. Additional parameters comprise the COSYNA standard observables. Observations are complemented by investigations of benthic species as well as sedimentological and granulometric analysis (laser diffraction) of the sediments sampled with grab samplers, box corers, and multi-corer equipment.

SedObs supports several applied and fundamental research projects, such as KÜNO NOAH (North Sea Observation and Assessment of Habitats). Until 2015, 11 ship surveys were carried out, and field data were collected and analysed at different reference sites in the German Bight with sedimentological and morphological characteristics that are representative of large areas of the German EEZ in the North Sea. A combination with other COSYNA seafloor observatories has produced consistent and extensive data sets on various physical and (micro-)biological properties of the domains (Krämer and Winter, 2016). Data are published at http://www.noah-project.de.

During some parts of the tidal cycle a periodic stratification of the water column has been observed in shallow areas of the German Bight forming distinct layers that move independently with a decoupled tidal ellipticity (Krämer and Winter, 2016; Kwoll et al., 2013, 2014; Ahmerkap et al., 2017). The difference in sea bed dynamics between fair weather conditions and storms is also investigated in the research area "Seafloor Dynamics" of the Deutsche Forschungsgemeinschaft (DFG, German Research Foundation) Research Center/Cluster of Excellence "The Ocean in the Earth System".

5.6.2 Lander NuSObs

The NuSObs (Nutrient and Suspension Observatory) benthic lander system was designed to quantify the exchange of nutrients and oxygen across the sediment–water interface and to sample surface sediments in situ (Oehler et al, 2015a, 2015b). The aim was to study the remineralisation of organic matter, the reflux of nutrients into the bottom water, the dissolution of biogenic silica (e.g. diatoms), and transport processes across the sediment–water transition zone, such as biologically mediated transport (e.g. bioirrigation) or wave-induced pore water advection. The target area was the North Sea. Three time-series sites were selected and revisited three to four times a year in order to identify seasonal variations.

NuSObs (Fig. 13) was equipped with two "Mississippi" type chambers (Witte and Pfannkuche, 2000). After the deployment of the lander, both chambers were moved slowly into the sediment by a motor, each enclosing a sediment area of $400 \, cm^2$ for typically 12–24 h. Each chamber was equipped with a syringe sampler (seven 50 mL glass syringes) to obtain water samples from the incubation chamber for subsequent chemical analysis. In addition, an oxygen optode and pH sensor were mounted in each chamber. The

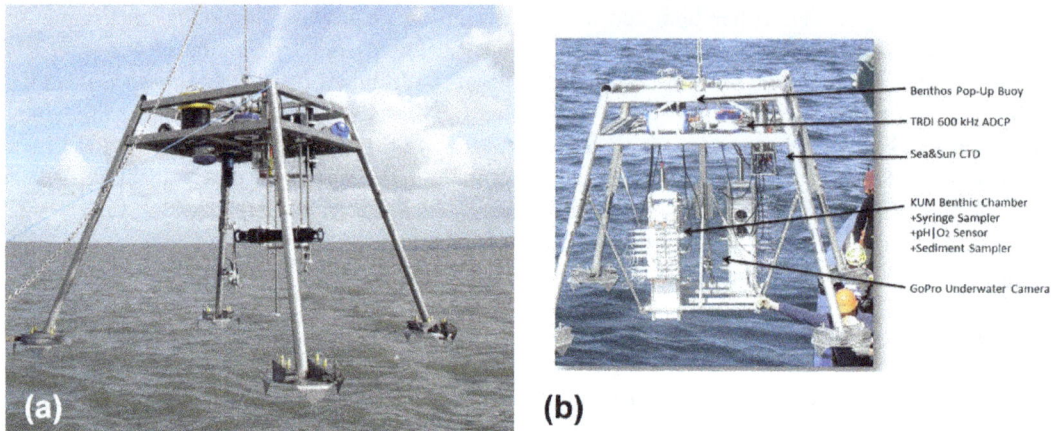

Figure 13. Deployment of landers SedObs **(a)** and NuSObs **(b)**.

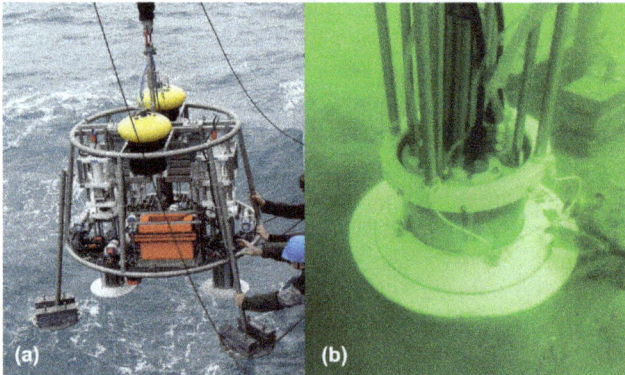

Figure 14. (a) Lander FLUXSO deployed for autonomous sampling in June 2015; **(b)** sampling chambers in mobile fine sand at 25 m depth.

syringe sampler was pre-programmed to obtain water samples from the chamber every 2–3 h, yielding time-series data of oxygen, nitrate, or silicic acid concentrations within the chambers.

5.6.3 Lander FLUXSO

The FLUXSO (Fluxes on Sands Observatory) benthic lander system was recently developed for studying in situ solute fluxes of nutrients, DIC, and oxygen in permeable consolidated sediments. The goal is to assess the importance of the seafloor as a sink or source of nutrients and benthic–pelagic coupling and to study advection-related processes in permeable shelf sediments. The lander was successfully applied on sandy sediments of the North Sea (Fig. 14; Friedrich et al., 2016; Neumann et al., 2016; Ahmerkamp et al., 2017).

The lander consists of a tripod base frame that is recovered from the seafloor using two pop-up buoys (Fig. 14). Power supply is provided by a deep-sea battery. The lander contains two wiggling chambers that are both equipped with

oxygen and CO_2 optodes, a pH sensor, and a conductivity sensor. A stirrer disk with variable speed and direction allows the simulation of advective or diffusive flow regimes in each chamber by creating rotationally symmetric pressure gradients between the center and the circumference of the enclosed sediment surface. The shape and magnitude of the pressure gradients closely resemble natural conditions. Two syringe samplers are used for tracer injection and sampling from the chambers. Outside water parameters are measured with a CTD with fluorescence and turbidity sensors, a PAR sensor, an oxygen optode and pH sensor, as well as a Doppler current sensor.

The FLUXSO lander can be deployed at the seafloor, where it autonomously measures solute fluxes between sediment and seawater using isolated sampling chambers. An innovative wiggling mechanism is used, permitting gentle and deep penetration of the chambers into consolidated sediments with minimum disturbance (Janssen et al., 2005).

5.7 Satellite oceanography

Satellite remote sensing is unique in providing a synoptic view over larger areas of the sea surface (Robinson, 2004). Standard algorithms are used widely to determine the optically dominant water constituents and the chlorophyll *a* concentration in clear oceanic waters (Carder et al., 1991; Lee et al., 1998; Gohin et al., 2002). These simple band-ratio algorithms, however, often fail in optically complex coastal waters. To gain concentrations of one coastal water constituent, other optically active substance categories have to be considered in the development of algorithms for the inversion of satellite spectral data. The correction of the atmospheric influence is more sensitive and complex as it accounts for 90 to 98 % of the radiance seen at the satellite. The algorithms for coastal waters developed by HZG and used in COSYNA are included in the ESA (European Space Agency) operational processing scheme for the sensors MERIS (MEdium Resolution Imaging Spectrometer) on ENVISAT (Doerffer

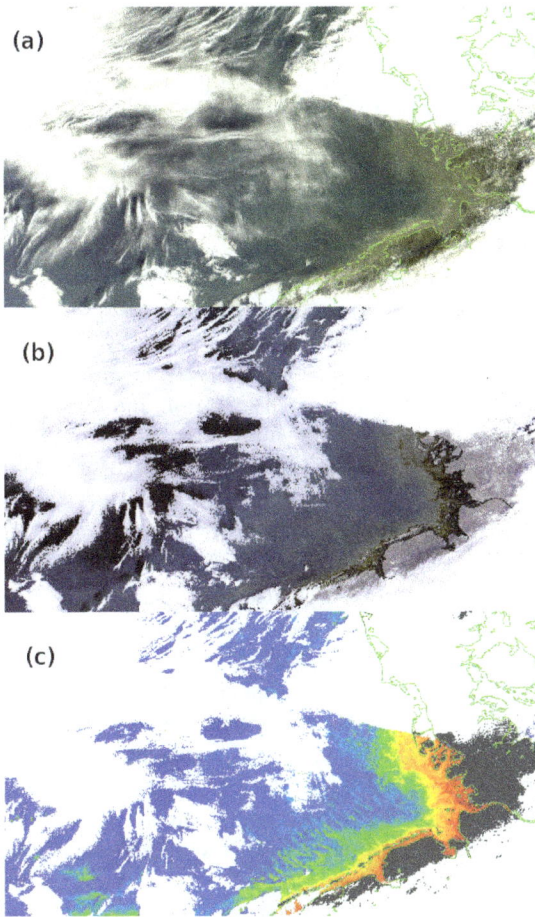

(a)

(b)

(c)

Figure 15. Satellite scene of the German Bight taken on 3 October 2012 by MERIS. **(a)** Radiance in the atmosphere; **(b)** reflectance at the bottom of the atmosphere (after atmospheric correction); **(c)** chlorophyll *a* concentration showing filaments of *phaeocystis* blooms along the West and East Frisian coasts.

Figure 16. Solar-powered GPS data logger attached to a tail of a Northern Gannet (photo: J. Dierschke).

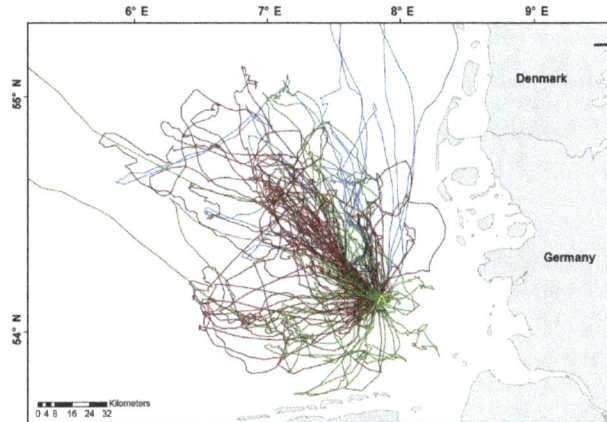

Figure 17. Foraging flights of three Northern Gannets (*Morus bassanus*) in 2015 starting from Helgoland.

and Schiller, 2007) and OLCI (Ocean and Land Colour Instrument) on Sentinel-3 providing chlorophyll *a* and total suspended matter (TSM) concentrations and the absorption by chromophoric dissolved organic matter (CDOM, "Gelbstoff").

MERIS provided COSYNA data (Fig. 15) for the North Sea until 2012, when ENVISAT failed, with the adaptation of the coastal algorithm to MODIS (on AQUA) and OLCI (Ocean and Land Colour Instrument) on Sentinel-3 providing chlorophyll *a* and total suspended matter (TSM) concentrations and the absorption of "yellow substances" (Gelbstoff) whose main part is chromophoric dissolved organic matter (CDOM).

5.8 Seabird tracking

Seabirds are top predators depending on marine resources. Their foraging behaviour may therefore indicate changes in their food resources which are often associated with vari-

ability in the marine environment (Furness and Camphuysen, 1997). In COSYNA, the Northern Gannet (*Morus bassanus*) has been selected as the target seabird species due to its size and large foraging range (Fig. 16; Garthe et al., 2017). Northern Gannets are widely distributed in the North Atlantic and breed in large colonies. Individual Northern Gannets were equipped with modern, lightweight GPS data loggers to track their flight patterns and foraging behaviour. In particular, information is collected on position, flight speed, altitude, and partly also on dive depth and water temperature. A strong feature of most modern data loggers is that they are powered by solar cells, thus enabling long-term tracking for several weeks, months, or even years. Furthermore, an increasing number of devices provide data transfer via UHF, satellite, and mobile phone networks (Wilson and Vandenabeele, 2012; Kays et al., 2015). A combination of the data collected by seabirds with environmental parameters from other COSYNA observations, such as salinity, sea surface temperature, or chlorophyll, facilitates the understanding of the seabirds' foraging behaviour, their likely food intake, and habitat choice (Fig. 17). On the other hand, the recorded spatial and temporal flight patterns and environmental parameters can help to characterise the environmental status of the North Sea.

5.9   In situ mapping of the COSYNA observation area

The regular operational observations in COSYNA primarily detect variables at the sea surface (currents observed with HF radar; chlorophyll a concentration, TSM, and SPMC observed with satellite remote sensing), at constant depths at fixed high-resolution time-series stations (Poles, FINO3 platform, MARNET stations), or at constant depth along regular ship routes (FerryBox transects). In order to observe the vertical distribution of key variables and their temporal development, these observations were complemented by extended in situ mapping of the North Sea during several research cruises and glider surveys. In situ observations taken with Wadden Sea poles, FINO3 platform, MARNET stations and Ferry-Box are also used in modelling (Stanev et al., 2016).

In particular, the surveys aimed at investigating the representativeness of single-point time-series observations, delivering larger-scale validation data for the COSYNA remote sensing systems and numerical models, testing the functioning of new sensors for permanent missions under North Sea conditions, and relating concentrations and characteristics of living and non-living water constituents to optical surrogate variables.

The regular COSYNA mapping grid covers estuarine, Wadden Sea, and open shelf seawater (Fig. 18). It consists of four east–west and four south–north cross-shore transects and touches the fixed COSYNA and MARNET stations covering the whole German EEZ. The land side is limited by a water depth of 10 m and its most seaward reach by the borders of the German EEZ.

From 2009 to 2013, up to four cruises per year were carried out with RV *Heincke*. The cruises took place between March and October to take seasonal variations into consideration. At a ship's speed of 6 to 8 knots, the grid was completed in less than a week. During this time, the water masses did not move substantially, as confirmed by model studies using Lagrangian tracers. The observations thus provide a good approximation of the spatial distribution of the observed variables.

Along the grid lines, an undulating towed Scanfish Mark II™ by EIVA was operated, yielding vertical profiles of oceanographic and bulk biogeochemical parameters at a vertical resolution of several centimetres and a horizontal resolution of 150 m at mid-water depth. A FerryBox system was used to analyse water continuously taken at a depth of 4 m with respect to the standard oceanographic parameters temperature, salinity, pH, chlorophyll fluorescence, turbidity, CDOM, nutrients, dissolved oxygen, and pCO_2. During the cruises, the FerryBox also served as a platform for test-

ing newly developed sensors. This includes a flow-through PSICAM (Point-Source Integrating Cavity Absorption Meter) for high-frequency hyperspectral absorption coefficient measurements (Wollschläger et al., 2013, 2014), a sequential injection analysis (SIA) approach for phosphate measurement (Frank and Schroeder, 2007), as well as high-precision spectrophotometric methods for the determination of pH and total alkalinity (Aßmann et al., 2011; Aßmann, 2012). Vertical current profiles were recorded with an ADCP. During two cruises, gliders were operated in parallel, enhancing the spatial observation density. At the cruise track crossing points, additional vertical profiles were taken and complemented with Secchi depth determination, light transmission, and scattering spectra taken from water samples.

As an example, the spatial distributions of σ_T (potential density – 1000 kg m^{-3}) and chlorophyll a fluorescence are shown for the cruise at the end of July 2010 (Fig. 18). Vertical density gradients at the 5 m thick pycnocline of up to 0.3 kg m^{-4} indicate a strong stratification typical of the summer months. In the outer reaches of the observation area, two pycnoclines can be discerned. In the presence of stratification, chlorophyll a shows a typical deep water maximum at the upper pycnocline. The sudden increase in oxygen saturation directly above this maximum can be attributed to photosynthesising phytoplankton. By coupling the observed vertical distribution of potential density and SPMC with a modelled turbulence parameter field, the spatial distribution of settling velocities in the COSYNA observation area was derived (März et al., 2016). Characteristic scales for the coupling of physical submesoscale and mesoscale processes and the distribution of chlorophyll a were identified by North et al. (2016) by applying wavelet analyses to Scanfish data.

6   Sensor and instrument development

In COSYNA, well-proven commercially available sensors and sensor systems are used. However, to automatically measure the main parameters that control and influence the North Sea and Arctic ecosystem, several novel, automated, and reliable sensors had to be developed and tested by the COSYNA partners. These are, in particular, sensors and samplers for biogeochemical and optical parameters as well as micropollutants. An overview is given in the following. For most of these sensors, the FerryBox was used as a test platform because it is protected from the environment, it provides a continuous seawater supply, and it offers high-frequency data acquisition and real-time data transmission.

6.1   pH sensor

pH can be used to estimate a system's state in terms of phytoplankton and primary production in regions of high biological activity, one of four parameters characterising the oceanic inorganic carbon system, and an indicator of the increasing

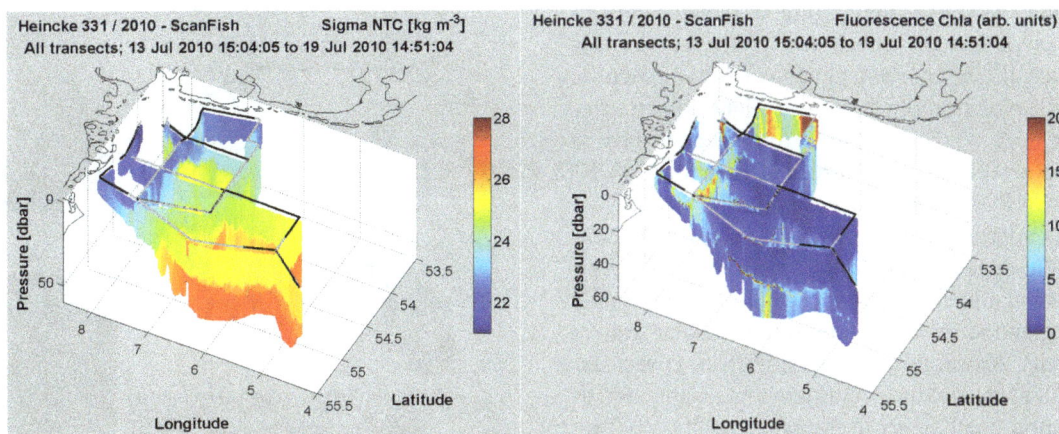

Figure 18. Spatial distribution of σ_T and chlorophyll a observed during RV *Heincke* cruise HE331 in July 2010.

acidification of seawater. In order to quantify the components of the carbon cycle in the context of climate change, a precise characterisation of the carbonate system is required.

In COSYNA, commercially available pH glass electrodes are routinely used. They are very sensitive to bio-fouling as bacterial biofilms on the electrodes changes the pH thus requiring cleaning and re-calibrating intervals of 7–10 days in summer. Although an accuracy of ±0.05–0.1 pH units can be achieved in FerryBox systems for several weeks due to their regular automatic cleaning procedures, a higher precision of < 0.01 pH units is necessary to detect the acidification process in coastal waters with a pH decrease of about 0.0019 pH units per year (Dore et al., 2009; Feely et al., 2009).

In COSYNA, a more precise sensor based on a spectrometric approach was developed (Aßmann, 2012) that detects the colour of a suitable indicator dye in a miniaturised flow-through system. A precision of ±0.0007 pH units with an offset of $+0.0081$ pH units to a certified standard buffer was achieved for several weeks. It is, however, not yet suitable for low-energy applications.

6.2 Alkalinity sensor

CO_2 flux estimates for the coastal ocean are subject to large uncertainties (Borges, 2005; Chen and Borges, 2009) due to strong seasonal variability. For a description of the carbonate system at least two of the following parameters have to be measured: pH, partial pressure of CO_2 (pCO_2), total alkalinity (AT), and total dissolved inorganic carbon (CT). Because a combination of pH and pCO_2 only yields a precision of about 1 %, a sensor for the additional measurement of alkalinity was developed that will allow to document the fast changing carbonate chemistry in the North Sea (Aßmann, 2012).

The approach for the photometric pH determination (Sect. 6.1) was modified for alkalinity, with the advantage that the same equipment can be used for both parameters. The chemical titration can be accomplished by using an

"open-cell technique" applying a simple seawater model as a calculation tool. The titration occurs at pH < 4.5, leading to a removal of all carbonate species by outgassing of CO_2. The precision is ±1.1 mol kg^{-1} with an accuracy of ±8 mol kg^{-1}. In a more complex "closed-cell technique" a broader pH range is used and no CO_2 escapes, yielding an accuracy of ±0.8 mol kg^{-1} with a precision of ±4.4 mol kg^{-1}.

6.3 Nutrient sensor

COSYNA uses commercially available nutrients analysers on FerryBoxes for long-term investigations of the nutrients ammonia, nitrite, nitrate, phosphate, and silicate, which are important parameters regarding eutrophication. However, as small-scale processes often require faster sensor response times, a flow-through system was developed for the fast determination of ammonia and phosphate based on sequential injection analysis (SIA), causing a chemical reaction of both species with a reagent that can be detected by fluorescence (Frank et al., 2006). The detection limits are 0.3 µmol L^{-1} for phosphate and 1 µmol L^{-1} for ammonia; 180 samples can be processed per hour and analysed.

This reliable analyser is especially useful for high-resolution surface mapping of ammonia and phosphate in coastal areas and for long-term monitoring due to the low amount of reagents used in this system (Frank and Schroeder, 2007). Nitrite and nitrate underway measurements were performed using ultraviolet absorption techniques with parallel temperature and salinity corrections, thus enabling application of this approach in coastal and estuarine waters (Zielinski et al., 2011; Frank et al., 2014).

6.4 Flow-through spectral light absorption measurements

One of the most important biogeochemical parameters for the assessment of the environmental status of the North Sea is the phytoplankton concentration. The standard method that is routinely used in COSYNA is the continuous in situ measurement of chlorophyll *a* fluorescence as a proxy for biomass estimation. Since fluorescence depends on factors such as plankton species, plankton physiology, or light climate, frequent sampling with subsequent lab analysis is necessary to reduce the large errors of up to 1 order of magnitude (UNESCO, 1980; SCOR Working Group, 1988).

Better suited to determine estimates of phytoplankton concentrations is the spectral absorption coefficient. To overcome the disturbing effects of the light scattering of inorganic and organic suspended matter, a flow-through Point-Source Integrating Cavity Absorption Meter (ft-PSICAM) was developed in COSYNA yielding continuous measurement of spectral absorption coefficients in the range of 400–710 nm with high temporal and spatial resolution. Additional useful information on CDOM/gelbstoff, algal pigments, and suspended matter can be obtained as well.

By using an integrating sphere, photons cannot get lost and the optical path length is increased, allowing the measurement of very clear waters. This PSICAM principle (Kirk, 1997; Lerebourg et al., 2002; Röttgers et al., 2005) was modified into a flow-through unit that can be used unattended on FerryBoxes or other platforms (Wollschläger et al., 2013, 2014). To reduce the contamination of the integrating sphere, it has to be cleaned automatically. The ft-PSICAM delivers data with a high temporal and spatial resolution.

6.5   Molecular observatory

Information on marine photosynthetic biomass distribution and biogeography with adequate temporal and spatial resolution is needed to better understand consequences of environmental change in marine ecosystems. Since COSYNA methods can only automatically measure proxy parameters for biomass, such as chlorophyll *a*, a method for the automatic determination of phytoplankton taxonomic composition is required. Molecular analyses, e.g. next-generation sequencing (NGS) or molecular sensors, are very well suited to provide comprehensive information on marine microbial or protist composition.

In COSYNA, the remotely controlled Automated Filtration System (AUTOFIM) for automated collection of samples for molecular analyses was developed. Resulting samples can either be preserved for later laboratory analyses or directly subjected to molecular surveillance of key species aboard the ship or at a monitoring side via quantitative polymerase chain reaction or an automated biosensor system (Metfies et al., 2017). The latter is based on an automated pre-treatment of the samples with an ultrasound sample preparation unit that was developed in COSYNA alongside AUTOFIM. The sampling system can either be deployed on a fixed monitoring platform or aboard a ship for near-real-time information on abundance and distribution of phytoplankton key species. Currently, two AUTOFIM systems

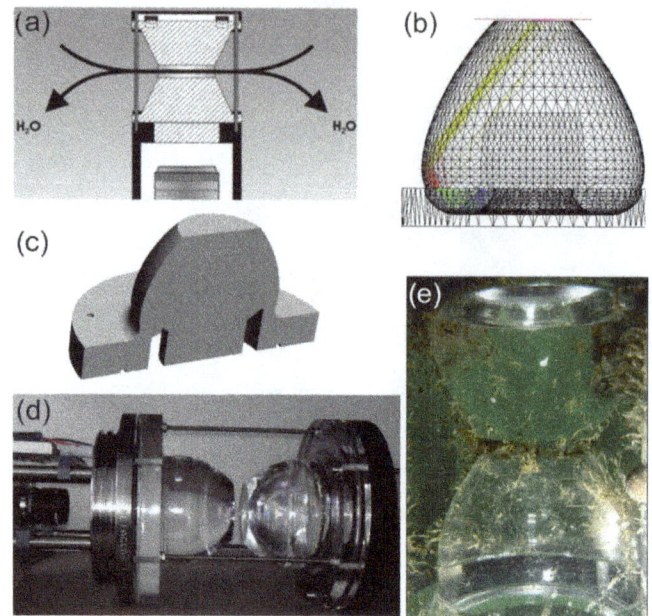

Figure 19. Design of the LOKI imaging head for moored operation. (**a**) Schematic overview. (**b**) Ray-tracing design model to investigate the best shape to increase efficiency. (**c**) Cross section of a 3-D model. The LEDs of the flash unit are positioned in the notch. (**d**) Imaging head with two optical cones: the right cone carries the circular flash unit, the left one the visual path of the camera's field of view. The camera is mounted on the left. (**e**) The system requires periodical cleaning in the field. The image shows bio-fouling after 5 weeks of operation in the North Sea.

are operating on Helgoland and aboard RV *Polarstern* in order to collect samples for molecular analyses.

6.6   Zooplankton sampling

In addition to phytoplankton distributions, the heterogeneities of the spatio-temporal zooplankton community assemblage are a key environmental parameter. Based on the established Lightframe On-sight Keyspecies Investigation technique (LOKI; Schulz et al., 2010), an imaging head for autonomous, moored operations was developed and attached to the COSYNA Underwater-Node System. A 360°-open flow chamber ensures optimal flow. The data are transferred to the shore in near real time.

LOKI combines several features bringing it close to the feasible borders set by the laws of optics (Schulz, 2013). These are an integrated flash unit providing sufficient light for short shutter times of $<30\,\mu s$ to avoid motion blurring, very high resolution of $<15\,\mu m\,pixel^{-1}$ to resolve fine taxonomical characteristics, and a depth of field of several millimetres. This was achieved by using two optical cones (Fig. 19e). The first one is attached to the camera housing and allows adjustment of the focal plane at a certain distance from the camera, while the tapering enhances water exchange in the flow chamber. The opposite cone houses a

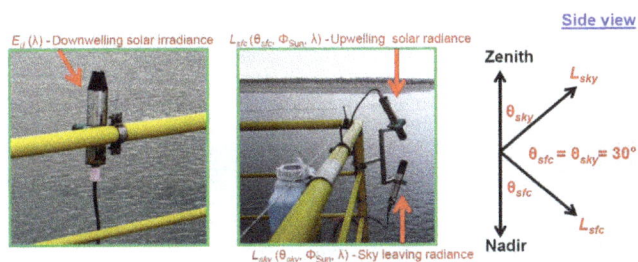

Figure 20. Set-up of RAMSES radiometers at the Spiekeroog Wadden Sea measurement pole.

high-power LED flash unit. The LEDs are arranged circular and off-axis to provide indirect and homogenous illumination resulting in high-resolution images of minute specimens and a large depth of field. The operation time is, however, limited by bio-fouling (Fig. 19).

6.7 Active and passive sampling tools

To determine the potential effects of micropollutants on the marine environment and biota, a set of integrative active and passive samplers has been developed. Suitable instruments for unattended use under the harsh conditions do not exist and pure concentration data of micropollutants are often not very meaningful.

For passive sampling, a Chemcatcher Metal (Petersen et al., 2015b) as well as DGTs have been used, while blue mussels (*mytilus edulis* sp.) have been applied as active sampling devices. After a deployment period of several weeks, the samples are analysed with conventional analytical laboratory methods. In contrast to spot sampling, passive samplers allow to measure the more representative time weighted average water concentrations (TWA). Passive sampling data also provide information about the biologically available trace element fraction of the analysed water body (Booij et al., 2016). Besides the measurement of contaminant body burdens, the application of mussels as active sampling devices allows also the analysis of potential biological effects induced by the contaminants present in the surrounding water. This is done with an analysis of the up and down regulation of specific proteins, whose expressions are related with certain detoxification mechanisms.

In COSYNA, two systems (Helmholz et al., 2015) have been developed featuring a modular design for the installation on different instrumental platforms, such as different passive sampling devices, SPM traps, and cages for biota deployment. An elevator enables the manual deployment and recovery of the experimental device at a fixed position approximately 3 m above the seafloor. The use of titanium reduces corrosion. The systems are deployed next to the FerryBox station in Cuxhaven at the mouth of the Elbe River and at the MARGate underwater testing site near Helgoland at a water depth of approximately 10 m.

A continuous flow box has been developed to overcome bio-fouling problems as well as to minimise effects of changing currents on the sampling rate, as it allows integration into FerryBox systems (Petersen et al., 2015b) for passive sampling, e.g. during ship cruises to obtain TWA contaminant data. Normally, the pumped water intake system is installed at the bow of the ship's hull several metres below the sea level, thus ensuring that the sampled water body is continuously exchanged due to the movement of the ship and the water is not contaminated by the metal construction of the ship. Alternatively, a metal-free pump system can be deployed on a crane several metres away from the ship's hull.

For the calculation of uptake rates, a calibration was carried out for Ni, Cu, Zn, Cd, Pb, Sc, Ti, Mn, Co, Ga, Sr, Y, Ba, U, and rare earth elements under different environmental conditions (Petersen et al., 2015a). Up to now, these calibrations have not been available for most elements of environmental concern besides Cu, Cd, Pb, Ni, and Zn. With these developments, a real multi-element analysis using passive sampling was possible for the first time.

6.8 Radiometric ocean colour measurements

The colour of the ocean is related to its optically active constituents and can be assessed with radiometric measurements within the water column and from above the water surface (Moore et al., 2009; Garaba and Zielinski, 2013a). The latter includes satellite and airborne platforms as well as measurement poles or vessels (Zielinski et al., 2009).

As part of COSYNA, the applicability of different low-altitude hyperspectral radiometer installations was investigated. Measurement poles at Spiekeroog (Fig. 20) and in Alfacs Bay (Ebro Delta, Mediterranean) were outfitted with TriOS RAMSES hyperspectral radiometers. Underway observations were performed from research vessels Otzum and Heincke, the latter with a permanent installation of a twin remote sensing reflectance set-up to account for different sun angles along the track.

One of the major challenges is the corruption of data from sun glint and white caps. It is therefore key for any operational observing system that robust automated quality assurance methods are applied, which is achieved by parallel image acquisition and analyses (Garaba et al., 2012) or from spectral feature utilisation (J. A. Busch et al, 2013; Garaba and Zielinski, 2013b). An ensemble of sun glint detection methods improves the flagging performance of the data quality algorithm (Garaba et al., 2015a). Remote sensing spectra of good quality are used to derive in-water constituents like chlorophyll, coloured dissolved organic matter, and suspended particulate matter along cruise tracks in the North West European Shelf Sea (Garaba et al., 2014b) and Arctic (Garaba et al., 2013a), and at a time-series station in the Wadden Sea (Garaba et al., 2014a). A very recent application is the calculation of the Forel–Ule Colour Index from reflectance spectra, which opens up the possibil-

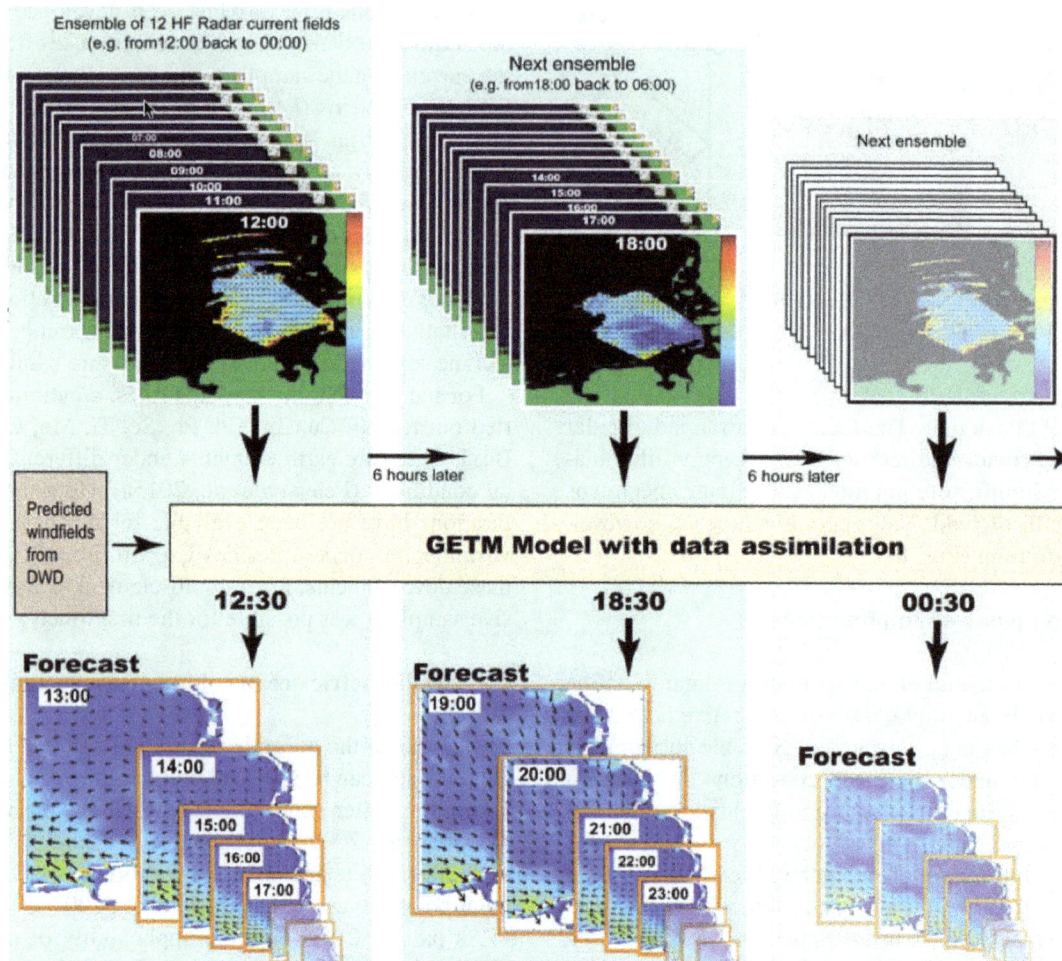

Figure 21. The functioning of data assimilation and forecasting in the pre-operational COSYNA system. HF radar system covering the German Bight. Radial current components are sent to the HZG data server, where current vectors are calculated and presented on the COSYNA data portal (Stanev et al., 2015).

ity of linking modern observations to long-term records and to involve citizens with smartphones in ocean colour measurements (J. A. Busch et al., 2016; Garaba et al., 2015b; http://www.eyeonwater.org).

6.9   Temperature sensor for sediments

To measure the exchange of heat and particulate matter between the German Bight and the Wadden Sea, the heat fluxes between the tidal flats and the water body have to be determined (Onken et al., 2007). As the stratification in the sediment is directly related to the heat content, the latter can easily be calculated and the heat flux between seabed and atmosphere or overlying water derived.

For these investigations, a vertical temperature sediment profiler was developed. The self-contained probe measures the temperature of intertidal sediments at depths of 0.02, 0.1, 0.2, 0.3, and 0.4 m. Two electrodes located about 2 cm above the sediment indicate whether the tidal flats are wet or dry.

The probe was deployed close to the Hörnum measurement pole (Sect. 5.1.1) where seawater temperatures were measured (Onken et al., 2010).

7   Modelling and data assimilation

Observations – and even automated observation networks – are limited by the fact that we cannot measure everywhere and at all times, which is in particular a challenge given the coastal ocean's strong variability. One of the distinguishing features of COSYNA lies therefore in the integration of observational data into models in order to close the spatial and temporal gaps of the observations and to calculate energy or matter fluxes (Stanev at al., 2016). Model studies are also essential for identifying regions with high sensitivity or variability in certain quantities that warrant the deployment of measurement devices. On the other hand, state-of-the-art numerical models of coastal dynamics require monitoring data

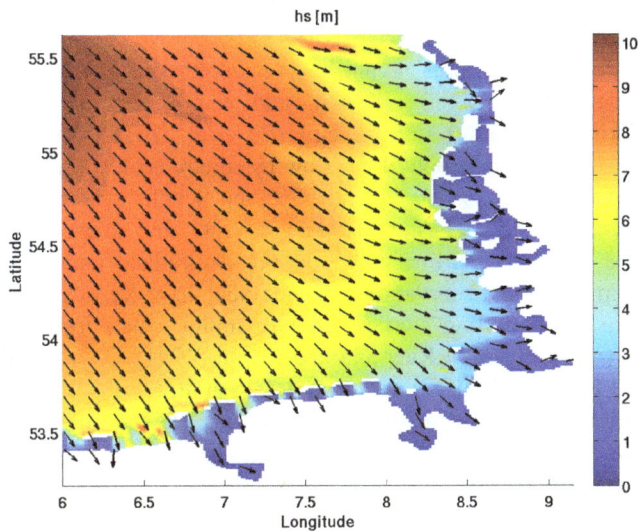

Figure 22. Significant wave height calculated for the German Bight on 1 November 2006 with the WAM wave model used in COSYNA.

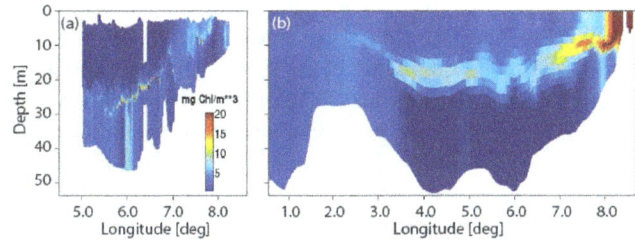

Figure 23. Chlorophyll transects around 55°15′ latitude in the German Bight (**a**) observed with a Scanfish in July 2010 (Sect. 5.9) and (**b**) as result of a coupled GETM and an adaptive ecosystem model showing a 1-week mean (Wirtz and Kerimoglu, 2017).

to reasonably manage large model uncertainties. The observations are used to bring models closer to the "real" state of the ocean, either by verifying model output or by assimilating them into models. These data sets should be representative and coherent. In order to continuously provide accurate pre-operational coastal ocean state estimates and forecasts, COSYNA integrates near-real-time measurements into numerical models in a pre-operational way that is meant to improve both historical model runs and forecasts.

In this context, COSYNA has explored different techniques to assimilate data into models. Satisfactory assimilation results were achieved when 2-D data fields were available, such as derived from HF radar or satellite observations (Stanev et al., 2015) providing a 12 h forecast. The assimilation of data from single locations or sections usually only influences the immediate vicinity of the locations where the observations were made and has limited value for greater spatial extensions (Grayek et al., 2011; Stanev et al., 2011). Data assimilation based on physical values is generally more easily achieved than with biogeochemical quantities. The successful assimilation products of COSYNA encompass surface currents, significant wave height, period and wave direction, as well as temperature.

For the assimilation of current observations, a nested 3-D hydrodynamic model is used. In situ current time series are measured with stationary ADCPs at the FINO-1 and FINO-3 research platforms. Remote sensing of surface currents is carried out with three HF radar systems installed in the German Bight (Sect. 5.3). For technical details of data processing and accuracy see Stanev et al. (2015). The flow of observational data including observing nodes, data management system, and data assimilation capabilities is streamlined toward

meeting the needs for high-quality operational data products in the German Bight (Fig. 21).

Although there are hundreds of HF radar systems installed worldwide, their operational use in numerical models, in particular at sub-tidal periods, is not well established. The assimilation of HF radar data is a challenge due to irregular data gaps in time and space, inhomogeneous observational errors, as well as inconsistencies between boundary forcing and observations. Furthermore, due to the high sampling frequency of typically several times per hour, it is difficult for the model to reach equilibrium between two time steps. Therefore, the Spatio-Temporal Optimal Interpolation (STOI) filter has been developed by Stanev et al. (2015). It enables a blending of model simulations from a free run and radar observations by extending the classical Kalman analysis method to time periods of at least one tidal cycle by using the Kalman analysis equation.

The modelling suite is based on the General Estuarine Transport Model (GETM; Burchard and Bolding, 2002) 3-D primitive equation. It is used in two configurations: a North Sea–Baltic Sea model of 5.6 km resolution and a one-way nested German Bight model with a horizontal resolution of about 1 km (Stanev et al., 2011). Both models use terrain-following equidistant vertical coordinates (s coordinates) with 21 non-intersecting layers.

The validation of the model and the physical interpretation of the results showed the good skills of STOI not only in the area covered by HF radar observations, but also outside it, revealing its upscaling capabilities (Stanev et al., 2015). By using HF radar data in the STOI system, homogeneous and continuous 2-D current fields were thus generated over the entire model area. The quality is superior to a free model run, demonstrating that data assimilation can enhance coastal ocean prediction capabilities by making use of observations and modelling, which is an essential aspect of an operational system. The combination of HF radar data and numerical model results can therefore also provide a deeper insight into the German Bight dynamics and provide useful indications of where further model developments (improvements) are needed.

COSYNA also provides a pre-operational wave forecast based on the WAM Cycle 4 wave model (release WAM 4.5.3; Komen et al., 1994; Günther et al., 1992). The computational system consists of a regional WAM for the North Sea with a spatial resolution of $\sim 5\,\text{km}$ and a nested grid with a spatial resolution of 900 m for the German Bight. Wind fields and boundary information are provided by the German Weather Service (DWD) derived from their regional wave model – EWAM. A number of wave parameters such as significant wave height, period, and total wave direction are calculated (Staneva et al., 2015). It has continuously provided hindcasts and forecasts since December 2009. Daily at 00:00 and 12:00 UTC, a 24 h regional forecast is issued for the North Sea and a local one for the German Bight. As an example, a typical wave height distribution with low values close to the coasts and higher values offshore is shown for the German Bight for 1 November 2006 (Fig. 22).

A combination of biogeochemical observational data and numerical models in COSYNA has been instrumental for a better understanding of material dynamics including steep cross-shore gradients ranging from shallow near-shore waters to the continental shelf, strong lateral gradients and mesoscale patchiness, as well as singular events, such as storms or ice winters. These processes are intimately linked to the functioning of coastal ecosystems but also affect efforts to maintain shipping pathways and coastal defense, as well as water quality.

A model- and data-based analysis (März et al., 2016) highlights a remarkable cross-shore separation of the coastal ocean with a maximum settling velocity of suspended material in the transition zone between the shallow Wadden Sea and the continental shelf, which modifies the traditional concept of continuous gradients. This acceleration of vertical deposition fluxes is likely due to enhanced particle aggregation induced by organic substances, which in turn are released by planktonic microorganisms (Su et al., 2015; Hofmeister et al., 2017). Enhanced deposition in the coastal transition zone is accounting for an effective trapping of lithogenic material within near-shore waters, while it may act as a barrier for offshore organic particles. Even higher variability at scales below the cross-shore gradients is evident in COSYNA lander observations (Sect. 5.6) of total benthic oxygen consumption.

Using an ecosystem model that includes turbidity fields, estimated from Scanfish observations (Sect. 5.9), and that accounts for the acclimation capacity of phytoplankton, spatial variability in chlorophyll a can be reproduced to a high degree (Fig. 23; Wirtz and Kerimoglu, 2017). Previous modelling attempts such as of van Leeuwen et al. (2013) or Schrum et al. (2006) do not capture the extreme vertical squeezing of chlorophyll a within thin layers, which may affect model-derived estimates of total primary production. Our new model results also reveal how reconstructed pelagic patterns decouple from benthic respiration patterns. Vertical deposition of freshly produced material greatly varies within the coastal ocean. In a few, mostly deeper regions, deposition prevails over resuspension, leading to depositional hotspots (Wirtz and Kerimoglu, 2017).

Vertical structures in nutrient concentrations are key to understanding whether, when, and where phytoplankton blooms form after storm events (Su et al., 2015). Vertical structures in chlorophyll a below the metre scale (thin layers) as recently observed by gliders and Scanfish (Sects. 5.2 and 5.9) as a persistent feature indicate that a considerable amount of primary production takes place unnoticed from satellite observations. To include these vertical patterns in modelling studies requires sophisticated formulations like those by Riegman and Colijn (1991), Behrenfeld and Falkowski (1997), and Behrenfeld et al. (2005). For the German Bight model validations, using COSYNA data can help to significantly improve estimates of total primary production.

8 Data management and data products

8.1 Data management

The COSYNA data management system (CODM) was established to make observational and model data publicly available in near-real time (Breitbach et al., 2016). The time between observations and the availability of data on CODM ranges from a few minutes for stationary measurements to about 24 h for data obtained from ships of opportunity and satellites.

Due to the various observational platforms and model output, it is a significant challenge to provide a comprehensive overview of the observations with their diverse data formats in terms of parameters, dimensionality, and observational methods. It is achieved by describing the data using metadata and by making all data available for different analyses and visualisations in a combined way independent of data dimensionality. This concerns in particular the presentation of different data types together in one plot, such as the mapping of the same variable derived from satellite imagery and in situ observations. Key for this is the harmonisation of parameter names. The various internally used parameter names for the same observed property are mapped to the corresponding Climate and Forecast (CF) standard name (Eaton et al., 2010).

Another important aspect of CODM is the use of standardised metadata that are adapted for the use in direct web service requests (Fig. 24). Two types of metadata are used in CODM: For observations, the first type describes an observational platform, its sensors, and observed properties, the second type describes the observed data.

The metadata are created automatically if the data sets have a distinct beginning and ending. Examples are ship or glider transects, or single satellite scenes. For stationary platforms, only one metadata record is created for the entire time series. For models, the first type of metadata describes the

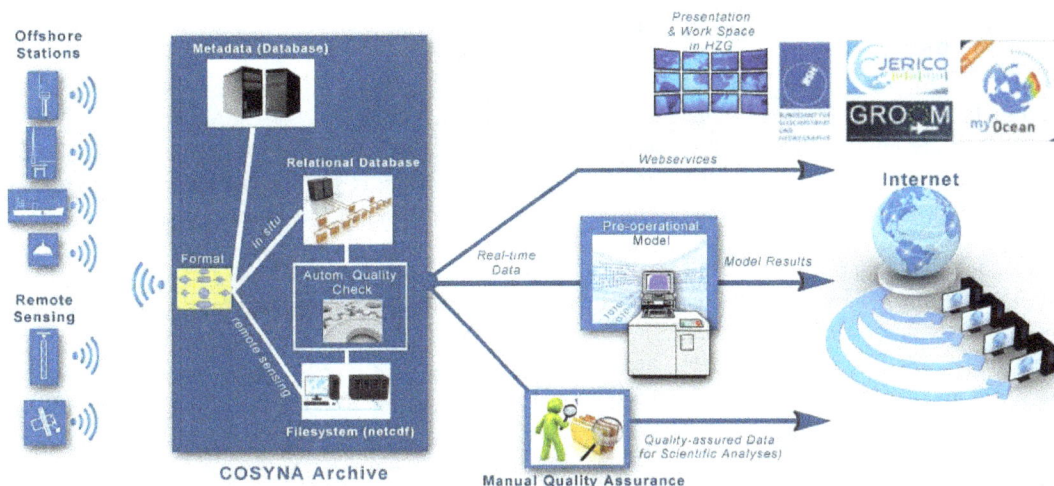

Figure 24. Data flow in COSYNA.

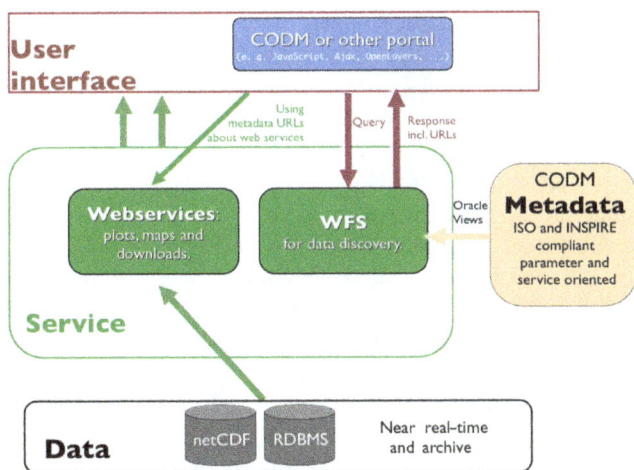

Figure 25. Data management architecture: the connection between the user interface on the one hand and data or metadata on the other hand is handled solely by web services like Web Feature Services (WFS) or Web Map Services (WMS).

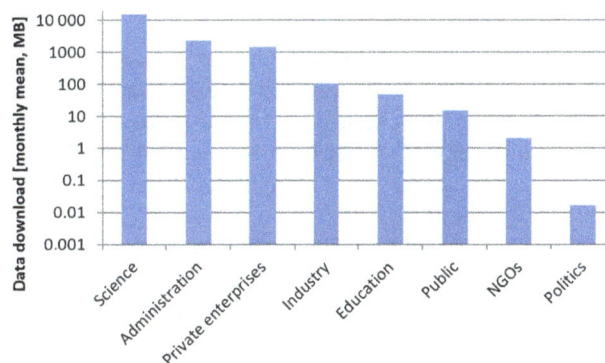

Figure 26. Mean monthly data use for different categories of users. Data are shown for the time period between November 2014, when the user registration started, and January 2016.

but CODM requires a basic user registration. Users are asked to provide country of origin, a user category, and the city. No other personal information is mandatory. Users are also asked to acknowledge COSYNA as a data source in their publications. The majority of users are in the science sector, followed by administration (Fig. 26).

8.2   Data products

COSYNA is monitoring the current state of the coastal system in the North Sea and is generating modelled pre-operational state reconstructions and forecasts. These routinely provided data can be grouped into four "product" categories.

a. High-resolution time series at fixed positions. Meteorological, oceanographic, water quality, and biological parameters are continuously observed at measuring poles (Sect. 5.1) Spiekeroog, Hörnum Deep, and Elbe, research platform FINO3 (Sect. 5.1.2), and at the station-

model itself, while the second type describes the model run. Data–metadata are ISO19115 and INSPIRE compliant (EC Directive, 2007) and contain all necessary information to access the data as download, plot, or map. The metadata themselves are also mapped to the Web Feature Service (Fig. 25).

The observational data have to pass a number of automated and supervised tests before they become publicly accessible in the data portal. Depending on the test results for range, stuck values, spikes, and – for some parameters – gradients, quality flags are assigned to the data. The procedures and quality flags are in line with international guidelines (Breitbach et al., 2016; SeaDataNet, 2010).

CODM is a publicly available Open Data portal. There are no restrictions or fees for downloading and using the data, ary FerryBox systems in Cuxhaven and on Helgoland (Sect. 5.4).

b. Repeated transects. Oceanographic and biogeochemical parameters are measured during regular ship and glider surveys (Sects. 5.2, 5.9) and with automated FerryBox systems on ships of opportunity (Sect. 5.4).

c. Remote sensing information. Regular maps of currents, chlorophyll distribution, and optical seawater properties are obtained with remote sensing by HF radar (Sect. 5.3) and satellites (Sect. 5.7). The data cover large areas of the German Bight and are integrated with observational in situ data.

d. Integrated COSYNA products. The automatically produced data fields of the German Bight are continuous in space and time and provide hindcast, nowcast, and short-term forecasts. The latter two are improved with data assimilation procedures (Sect. 7).

The COSYNA product, "Surface Current Fields", provides data fields and maps of tidal hindcasts and forecasts of sea surface currents in the German Bight. The fields are updated every 30 min. They are created by assimilating regular HF radar measurements into a 3-D circulation model (Stanev et al., 2011, 2015; Sect. 7).

The pre-operational COSYNA wave forecast model system runs twice a day and provides a 72 h forecast on the regional scale for the North Sea and on the local scale for the German Bight. Significant wave height, period, and total wave direction are calculated (Staneva et al., 2014).

In order to provide the spatial distribution of sea surface temperature and salinity in the North Sea, FerryBox observations taken along ship tracks are extrapolated to larger areas combining them with information from numerical models. Data from the Cuxhaven–Immingham route are assimilated into a 3-D circulation model every 24 h (Grayek et al., 2011).

9   Outreach and stakeholder interaction

COSYNA aims to make scientific data, results, and data products publicly available by reaching out to different target groups and users, such as the scientific community, potential users in business enterprises and authorities, and the general public. To serve this purpose, COSYNA publishes several print products in German and English that are publicly available for download at the COSYNA website, or that can be ordered. Flyers and more comprehensive brochures provide an overview of the goals, approaches, activities, and results of COSYNA. The annual progress reports are intended for COSYNA partners and users and describe selected results and activities of the various working groups and subprojects within COSYNA. Newsletter and product fact sheets provide COSYNA partners and users as well as interest groups or the general public with information on activities, events, or data products.

COSYNA maintains the website http://www.cosyna.de that informs about motivation, approach, observations, modelling, products, and outreach activities. The COSYNA data portal is linked to that website and provides access to data download and visualisation. On average, the COSYNA website has been visited by more than 500 different external visitors per month.

Furthermore, COSYNA has developed an interactive app with versions for iPad and other tablet PCs as well as Android- and iOS-based smartphones. The app provides explanatory texts and pictures describing the observing systems, instruments, models, and products, as well as the COSYNA partners. Near-real-time data for several platforms are available. COSYNA also presents the app in permanent exhibits in museums, or temporarily at public events or trade shows.

It is one of the main goals of COSYNA to bridge the gap between operational oceanography and the users of marine data in local authorities, non-governmental organisations, science, and industry. In order to ensure that products are applicable, COSYNA has been initiating a dialogue with stakeholders, allowing for direct feedback and input to COSYNA. In the initial phase of COSYNA, a national and an international survey showed that the COSYNA data products are useful to a great number of users from different sectors and fit into the international context. Follow-up workshops and an external evaluation of the Surface Current Fields integrated COSYNA product have clearly improved COSYNA products and their usability. To explore the streamlining of COSYNA products for the offshore wind energy industry, several workshops were held to pave the way for future co-operation with offshore wind energy companies (Eschenbach, 2017).

10   Conclusions and outlook

COSYNA was established with its sights on understanding the state and variability of complex interdisciplinary processes in the North Sea and the Arctic. During its first years, work concentrated on establishing the observational network, developing sensors and numerical models, testing and applying data assimilation techniques, building a data management system and testing outreach strategies. Now that the core of what had been envisioned in the original concepts is operational and functioning, COSYNA will expand into new areas, spatially as well as scientifically.

Currently, COSYNA is being extended to the western part of the Baltic Sea (in cooperation with a new partner, GEOMAR, Helmholtz Centre for Ocean Research) by installing an Underwater-Node System in spring 2016 in the Eckernförde Bight near the location of GEOMAR's long established Boknis Eck time-series station (Lennartz et al., 2014). COSYNA already contributes to observations of other coastal areas in the world, such as the Lena delta, the Bohai Sea in China, or with instruments on research vessels and cruise ships operating in various parts of the world ocean. In the long run, COSYNA will be part of HZG's Global Coast project that aims at identifying representative coastal regions worldwide that will help evaluate the role of coastal areas for global processes, while using a global context for understanding regional and coastal processes.

To this end and for use in large national and international research projects, COSYNA plans to develop mobile observing systems with high-resolution capabilities in space and time that have very short deployment times in order to be able to react to extreme events such as storms and floods. As the focus of research projects will be shifting more and more to an integrated understanding of complex systems, this approach will require cooperation with partners in the atmospheric and terrestrial research communities. In the future, COSYNA will be closely interlinked with the Elbe River supersite of DANUBIUS, the most recent European ESFRI Roadmap project studying river–delta–sea systems, and will be part of the Helmholtz Association's MOSES (Modular Observing System for the Earth System) research infrastructure.

Intensified modelling efforts, especially regarding biogeochemical models and data assimilation, are needed to put the COSYNA observations in a broad context and help understand coastal systems. This will also yield future data products including wind fields, ship detection, and biogeochemical parameters. Chlorophyll maps and maps of suspended particulate matter will be obtained from satellites on a regular basis. The assimilation of other quantities is a work in progress and will be published when they become available.

The successful technology development of underwater nodes will continue. Currently, experiments with smaller, more flexible units are underway. Alternative forms of power supplies, such as fuel cells, are being tested and may allow for a flexible network of nodes.

New partners are joining COSYNA: GEOMAR in Kiel and the Franzius Institute for Hydraulic, Estuarine, and Coastal Engineering at the University of Hannover have recently agreed to become COSYNA partners. For the future, discussions with international partners will be sought and international cooperation will be intensified – in particular with the countries bordering the North Sea.

While COSYNA has evolved into a well-established integrated pre-operational observing system, research will become more central to defining COSYNA's endeavours. Utilising the combined expertise of its various partner institutions, COSYNA's science foci will include biogeochemical cycles from rivers to the North Sea and the North Atlantic, the role of wind farms for physical, biogeochemical, and biological processes in the coastal ocean as well as asso-ciated engineering questions, land–Wadden Sea–North Sea exchange processes with an extensive experiment spanning from the Netherlands along the German coast to Denmark involving physics and biogeochemistry, and exploration of the possibilities and challenges associated with citizen science.

Competing interests. The authors declare that they have no conflict of interest.

Acknowledgements. COSYNA was implemented between 2010 and 2014. The infrastructure and instrumentation of COSYNA was funded by the German Federal Ministry of Education and Research through the Helmholtz Association. COSYNA is developed and operated jointly between 11 partner institutions entirely contributing personnel and funding for development, operation, and maintenance. COSYNA is coordinated by the Helmholtz-Zentrum Geesthacht. Currently, funding is available through the Advanced Remote Sensing – Ground Truth Demo and Test Facilities (ACROSS) infrastructure of the Helmholtz Association.

The implementation and operation of COSYNA would not have been possible without the competent and tireless efforts of the technical staff at all involved institutions. We thank the master and the crew of RV *Heincke* for their help and support during the ship cruises. The cruises were conducted under grant numbers AWI_HE298_00, AWI_HE303_00, AWI_HE308_00, AWI_HE312_00, AWI_HE319_00, AWI_HE325_00, AWI_HE331_00, AWI_HE336_00, AWI_HE353_00, AWI_HE359_00, AWI_HE365_00, AWI_HE371_00, AWI_HE391_00, AWI_HE397_00, AWI_HE407_00, AWI_HE412_00 AWI_HE417_00, AWI_HE441_00, and AWI_HE447_00.

Edited by: P. Testor

References

ABARE: Economics of Australia's sustained ocean observing system, benefits and rationale for public funding, The Australian Bureau of Agricultural and Resource Economics and Economic Consulting Services, Report 44, 2006.

Ahmerkamp, S., Winter, C., Janssen, F., Kuypers, M. M. M., and Holtappels, M.: The impact of bedform migration on benthic oxygen fluxes, J. Geophys. Res.-Biogeo., 120, 2229–2242, doi:10.1002/2015JG003106, 2015.

Ahmerkamp, S., Winter, C., Krämer, K., de Beer, D., Janssen, F., Kuypers, M. M. M., and Holtappels, M.: Regulation of benthic oxygen fluxes in permeable sediments of the coastal ocean, Limnol. Oceanogr., doi:10.1002/?lno.10544, 2017.

Amirshahi, S. M., Winter, C., and Kwoll, E.: Characteristics of instantaneous turbulent events in southern German Bight, in: River Flow, 2016.

Aßmann, S.: Entwicklung und Qualifizierung autonomer Messsysteme für den pH-Wert und die Gesamtalkalinität von Meerwasser, PhD-Thesis, Christian-Albrechts-University, Kiel, 2012.

Aßmann, S., Frank, C., and Körtzinger, A.: Spectrophotometric high-precision seawater pH determination for use in underway measuring systems, Ocean Sci., 7, 597–607, doi:10.5194/os-7-597-2011, 2011.

Badewien, T. H., Zimmer, E., Bartholoma, A., and Reuter, R.: Towards continuous long-term measurements of suspended particulate matter (SPM) in turbid coastal waters, Ocean Dynam., 59, 227–238, doi:10.1007/S10236-009-0183-8, 2009.

Behrenfeld, M. J. and Falkowski, P. G.: Photosynthetic Rates Derived from Satellite-Based Chlorophyll Concentration, Limnol. Oceanogr., 42, 1–20, 1997.

Behrenfeld, M. J., Boss, E., Siegel, D. A., and Shea, D. M.: Carbon-based ocean productivity and phytoplankton physiology from space, Global Biogeochem. Cy., 19, GB1006, doi:10.1029/2004GB002299, 2005.

Bell, P. S. and Thorne, P. D.: Application of a high resolution acoustic scanning system for imaging sea bed microtopography, Seventh International Conference on Electronic Engineering in Oceanography, Tech. Trans. Res. Ind., 128–133, doi:10.1049/cp:19970673, 1997.

Booij, K., Robinson, C. D., Burgess, R. M., Mayer, P., Roberts, C. A., Ahrens, L., Allan, I. J., Brant, J., Jones, L., Kraus, U. R., Larsen, M. M., Lepom, P., Petersen, J., Pröfrock, D., Roose, P., Schäfer, S., Smedesp, F., Tixier, C., Vorkamp, K., and Whitehouse, P.: Passive Sampling in Regulatory Chemical Monitoring of Nonpolar Organic Compounds in the Aquatic Environment, Environ. Sci. Technol., 50, 3–17, doi:10.1021/acs.est.5b04050, 2016.

Borges, A. V.: Do we have enough pieces of the jigsaw to integrate CO_2 fluxes in the coastal ocean?, Estuaries, 28, 3–27, doi:10.1007/BF02732750, 2005.

Brand, M. and Fischer, P.: Species composition and abundance of the shallow water fish community of Kongsfjorden, Svalbard, Polar Biol., 39, 1–13, doi:10.1007/s00300-016-2022-y, 2016.

Breitbach, G., Krasemann, H., Behr, D., Beringer, S., Lange, U., Vo, N., and Schroeder, F.: Accessing diverse data comprehensively – CODM, the COSYNA data portal, Ocean Sci., 12, 909–923, doi:10.5194/os-12-909-2016, 2016.

Bruno, L., Braca, P., Horstmann, J., and Vespe, M.: Experimental Evaluation of the Range-Doppler Coupling on HF Surface Wave Radar, IEEE Geosci. Remote S., 10, 4, 850–854, doi:10.1109/LGRS.2012.2226203, 2013.

Burchard, H. and Badewien, T. H.: Thermohaline circulation of the Wadden Sea, Ocean Dynam., 65, 1717–1730, doi:10.1007/s10236-015-0895-x, 2015.

Burchard, H. and Bolding, K.: GETM: A general estuarine transport model. European Commission Joint Research Centre Tech. Rep. EUR 20253 EN, 157, 2002.

Burchard, H., Flöser, G., Staneva, J. V., Badewien, T. H., and Riethmüller, R.: Impact of Densitiy Gradients on Net Sediment Transport into the Wadden Sea, J. Phys. Oceanogr., 38, 566–587, doi:10.1175/2007JPO3796.1, 2008.

Busch, J., Price, I., Jeauson, E., Zielinski, O., and v.d. Woerd, H. J.: Citizens and satellites: Assessment of phytoplankton dynamics in a NW Mediterranean aquaculture zone, Int. J. Appl. Earth Obs., 47, 40–49, doi:10.1016/j.jag.2015.11.017, 2016.

Busch, J. A., Hedley, J. D., and Zielinski, O.: Correction of hyperspectral reflectance measurements for surface objects and direct sun reflection on surface waters, Int. J. Remote Sens., 34.2013, 19, 6651–6667, doi:10.1080/01431161.2013.804226, 2013.

Busch, M., Kannen, A., Garthe, S., and Jessopp, M.: Consequences of a cumulative perspective on marine environmental impacts: offshore wind farming and seabirds at North Sea scale in context of the EU Marine Strategy Framework Directive, Ocean Coast. Manage., 71, 213–224, doi:10.1016/j.ocecoaman.2012.10.016, 2013.

Buschbaum, C., Lackschewitz, D., and Reise, K.: Nonnative macrobenthos in the Wadden Sea ecosystem, Ocean Coast. Manage., 68, 89–101, doi:10.1016/j.ocecoaman.2011.12.011, 2012.

Carder, K. L., Hawes, D. K., Baker, K. A., Smith, R. C., Steward, R. G., and Mitchell, B. G.: Reflectance model for quantifying chlorophyll a in the presence of productivity degradation products, J. Geophys. Res., 96, 20599–20611, doi:10.1029/91JC02117, 1991.

Carpenter, J. R., Merckelbach, L., Callies, U., Clark, S., Gaslikova, L., and Baschek, B.: Potential Impacts of Offshore Wind Farms on North Sea Stratification, PLoS ONE, 11, e0160830, doi:10.1371/journal.pone.0160830, 2016.

Chen, C. T. A. and Borges, A. V.: Reconciling opposing views on carbon cycling in the coastal ocean: continental shelves as sinks and near-shore ecosystems as sources of atmospheric CO_2, Deep-Sea Res. Pt. II, 56, 8–10, 578–590, doi:10.1016/j.dsr2.2009.01.001, 2009.

Colijn, F. and de Jonge, V. N.: Primary production of microphytobenthos in the Ems_Dollard estuary, Mar. Ecol.-Progr. Ser., 14, 185–196, 1984.

Constantinescu, G., Garcia, M., and Hanes, D. (Eds.): Taylor & Francis Group, Boca Raton, FL CRC Press 2016, 175–182, ISBN: 978-1-138-02913-2, ISBN: 978-1-317-28912-8, 2016.

Cottier, F., Tverberg, V., Inall M., Svendsen, H., Nilsen, F., and Griffiths, C.: Water mass modification in an Arctic fjord through cross-shelf exchange: The seasonal hydrography of Kongsfjorden, Svalbard, J. Geophys. Res., 110, C12005, doi:10.1029/2004JC002757, 2005.

Dittmann, S. (Ed.): The Wadden Sea Ecosystem: stability properties and mechanism, Springer Berlin Heidelberg, ISBN: 978-3-642-64256-2, doi:10.1007/978-3-642-60097-5, 1999.

Doerffer, R. and Schiller, H.: The MERIS Case 2 water algorithm, Int. J. Remote Sens., 28, 517–535, doi:10.1080/01431160600821127, 2007.

Dore, J. E., Lukas, R., Sadler, D. W., Church, M. J., and Karl, D. M.: Physical and biogeochemical modulation of ocean acidification in the central North Pacific, P. Natl. Acad. Sci. USA, 106, 12235–12240, doi:10.1073/pnas.0906044106, 2009.

Drücker, S., Steglich, D., Merckelbach, L., Werner, A., and Bargmann, S.: Finite element damage analysis of an underwater glider–ship collision, J. Mar. Sci. Technol., 1–10, doi:10.1007/s00773-015-0349-7, 2015.

Dzvonkovskaya, A., Gurgel, K.-W., Rohling, H., and Schlick, T.: Low power high frequency surface wave radar application for ship detection and tracking, Proc. Int. Conf. Radar, Adelaide, SA, 627–632, doi:10.1109/RADAR.2008.4653998, 2008.

Eaton, G., Drach, T., and Hankin, S.: netCDF Climate and Forecast (CF) Metadata Conventions, Version 1.5, available at: http://cfconventions.org/Data/cf-conventions/cf-conventions-1.5/build/cf-conventions.html (last access: May 2017), 2010.

EC Directive: Directive 2007/2/EC of the European Parliament and of the Council of 14 March 2007 establishing an Infrastructure for Spatial Information in the European Community, available at: http://eurlex.europa.eu/LexUriServ/LexUriServ.do?uri=OJ:L:2007:108:0001:0014:EN:PDF (last access: May 2017), 2007.

Emeis, K.-C., van Beusekom, J., Callies, U., Ebinghaus, R., Kannen, A., Kraus, G., Kröncke, I., Lenhart, H., Lorkowski, I., Matthias, V., Möllmann, C., Pätsch, J., Scharfe, M., Thomas, H., Weisse, R., and Zorita, E.: The North Sea – A shelf sea in the Anthropocene, J. Mar. Syst., 141, 18–33, doi:10.1016/j.jmarsys.2014.03.012, 2015.

Eschenbach, C. A.: Bridging the gap between observational oceanography and users, Ocean Sci., 13, 161–173, doi:10.5194/os-13-161-2017, 2017.

Feely, R. A., Doney, S. C., and Cooley, S. R.: Ocean acidification: Present conditions and future changes in a high-CO_2 world, Oceanogr., 22, 36–47, doi:10.5670/oceanog.2009.95, 2009.

Fischer, P., Schwanitz, M., Loth, R., Posner, U., Brand, M., and Schröder, F.: First year of practical experiences of the new Arctic AWIPEV-COSYNA cabled Underwater Observatory in Kongsfjorden, Spitsbergen, Ocean Sci., 13, 259–272, doi:10.5194/os-13-259-2017, 2017.

Flöser, G., Burchard, H., and Riethmüller, R.: Observational evidence for estuarine circulation in the German Wadden Sea, Cont. Shelf Res., 31, 1633–1639, doi:10.1016/j.csr.2011.03.014, 2011.

Frank, C. and Schroeder, F.: Using Sequential Injection Analysis to Improve System and Data Reliability of Online Methods: Determination of Ammonium and Phosphate in Coastal Waters, J. Autom. Method. Manag., 49535, doi:10.1155/2007/49535, 2007.

Frank, C., Schroeder, F., Ebinghaus, R., and Ruck, W.: A Fast Sequential Injection Analysis System for the Simultaneous Determination of Ammonia and Phosphate, Microchim. Ac., 154, 31–38, doi:10.1007/s00604-006-0496-y, 2006.

Frank, C., Meier, D., Voss, D., and Zielinski, O.: Computation of nitrate concentrations in coastal waters using an *in situ* ultraviolet spectrophotometer: Behavior of different computation methods in a case study a steep salinity gradient in the southern North Sea, Meth. Oceanogr., 9, 34–43, doi:10.1016/j.mio.2014.09.002, 2014.

Friedrich, J., van Beusekom, J. E. E., Neumann, A., Naderipour, C., Janssen, F. Ahmerkamp, S., Holtappels, M., and Krämer, K.: Long-term impact of bottom trawling on pelagic-benthic coupling inthe southern North Sea (German Bight), EGU2016-15791, 2016.

Furness, R. W. and Camphuysen, C. J.: Seabirds as monitors of the marine environment, ICES J. Mar. Sci., 54, 726–737, doi:10.1006/jmsc.1997.0243, 1997.

Garaba, S. P. and Zielinski, O.: Comparison of remote sensing reflectance from above-water and in-water measurements west of Greenland, Labrador Sea, Denmark Strait, and west of Iceland, Opt. Expr., 21, 15938–15950, doi:10.1364/OE.21.015938, 2013a.

Garaba, S. P. and Zielinski, O.: Methods in reducing surface reflected glint for shipborne above-water remote sensing, J. Eur. Opt. Soc.-Rapid, 8, 1–8, doi:10.2971/jeos.2013.13058, 2013b.

Garaba, S. P., Schulz, J., Wernand, M. R., and Zielinski, O.: Sunglint detection for unmanned and automated platforms, Sensors 12, 12545–12561, doi:10.3390/s120912545, 2012.

Garaba, S. P., Badewien, T. H., Braun, A., Schulz, A.-C., and Zielinski, O.: Using ocean colour remote sensing products to estimate turbidity at the Wadden Sea time series station Spiekeroog, J. Eur. Opt. Soc.-Rapid, 9, 1–6, doi:10.2971/jeos.2014.14020, 2014a.

Garaba, S. P., Voß, D., and Zielinski, O.: Physical, bio-optical state and correlations in North–Western European Shelf Seas, Remote Sens., 6, 5042–5066, doi:10.3390/rs6065042, 2014b.

Garaba, S. P., Friedrichs, A., Voss, D., and Zielinski, O.: Classifying natural waters with Forel-Ule Colour Index System: Results, Applications, Correlations and Crowdsourcing, Intl. J. Environ. Res. Publ. Health, 12, 16096–16109, doi:10.3390/ijerph121215044, 2015a.

Garaba, S. P., Voß, D., Wollschläger, J., and Zielinski, O.: Modern approaches to shipborne ocean colour remote sensing, Appl. Opt., 54, 3602–3612, doi:10.1364/AO.54.003602, 2015b.

Garthe, S. and Hüppop, O.: Scaling possible adverse effects of marine wind farms on seabirds: developing and applying a vulnerability index, J. Appl. Ecol., 41, 724–734, doi:10.1111/j.0021-8901.2004.00918.x, 2004.

Garthe, S., Peschko, V., Kubetzki, U., and Corman, A.-M.: Seabirds as samplers of the marine environment – a case study in Northern Gannets, Ocean Sci., submitted, 2017.

Gohin, F., Druon, J. N., and Lampert, L.: A five channel chlorophyll concentration algorithm applied to SeaWiFS data processed by SeaDAS in coastal waters, Int. J. Remote Sens., 23, 8, 1639–1661, doi:10.1080/01431160110071879, 2002.

Gollasch, S., Haydar, D., Minchin, D., Wolff, W. J., and Reise, K.: Introduced aquatic species of the North Sea coasts and adjacent brackish waters, in: Biological Invasions in Marine Ecosystems, edited by: Rilov, G. and Crooks, J. A., Ecological Studies, 204, 507–528, doi:10.1007/978-3-540-79236-9_29, 2009.

Grashorn, S., Lettmann, K. A., Wolff, J.-O., Badewien, T. H., and Stanev, E. V.: East Frisian Wadden Sea hydrodynamics and wave effects in an unstructured-grid model, Ocean Dynam., 65, 419–434, doi:10.1007/s10236-014-0807-5, 2015.

Grayek, S., Staneva, J., Schulz-Stellenfleth, J., Petersen, W., and Stanev, E. V.: Use of FerryBox surface temperature and salinity measurements to improve model based state estimates for the German Bight, J. Mar. Syst., 88, 45–59, doi:10.1016/j.jmarsys.2011.02.020, 2011.

Gurgel, K.-W., Antonischki, G., Essen, H.-H., and Schlick, T.: Wellen Radar (WERA): A new ground-wave based HF radar for ocean remote sensing, Coast. Eng., 37, 219–234, 1999.

Günther, H., Hasselmann, S. and Janssen, P. A. E. M.: The WAM model cycle 4, Report Deutsches Klimarechenzentrum, DKRZ-TR-4, 26, 1, 1992.

Haller, M., Janssen, F., Siddorn, J., Petersen, W., and Dick, S.: Evaluation of numerical models by FerryBox and fixed platform in situ data in the southern North Sea, Ocean Sci., 11, 879–896, doi:10.5194/os-11-879-2015, 2015.

Hegseth, E. N. and Tverberg, V.: Effect of Atlantic water inflow on timing of the phytoplankton spring bloom in a high Arctic fjord (Kongsfjorden, Svalbard), J. Mar. Syst., 113–114, 94–105, doi:10.1016/j.jmarsys.2013.01.003, 2013.

Helmholz, H., Lassen, S., Ruhnau, C., Pröfrock, D., Erbslöh, H.-B., and Prange, A.: Investigation on the proteome response of transplanted Blue mussel (*Mytilus sp.*) during a long term exposure experiment at differently impacted field stations in the German Bight (North Sea), Mar. Environ. Res., 110, 69–80, 2015.

Hodapp, D., Meier, S., Muijsers, F., Badewien, T. H., and Hillebrand, H.: Structural equation modeling approach to the diversity-productivity relationship of Wadden Sea phytoplankton, Mar. Ecol.-Progr. Ser., 523, 31–40, doi:10.3354/meps11153, 2015.

Holinde, L., Badewien, T. H., Freund, J. A., Stanev, E. V., and Zielinski, O.: Processing of water level derived from water pressure data at the Time Series Station Spiekeroog, Earth Syst. Sci. Data, 7, 289–297, doi:10.5194/essd-7-289-2015, 2015.

Hofmeister, R., Flöser, G., and Schartau, M.: Estuarine circulation sustains horizontal nutrient gradients in freshwater-influenced coastal systems, Geo-Marine Letters, under revision, 2017.

Hop, H., Pearson, T., Hegseth, E. N., Kovacs, K. M., Wiencke, C., Kwasniewski, S., Eiane, K., Mehlum, F., Gulliksen, B., Wlodarska-Kowalczuk, M., Lydersen, C., Weslawski, J. M., and Cochrane, S.: The marine ecosystem of Kongsfjorden, Svalbard, Polar Res., 21, 167–208, doi:10.1111/j.1751-8369.2002.tb00073.x, 2002.

Hop, H., Wiencke, C., Vögele, B., and Kovaltchouk, N. A.: Species composition, zonation, and biomass of marine benthic macroalgae in Kongsfjorden, Svalbard, Bot. Mar., 55, 399–414, doi:10.1515/bot-2012-0097, 2012.

Howarth, M. J.: North Sea circulation, Encyclopedia of Ocean Sciences, 1, 1912–1921, 2001.

Intergovernmental Oceanographic Commission (of UNESCO): The Case for GOOS. Report of the IOC Blue Ribbon Panel for a Global Ocean Observing System (GOOS), IOC/INF-915, Paris, 62, 27 January 1993.

Intergovernmental Oceanographic Commission: GOOS Coastal Module Planning Workshop Report, Scientific Committee on Oceanic Research of ICSU for J-GOOS, Workshop Report No. 131, GOOS No. 35, Miami, 57, 24–28 February 1997.

IPCC: Climate Change 2014: Synthesis Report, Contribution of Working Groups I, II and III to the Fifth Assessment Report of the Intergovernmental Panel on Climate Change, Core Writing Team, edited by: Pachauri, R. K. and Meyer, L. A., IPCC, Geneva, Switzerland, 151 pp., 2014.

Janssen, F., Färber, P., Huettel, M., Meyer, V., and Witte, U.: Porewater advection and solute fluxes in permeable marine sediments (I): Calibration and performance of the novel benthic chamber system Sandy, Limnol. Oceanogr., 50, 768–778, 2005.

Jones, C., Creed, E., Glenn, S., Kerfoot, J., Kohut, J., Mudgal, C., and Schofield, O.: Slocum gliders – a component of operational oceanography, Proc. UUST, Autonomous Undersea Systems Institute, 2005.

Kays, R., Crofoot, M. C., Jetz, W., and Wikelski, M.: Terrestrial animal tracking as an eye on life and planet, Science, 348, aaa2478, doi:10.1126/science.aaa2478, 2015.

Kirk, J. T. O.: Point-source integrating-cavity absorption meter: theoretical principles and numerical modeling, Appl. Opt., 36, 6123–6128, doi:10.1364/AO.36.006123, 1997.

Komen, G. J., Cavaleri, L., Donelan, M., Hasselmann, K., Hasselmann, S., and Janssen, P.: Dynamics and Modelling of Ocean Waves, Cambridge University Press, Cambridge, UK, 560, 1994.

Koschinsky, S., Culik, B., Henriksen, O. D., Tregenza, N., Ellis, G., Jansen, C., and Kathe, G.: Behavioral reactions of free-ranging porpoises and seals to the noise of a simulated 2 MW windpower generator, Mar. Ecol.-Progr. Ser., 265, 263–273, doi:10.3354/meps265263, 2003.

Kwoll, E., Winter, C., and Becker, M.: Intermittent suspension and transport of fine sediment over natural tidal bedforms, in: Coherent Structures in Flows at the Earth's Surface, edited by: Venditti, J. G., Best, J., Church, M. and Hardy, R. J., Wiley-Blackwell, London, doi:10.1002/9781118527221.ch15, 2013.

Kwoll, E., Becker, M., and Winter, C.: With or against the tide: the influence of bedform asymmetry on the formation of macroturbulence and suspended sediment patterns, Water Resour. Res., 50, 1–16, doi:10.1002/2013WR014292, 2014.

Krämer, K. and Winter, C.: Predicted ripple dimensions in relation to the precision of in situ measurements in the southern North Sea, Ocean Sci., 12, 1221–1235, doi:10.5194/os-12-1221-2016, 2016.

Lass, H. U., Mohrholz, V., Knoll, M., and Prandke, H.: Enhanced Mixing Downstream of a Pile in an Estuarine Flow, J. Mar. Syst., 74, 505–527, doi:10.1016/j.jmarsys.2008.04.003, 2008.

Lee, Z. P., Carder, K. L., Steward, R. G., Peacock, T. G., Davis, C. O., and Patch, J. S.: An Empirical Algorithm for Light Absorption by Ocean Water Based on Colour, J. Geophys. Res.-Oceans, 103, 27967–27978, doi:10.1029/98JC01946, 1998.

Lennartz, S. T., Lehmann, A., Herrford, J., Malien, F., Hansen, H.-P., Biester, H., and Bange, H. W.: Long-term trends at the Boknis Eck time series station (Baltic Sea), 1957–2013: does climate change counteract the decline in eutrophication?, Biogeosciences, 11, 6323–6339, doi:10.5194/bg-11-6323-2014, 2014.

Lerebourg, C. J. Y., Pilgrim, D. A., Ludbrook, G. D., and Neal, R.: Development of a point source integrating cavity absorption meter, J. Opt. A Pure Appl. Opt., 4, 4 S56–S65, 2002.

Lettmann, K. A., Wolff, J. O., and Badewien, T. H.: Modeling the impact of wind and waves on suspended particulate matter fluxes in the East Frisian Wadden Sea (southern North Sea), Ocean Dynam., 59, 239–262, doi:10.1007/S10236-009-0194-5, 2009.

Loewe, P. (Ed.): System Nordsee — Zustand 2005 im Kontext langzeitlicher Entwicklungen, Berichte des BSH, 44, Bundesamt für Seeschifffahrt und Hydrographie, Hamburg und Rostock, 270, 2009.

Lorkowski, I., Paetsch, J., Moll, A., and Kuehn, W.: Interannual variability of carbon fluxes in the North Sea from 1970 to 2006 – competing effects of abiotic and biotic drivers on the gas exchange of CO_2, Estuarine, Coastal and Shelf Science 100, 38–57, doi:10.1016/j.ecss.2011.11.037, 2012.

Ludewig, E.: On the Effect of Offshore Wind Farms on the Atmosphere and Ocean Dynamics, Hamburg Studies on Maritime Affairs 31, Springer International Publishing Switzerland, doi:10.1007/978-3-319-08641-5_1, 2015 ISBN: 978-3-319-08640-8 (Print) 978-3-319-08641-5, 2015.

Maerz, J., Hofmeister, R., van der Lee, E. M., Gräwe, U., Riethmüller, R., and Wirtz, K. W.: Maximum sinking velocities of suspended particulate matter in a coastal transition zone, Biogeosciences, 13, 4863–4876, doi:10.5194/bg-13-4863-2016, 2016.

Maresca, S., Braca, P., Horstmann, J., and Grasso, R.: Maritime surveillance using multiple high-frequency surface-wave radars, IEEE T. Geosci. Remote, 52, 5056–5071, 2014.

Meier, S., Muijsers, F., Beck, M., Badewien, T. H., and Hillebrand, H.: Dominance of the non-indigenous diatom Mediopyxis helysia in Wadden Sea phytoplankton can be linked to broad tolerance to different Si and N supplies, J. Sea. Res., 95, 36–44, doi:10.1016/J.Seares.2014.10.001, 2015.

Merckelbach, L.: On the probability of underwater glider loss due to collision with a ship, J. Mar. Sci. Techonol., 18, 75–86, doi:10.1007/s00773-012-0189-7, 2013.

Merckelbach, L.: Depth-averaged instantaneous currents in a tidally dominated shelf sea from glider observations, Biogeosciences, 13, 6637–6649, doi:10.5194/bg-13-6637-2016, 2016.

Metfies, K., Schroeder, F., Hessel, J., Wollschläger, J., Micheller, S., Wolf, C., Kilias, E., Sprong, P., Neuhaus, S., Frickenhaus, S., and Petersen, W.: High-resolution monitoring of marine protists based on an observation strategy integrating automated on-board filtration and molecular analyses, Ocean Sci., 12, 1237–1247, doi:10.5194/os-12-1237-2016, 2016.

Moore, C., Barnand, A., Fietzek, P., Lewis, M., Sosik, H., White, S., and Zielinski, O.: Optical tools for ocean monitoring and research, Ocean Sci., 5, 661–684, doi:10.5194/os-5-661-2009, 2009.

Neumann, A., Friedrich, J., van Beusekom, J. E. E., and Naderipour, C.: Spatial and temporal patterns in oxygen and nutrient fluxes in sediment of German Bight (North Sea), EGU 2016-14238, 2016.

North, R. P., Riethmüller, R., and Baschek, B.: Detecting small–scale horizontal gradients in the upper ocean using wavelet analysis, Estuarine, Coast. Shelf Sci., 180, 221–229, doi:10.1016/j.ecss.2016.06.031, 2016.

NOSCCA: North Sea Climate Change Assessment, edited by: Quante, M. and Colijn, F., Springer, 528 pp., doi:10.1007/978-3-319-39745-0, 2016.

Oehmcke, S., Zielinski, O., and Kramer, O.: Event detection in marine time series data. Proceedings of KI 2015: Advances in Artifical Intelligence, Lecture Notes Comp. Sci., 9324, 279–286, doi:10.1007/978-3-319-24489-1_24, 2015.

Onken, R. and Riethmüller, R.: Determination of the freshwater budget of tidal flats from measurements near a tidal inlet, Cont. Shelf Res., 30, 924–933, doi:10.1016/j.csr.2010.02.004, 2010.

Onken, R., Callies, U., Vaessen, B., and Riethmüller, R.: Indirect determination of the heat budget of tidal flats, Cont. Shelf Res., 27, 1656–1676, doi:10.1016/j.csr.2007.01.029, 2007.

Onken, R., Garbe, H., Schröder, S., and Janik, M.: A new instrument for sediment temperature measurements, J. Mar. Sci. Technol., 15, 427–433, doi:10.1007/s00773-010-0096-8, 2010.

Oehler, T., Martinez, R., Schückel, U., Winter, C., Kröncke, I., and Schlüter, M.: Seasonal and spatial variations of benthic oxygen and nitrogen fluxes in the Helgoland Mud Area (southern North Sea), Cont. Shelf Res., 106, 118–129, doi:10.1016/j.csr.2015.06.009, 2015a.

Oehler, T., Schlüter, M., and Schückel, U.: Seasonal dynamics of the biogenic silica cycle in surface sediments of the Helgoland Mud Area (southern North Sea), Cont. Shelf Res., 107, 103–114, doi:10.1016/j.csr.2015.07.016, 2015b.

Örek, H., Doerffer, R., Röttgers, R., Boersma, M., and Wiltshire, K. H.: Contribution to a bio-optical model for remote sensing of Lena River water, Biogeosciences, 10, 7081–7094, doi:10.5194/bg-10-7081-2013, 2013.

OSPAR: Second OSPAR Integrated Report on the Eutrophication Status of the OSPAR Maritime Area, OSPAR Commission, 2008.

Otto, L., Zimmerman, J. T. F., Furnes, G. K., Mork, M., Saetre, R., and Becker, G.: Review of the physical oceanography of the North Sea, Netherlands, J. Sea Res., 26, 161–238, 1990.

Petersen, J., Pröfrock, D., Paschke, A., Broekaert, J. A. C., and Prange, A.: Laboratory calibration and field testing of the Chemcatcher Metal for trace levels of rare earth elements in estuarine waters, Environ. Sci. Water Res. Technol., 22, 16051–16059, doi:10.1007/s11356-015-4823-x, 2015a.

Petersen, J., Pröfrock, D., Paschke, A., Langhans, V., Broekaert, J. A. C., and Prange, A.: Development and field test of a mobile continuous flow system utilizing Chemcatcher for monitoring of rare earth elements in marine environments, Environ. Sci. Water Res. Technol., doi:10.1039/c5ew00126a, 2015b.

Petersen, W.: FerryBox Systems: State-of-the-Art in Europe and Future Development, J. Mar. Syst., 140A, 4–12, doi:10.1016/j.jmarsys.2014.07.003, 2014.

Petersen, W., Schroeder, F., and Bockelmann, F.-D.: FerryBox – Application of continuous water quality observations along transects in the North Sea, Ocean Dynam., 61, 1541–1554, doi:10.1007/s10236-011-0445-0, 2011.

Postma, H.: Introduction to the symposium on organic matter in the Wadden Sea, Neth. Inst. Sea Res. Publ. Ser., 10, 15–22, 1984.

Reuter, R., Badewien, T. H., Bartholoma, A., Braun, A., Lübben, A., and Rullkötter, J.: A hydrographic time series station in the Wadden Sea (southern North Sea), Ocean Dynam., 59, 195–211, doi:10.1007/S10236-009-0196-3, 2009.

Riegman, R. and Colijn, F.: Evaluation of measurements and calculation of primary production in the Doggerbank area (North Sea) in summer 1988, Mar. Ecol.-Progr. Ser., 69, 125–132, 1991.

Robinson, I. S.: Measuring the oceans from space: the principles and methods of satellite oceanography, Berlin, Germany, Springer/Praxis Publishing, 669, 2004.

Röttgers, R., Schönfeld, W., Kipp, P. R., and Doerffer, R.: Practical test of a point-source integrating cavity absorption meter: the performance of different collector assemblies, Appl. Opt., 44, 5549–5560, doi:10.1364/AO.44.005549, 2005.

Rullkötter, J.: The back-barrier tidal flats in the southern North Sea – a multidisciplinary approach to reveal the main driving forces shaping the system, Ocean Dynam., 59, 157–165, 2009.

Schaap, D. and Lowry, R. K.: SeaDataNet—Pan-European Infrastructure for Marine and Ocean Data Management: Unified Access to Distributed Data Sets, Int. J. Digital Earth, 3, S1, 50–69, doi:10.1080/17538941003660974, 2010.

Schulz, A. C., Badewien, T. H., and Zielinski, O.: Impact of currents and turbulence on turbidity dynamics at the time series station Spiekeroog (Wadden Sea, Southern North Sea). Current, waves and turbulence measurement (CWTM), 11th IEEE/OES, 2015, doi:10.1109/CWTM.2015.7098095, 2015.

Schulz, J.: The geometric optics of subsea imaging. Invited chapter for the book "Subsea optics and imaging", Woodhead Publishing, Editors: Watson and Zielinski, doi:10.1533/9780857093523.3.243, 2013.

Schulz, J., Barz, K., Ayon, P., Lüdtke, A., Zielinski, O., Mengedoht, D., and Hirche, H.-J.: Imaging of plankton specimens with the Lightframe On-sight Keyspecies Investigation (LOKI) system, Journal of the European Optical Society – Rapid publications 10017s, doi:10.2971/jeos.2010.10017s, 2010.

Schulz, M., von Beusekom, J., Bigalke, K., Brockmann, U. H., Dannecker, W., Gerwig, H., Grassl, H., Lenz, C.-J., Michaelsen, K., Niemeier, U., Nitz, T., Plate, E., Pohlmann, T., Raabe, T., Rebers, A., Reinhardt, V., Schatzmann, M., Schlünzen, K. H., Schmidt-Nia, R., Stahlschmidt, T., Steinhoff, G., and von Salzen, K.: The atmospheric impact on fluxes of nitrogen, POPs and energy in the German Bight, German J. Hydrogr., 51, 133–154, doi:10.1007/BF02764172, 1999.

Schrum, C., Alekseeva, I., and St. John, M.: Development of a coupled physical–biological ecosystem model ECOSMO: part I: model description and validation for the North Sea, J. Mar. Syst., 61.1, 79–99, 2006.

SCOR Working Group 75: Methodology for oceanic CO_2 measurements, UNESCO technical papers in marine science, 65, 44, 1988.

SeaDataNet: Standards for Data Quality Control, available at: https://www.seadatanet.org/Standards (last access: May 2017), 2010.

Stanev, E. V., Schulz-Stellenfleth, J., Staneva, J., Grayek, S., Seemann, J., and Petersen, W.: Coastal observing and forecasting system for the German Bight – estimates of hydrophysical states, Ocean Sci., 7, 569–583, doi:10.5194/os-7-569-2011, 2011.

Stanev, E. V., Ziemer, F., Schulz-Stellenfleth, J., Seemann, J., Staneva, J., and Gurgel, K. W.: Blending Surface Currents from HF Radar Observations and Numerical Modelling: Tidal Hindcasts and Forecasts, J. Atmos. Ocean Techn., 32, 256–281, doi:10.1175/JTECH-D-13-00164.1, 2015.

Stanev, E. V., Schulz-Stellenfleth, J., Staneva, J., Grayek, S., Grashorn, S., Behrens, A., Koch, W., and Pein, J.: Ocean forecasting for the German Bight: from regional to coastal scales, Ocean Sci., 12, 1105–1136, doi:10.5194/os-12-1105-2016, 2016.

Staneva, J., Stanev, E. V., Wolff, J. O., Badewien, T. H., Reuter, R., Flemming, B., Bartholoma, A., and Bolding, K.: Hydroynamics and sediment dynamics in the German Bight. A focus on observations and numerical modelling in the East Frisian Wadden Sea, Cont. Shelf Res., 29, 302–319, doi:10.1016/J.Csr.2008.01.006, 2009.

Staneva, J., Behrens, A., and Groll, N.: Recent Advances in Wave Modelling for the North Sea and German Bight, Die Küste, 81, 1–586, 2014.

Stempniewicz, L., Błachowiak-Samołyk, K., and Wesławski, J. M.: Impact of climate change on zooplankton communities, seabird populations and arctic terrestrial ecosystem–A scenario, Deep-Sea Res. Pt. II, 54, 2934–2945, doi:10.1016/j.dsr2.2007.08.012, 2007.

Stroeve, J., Holland, M. M., Meier, W., Scambos, T., and Serreze, M.: Arctic sea ice decline: Faster than forecast, Geophys. Res. Lett., 34, L09501, doi:10.1029/2007GL029703, 2007.

Su, J., Tian, T., Krasemann, H., Schartau, M., and Wirtz, K.: Response patterns of phytoplankton growth to variations in resuspension in the German Bight revealed by daily MERIS data in 2003 and 2004, Oceanologia, 57, 328–341, doi:10.1016/j.oceano.2015.06.001, 2015.

Sündermann, J., Hesse, K.-J., and Beddig, S.: Coastal mass and energy fluxes in the southeastern North Sea, German J. Hydrogr., 51, 113–132, doi:10.1007/BF02764171, 1999.

Svendsen, H., Beszczynska-Møller, A., Hagen, J. O., Lefauconnier, B., Tverberg, V., Gerland, S., Ørbæk, J. B., Bischof, K., Papucci,

C., Zajaczkowski, M., Azzolini, R., Bruland, O., and Wiencke, C.: The physical environment of Kongsfjorden/Krossfjorden, an Arctic fjord system in Svalbard, Polar Res., 21, 133–166, doi:10.1111/j.1751-8369.2002.tb00072.x, 2002.

Testor, P., Meyers, G., Pattiaratchi, C., Bachmayer, R., Hayes, D., Pouliquen, S., de la Villeon, L. P., Carval, T., Ganachaud, A., Gourdeau, L., Mortier, L., Claustre, H., Taillandier, V., Lherminier, P., Terre, T., Visbeck, M., Krahman, G., Karstensen, J., Alvarez, A., Rixen, M., Poulain, P., Osterhus, S., Tintore, J., Ruiz, S., Garau, B., Smeed, D., Griffiths, G., Merckelbach, L., Sherwin, T., Schmid, C., Barth, J., Schofield, O., Glenn, S., Kohut, J., Perry, M., Eriksen, C., Send, U., Davis, R., Rudnick, D., Sherman, J., Jones, C., Webb, D., Lee, C., Owens, B., and Fratantoni, D.: Gliders as a component of future observing systems, OceanObs'09: Sustained Ocean Observations and Information for Society, edited by: Hall, J., Harrison, D., and Stammer, D., 2 of OceansObs'09, Venice, Italy, eSA Publication WPP-306, doi:10.5270/OceanObs09.cwp.89, 2010.

UNESCO: Determination of chlorophyll in seawater: Report of intercalibration tests, UNESCO technical papers in marine science 35, 21, 1980.

van Beusekom, J. E. E. and de Jonge, V. N.: Long-term changes in Wadden Sea nutrient cycles: importance of organic matter import from the North Sea, Hydrobiol., 475/476, 185–194, doi:10.1023/A:1020361124656, 2002.

van Beusekom, J. E. E., Brockmann, U. H., Hesse, K.-J., Nickel, W., Pormeba, K., and Tillmann, U.: The importance of sediments in the transformation and turnover of nutrients and organic matter in the Wadden Sea and German Bight, German J. Hydrogr., 51, 245–266, doi:10.1007/BF02764176, 1999.

van Beusekom, J. E. E., Buschbaum, C., and Reise, K.: Wadden Sea tidal basins and the mediating role of the North Sea in ecological processes: scaling up of management?, Ocean Coast. Manage., 68, 69–78, doi:10.1016/j.ocecoaman.2012.05.002, 2012.

van De Poll, W. H., Maat, D. S., Fischer, P., Rozema, P. D., Daly, O. B., Koppelle, S., Visser, R. J. W., and Buma, A. G. J.: Atlantic Advection Driven Changes in Glacial Meltwater: Effects on Phytoplankton Chlorophyll-a and Taxonomic Composition in Kongsfjorden, Spitsbergen, Front. Mar. Sci., 3, 60, doi:10.3389/fmars.2016.00200, 2016.

van Leeuwen, S. M., van der Molen, J., Ruardij, P., Fernand, L., and Jickells, T.: Modelling the contribution of deep chlorophyll maxima to annual primary production in the North Sea, Biogeochemistry, 113, 137–152, 2013.

Vivone, G., Braca, P., and Horstmann, J.: Knowledge-Based Multi-Target Ship Tracking for HF Surface Wave Radar Systems, IEEE T. Geosci. Remote, 53, 3931–3949, doi:10.1109/TGRS.2014.2388355, 2015.

Voynova, Y. G., Brix, H., Petersen, W., Weigelt-Krenz, S., and Scharfe, M.: Extreme flood impact on estuarine and coastal biogeochemistry: the 2013 Elbe flood, Biogeosciences, 14, 541–557, doi:10.5194/bg-14-541-2017, 2017.

Wahl, T., Haigh, I. D., Woodworth, P. L., Albrecht, F., Dillingh, D., Jensen, J., Nicholls, R. J., Weisse, R., and Wöppelmann, G.: Observed mean sea level changes around the North Sea coastline from 1800 to present, Earth-Sci. Rev., 124, 51–67, doi:10.1016/j.earscirev.2013.05.003, 2013.

Wehkamp, S. and Fischer, P.: Impact of hard-bottom substrata on the small-scale distribution of fish and decapods in shallow subtidal temperate waters, Helgoland Mar. Res., doi:10.1007/s10152-012-0304-5, hdl:10013/epic.39252, 2012.

Wehkamp, S. and Fischer, P.: The impact of coastal defence structures (tetrapods) on decapod crustaceans in the southern North Sea, Mar. Environ. Res., 92, 52–60, doi:10.1016/j.marenvres.2013.08.011, hdl:10013/epic.47436, 2013a.

Wehkamp, S. and Fischer, P.: Impact of coastal defence structures (tetrapods) on a demersal hard-bottom fish community in the southern North Sea, Mar. Environ. Res., doi:10.1016/j.marenvres.2012.10.013, hdl:10013/epic.40445, 2013b.

Wehkamp, S. and Fischer, P.: A practical guide to the use of consumer-level digital still cameras for precise stereogrammetric *in situ* assessments in aquatic environments, Underw. Tech., 32, 111–128, doi:10.3723/ut.32.111, hdl:10013/epic.43817, 2014.

Willis, K., Cottier, F., Kwasniewski, S., Wold, A., and Falk-Petersen, S.: The influence of advection on zooplankton community composition in an Arctic fjord (Kongsfjorden, Svalbard), J. Marine Syst., 61, 39–54, doi:10.1016/j.jmarsys.2005.11.013, 2006.

Wilson, R. P. and Vandenabeele, S. P.: Technological innovation in archival tags used in seabird research, Mar. Ecol.-Prog. Ser., 451, 245–262, doi:10.3354/meps09608, 2012.

Wirtz, K. and Kerimoglu, O.: Optimality and variable co-limitation controls autotrophic stoichiometry, Front. Mar. Sci., submitted, 2017.

Witte, U. and Pfannkuche, O.: High rates of benthic carbon remineralization in the abyssal Arabian Sea, Deep-Sea Res. Pt. II, 47, 2785–2804, doi:10.1016/S0967-0645(00)00049-7, 2000.

Wollschläger, J., Grunwald, M., Röttgers, R., and Petersen, W.: Flow-through PSICAM: a new approach for determining water constituents absorption continuously, Ocean Dynam., 63, 761–775, doi:10.1007/s10236-013-0629-x, 2013.

Wollschläger, J., Röttgers, R., Petersen, W., and Wiltshire, K. H.: Performance of absorption coefficient measurements for the *in situ* determination of chlorophyll *a* and total suspended matter, J. Exp. Mar. Biol. Ecol., 453, 138–147, doi:10.1016/j.jembe.2014.01.011, 2014.

Zielinski, O., Busch, J. A., Cembella, A. D., Daly, K. L., Engelbrektsson, J., Hannides, A. K., and Schmidt, H.: Detecting marine hazardous substances and organisms: sensors for pollutants, toxins, and pathogens, Ocean Sci., 5, 329–349, doi:10.5194/os-5-329-2009, 2009.

Zielinski, O., Voss, D., Saworski, B., Fiedler, B., and Koertzinger, A.: Computation of nitrate concentrations in turbid coastal waters using an *in situ* ultraviolet spectrophotometer, J. Sea Res., 65, 456–460, 2011.

9

Recurrence intervals for the closure of the Dutch Maeslant surge barrier

Henk W. van den Brink[1] **and Sacha de Goederen**[2]

[1]KNMI, Utrechtseweg 297, De Bilt, the Netherlands
[2]Rijkswaterstaat, Boompjes 200, Rotterdam, the Netherlands

Correspondence to: Henk W. van den Brink (henk.van.den.brink@knmi.nl)

Abstract. The Dutch Maeslant Barrier, a movable surge barrier in the mouth of the river Rhine, closes when there is a surge in the North Sea and the water level in the river at Rotterdam exceeds 3 m above mean sea level. An important aspect of the failure probability is that the barrier might get damaged during a closure and that, within the time needed for repair, a second critical storm surge may occur. With an estimated closure frequency of once in 10 years, the question of how often the barrier has to be closed twice within one month arises.

Instead of tackling this problem by the application of statistical models on the (short) observational series, we solve the problem by combining the surge model WAQUA/DCSMv5 with the output of all seasonal forecasts of the European Centre of Medium-Range Weather Forecasting (ECMWF) in the period 1981–2015, whose combination cumulates in a pseudo-observational series of more than 6000 years.

We show that the Poisson process model leads to wrong results as it neglects the temporal correlations that are present on daily, weekly and monthly timescales.

By counting the number of double events over a threshold of 2.5 m and assuming that the number of events is exponentially related to the threshold, it is found that two closures occur on average once in 150 years within a month, and once in 330 years within a week. The large uncertainty in these recurrence intervals of more than a factor of two is caused by the sensitivity of the results to the Gumbel parameters of the observed record, which are used for bias correction.

Sea level rise has a significant impact on the recurrence time for both single and double closures. The recurrence time of single closures doubles with every 18 cm mean sea level rise (assuming that other influences remain unchanged) and double closures double with every 10 cm rise. This implies a 3–14 times higher probability of a double closure for a 15–40 cm sea level rise in 2050 (according to the KNMI climate scenarios).

1 Introduction

In 1953, a large part of south-west Netherlands was flooded by the sea, with over 1800 casualties. After these floods it was decided to shorten the Dutch coastline by approximately 700 km by building both closed and permeable dams between the isles in the south-west of the country. In this way not all dikes had to be made higher.

In 1987 it was decided to build a movable surge barrier in the so-called New Waterway (in Dutch: "Nieuwe Waterweg", which is the artificial mouth of the river Rhine into the North Sea, located at 20 km downstream from the city of Rotterdam), which only has to be closed during dangerous situations. In this way, the Rotterdam harbour can remain accessible for sea shipping. This barrier, called the Maeslant Barrier (in Dutch: "Maeslantkering"), has been operational since 1997. When the forecasted water level in Rotterdam exceeds 3 m above Normaal Amsterdams Peil (NAP; "Amsterdam Ordnance Datum", which is approximately equal to mean sea level), the barrier is closed.[1] This situation was expected to happen once in approximately 10 years. In the pe-

---

[1]Formally, the barrier also closes if the level in Dordrecht exceeds 2.90 m above NAP. However, it is very unlikely that the water level exceeds the threshold in Dordrecht but not in Rotterdam.

riod 1997–2016, the barrier has been closed once in storm conditions: this event happened on 8–9 November 2007.

In order to guarantee the required safety level for the hinterland, the failure probability of the Maeslant Barrier is required to be maximally 0.01, i.e. it has to close correctly in 99 out of 100 cases (Rijkswaterstaat, 2013). An important aspect of the failure probability is the scenario that the barrier gets damaged during a closure and that, within the time needed for repair, a second critical storm surge occurs.

The time that the barrier can not be closed due to repair depends, naturally, on the complexity of the breakdown. Therefore, we explore the frequency of all succeeding closures with an inter-arrival time from one day to one month.

For the estimation of the probability of two closures within a given short time interval (which we will call a double closure here), the observational record of the single event that did occur obviously does not provide any information about inter-arrival times. Nevertheless, in order to derive information about the double closures from the observations, one possibility is to explore how often some threshold lower than 3 m above NAP has been exceeded and then to scale these probabilities to the required level. A different approach might be to regard the closures to be independent, which leads to a Poisson distribution for the inter-arrival times (see Sect. 3). Assuming that the average return period is about 10 years, an estimate can be obtained of how often the recurrence time is 1 week or 1 month. However, the result of this approach is very sensitive to the estimated recurrence period, and is biased due to the neglect of temporal correlations in the atmosphere.

We therefore used an alternative approach (Van den Brink et al., 2005b), i.e. by combining the seasonal forecasts (Vialard et al., 2005) of the European Centre of Medium-Range Weather Forecasting (ECMWF) into a large dataset, representing the current climate with more than 6000 independent years (up until December 2015). Thereafter we calculated the surges from the winds and pressures from this dataset, resulting in a high-frequency time series of water levels with the same length as the ECMWF dataset. From this dataset of water levels, the required inter-arrival times in a stationary climate can be counted and analysed.

The paper is structured as follows: the meteorological and hydrological models, and the observational dataset are described in Sect. 2. Section 3 explains the applied methodology, Sect. 4 shows the validation of the model outcomes and Sect. 5 describes the results. The conclusions are presented in Sect. 6.

2 Models and observations

The seasonal forecasts of the ECMWF are used to drive the surge model WAQUA/DCSMv5, which outputs (among others) the water level at the coastal station Hoek van Holland (see Fig. 4 for its location). The city of Rotterdam is located about 25 km from Hoek van Holland upstream of the river Rhine. Although the height of the water level in Rotterdam is mainly determined by the water level at Hoek van Holland, it is also influenced by the discharge of the river Rhine. A simple analytical relation is therefore used to simulate the effect of the Rhine discharge on the water levels in Rotterdam. All three models are briefly described below and the observational record is also described.

2.1 ECMWF seasonal model runs

From November 2011 onward the ECMWF produces every month an ensemble of 51 global seasonal forecasts up to 7 months ahead, i.e. amply surpassing the 2 week horizon of weather predictability from the atmospheric initial state. Over the period 1981–2011, re-forecasts with smaller ensembles have been performed to calibrate the system. The forecast system consists of a coupled atmosphere–ocean model. The atmospheric component has a horizontal resolution of T255 (80 km) and 91 levels in the vertical (Molteni et al., 2011). The ocean component NEMO has a resolution of 1 degree and 29 vertical levels (Madec, 2008). The wave model WAM (Janssen, 2004) allows for the two-way interaction of wind and waves with the atmospheric model. All forecasts are generated by the so-called System 4 (Molteni et al., 2011).

The ECMWF dataset provides, among other fields, global fields of 6-hourly wind and sea-level pressures (SLP). We have regridded the data to a regular grid of 0.5°.

From every 7-month forecast, we skipped the first month in order to remove dependence between the perturbed members due to the correlation in the initial meteorological states. Van den Brink et al. (2005a) show that the correlation of the North Atlantic Oscillation (NAO) index approaches zero for the forecasts after 1 month. We combined two forecasts that differ by 6 months in start time to construct a full calendar year. The total number of forecasts that have been combined to full years is 12 556, resulting in 6282 independent calendar years.

Table 1 clarifies how the individual members are combined to construct the 6282-year time series. It shows that the first year is constructed from the combination of ensemble member 0 starting in January 1981 with ensemble member 0 starting in July 1981. As the first months are skipped, together they cover the period 1 February 1981 to 31 January 1982. The next year continues with 1 February 1982.

Although the thus obtained dataset is as continuous as possible, several peculiarities are left. First, there is a discontinuity at every concatenation point, which aborts the temporal correlation in the meteorological situation. The correlation in the astronomical tide is however preserved. As the concatenation follows the historical order for every perturbation number, possible low-frequency variability (e.g. due to the sea surface temperature) is maintained (Graff and LaCasce, 2012). In this way, the 18.6-year lunar nodal cycle is also in-

Table 1. Combination of individual forecast members to construct the 6282-year time series. The numbers indicate the start year (1981–2015) followed by the perturbation number (0–50). See the text for further explanation.

Year	First half year	+	Second half year
1	Jan 1981-0	+	Jul 1981-0
2	Jan 1982-0	+	Jul 1982-0
⋮			
35	Jan 2015-0	+	Jul 2015-0
36	Jan 1981-1	+	Jul 1981-1
⋮			
669	Jan 2015-50	+	Jul 2015-50
670	Feb 1981-0	+	Aug 1981-0
⋮			
6282	Jun 2015-50	+	Dec 2015-50

corporated. The only discontinuities in the initial states occur when 2015 is reached and the next year starts again in 1981 (from year 35 to 36 in Table 1). Discontinuities in the calendar years are made when the perturbation number jumps back from 50 to 0. In that case, one calendar month is skipped (from year 669 to 670 in Table 1). These few discontinuities have negligible influence on the outcomes.

2.2 WAQUA/DCSMv5 model

To infer surge heights in the North Sea from the ECMWF output we use WAQUA/DCSMv5 (Gerritsen et al., 1995). This model solves the two-dimensional shallow-water equations on a $1/12 \times 1/8$ (approximately 8×8 km) grid on the north-west European shelf region. It is operationally used at KNMI to predict the water levels along the Dutch coast. Meteorological input are SLP and 10 m wind. The latter is translated into wind stress using a drag coefficient based on the parameterisation of Charnock (1955), with a Charnock parameter of 0.032. The astronomical tide is prescribed at the open boundaries in ten harmonic constituents ($O_1, K_1, N_2, M_2, S_2, K_2, Q_1, P_1, \nu_2$ and L_2) and propagates from there into the model domain. The model output consists of total water level and the height of the astronomical tide in the absence of meteorological forcing.

We analyse the model results in terms of total water level as this is the quantity relevant for the closure of the Maeslant Barrier.

2.3 Rhine discharge model

The water level at Rotterdam is influenced both by the sea level at Hoek van Holland and the river discharge. Based on calculations by Rijkswaterstaat (De Goederen, 2013, p. 25), the water level at Rotterdam L_R can be approximated by the following:

$$L_R - L_{HvH} = 4.08 \times 10^{-5}(Q_L - 1750), \qquad (1)$$

in which L_{HvH} is the level at Hoek van Holland and Q_L is the river Rhine discharge at Lobith (where the Rhine enters the Netherlands, see Fig. 4) in $m^3 s^{-1}$.

In order to take the effect of the river discharge into account, we applied the right-hand side of Eq. (1) to the historical 1901–2000 daily record of discharges at Lobith. To every WAQUA/DCSMv5 ensemble member time series for Hoek van Holland we added a 6-month period starting with the same date as the ECMWF ensemble member starts with, randomly selected from the Lobith record. In this way the seasonal variation of the river discharge is maintained. This approach implies that there is no correlation between high sea surges and river discharges, which is approximately true (Van den Brink et al., 2005a; Kew et al., 2013)[2].

2.4 Water level observations

The observational record of water levels at Hoek van Holland starts in 1864 (Holgate et al., 2013; PSMSL, 2017). Accurate readings of the water level start in August 1887. We used the data from 1888 onward. The data before 1987 are obtained visually from (digitised) charts and afterwards 10-min average values are used.

Due to sea level rise and land subsidence, the observational water levels in the historical record have to be corrected for these influences. Figure 1 shows how the observations taken in Hoek van Holland from 1880 to 2015 have to be adjusted to be representative for the year 2009. The correction varies from 31 cm for 1888 to −2 cm for 2015.

The rapid change around 1965 and the change in the slope can be attributed to the extension of the Rotterdam harbour to the west (Dillingh et al., 1993; Hollebrandse, 2005; Becker et al., 2009).

The annual maxima of the water level in Hoek van Holland are shown in Fig. 2, both uncorrected and corrected.

The observational record contains (after correction) 10 events that exceed 3 m in Hoek van Holland, the smallest inter-arrival time being 1.2 years (in 1953 and 1954). This makes direct derivation of the recurrence intervals of inter-arrival times less than 1 year impossible.

3 Methodology

In this section the methodology that we use to derive the recurrence times of double closures within 1 month is pre-

[2]In the case that extreme discharge and extreme water levels are correlated, the most promising solution – in line with the topic of this paper – is to use the precipitation amounts, the temperature and snow melt in the Rhine basin as input for a hydrological model to calculate the Rhine discharge. In this way no explicit assumptions about the correlation have to be made. This is however outside the scope of this paper.

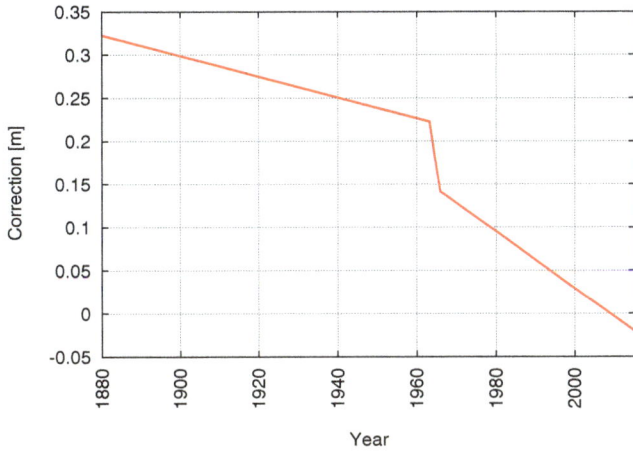

Figure 1. Adjustment of the observed sea levels in Hoek van Holland (1880–2015) to correct for sea level rise and land subsidence. The observational record is adjusted to be representative for the situation in 2009.

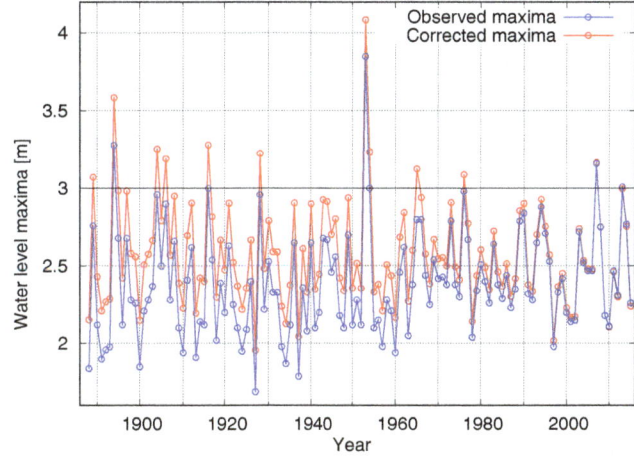

Figure 2. Annual maxima of the observed water levels in Hoek van Holland, 1887–2015 (blue). Correction according to Fig. 1 leads to the red plot.

sented. As mentioned earlier, the observational record does not contain double closures within one year, which makes direct derivation of the required inter-arrival times impossible. We therefore combine the observations with information from the ECMWF-WAQUA/DCSMv5 dataset. The first step in the evaluation of the quality of this dataset is to check whether the annual maxima of the water level in Hoek van Holland as derived from the ECMWF and WAQUA/DCSMv5 dataset has the same distribution as the observational dataset. The theory of the annual maxima, which is applied for the intercomparison, is described in Sect. 3.1.

The next step is to check whether the recurrence times of double closures can be described by a Poisson distribution, which would occur if the events are mutually independent (Sect. 3.2). Section 5 shows that the events are not mutually independent, which hinders the application of the Poisson distribution, either to the observations or to the ECMWF-WAQUA/DCSMv5 dataset.

The huge size of the ECMWF-WAQUA/DCSMv5 dataset invites us to explore whether the recurrence times of the events that exceed the threshold of 3.0 m can be derived directly from the empirical density function (EDF) of the dataset. This EDF is introduced in Sect. 3.3. Section 5 shows, however, that even the 6282-year dataset is not long enough for an accurate estimation of the desired recurrence times by counting the intervals. We introduce an extra step by using a lower threshold than 3.0 m, and by deriving the relation between the threshold and the number of inter-arrival times (Eq. 11).

Section 4.2 shows that a small bias correction of the ECMWF-WAQUA/DCSMv5 dataset is necessary (Eq. 10). A short analysis of the uncertainty analysis introduced by this

bias correction, as well as by the use of a lower threshold, is given in Sect. 3.4.

3.1 Extreme value analysis

To determine the extreme water levels that occur on average once in a given period (the return period), annual maxima are fitted to a generalized extreme value (GEV) distribution, which is the theoretical distribution for block maxima (e.g. Coles, 2001).

$$G(y) = \exp\{-[1 + \xi(\frac{y - \mu}{\sigma})]^{-1/\xi}\} \tag{2}$$

Here, μ, σ and ξ are called the location, scale and shape parameter, respectively, and y is the sea water height. If $|\xi| \to 0$, Eq. (2) can be written as

$$G(y) = \exp\{-\exp[-\frac{y - \mu}{\sigma}]\}, \tag{3}$$

which is called the Gumbel distribution.

The return period T_s, which is the average recurrence time of a single exceedence of level y, is defined by

$$T_s = \frac{1}{1 - G(y)}. \tag{4}$$

For large return periods, the combination of Eqs. (3) and (4) can be approximated by the following:

$$y \approx \mu + \sigma \log(T_s). \tag{5}$$

The distributions of the annual extremes are presented in this paper in the form of Gumbel plots, in which the annual maxima (or minima) are plotted as a function of the Gumbel variate $x = -\ln(-\ln(G(y)))$. In the case of a Gumbel distribution this results in a straight line. Using Eq. (4) the Gumbel variate is directly related to the return period, which we label on the upper horizontal axis of the plots.

The parameters are derived by maximum likelihood estimation.

3.2 Inter-arrival times

The inter-arrival times of independent events can be described as a Poisson process. If N_t is the number of events that occurs before time t and $1/\lambda$ is the average recurrence time, then

$$P(N_t = k) = \frac{(\lambda t)^k}{k!} e^{-\lambda t}. \tag{6}$$

We are interested in the probability that the time until the next event ΔT is larger than a given value t. This means that no events occurred before time t, i.e. $k = 0$. It thus follows that

$$P(\Delta T > t) = P(N_t = 0) = e^{-\lambda t}, \tag{7}$$

which states that the inter-arrival time between independent events is exponentially distributed.

For small inter-arrival times ($\Delta T \ll 1/\lambda$), Eq. (7) can be rewritten as

$$T_d \approx \frac{1}{\lambda \Delta T}, \tag{8}$$

in which T_d is the recurrence time of a double event. An average recurrence time of 10 years thus implies that a double closure within a month occurs once in 120 years if independence is assumed.

3.3 Empirical distribution function

Let $x_1 \leq x_2 \ldots \leq x_n$ be n observations from distribution F, then the empirical distribution function (EDF) $\hat{F}$ (e.g. Buishand and Velds, 1980) is given by the following:

$$\hat{F}(x_i) = \frac{i}{n+1}. \tag{9}$$

Equation (9) states that $F(x)$ can be estimated from the number of observations lower than x. The advantage of Eq. (9) is that it requires no assumptions about F.

3.4 Uncertainty analysis

We consider two contributions to the uncertainty in the estimation of the recurrence intervals. The first one is the bias correction in the location and scale parameter of the GEV distribution (Eq. 10). As the ECMWF-WAQUA/DCSMv5 dataset is 49-times longer than the observational record, the uncertainty in the bias correction will be dominated by the uncertainty in the Gumbel fit to the observations. A first-order estimation of the 95 % uncertainty range due

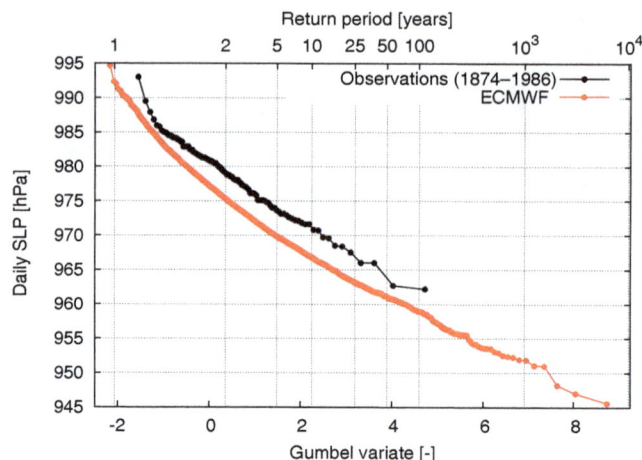

Figure 3. Gumbel plot of observed annual minimum sea level pressure (SLP) in Nordby (black) and as simulated at the nearest ECMWF grid point (red). The location of Nordby is indicated in Fig. 4.

to the bias correction is made by keeping the ECMWF-WAQUA/DCSMv5 unchanged in Eq. (10) and replacing μ_{obs} and σ_{obs} with $\mu_{obs} \pm 2\Delta\mu_{obs}$ and $\sigma_{obs} \pm 2\Delta\sigma_{obs}$, respectively. Here, $\Delta\mu_{obs}$ and $\Delta\sigma_{obs}$ are the standard errors in the location and scale parameter as derived from the observations. Redoing the calculations with those adjusted bias corrections gives a good indication of the uncertainty range in the estimated recurrence intervals.

The second contribution to the uncertainty in the estimation of the recurrence intervals is the choice of the threshold and its scaling to the threshold of 3.0 m (Eq. 11). The range of thresholds is varied from 2.3 m (high enough to resemble extreme conditions) and 2.7 m (low enough to reduce statistical uncertainty). The variation of the recurrence intervals for different thresholds gives an indication of the sensitivity of the estimated recurrence intervals for the choice of threshold.

4 Evaluation

4.1 ECMWF wind and pressure fields

In order to model extreme surge events correctly, in particular the wind and pressure should be well represented by the model. Due to the sensitivity of model wind to the drag parameterisation, it is difficult to verify the model winds directly. Instead, we validated the SLP. This direct model parameter can be compared more easily with observations than wind data and is a good measure of the capability of the model to produce deep depressions (see also Sterl et al., 2009).

Figure 3 shows the annual minimum daily-mean SLP for the observations in Nordby, Denmark (8.4° E, 55.45° N) for

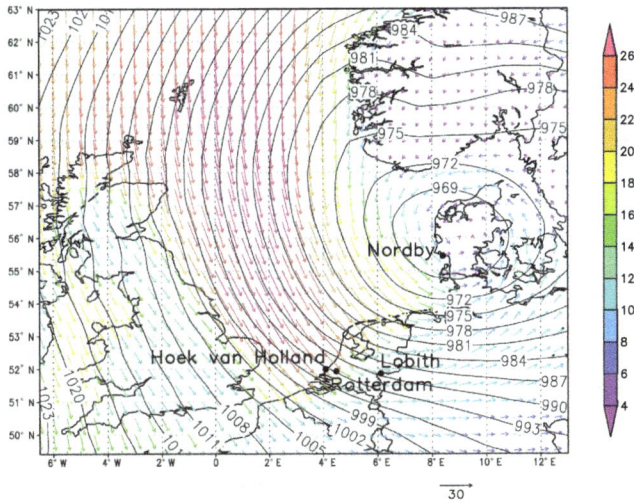

Figure 4. Wind and pressure fields for the situation leading to the highest surge in Hoek van Holland that occurred in the ECMWF-WAQUA/DCSMv5 ensemble. The locations of Hoek van Holland, Rotterdam, Lobith and Nordby are indicated.

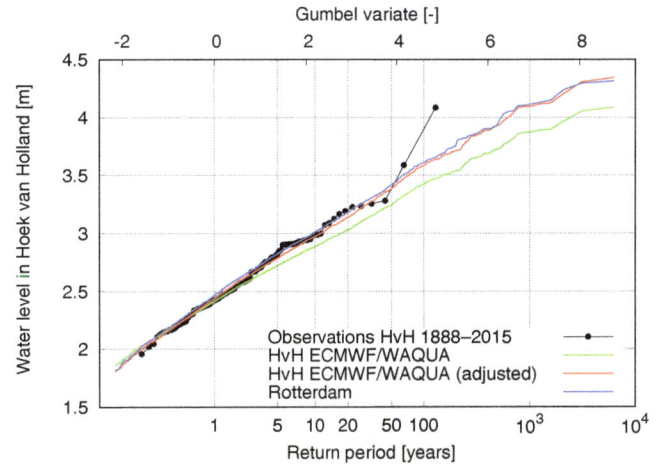

Figure 5. Gumbel plot of the annual maximum water levels in Hoek van Holland according to the observations (black) and the ECMWF-WAQUA/DCSMv5 ensemble (red). Adjusting the Gumbel parameters of the ECMWF-WAQUA/DCSMv5 ensemble to match the observed Gumbel parameters (according to Eq. 10) results in the blue distribution.

the 1874–1986 period, and the ECMWF data for the nearest grid point. This location was chosen because a pressure minimum in this area leads to long north–west oriented wind fetches over the North Sea and therefore to high surges at the Dutch coast. This is illustrated in Fig. 4, which depicts the pressure and the wind field related to the highest surge of 4.29 m in Hoek van Holland that occurred on 21 January 1988, in ensemble member 17 that started on 1 August 1987. The figure also depicts the location of Nordby.

In the Gumbel plot (Fig. 3) observed and simulated values yield parallel curves. Modelled pressures are slightly lower than the observed ones but the model has the same relation between intensity and frequency of low pressures as the observations have. There is no sign of an artificial lower limit on pressure in the model. From Fig. 3 we conclude that the ECMWF dataset is appropriate to drive a surge model for water level calculations.

4.2 WAQUA/DCSMv5 surge model

Figure 5 shows the Gumbel plot of the annual maxima for the observations (black) and the ECMWF-WAQUA/DCSMv5 ensemble (green) for Hoek van Holland. The once-a-year extreme water level of 2.43 ± 0.02 m (represented by the Gumbel location parameter) is reproduced within 0.6 % (2.36 ± 0.004). The scale parameter of the Gumbel distribution (0.264 ± 0.018) is slightly underestimated (0.252 ± 0.002). It is likely that this underestimation is caused by the fact that WAQUA/DCSMv5 uses a fixed Charnock parameter, whereas the ECMWF uses a time-varying Charnock parameter. For high winds, the ECMWF Charnock parameter exceeds the value of 0.032 used by WAQUA/DCSMv5, which leads to an underestimation of

high surges (Zweers et al., 2010; Van Nieuwkoop et al., 2015).

In order to correct for this feature, we applied the following correction to the ECMWF-WAQUA/DCSMv5 water levels at Hoek van Holland:

$$L_{\text{adj}} = \mu_{\text{obs}} + \sigma_{\text{obs}} \frac{L_{\text{org}} - \mu_{\text{EW}}}{\sigma_{\text{EW}}}, \qquad (10)$$

in which L_{org} is the original water level as calculated by ECMWF-WAQUA/DCSMv5 and L_{adj} is the adjusted water level. The subscripts obs and EW refer to the Gumbel parameters of the observations and ECMWF-WAQUA/DCSMv5, respectively. The quantile mapping of Eq. (10) ensures that the ECMWF-WAQUA/DCSMv5 water levels have the same Gumbel location and scale parameter as the observations have. The results presented in this paper are based on L_{adj}.

Correction according to Eq. (10) results in the red line of Fig. 5. For the once-in-10-years return level this implies a correction of only 3 % in the water level and 5 % in the surge (taking the average astronomical high tide of 1.21 m into account). We conclude that, although a correction is necessary, this correction is small enough to trust the water levels of the ECMWF-WAQUA/DCSMv5 ensemble for determining the closure frequencies.

4.3 Rhine discharge

Figure 6 shows the histogram of the effect of the Rhine discharge on the water level at Rotterdam.

In the observational record the maximum effect of the river discharge to the water level in Rotterdam (according to Eq. 1) is 0.43 m for the highest observed discharge of 12 280 m³ s⁻¹ at Lobith. The yearly averaged addition of the river discharge

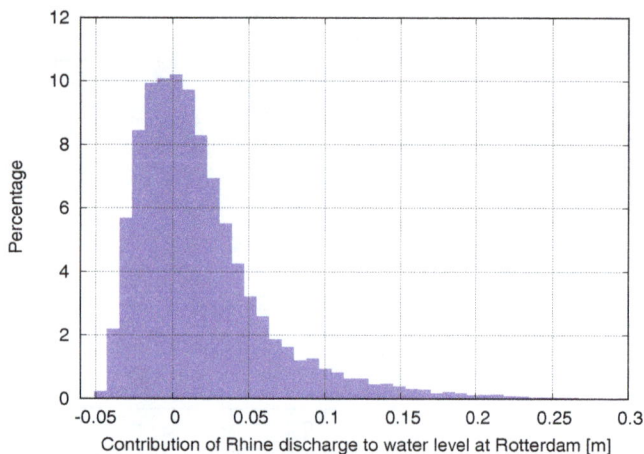

Figure 6. Histogram of the effect of the Rhine discharge on the water level at Rotterdam. The bin width is 0.5 cm.

to the water level in Rotterdam is 0.02 m and the average effect to the annual maxima at Rotterdam is 0.03 m. This means that the Gumbel plot of the annual maximum water levels at Rotterdam is about 0.03 m higher than that of Hoek van Holland (see blue line in Fig. 5). We note that the equation used by Zhong et al. (2012) to model the effect of the Rhine discharge on the water level in Rotterdam gives identical results.

We conclude that the effect of the Rhine discharge on the water level in Rotterdam can be substantial but that the average effect on the extreme levels is only a few centimetres.

5 Results

Figure 7 shows the distribution of the inter-arrival times for two thresholds: 2 m (panel a) and 3 m (panel b). The insets show the distributions for the first 14 days (a) and 30 days (b). Note that the unit of the horizontal axis of the upper panel is in days and of the lower panel is in years. Figure 7 illustrates the following six items.

First, in the 6282-year dataset, the maximum recurrence time for a level of 2 m is 1095 days and 135 years for a level of 3 m. Second, the deviations from the straight black line (which represents an exponential distribution on the logarithmic vertical scale) indicate that the distribution of recurrence times is not a Poisson process – neither for the 2 m nor for the 3 m threshold. Especially in Fig. 7(a) the seasonal variation in the recurrence times is clearly visible by the oscillation around the black line. This is the result of the low probability in summer and higher probability in winter that a 2 m event occurs. Third, as illustrated by the insets, the probability of a recurrence time of less than 5 days is considerably higher than independence between the inter-arrival times would indicate. Apparently, there is clustering of extreme events up to inter-arrival times of 5 days. This is in agreement with Mailier et al. (2006) who quantify the clus-

tering of extra-tropical cyclones for the area of interest. Also, the influence of spring tide will increase the probability of double closures (Van den Hurk et al., 2015). Fourth, the inset of Fig. 7(a) also shows the influence of the deterministic astronomical tide: the figure shows a 12.5-hourly oscillation caused by the fact that all exceedances of the thresholds occur at high tide. Fifth, the inter-arrival times for very high thresholds (as shown in Fig. 7b for the 3 m threshold) are distributed according to Eq. (7), which implies that the inter-arrival times on an annual scale can be considered to be independent.[3] Sixth, the ECMWF-WAQUA/DCSMv5 shows a good agreement with the observations for the 2 m threshold (blue line in Fig. 7a). As there are only 10 exceedances above the 3 m threshold in the observational record (blue line in lower panel), it is hard to verify the distribution for the 3 m threshold. None of the 3 m exceedances in the observations are within a month of each other.

From Fig. 7 it can be concluded that the assumption of independence for the occurrence of (extreme) water levels is violated on a daily scale by the astronomical tide, on a weekly scale by the clustering of extra-tropical cyclones and spring tide, and on a monthly scale by the seasonal variation in the storm intensity and frequency. Only at annual scales are the inter-arrival times exponentially distributed, and thus can be considered to be independent. This means that we cannot assume independence in order to calculate recurrence times for short intervals, and thus cannot apply Eq. (7) to estimate the probability of a double closure within a week or month.

Instead of assuming independence, we could directly count the inter-arrival times between the events that exceed the 3 m water level and construct an EDF from them (see Sect. 3.3). However, even the 6282-year dataset is too short for this approach as the dataset contains only 30 inter-arrival times that are less than a month. Direct derivation of the EDF is therefore not possible.

In order to bypass this problem, we explore how the required EDF for 3 m relates to the EDF for lower thresholds. Figure 8 shows that the number of occurrences in which the threshold is exceeded twice within 1 week is exponentially related to the threshold (blue line and points). The same holds for inter-arrival times of 2 and 4 weeks (green and red resp.).

Fitting the line

$$\ln(N) = N_0 - y/\beta, \qquad (11)$$

where N is the count, y the threshold and β the slope yields a value of N_0 that depends on the time window and $\beta = 0.145 \pm 2\%$ for all three lines.

The figure indicates that we can base our desired EDF on a lower threshold than 3 m and transform those results to the required EDF for a threshold of 3 m by a simple multiplica-

---

[3] Although the forecast runs are only 7 months in length, they are combined in such a way that a possible oceanic influence remains present; see Table 1.

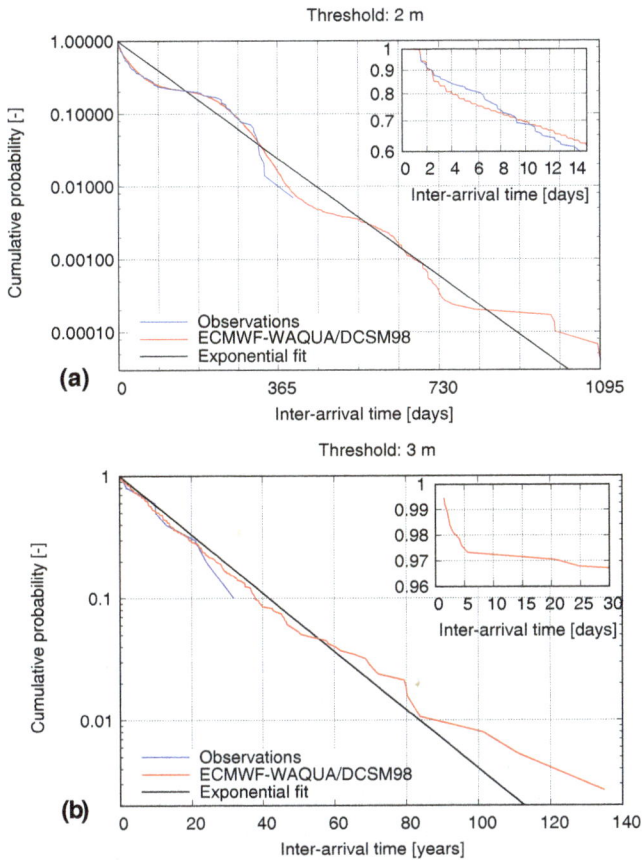

Figure 7. Distributions of the recurrence times of the water level in Hoek van Holland for a threshold of 2 m **(a)** and 3 m **(b)**. The insets show the distribution for the first 14 and 30 days, respectively. The black line represents an exponential fit to the whole dataset. The vertical axes are logarithmic. The blue lines represent the observations.

tion. The value of that multiplication factor M for the probabilities of the 3 m threshold is given by the following:

$$M = \frac{N_1}{N_2} = \exp\left(\frac{y_2 - y_1}{\beta}\right), \tag{12}$$

in which N_1 is the counted number of occurrences for which the water level exceeds y_1 metres twice within the given time window and N_2 is the number of double closures for the level y_2 (3.0 m) we are looking for.

The fact that the three lines in Fig. 8 are parallel indicates that this multiplication factor is virtually independent of the time window.

We chose to derive the EDF on a threshold of y_1 equal to 2.5 m as this threshold gives a good compromise between the number of occurrences (we then have 1228 events that occur within 4 weeks of the previous event and 601 events that occur within 1 week) and the extremity of the threshold (the 2.5 m threshold is exceeded on average once in 2 years; see Fig. 5). According to Eq. (12), the probabilities of the

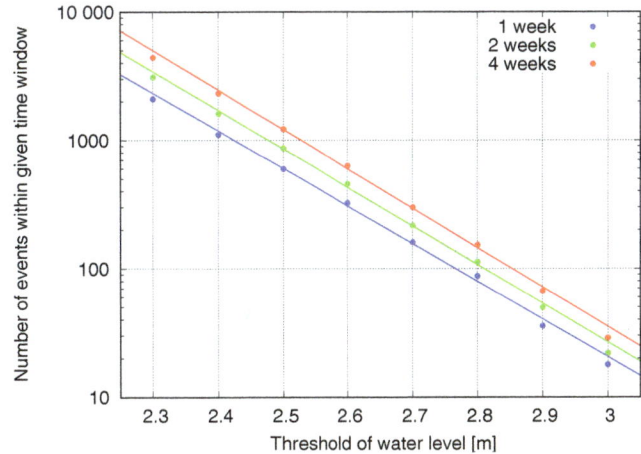

Figure 8. Number of occurrences N in which the threshold y is exceeded twice within 1, 2 or 4 weeks (blue, green and red respectively). The vertical axis is logarithmic.

2.5 m threshold have then to be multiplied by 0.032 to be transformed to the 3.0 m threshold.

5.1 Recurrence times

Figure 9 shows the recurrence times as a function of the inter-arrival times for a threshold of 3 m.

The green shading in the figure illustrates the effect on the recurrence time if the EDF is based on thresholds in the range of 2.3–2.7 m. Apparently, the outcome varies about 10 % with the choice of the threshold. The blue shading represents the uncertainty due to the bias correction of Eq. (10), in which $\mu_{obs} = 2.43$ and $\sigma_{obs} = 0.264$ are replaced by $\mu \pm 2\Delta\mu_{obs}$ (2.38, 2.47) and $\sigma \pm 2\Delta\sigma_{obs}$ (0.23, 0.30), respectively. The figure shows that the uncertainty due to the bias correction is much larger than that due to the threshold selection.

From Fig. 9 it can be seen that in the current climate the Maeslant Barrier has to be closed twice in a month due to exceedence of the 3.0 m threshold once in about 150 years (95 % uncertainty range is 70–390 years). Once in about 330 years (95 % uncertainty range is 150–810 years) it has to be closed twice in a week. The 95 % uncertainty ranges show that the uncertainty in the recurrence time is slightly more than a factor of two.

The oscillations in the graph of Fig. 9 are caused by the fact that exceedences of the threshold always occur at high tide, i.e. the inter-arrival times are always a multiple of 12.5 h.

The dashed line represents the Poisson distribution (Eq. 7) with $\lambda = 0.10$ per year. It shows that the assumption of independence leads to considerable deviations in the estimation of the recurrence times of double events.

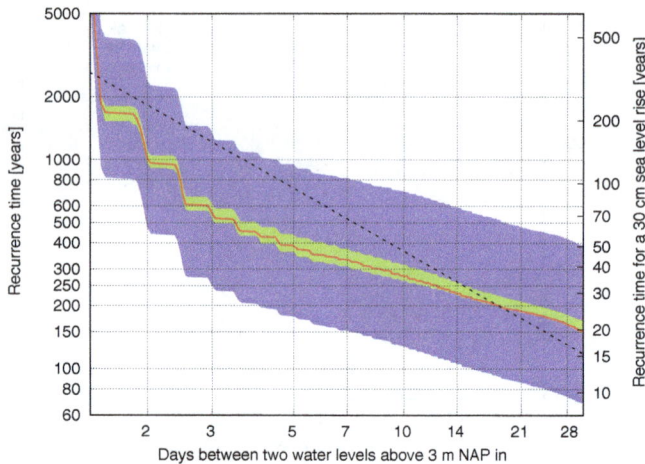

Figure 9. Recurrence times in years as a function of the inter-arrival time for a threshold of 3.0 m. The line is determined by counting the events for a threshold of 2.5 m and multiplying the probabilities with 0.032 (according to Eq. 12.). The green shading illustrates the effect on the recurrence time if the EDF is based on thresholds in the range of 2.3–2.7 m. The blue shading is the 95 % confidence interval due to the bias correction of Eq. (10). The dashed line represents the Poisson distribution (Eq. 7) with $\lambda = 0.10$ per year. The right axis shows the recurrence times for a sea level rise of 30 cm. All axes are logarithmic.

5.2   Effect of sea level rise

In order to estimate the first-order effect of sea level rise on the closure frequency we assume no changes in the wind climate, no change in river discharge and no effect of sea level rise on the surge and astronomical tides (which is approximately true; see e.g. Lowe et al., 2001; Sterl et al., 2009). In that case, the effect of sea level rise can be incorporated by considering the probabilities of a threshold that is accordingly lower. A sea level rise of 0.3 m will thus lead to the situation as if closure takes place at 2.7 m instead of 3.0 m.

5.2.1   Effect of sea level rise on single closures

The effect of sea level rise on the number of single closures can be derived by calculating $\partial T / \partial y$ from Eq. (5). It easily follows that

$$\ln\left(\frac{T_{s,2}}{T_{s,1}}\right) \approx \frac{y_2 - y_1}{\sigma}, \tag{13}$$

in which return periods $T_{s,1}$ and $T_{s,2}$ belong to water levels y_1 and y_2, respectively. Here, $\sigma = 0.26$ is the Gumbel scale parameter. It directly follows from Eq. (13) that a 0.18 m sea level rise doubles the closure frequency. With the expected sea level rise of 0.15–0.40 m in 2050 with respect to 1981–2010 (Van den Hurk et al., 2014), the closure frequency will increase by a factor of 1.8–4.6.

5.2.2   Effect of sea level rise on double closures

The effect of sea level rise on the probability of two closures within a time window can directly be derived from Fig. 8 and Eq. (12). In a way similar to Eq. (13) it follows from Eq. (11) that

$$\ln\left(\frac{N_1}{N_2}\right) = \frac{y_2 - y_1}{\beta}. \tag{14}$$

Equation (14) shows that approximately every 0.10 m, sea level rise doubles the probability that two closures occur within a given time window. The expected sea level rise of 0.15–0.40 m in 2050 results in recurrence times that are 2.8–16 times more frequent than in the reference situation. These recurrence times for 0.3 m sea level rise are indicated on the right axis of Fig. 9.

6   Conclusions

The seasonal forecasts of the ECMWF, with a total length of more than 6000 years, represent the current wind climatology over the North Sea area very accurately. Combination of the ECMWF output with the surge model WAQUA/DCSMv5 results in a 6282-year dataset of water levels that (after a small correction) are well suited for many research objectives.

In this paper we apply the dataset in order to estimate how often the movable Maeslant Barrier in the New Waterway (which is the artificial mouth of the river Rhine) has to be closed twice in a short time interval – varying between days up until a month. This is of importance as the barrier might get damaged during the first closure and the barrier can not be closed during the repair time.

Assuming independence between two closures leads to wrong estimates of the double closures. Independence is violated by the deterministic component of the astronomical tide on the daily scale, by clustering of depressions and by spring tide on the weekly scale, and by seasonality on the monthly scale.

By counting the number of double events over a threshold of 2.5 m, and assuming that the number of events is exponentially related to the threshold, it is found that the barrier has to be closed within a month approximately once in 150 years, and once in 330 years within a week. The large uncertainty in these recurrence intervals of more than a factor of two is caused by the sensitivity of the results to the Gumbel parameters of the observed record, which are used for bias correction.

Sea level rise has a large impact on the frequency of single and double closures. Every 10 cm sea level rise doubles the probability of double closures, resulting in 2.7–14 times more double closures for the 0.15–0.40 m expected sea level rise in 2050.

Competing interests. The authors declare that they have no conflict of interest.

Acknowledgements. This work was initiated and funded by Rijkswaterstaat.

Edited by: Mario Hoppema

References

Becker, M., Karpytchev, M., Davy, M., and Doekes, K.: Impact of a shift in mean on the sea level rise: Application to the tide gauges in the Southern Netherlands, Cont. Shelf Res., 29, 741–749, https://doi.org/10.1016/j.csr.2008.12.005, 2009.

Buishand, T. and Velds, C.: Neerslag en Verdamping (Eng: "Precipitation and Evaporation"), Staatsdrukkerij, 's Gravenhage, NL, 1980.

Charnock, H.: Wind stress on a water surface, Q. J. Roy. Meteor. Soc., 81, 639–640, https://doi.org/10.1002/qj.49708135027, 1955.

Coles, S.: An Introduction to Statistical Modelling of Extreme Values, Springer-Verlag, London, 2001.

De Goederen, S.: Betrekkingslijnen Rijn-Maasmonding 2006 (Eng: "Isolines of water levels for the Rhine-Meuse mouth"), available at: http://publicaties.minienm.nl/documenten/betrekkingslijnen-rijn-maasmonding-2006 (last access: 4 July 2017), 2013.

Dillingh, D., de Haan, L., Helmers, R., Können, G., and van Malde, J.: De basispeilen langs de Nederlandse kust, Statistisch onderzoek (Eng: "Critical return values along the Dutch coast; Statistical research"), Tech. Rep. DGW-93.023, Ministerie van Verkeer en Waterstaat, Directoraat-Generaal Rijkswaterstaat, 1993.

Gerritsen, H., de Vries, H., and Philippart, M.: The Dutch Continental Shelf Model, in: Quantitative Skill Assessment for Coastal Ocean Models, edited by: Lynch, D. and Davies, A., vol. 47 of Coastal and Estuarine Studies, American Geophysical Union, 1995.

Graff, L. S. and LaCasce, J. H.: Changes in the Extratropical Storm Tracks in Response to Changes in SST in an AGCM, J. Climate, 25, 1854–1870, https://doi.org/10.1175/JCLI-D-11-00174.1, 2012.

Holgate, S. J., Matthews, A., Woodworth, P. L., Rickards, L. J., Tamisiea, M. E., Bradshaw, E., Foden, P. R., Gordon, K. M., Jevrejeva, S., and Pugh, J.: New Data Systems and Products at the Permanent Service for Mean Sea Level, J. Coastal Res., 3, 493–504, https://doi.org/10.2112/JCOASTRES-D-12-00175.1, 2013.

Hollebrandse, F.: Temporal development of the tidal range in the southern NorthSea, PhD thesis, Faculty of Civil Engineering and Geosciences, Hydraulic Engineering, TUDelft, 2005.

Janssen, P.: The interaction of ocean waves and wind, Cambridge University Press, Cambridge, 2004.

Kew, S. F., Selten, F. M., Lenderink, G., and Hazeleger, W.: The simultaneous occurrence of surge and discharge extremes for the Rhine delta, Nat. Hazards Earth Syst. Sci., 13, 2017–2029, https://doi.org/10.5194/nhess-13-2017-2013, 2013

Lowe, J. A., Gregory, J. M., and Flather, R. A.: Changes in the occurrence of storm surges around the United Kingdom under a future climate scenario using a dynamic storm surge model driven by the Hadley Centre climate models, Clim. Dynam., 18, 179–188, https://doi.org/10.1007/s003820100163, 2001.

Madec, G.: NEMO reference manual, ocean dynamics component: NEMO-OPA, available at: http://www.nemo-ocean.eu/content/download/180742/735839/file/NEMO_book_3.6_STABLE.pdf (last access: 15 December 2016), 2008.

Mailier, P., Stephenson, D. B., Ferro, C. A. T., and Hodges, K. I.: Serial Clustering of Extratropical Cyclones, Mon. Weather Rev., 134, 2224–2240, https://doi.org/10.1175/MWR3160.1, 2006.

Molteni, F., Stockdale, T., Balmaseda, M. A., Balsamo, G., Buizza, R., Ferranti, L., Magnusson, L., Mogensen, K., Palmer, T., and Vitart, F.: The new ECMWF seasonal forecast system (System 4), Technical Memorandum, 656, 49 pp., 2011.

PSMSL: Permanent Service for Mean Sea Level, Tide Gauge Data, available at: http://www.psmsl.org/data/obtaining/, last access: 5 June 2017.

Rijkswaterstaat: Maeslantkering, stormvloedkering in de Nieuwe Waterweg (Eng: Maeslant barrier, storm surge barier in the New Waterway), available at: https://staticresources.rijkswaterstaat.nl/binaries/FactsheetMaeslantkering_tcm21-65750.pdf (last access: 8 September 2016), 2013.

Sterl, A., van den Brink, H., de Vries, H., Haarsma, R., and van Meijgaard, E.: An ensemble study of extreme storm surge related water levels in the North Sea in a changing climate, Ocean Sci., 5, 369–378, https://doi.org/10.5194/os-5-369-2009, 2009.

Van den Brink, H., Können, G., Opsteegh, J., Van Oldenborgh, G., and Burgers, G.: Estimating return periods of extreme events from ECMWF seasonal forecast ensembles, Int. J. Climatol., 25, 1345–1354, https://doi.org/10.1002/joc.1155, 2005a.

Van den Brink, H., Können, G., Opsteegh, J., van Oldenborgh, G., and Burgers, G.: Estimating return periods of extreme events from ECMWF seasonal forecast ensembles, Int. J. Climtol., 25, 1345–1354, https://doi.org/10.1002/joc.1155, 2005b.

Van den Hurk, B., Siegmund, P., Klein Tank, A., Attema, J., Bakker, A., Beersma, J., Bessembinder J., Boers R., Brandsma T., Van den Brink, H., Drijfhout, S., Eskes, H., Haarsma, R., Hazeleger W., Jilderda, R., Katsman, C., Lenderink, G., Loriaux, J., Van Meijgaard, E., Van Noije, T., Van Oldenborgh, G. J., Selten, F., Siebesma, P., Sterl, A., De Vries, H., Van Weele, M., De Winter, R., and Van Zadelhoff, G. J.: KNMI-14: Climate Change scenarios for the 21st Century – A Netherlands perspective, available at: http://bibliotheek.knmi.nl/knmipubWR/WR2014-01.pdf (last access: 6 december 2016), 2014.

Van den Hurk, B., van Meijgaard, E., de Valk, P., van Heeringen, K.-J., and Gooijer, J.: Analysis of a compounding surge and precipitation event in the Netherlands, Environ. Res. Lett., 10, https://doi.org/10.1088/1748-9326/10/3/035001, 2015.

Van Nieuwkoop, J., Baas, P., Caires, S., and Groeneweg, J.: On the consistency of the drag between air and water in meteorological, hydrodynamic and wave models, Ocean Dynam., 65, 989–1000, https://doi.org/10.1007/s10236-015-0849-3, 2015.

Vialard, J., Vitart, F., Balmaseda, M., Stockdale, T., and Anderson, D. L. T.: An ensemble generation method for seasonal forecasting with an ocean-atmosphere coupled model, Mon. Weather Rev., 133, 441–453, 2005.

10

Large-scale forcing of the European Slope Current and associated inflows to the North Sea

Robert Marsh[1], **Ivan D. Haigh**[1], **Stuart A. Cunningham**[2], **Mark E. Inall**[2], **Marie Porter**[2], and **Ben I. Moat**[3]

[1]Ocean and Earth Science, University of Southampton, National Oceanography Centre, Southampton, European Way, Southampton SO14 3ZH, UK
[2]Scottish Association for Marine Science, Scottish Marine Institute, Oban, Argyll PA37 1QA, UK
[3]National Oceanography Centre, European Way, Southampton SO14 3ZH, UK

Correspondence to: Robert Marsh (rma@noc.soton.ac.uk)

Abstract. The European "Slope Current" provides a shelf-edge conduit for Atlantic Water, a substantial fraction of which is destined for the northern North Sea, with implications for regional hydrography and ecosystems. Drifters drogued at 50 m in the European Slope Current at the Hebridean shelf break follow a wide range of pathways, indicating highly variable Atlantic inflow to the North Sea. Slope Current pathways, timescales and transports over 1988–2007 are further quantified in an eddy-resolving ocean model hindcast. Particle trajectories calculated with model currents indicate that Slope Current water is largely recruited from the eastern subpolar North Atlantic. Observations of absolute dynamic topography and climatological density support theoretical expectations that Slope Current transport is to first order associated with meridional density gradients in the eastern subpolar gyre, which support a geostrophic inflow towards the slope. In the model hindcast, Slope Current transport variability is dominated by abrupt 25–50 % reductions of these density gradients over 1996–1998. Concurrent changes in wind forcing, expressed in terms of density gradients, act in the same sense to reduce Slope Current transport. This indicates that coordinated regional changes of buoyancy and wind forcing acted together to reduce Slope Current transport during the 1990s. Particle trajectories further show that 10–40 % of Slope Current water is destined for the northern North Sea within 6 months of passing to the west of Scotland, with a general decline in this percentage over 1988–2007. Salinities in the Slope Current correspondingly decreased, evidenced in ocean analysis data. Further to the north, in the Atlantic Water conveyed by the Slope Current through the Faroe–Shetland Channel (FSC), salinity is observed to increase over this period while declining in the hindcast. The observed trend may have broadly compensated for a decline in the Atlantic inflow, limiting salinity changes in the northern North Sea during this period. Proxies for both Slope Current transport and Atlantic inflow to the North Sea are sought in sea level height differences across the FSC and between Shetland and the Scottish mainland (Wick). Variability of Slope Current transport on a wide range of timescales, from seasonal to multi-decadal, is implicit in sea level differences between Lerwick (Shetland) and Tórshavn (Faroes), in both tide gauge records from 1957 and a longer model hindcast spanning 1958–2012. Wick–Lerwick sea level differences in tide gauge records from 1965 indicate considerable decadal variability in the Fair Isle Current transport that dominates Atlantic inflow to the northwest North Sea, while sea level differences in the hindcast are dominated by strong seasonal variability. Uncertainties in the Wick tide gauge record limit confidence in this proxy.

1 Introduction

The European Slope Current that lies to the west and north of Scotland exerts considerable influence on the physical and biogeochemical conditions on the adjacent western European shelf seas (Huthnance et al., 2009), with Atlantic Water prevalent across much of the shelf (Inall et al., 2009). Located above the topographic slope at the eastern boundary, the Slope Current is associated with large-scale density gradients and wind forcing (Huthnance, 1984). Sea surface height drops in the northward direction, while prevailing wind stress is oriented from southwest to northeast. Density gradients and winds together drive eastward flows towards the slope that are diverted poleward as an intensified geostrophic flow along the slope. The barotropic transport of the Slope Current may be considered buoyancy-forced to first order, modified by frictional influences, with much of the seasonal variability in transport attributed to wind forcing (Huthnance, 1984).

The Slope Current is part of a greater inflow of Atlantic Water through the Faroe–Shetland Channel (FSC) (Sherwin et al., 2008; Richter et al., 2012; Berx et al., 2013) that also includes some recirculation of the Faroe branch of Atlantic inflow, north of the Faroe Islands, that turns to flow southwestward, and an additional flow that has negotiated the Faroe Bank and Wyville Thomson Ridge. Sherwin et al. (2008) identify a long-term mean barotropic transport of 2.1 Sv over the upper part of the slope region of the Shetland shelf. Richter et al. (2012) refer to the flow between the Faroe and Shetland Islands as the Shetland Current, and use a range of tide gauge data to reconstruct transports in the region. However, they are unable to reconstruct Shetland Current transports, an issue that we return to in the discussion. Over 1995–2009, Berx et al. (2013) estimate an average net Atlantic inflow of 2.7 ± 0.5 Sv through the FSC. Calibrating sea level height with transport fluctuations, Berx et al. (2013) further use satellite altimetry to reconstruct volume transport since 1992, revealing a seasonal variation of 0.7 Sv in Atlantic inflow, becoming warmer and more saline since 1994, but with no trend in volume transport. However, this method detects the net inflow, not just that part associated with the Slope Current. As reviewed and discussed in Berx et al. (2013), issues remain with both the "altimetry transport" and the transport estimates based on ADCP and hydrography data, due to under-sampling of variability in time and space, and the extent to which either estimate is able to represent the net volume transport.

Beyond the FSC, most of the Atlantic inflow progresses beyond the Greenwich meridian to the Nordic seas, with a small fraction branching southward along the western flank of the Norwegian trench. Upstream of the FSC, Atlantic Water migrates up-slope and onto the shelf through several processes, including wind forcing, frictional effects and flow instability related to topographic features (see Inall et al., 2009 and references therein). Major flow instability is associated with the Wyville Thomson Ridge, which presents a transverse obstacle to the Slope Current, bringing a substantial quantity of Atlantic Water onto the Shetland shelf (Souza et al., 2001), augmenting the on-shelf flows derived from further south (see Fig. 1 in Inall et al., 2009). The majority of this shelf flow turns into the North Sea between Orkney and Shetland as the Fair Isle Current (Dooley, 1974). The Fair Isle Current and southward flow along the flank of the Norwegian trench together comprise the Atlantic inflow to the North Sea,

providing a relatively warm influence on the northern North Sea in winter. The North Sea as a whole has warmed considerably since the late 1980s, to an extent considered unprecedented in the historical record (MacKenzie and Schiedek, 2007). Recent warming of the North Sea follows a wider pattern of warming across Europe, and increasingly mild winters in particular (MacKenzie and Schiedek, 2007), although major inflows of warm Atlantic Water in 1988 and 1998 are also believed to have contributed to the warming (Reid et al., 2001), evident also in changes of zooplankton (Reid et al., 2003).

Episodic changes in Atlantic inflow have been attributed to anomalous wind forcing; hence wind-driven changes in Atlantic inflow and the associated warming have been a focus of recent model studies. Hjøllo et al. (2009) use a numerical model of the North Sea region to investigate changes of heat content over 1985–2007, in particular a long-term warming of 0.62 °C. Dividing the North Sea into northern and southern circulation regimes, they find that inflows at the northern boundary are strongly influenced by large-scale atmospheric forcing associated with the North Atlantic Oscillation, but that variable inflow at open boundaries has a limited direct influence on heat content variability. Winther and Johannessen (2006) relate changes of Atlantic inflow to wind forcing, but also emphasize the dilution of Atlantic Water as it circulates the North Sea before leaving in the Norwegian Coastal Current.

Interannual variability in the European Slope Current has recently been explored using altimetry data over a 20-year period, revealing a peak in poleward flow along much of the continental slope from Portugal to Scotland during 1995–1997, and a long-term decreasing trend of $\sim 1\%$ per year (Xu et al., 2015). Here, we consider the extent to which changes in the Atlantic inflow to the North Sea are associated with variability of the Slope Current driven by changing large-scale meridional density gradients and winds. We use a wide range of observations and eddy-resolving model hindcast data to examine the Slope Current, large-scale forcing mechanisms, and Atlantic inflow to the North Sea.

The paper is organized as follows. In Sect. 2, we outline the variety of data and methods used. In Sect. 3, we evaluate simulated Slope Current drift over 1995–1997 using archived drifter data (Sect. 3.1). We then characterize Slope Current pathways, timescales and transports in a model hindcast spanning 1988–2007 (Sect. 3.2). With considerable Slope Current variability evident in the hindcast, we consider the influence of two large-scale driving mechanisms, meridional density gradients and wind forcing (Sect. 3.3), in both the model and observations. Finally, we explore the evidence for variable Atlantic inflow to the North Sea (Sect. 3.4), and sea level differences as proxies for Slope Current transport and this inflow (Sect. 3.5). In the Discussion and conclusion section (Sect. 4), we suggest that major variations in Slope Current transports and the Atlantic Water influence on the North Sea are primarily linked to variable meridional density gradients in the eastern subpolar gyre that are attributed to the combined (reinforcing) effects of wind and buoyancy forcing.

2 Datasets and methodology

In Sect. 2.1, we summarize the available time series data that record variability in the Slope Current and the Atlantic inflow to the North Sea over recent decades. In Sect. 2.2, we introduce the drifter data used to provide an observational perspective on the Slope Current system, and for preliminary evaluation of model currents. We then introduce the observations used to explore forcing mechanisms: mean absolute dynamic topography and climatological density data (Sect. 2.3), and wind-stress reanalysis data (Sect. 2.4). In Sect. 2.5 we introduce the tide gauge data and analysis used to explore Slope Current transport between Scotland, Shetland and the Faroes. Finally, we outline the model hindcasts used to characterize variability of the Slope Current system (Sect. 2.6) and the Lagrangian diagnosis of hindcast data using the ARIANE methodology for calculation of particle trajectories based on velocity fields, and the accompanying statistical analyses (Sect. 2.7).

2.1 Time series data

The following data are provided as part of the ICES Report on Ocean Climate (IROC), available at http://ocean.ices.dk/iroc:

- depth-averaged inflow and outflow to/from the North Sea, centred on 59° N, 1° E, as modelled volume transport between Orkney (Scotland) and Utsira (Norway), monthly averaged from January 1985;

- salinity in the Fair Isle Current, centred on 59° N, 2° W (first two stations on the JONSIS line), averaged over the depth range 0–100 m, irregularly sampled and annually averaged from 1960;

- salinity for the FSC Shetland Shelf, centred on 61° N, 3° W, the maximum in the upper layer high salinity core, sampled 3 times per year (April–May, September–October and December) from 1950.

We further sample monthly-mean salinity in the NCEP Global Ocean Data Assimilation System (GODAS) analysis fields spanning 1980–2016 (NOAA Climate Prediction Center; see http://www.cpc.ncep.noaa.gov/products/GODAS/). These time series data are used collectively to evaluate time series of similar quantities in the model hindcast and derived Lagrangian data.

2.2 Drifter data

As part of the Land–Ocean Interaction Study (LOIS), the Shelf Edge Study (SES) was undertaken in the mid-1990s.

LOIS-SES included two Slope Current drifter experiments within which drifters were released in three groups of seven in an east–west line 20 km long across the continental shelf west of Scotland near 56.25° N, on 5 December 1995 and on 5–9 May 1996 (Burrows and Thorpe, 1999; Burrows et al., 1999), to characterize winter and summer conditions. The drifters were drogued at a depth of 50 m and tracked for up to 240 days, to study the regional circulation and dispersion. The archived drifter positions from both experiments are used to provide some context for the study, and for a basic evaluation of corresponding circulation in the model (see below).

2.3 Mapped absolute dynamic topography (MADT) and climatological density

Daily global absolute sea-surface dynamic topography distributions with a spatial resolution of 0.25° are produced by Centre National d'Etudes Spatiales (CNES), and distributed through AVISO+ (http://www.aviso.altimetry.fr). Here, we use "Delayed Time" data from the SSALTO/DUACS system (AVISO+, 2016), which provides a homogeneous, intercalibrated time series of sea-level anomaly. Absolute sea surface dynamic topography is the sum of sea level anomalies and a mean dynamic topography, both referenced over a 20-year period (1993–2012). Key improvements in this new dataset are the use of a new mean dynamic topography (MDT CNES_CLS13) calculated from GOCE satellite data, increased use of in situ observations over the longer reference period, and more accurate mapping of the mesoscale (Rio et al., 2011). The geoid model developed from the GOCE satellite data has a horizontal resolution of 125 km. Multivariate objective analysis (including wind and in situ data) is used to improve the large-scale solution, resulting in a final gridded horizontal resolution of 0.25°.

Monthly estimates of ocean temperature and salinity spanning the same period are available as objectively analysed gridded fields from the EN4 dataset provided by the UK Met Office Hadley Centre (Good et al., 2013). EN4 comprises global gridded fields of potential temperature and salinity at 1° resolution with 42 vertical levels. From 2002, the Argo float programme significantly improved EN4 data coverage in the northeast Atlantic (Good et al. 2013). The gridded temperature and salinity estimates are used to calculate climatological potential density referenced to the surface (σ_0), at selected depth levels. Of specific relevance to the Slope Current, and the present study, are meridional gradients of mapped absolute dynamic topography (MADT) and potential density.

2.4 Wind-stress data

For the study period 1988–2007, we obtain 10 m winds from the ERA-interim 12-hourly, 0.75° × 0.75° resolution reanalysis datasets (Dee et al., 2011). We calculate wind stress following the methods of Large and Pond (1981). To address wind forcing of the Slope Current, a slope-based subset is extracted from the wind-stress field between the 200 and 1000 m contours (bathymetry from ETOPO1, Amante and Eakins, 2009) in the latitude range 48–60° N. These wind-stress vectors are rotated into a coordinate system parallel to the 500 m contour and then averaged to obtain annual-mean averages over 1988–2007.

2.5 Tide gauge data

Monthly mean sea level records were obtained from the Permanent Service for Mean Sea Level (PSMSL; http://www.psmsl.org) for tide gauges at Wick (mainland Scotland; 3.09° W, 58.44° N), Lerwick (Shetland; 1.14° W, 60.15° N) and Tórshavn (Faroes; 6.77° W, 62.01° N) (Holgate et al., 2013; PSMSL, 2017). The Wick record spans the period 1965–2014 and is 91.5 % complete. The Lerwick record spans the period 1957–2014 and is 91 % complete. The Tórshavn record spans the period 1957–2006, and is 84 % complete. Records have been corrected for the effects of glacial isostatic adjustment, using results from the ICE-5G model of post-glacial relative sea level history. We calculate Lerwick–Tórshavn and Wick–Lerwick sea level differences as proxies for Slope Current transport and Fair Isle Current transport respectively. Note that the tide gauge records are referenced to different local datums and within the scope of this study it has not been possible to directly tie these together. However, as we are interested in transport variability, on seasonal to decadal timescales, we focus on the relative difference in sea level recorded by each tide gauge.

2.6 Model hindcasts

NEMO (Madec, 2008) is a state-of-the-art, portable ocean modelling framework developed by a consortium of European institutions. We sample currents and hydrographic data (temperature, salinity) from the northeast Atlantic region of an eddy-resolving (1/12°) global ocean model hindcast, the ORCA12 configuration of NEMO, for the period 1988–2007 (see Blaker et al., 2015), henceforth ORCA12-N01. With the barotropic Rossby radius at 55° N ranging from ∼ 375 km (water depth 200 m) to ∼ 1200 km (water depth 2000 m), the horizontal resolution of ORCA12 will comfortably resolve large instabilities and eddies associated with the Slope Current, although with corresponding baroclinic Rossby radii in the range 5–10 km, smaller-scale variability cannot be resolved. In the vertical dimension, there are 75 vertical levels, with 46 in the upper 1000 m, resolving the surface and bottom boundary layers that play an important role in Slope Current dynamics. The advantage of using fields from a global model is that large-scale influences on the Slope Current are fully represented, rather than being prescribed at the boundaries in a regional model, which can be problematic.

We use results for the hindcast period to simulate region-typical patterns of particle drift and dispersal (see Sect. 2.7). Our choice of this hindcast is guided by evidence that eddy-resolving simulations can faithfully reproduce the global EKE field observed with satellite altimetry (Petersen et al., 2013), while lower-resolution eddy-permitting simulations are known to substantially underestimate EKE (McClean et al., 2002; Hecht and Smith, 2013). NEMO is forced with 6-hourly winds supplied by the DFS4.1 (1988–2006) and DFS5.1.1 (2007–2010) datasets (Brodeau et al., 2010). The hindcast provides 5-day averages of currents and tracers (temperature, salinity), a time window appropriate to model realistically and with high precision the advection of an ensemble of particles representative of Slope Current transport, and associated variability on an eddy timescale of order 1 month. Most recently, a longer hindcast simulation with ORCA12 became available (e.g. Moat et al., 2016), henceforth ORCA12-N06, and we use diagnostics from this experiment to extend our analysis to the longer period 1958–2012. While it would be instructive to also calculate particle drift and dispersal with the longer hindcast, such calculations are not straightforward with the remotely archived ORCA12-N06 datasets.

2.7 Lagrangian model diagnostics

We use the ARIANE particle-tracking software (Blanke and Raynaud, 1997) to track ensembles of particles that are "seeded" in the northward-flowing Slope Current. We release 630 particles, at 30 model levels from 9.85 m down to 371.22 m, and at 21 equally spaced locations across a short section on the ORCA12 mesh (9.46° W 55.83° N to 9.28° W 55.82° N). This section is close to where floats have been deployed as part of the UK NERC project FASTNEt (http://www.sams.ac.uk/fastnet), henceforth the "FASTNET release section", and co-located with the location of the Slope Current in ORCA12, identified as a narrow band of high velocity ($> 10\,\mathrm{cm\,s^{-1}}$) in 5-day mean fields. Particles are released on 1 January or 1 July, to sample the two halves of the seasonal cycle, with location, depth and ambient water properties (temperature, salinity) recorded every 24 h for 183 days. We also use ARIANE in "backward" mode, which simply reverses (in time) the analytical calculation of particle progress through grid cells, to examine the source of particles recruited to the Slope Current.

Particle locations are statistically analysed to obtain a measure of particle density, dividing the number of particle occurrences in a limited longitude–latitude range by the total number of particle occurrences during the tracking period. We use a 0.5° × 0.5° mesh to sample for particle occurrence, optimal for both the resolution of the Slope Current and sampling of sufficient particles from a statistical perspective. This quantifies interannual variation in pathways, broadly distinguishing between years of low and high influence of the Slope Current on the northern North Sea,

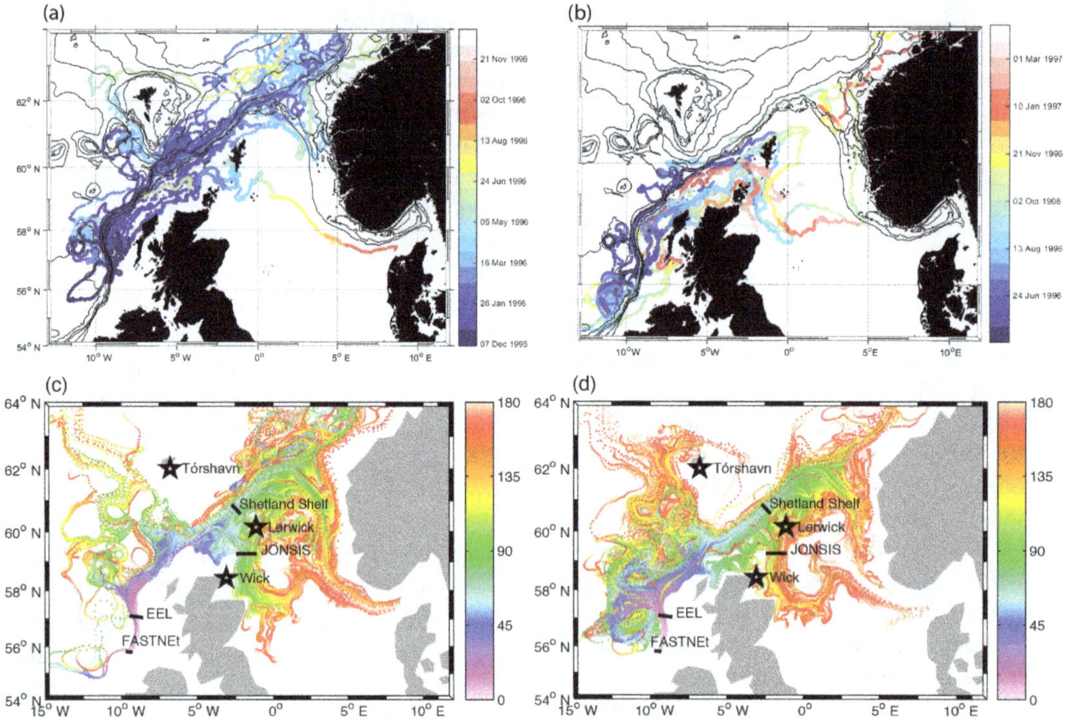

Figure 1. Drogued drifters released in **(a)** December 1995 and **(b)** May 1996; model particle trajectories spanning 6 months, released on **(c)** 1 January 1996 and **(d)** 1 July 1996 (630 model trajectories are plotted in each case). Drifters are colour-coded by calendar date. Model particles are colour-coded by age (days). Note that the drifter data span slightly different durations: up to 11 months, December 1995–November 1996 **(a)**; up to 10 months, May 1996–March 1997 **(b)**. In **(c)** and **(d)**, we also indicate the FASTNEt, EEL and Shetland Shelf sections where we sample the Slope Current, the western JONSIS line where we sample the Fair Isle Current, and the locations of Wick, Lerwick and Tórshavn where we take the sea surface height records used to develop proxies for variability of transport in the Slope Current and the Fair Isle Current.

where we further record the presence of particles reaching the "NW North Sea" (south of 59° N, bounded by longitudes 4.5° W and 1.5° E) and the "NE North Sea" (east of 1.5° E, south of 62° N). Alongside particle density, we also obtain an average particle age (since release), depth and salinity, per $0.5° \times 0.5°$ grid cell.

3 Results

We begin with a broad perspective of Slope Current pathways, moving on to examine time series of transport variability at selected locations. We then consider the drivers of transport variability, introducing a theoretical framework and applying this in an evaluation of the changes evident in our time series. We conclude the results section with a consideration of Atlantic inflow to the North Sea, and re-visit the prospects for monitoring regional transports with sea level observations.

3.1 Drifter observations and ORCA12 simulations of Slope Current pathways, 1995–1997

To provide some context for the study, and a basic evaluation of corresponding model drift, in Fig. 1 we show LOIS-SES drifter data alongside example model particle trajectories, with the caveat that variability on length scales below ~ 10 km and timescales shorter than ~ 10 days are unresolved in the latter. LOIS-SES drifter deployments in December 1995 (Fig. 1a) and May 1996 (Fig. 1b) reveal somewhat different pathways in and around the Slope Current, with a tendency for more extensive drift in winter releases, compared to summer releases. Most drifters follow the Slope Current for several hundred kilometres following release in December 1995. Several drifters enter the northeast sector of the North Sea by late winter or early spring, a travel timescale of 2–3 months. In contrast, several drifters released in May 1996 move onto the shelf and directly to the northwest North Sea, but on a wide range of timescales due to highly variable shelf currents.

Model particle trajectories start on 1 January and 1 July 1996 (Fig. 1c and d) at the FASTNEt release section (see Sect. 2.7), and are tracked forwards for 6 months.

Particles released on 1 January tend to disperse more widely than those released on 1 July, with a larger number reaching the northwest and northeast sectors of the North Sea within 6 months. There are limitations to the direct comparison of the drifters and particle trajectories, as the former are subject to sub-mesoscale processes and tides that are not represented in the model. Given the chaotic nature of mesoscale variability, we further note that pathways inferred from a more limited number of drifters are less statistically significant. However, in broad terms, model particle trajectories indicate drift pathways and timescales similar to the drifters, and suggest more extensive drift in the first half of 1996, compared to the second half, that is consistent with the observations and indicative of known seasonality in Slope Current dynamics.

3.2 Characterizing Slope Current pathways, timescales and transports in the ORCA12 hindcast of 1988–2007

As outlined in Sect. 2, particle density and mean age maps are obtained on a $0.5° \times 0.5°$ mesh for each year from 1988 to 2007 (see Figs. S1, S2, S4 and S5 in the Supplement for 1 July starts, tracking forwards and backwards). The statistics for each set of 20 ensembles (January–July releases, tracked forwards and backwards) are further averaged to obtain the "grand ensemble" results shown in Figs. 2 and 3, respectively.

Tracking forwards, Fig. 2 shows the "grand mean" of particle density, age, depth and salinity, for particles released on 1 January (left panels) and 1 July (right panels). Ages are expressed as days since 1 January or 1 July. Particle density is simply the fraction of all particle positions in each $0.5° \times 0.5°$ grid square, in relation to all particle positions. With variation across 3 orders of magnitude, we use a log scale to highlight the distribution of density and depth statistics. Highest particle density (~ 0.1) and youngest age (0–20 days) is naturally located near the release section. Relatively high particle density in Fig. 2a and b otherwise traces the Slope Current pathway, characteristically following the shelf break (see Fig. 1), but bifurcating to the northeast of Scotland and just west of Norway. Only a small fraction of particles are tracked further to the northwest and a destination in the Norwegian Sea. Back upstream, particles can also follow a minor pathway offshore to the north of the release section, turning westward to the south of Iceland and then southward along the Reykjanes Ridge. Another minor pathway involves almost immediate recirculation to the west of Ireland. Within 6 months, a few particles reach domain boundaries, to the north, east and west. With the focus of this study on the northern North Sea, these boundary terminations are not problematic.

Turning to the mean age of particles (Fig. 2c and d), this correspondingly increases to 140–180 days at locations most remote from the release section. There are relatively small differences between the January and July releases, although

it appears that particles released in January reach the northwest North Sea more quickly, and in larger numbers. Mean ages for January releases are younger by ~ 10 days at many locations, suggestive of a more vigorous circulation during the first half of the year and more extensive shelf edge exchange, with higher on-shelf particle densities in particular.

Mean depths (Fig. 2e and f) are around 50 m in the North Sea inflow, which likely reflects the initial vertical distribution of particles in the Slope Current, but is consistent with residence of Atlantic Water in a sub-surface layer, below the fresh surface layer which is dominated by Baltic outflow. January releases reach slightly greater depths in the North Sea, compared to July releases, consistent with stronger southwesterly winds (aligned with the slope) and downwelling in winter–spring. Particles that leave the Slope Current system for an Atlantic fate are subducted across a wide range of depths, up to around 1000 m, for both July and January releases. Particles that persist in the Atlantic inflow as far as the southern Nordic seas also descend on average, with mean depth of around 300 m.

Salinity in the Slope Current (Fig. 2g and h) is generally higher than surrounding waters. We consider the mean salinity of forward trajectories to trace high-salinity Atlantic Water through the Slope Current system. In the FSC, maximum salinity averages around 35.5 psu. Water of this salinity corresponds to the "North Atlantic Water" of salinity around 35.42 psu that dominates the upper 200 m on the Shetland side of the FSC, recently identified by McKenna et al. (2016). Moving onto the Shetland shelf, upstream of FSC, mean salinity declines to around 35.3 psu. This is consistent with mixing of Atlantic Water and relatively fresher water on the shelf, for which salinity ranges 35.00–35.25 psu at the Ellett line (e.g. Fig. 14 in Inall et al., 2009).

Tracking backwards, Fig. 3 shows the corresponding particle density, age and depth distributions for flows feeding transport across the FASTNEt release section. Similar to the results for forward tracking, highest particle densities are located adjacent to this section. In contrast to the forward tracking, particle density is generally lower across a broader area of the eastern subpolar gyre, indicative of a widespread inflow across the approximate latitude range 48–60° N. As for forward tracking, a few back trajectories reach the western domain boundaries in a little under 180 days. There is some evidence for a southward continuation of the Slope Current, along the shelf break, to around 13° W, 48° N. This is more evident in back-trajectories that span the second half of the year (i.e. reaching the FASTNEt release section in January). This "upstream" branch of the Slope Current is slower than the "downstream" branch represented in Fig. 2, consistent with downstream strengthening of Slope Current transport through progressive inflow from the west. Mean depth across the catchment area generally increases to the south, with an impression that the upper ~ 100 m layer of the Slope Current is recruited from the southeast subpolar gyre, while deeper layers (below 100 m) are recruited from the northeast sub-

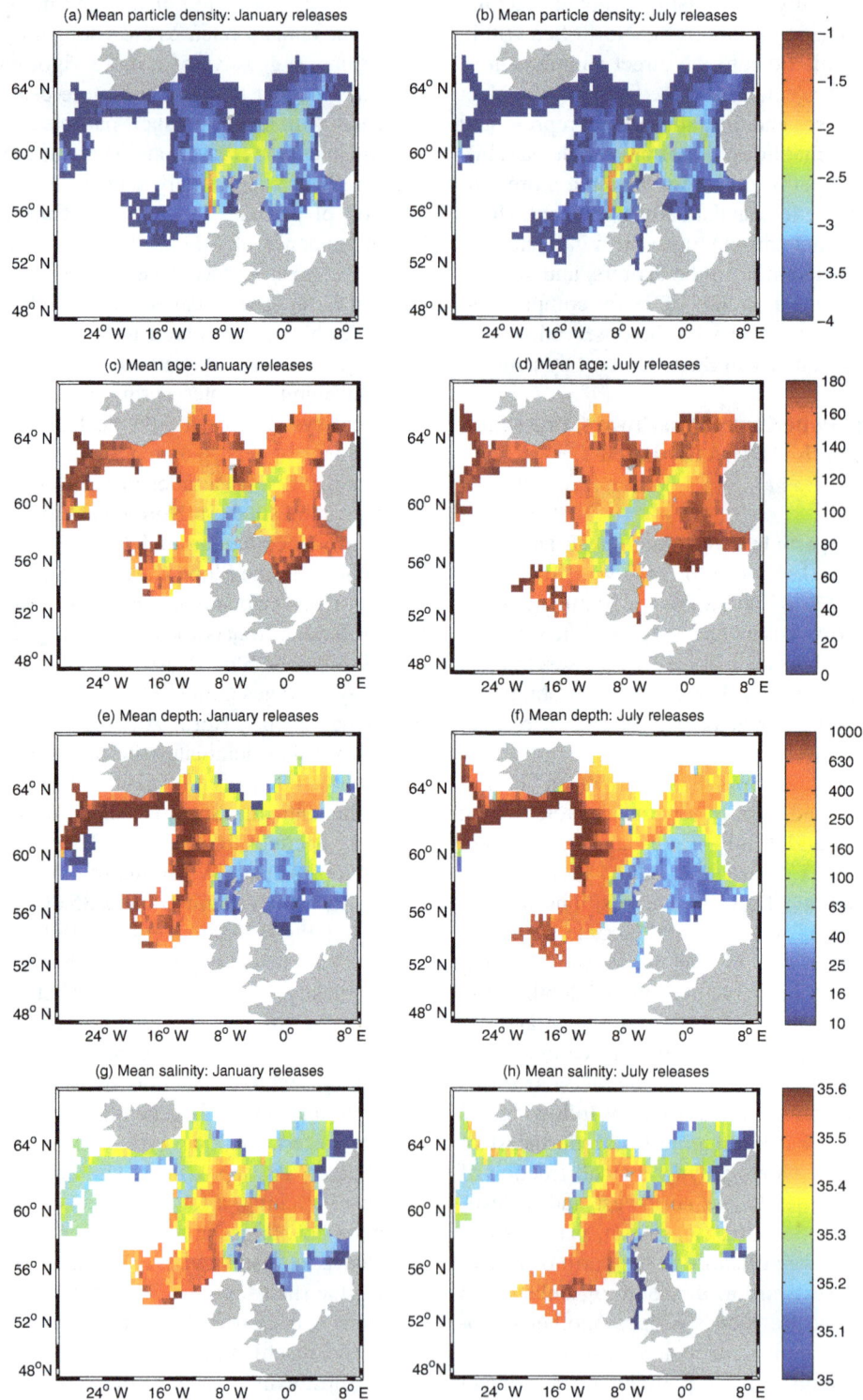

Figure 2. Forward trajectories from the FASTNEt release section in ORCA12-N01, for particles released on 1 January and 1 July: **(a, b)** mean particle density; **(c, d)** mean particle age (days since 1 January or 1 July); **(e, f)** mean particle depth (m); **(g, h)** mean particle salinity (psu). Averages are for 1988–2007, and values are binned at 0.5° × 0.5° resolution. Particle density is expressed as a fraction, obtained as the number of particle occurrences per 0.5° × 0.5° grid cell divided by the total number of particle occurrences. The logarithmic scale for density, ranging from −4 to −1, equates to 0.01–10 % of all particle positions.

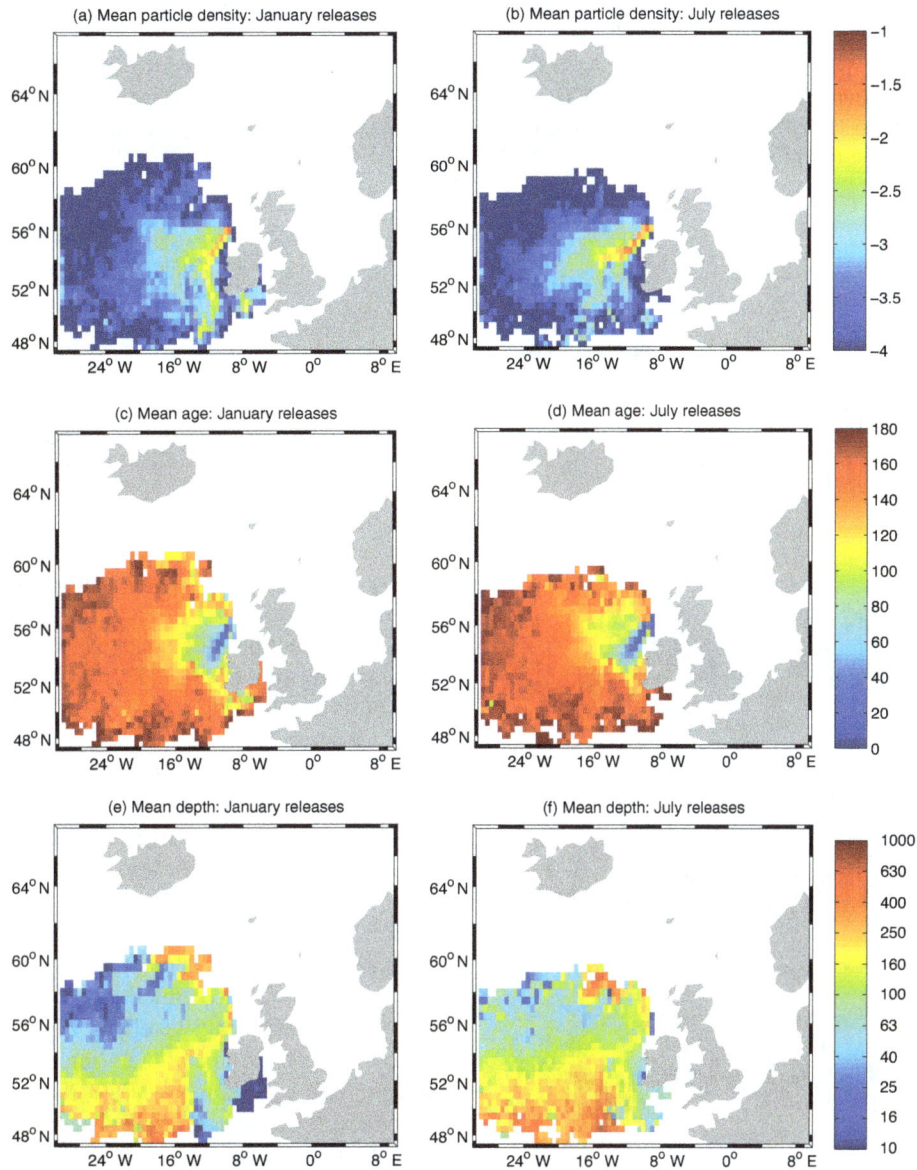

Figure 3. As in Fig. 2a–f, for backward trajectories.

tropical gyre. Particles arriving in the Slope Current in January originate from a depth range 10–50 m in the subpolar gyre, while particles arriving in July arrive from greater depths in this region (around 50–100 m), indicating stronger upwelling of particles recruiting to the Slope Current during the first half of the year.

We now consider the corresponding changes in Slope Current transport at selected locations along the shelf break. Figure 4 shows Slope Current transport every 5 days over 1988–2007 at a somewhat longer FASTNEt section (from 9.74° W, 55.82° N to 9.28° W, 55.79° N), and at two further sections, EEL and Shetland Slope (see Fig. 1c and d). Endpoints for the EEL section (from 9.48° W, 57.11° N to 8.57° W, 57.05° N) and the Shetland Slope section (from 2.72° W,

60.87° N to 2.15° W, 60.57° N) are based on particle trajectories (see Fig. 1), rather than strictly delimited by the same isobaths (model bathymetry) spanned at the FASTNEt section (181 to 1516 m). Hence the EEL section spans a depth range 120–1046 m, while the Shetland Slope section spans 133–412 m. On this basis, we diagnose Slope Current transports at the upper end of observed ranges (e.g. Sherwin et al., 2008), and considerably higher than "Atlantic inflow" estimates of 2.7 ± 0.5 Sv in the FSC (Berx et al., 2013). Between the FASTNEt (EEL) and Shetland Slope sections, long-term mean transport increases, while the standard deviation decreases, from 4.62 ± 3.34 (4.17 ± 2.71) Sv to 7.04 ± 2.18 Sv (Table 1). The Fair Isle Current, an inshore component of Slope Current transport, amounts to 1.17 ± 1.11 Sv. The rel-

Table 1. Long-term mean and standard deviation (SD) of transport, decadal means for 1988–1997 and 1998–2007, and corresponding age statistics at the three selected sections along the continental shelf break, and across the Fair Isle Current branch of Atlantic inflow to the northwest North Sea (positive southward). The FASTNEt section extends from 9.74° W, 55.82° N (1516 m) to 9.28° W, 55.79° N (181 m). The EEL section extends from 9.48° W, 57.11° N (1046 m) to 8.57° W, 57.05° N (120 m). The Shetland Slope section extends from 2.72° W, 60.87° N (412 m) to 2.15° W, 60.57° N (133 m). The Fair Isle Current is associated with flow between 1.1 and 2.46° W at 59.27° N, approximately the western portion of the JONSIS line.

Section	Transport				Travel time (days)			
	(Sv)	SD	Mean	Mean	Since 1 January		Since 1 July	
	Mean		(1988–1997)	(1998–2007)	Mean	SD	Mean	SD
FASTNEt	4.62	3.34	5.94	3.30	0	n/a	0	n/a
EEL	4.17	2.71	5.33	3.00	19.4	7.7	28.7	20.4
Shetland Slope	7.04	2.18	7.59	6.49	90.4	23.7	98.6	32.3
Fair Isle Current	1.17	1.11	1.22	1.13	123.6	20.0	145.6	14.6

n/a: not applicable.

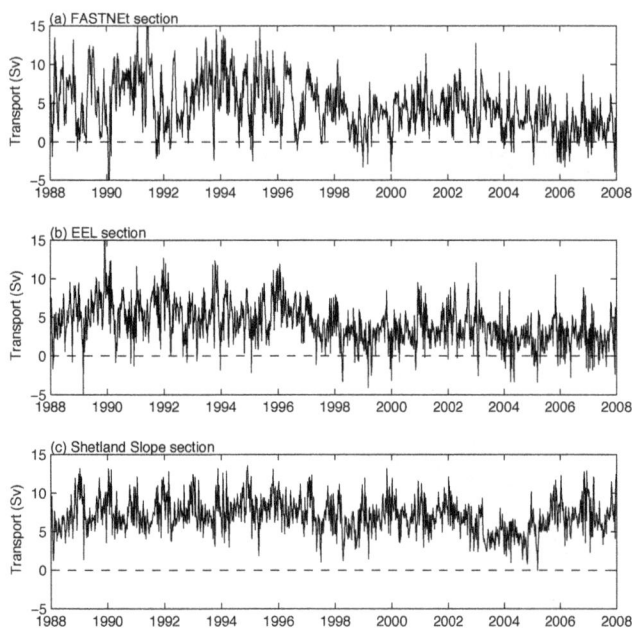

Figure 4. Slope Current transport at FASTNEt, EEL and Shetland Slope sections, 5-day averages over 1988–2007 in ORCA12-N01.

atively large standard deviation results from strong seasonality, with peak inflow in winter (see Fig. 9b).

As a metric of seasonal variations in transport, we sample the ensemble-mean particle ages in Fig. 2 for travel times between sections. These are shorter and less variable in the first half of the year: 19.4 ± 7.7 days (January releases) compared to 28.7 ± 20.4 days (July releases) between FASTNEt and EEL sections; 90.4 ± 23.7 days (January releases) compared to 98.6 ± 32.3 days (July releases) between FASTNEt and Shetland Slope sections; 123.6 ± 20.0 days (January releases) compared to 145.6 ± 14.6 days (July releases) between FASTNEt and the Fair Isle Current sections (Table 1). This is indicative of a somewhat more vigorous circulation

during January–June, although monthly-mean transports at the three Slope Current sections and the JONSIS section (Fig. S6) do not provide conclusive evidence for this.

Regarding the 1988–2007 variability in Figs. 4 and 9b, transports are weaker by 9–45 % in the second decade of the hindcast at all sections (see Table 1), with a most striking shift to weaker transport at the EEL section over 1996–1998. To investigate the extent to which Slope Current transport variability (including the seasonal cycle) is instantaneously correlated at the three sections, we compute correlation coefficients between the 5-day averaged transports in Fig. 4, confirming that variability along the shelf break is coordinated to a large extent. This is most evident between the FASTNEt and EEL sections, for which the correlation is 0.64 (significant at 99 % confidence level), as might be expected given the relatively short distance separating these two sections. More striking is a correlation of 0.43 (significant at 99 % confidence level) between transports at the widely separated EEL and Shetland Slope sections.

3.3 Mechanisms driving Slope Current variability

We now consider in turn the influences of meridional density gradients to the west of the shelf break, and the local winds along the shelf break, in driving the variability in Slope Current transport evident in the ORCA12-N01 hindcast.

3.3.1 Meridional density gradients

To first order, the Slope Current is driven by the deep ocean meridional density gradient. We can accommodate this in the geostrophic momentum balance, as presented in Simpson and Sharples (2012) and reproduced here. First consider the zonal momentum equation, given reference density ρ_0 and Coriolis parameter f. We use the hydrostatic balance, whereby pressure $p = \rho g(z + \eta)$, given density ρ, gravitational acceleration g, arbitrary ocean depth z and sea surface elevation η, and we assume that the zonal density gradient is zero. The right-hand side thus simplifies to a zonal gradient in sea surface height:

$$-fv = -\frac{1}{\rho_0}\frac{\partial p}{\partial x} = -\frac{1}{\rho_0}\frac{\partial(\rho g(z+\eta))}{\partial x} = -g\frac{\partial \eta}{\partial x}. \quad (1)$$

Vertically integrating in the depth range $-h$ and η, defining meridional transport, $V = \int_{-h}^{\eta} v\,dz$, and assuming $h \gg \eta$, the depth-integrated zonal momentum balance is as follows:

$$-fV = -g\frac{\partial \eta}{\partial x}h. \quad (2)$$

Considering the y-momentum equation, we follow the same approach, noting that the meridional density gradient is non-zero, so the right-hand side now includes an extra term:

$$fu = -\frac{1}{\rho_0}\frac{\partial p}{\partial y} = -\frac{1}{\rho_0}\frac{\partial(\rho g(z+\eta))}{\partial y} = -\frac{g}{\rho_0}\frac{\partial \rho}{\partial y}z - g\frac{\partial \eta}{\partial y}. \quad (3)$$

Vertically integrating again, defining zonal transport, $U = \int_{-h}^{\eta} u\,dz$, the meridional momentum balance is as follows:

$$fU = -\frac{g}{2\rho_0}\frac{\partial \rho}{\partial y}h^2 - g\frac{\partial \eta}{\partial y}h. \quad (4)$$

Cross-differentiating Eqs. (2) and (4) for $\partial U/\partial x$ and $\partial V/\partial y$, and given vertically integrated continuity of volume, $\partial U/\partial x + \partial V/\partial y = 0$, we obtain an expression for the meridional gradient in sea surface elevation as a function of local depth (h) and the meridional density gradient:

$$\frac{\partial \eta}{\partial y} = -\frac{h}{\rho}\frac{\partial \rho}{\partial y}. \quad (5)$$

Following Simpson and Sharples (2012), we further distinguish between the shelf (depth $h = h_\mathrm{s}$) and the deep ocean ($h = H$):

$$\text{Shelf}: \left(\frac{\partial \eta}{\partial y}\right)_{h_\mathrm{s}} = -\frac{h_\mathrm{s}}{\rho}\frac{\partial \rho}{\partial y}, \quad (6)$$

$$\text{Deep ocean}: \left(\frac{\partial \eta}{\partial y}\right)_{H} = -\frac{H}{\rho}\frac{\partial \rho}{\partial y}. \quad (7)$$

Applying this theoretical framework, we use MADT and climatological temperature and salinity observations to evaluate the meridional gradients of Eq. (7) in the eastern subpolar North Atlantic, where the Slope Current originates (see Fig. 3). Fields of MADT and mean σ_0 at 500 m (Fig. S7a and b) are broadly characterized by negative and positive meridional gradients respectively (Fig. S7c and d). Dividing $\partial \eta/\partial y$ by $-\rho^{-1}\partial \rho/\partial y$, we obtain an estimate of the deep ocean depth scale H, plotted in Fig. 5. Over large areas of the region, H is thus predicted in the range 500–2500 m, representative of the deep ocean.

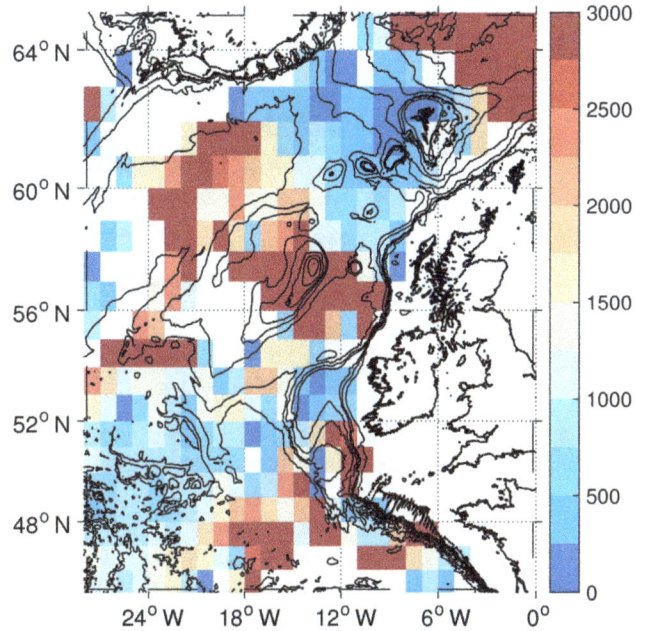

Figure 5. The depth scale H of Eq. (7), predicted from the meridional gradients of density and sea surface height; H is set to the local water depth, where H exceeds that depth; white areas inside the bold black contour indicate where H is negative (undefined).

Since $H \gg h_\mathrm{s}$, Eqs. (6) and (7) predict that the (downward) meridional gradient in η will be greater over the deep ocean than over the shelf, so the cross-slope (downward) gradient in η increases with latitude (see also equation pair 10.11 in Simpson and Sharples, 2012). This cross-slope difference in sea surface elevation will result in a geostrophic current parallel to the isobaths. At the same time, the momentum balance in the meridional direction implies a geostrophic transport in deep water towards the slope, predicted by substituting Eq. (7) into Eq. (4):

$$U = \frac{gH^2}{2\rho_0 f}\frac{\partial \rho}{\partial y}. \quad (8)$$

As this zonal flow reaches the slope, it turns to the north and joins the meridional current. This current increases with latitude as the zonal difference in height across the sloping seabed, which increases likewise (see also equation 10.12 in Simpson and Sharples, 2012). Using Eq. (8) with $g = 9.81\,\mathrm{m}^2\,\mathrm{s}^{-1}$, $\rho_0 = 1025\,\mathrm{kg}\,\mathrm{m}^{-3}$, representative latitude 55° N ($f \sim 1.19 \times 10^{-4}\,\mathrm{s}^{-1}$), and $H = 1000\,\mathrm{m}$ as a depth scale appropriate for the inflow (from Fig. 5), we obtain $U \sim 8 \times 10^7\,\partial \rho/\partial y$ ($\mathrm{m}^3\,\mathrm{s}^{-1}$ per m along slope).

Considering the Slope Current to be thus "fed from the west" by geostrophic inflows that are supported by a meridional density gradient, we now investigate how this large-scale pattern may have changed over the hindcast period. Figures S8 and S9 show maps of potential density (σ_0) for 1–5 January, biennially over 1988–2006, at two depth levels

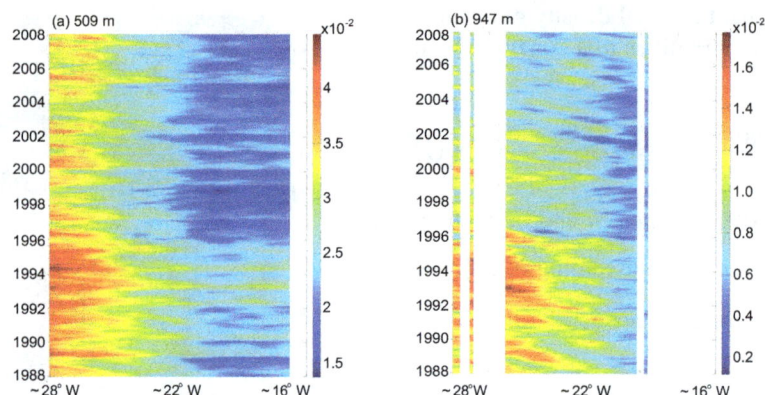

Figure 6. The northward trend of density (10^{-2} kg m^{-3} °$^{-1}$, regressed in the latitude range 45–62° N) in the northeast Atlantic (in the approximate longitude range 16–28° W), 5-day averages for 1988–2007: **(a)** at 509 m; **(b)** at 947 m.

– 500 and 947 m – which are representative of the inflow. At 500 m, it is evident that density to the south of Iceland progressively decreases over 1988–2006; at 947 m there is a progressive increase of density at mid-latitudes over the study period, with little change further to the north. Both changes result in a reduction of the northward density gradient.

Variability of $\partial\rho/\partial y$ is more explicitly shown in Fig. 6 as northward trends of σ_0 over the latitude range from 45 to 62° N, in the northeast Atlantic (across 15–28° W), 5-day averages for 1988–2007, at 500 m (Fig. 6a) and at 947 m (Fig. 6b). Reductions in the trend are evident throughout most of the period, with particularly abrupt reductions over 1995–1997 and overall reductions in the range 25–50 %, with the strongest percentage reductions in the east (around 15° W). These abrupt changes are consistent with the sharp reduction of Slope Current transport at the FASTNEt and EEL sections (Fig. 4a and b).

As a metric for density forcing of the Slope Current, we average the 45–62° N density gradients across 15–28° W and annually, and then take annual anomalies relative to the 20-year mean. Figure 7 shows 1988–2007 time series of these metrics for density gradients at 500 m (Fig. 7a) and at 947 m (Fig. 7b). The annual index clearly shows how the density gradients weakened around the mid-1990s. Taking a change in $\partial\rho/\partial y$ of 2–4×10^{-3} kg m^{-3} °$^{-1}$ (111 km), Eq. (8) suggests a change of inflow, $\Delta U \sim 8 \times 10^7 \times [2$–$4] \times 10^{-3} \times (111 \times 10^3)^{-1} = 1.42$–$2.9$ m^2 s^{-1}. Across 17° of latitude, this amounts to a change (decrease) of total inflow in the range 2.75–5.5 Sv, broadly consistent with the abrupt drop in transport, over 1996–1998, at the EEL section in particular (Fig. 4b).

In Fig. 7c and d, we plot these metrics against annual-mean Slope Current transports at the EEL section. We find strong and significant correlations, of up to 0.67 and 0.75 (both significant at 99 % confidence level) between Slope Current transport at the EEL section and the density gradient indices, at 509 and 947 m respectively. The strong correlations are associated with a degree of bimodal scatter in Fig. 7c and d,

associated in turn with abrupt declines of density gradients (Fig. 7a and b) and Slope Current transport (Fig. 4b) in the mid-1990s. Similar correlations, also significant at 99 % confidence level, are obtained between the density gradient indices and annual-mean transports at the FASTNEt section (0.67 at 509 m; 0.71 at 947 m) and at the Shetland Slope section (0.61 at 509 m; 0.60 at 947 m).

3.3.2 Wind forcing

While density gradients do indeed appear to exert a leading control on barotropic Slope Current transport, wind forcing is also likely to play an important role on short timescales. In particular, strong wintertime winds likely explain strongest Slope Current transport at that time of the year (Huthnance, 1984). For a circular basin with a sloping margin and wind-stress forcing, Huthnance (1984) uses scaling arguments applied to incompressible, hydrostatic momentum equations (with horizontal flow scales > topographic slope scales and vertical scales; $w \ll u$) to demonstrate that in a steady state, any component of wind stress parallel to a steep slope (τ^s) will induce a downwind current along the continental shelf and slope, with a speed given by $\tau^s/(\overline{\rho}k)$, where $\overline{\rho}$ is a depth mean density and k is a linearized friction coefficient. Additionally, applying a uniform azimuthal density gradient, and similar scaling arguments, a slope or shelf current (in the direction of decreasing sea surface height) will have strength comparable to the wind-stress-induced current if

$$\frac{\partial\rho}{\partial s} = \frac{A}{hHg}\left|\tau^s\right|, \tag{9}$$

where $\partial\rho/\partial s$ expresses the depth-mean along-slope density gradient, $A = 2$ to 4 (dependent on scaling assumptions) and h and H are the local and maximum water depths (see Huthnance, 1984, un-numbered equation, end of p. 799).

In interpreting these scaling arguments as applied to the eastern margin of the North Atlantic, the underlying physics is such that in a steady sense both the eastward geostrophic

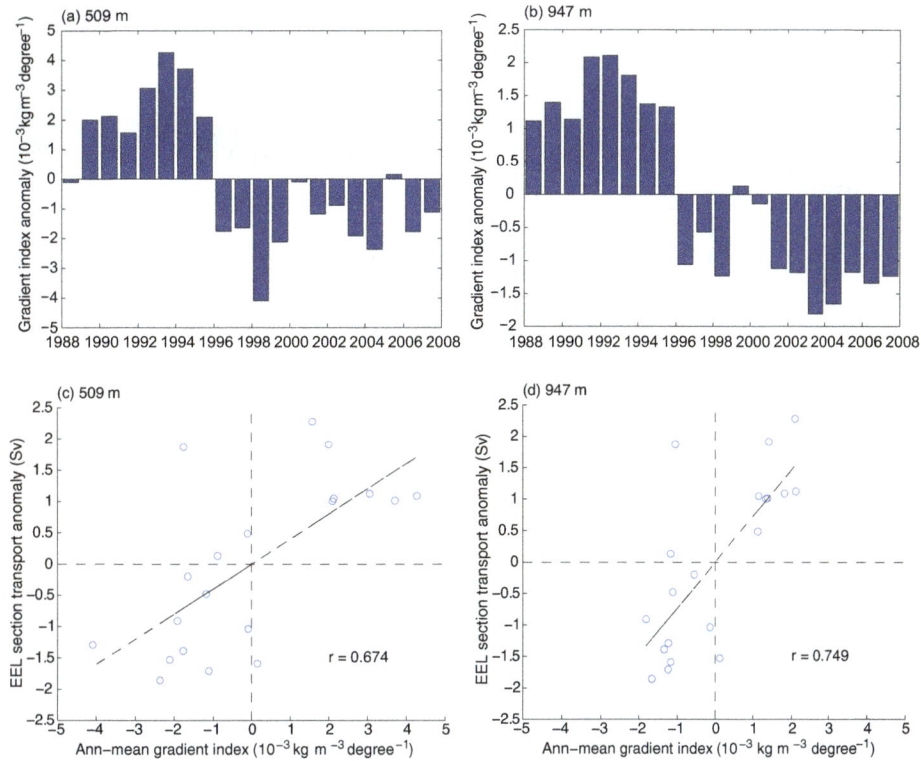

Figure 7. Time series over 1988–2007, in ORCA12-N01, of anomalies in density gradients averaged across 15–28° W and annually, for 509 m (**a**) and 947 m (**b**)–(**d**), plotted against annual-mean Slope Current transports at the EEL section.

flow (derived from the meridional density gradient) and an eastward surface Ekman response (derived from a northward wind stress) drive water initially towards the closed eastern boundary. These eastward flows raise the sea level near the eastern boundary, and result in a geostrophically balanced northward flow, with friction balancing the down-wind and down-pressure gradient accelerations.

Although unimportant when making a relative comparison between wind stress and buoyancy forcing, the linearized friction coefficient determines the absolute strength of the northward current (for given forcing). Expressed as $k \propto U_{M2}/h$ (where U_{M2} is a magnitude for the semi-diurnal tidal current speed), friction therefore determines the zonal structure of a Slope Current (moving across the slope). In the setting of the shallow and strongly tidal northwest European shelf, friction is greatest on the shelf, and the strongest northward flows are therefore concentrated over the slope.

The effects of seasonality in a northeast Atlantic setting are noteworthy. In winter, the surface Ekman layer will be deeper than the shelf break, and the eastward Ekman mass convergence will manifest at least in part over the continental slope. Both eastward Ekman and eastward geostrophic flow (in balance with the meridional density gradient) therefore have co-located convergence over the slope in winter, forcing a strong and mostly barotropic Slope Current. In summer, by contrast, the surface Ekman layer will be consider-

ably shallower than the shelf break, with convergence occurring more towards the coast. This leads to greater surface water exchange onto the shelf, and a more spatially diffuse northward flow over the slope and shelf, since the effects of wind-stress and meridional buoyancy forcing are no longer co-located over the slope.

Here, we focus on interannual changes in wind forcing, considering the influence on Slope Current transport of anomalies in along-slope wind stress. As a metric for the wind-stress forcing of the Slope Current, and for direct comparison with the density gradient metric developed in Sect. 3.3.1, we evaluate the right-hand side of Eq. (9) annually over the continental slope between 48 and 62° N, where $|\tau_s|$ is obtained from reanalysis 10 m winds, as outlined in Sect. 2.4. Expressed in units of northward density gradient (10^{-3} kg m^{-3} °$^{-1}$, regressed in the latitude range 48–62° N), annual anomalies of this metric are shown in Fig. 8. The anomalies are generally smaller in magnitude compared to the large-scale anomalies in Fig. 7, but there is a similar tendency for positive (negative) anomalies before (after) 1996 (as Fig. 7). In Sect. 4, we discuss the combined influences of wind and buoyancy forcing on variable Slope Current transport, as diagnosed with this common framework.

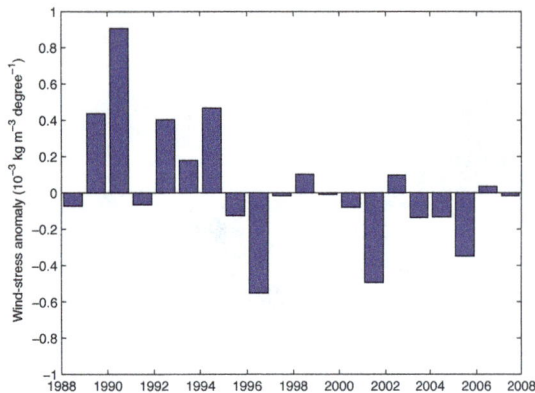

Figure 8. The wind forcing metric, expressed in units of northward density gradient (10^{-2} kg m^{-3} $\circ{-1}$, regressed in the latitude range 48–62° N) – see text for details.

3.4 Atlantic inflow to the North Sea

Considering the Slope Current and associated flows through the FSC, we identify that part of the flow diverted as Atlantic inflow to the North Sea. Time series of the associated transports in ORCA12-N01 are presented in Fig. 9. Net transports between Faroes and the Scottish mainland at Wick (not shown) are very highly correlated with net transport between the Faroes and Shetland ($r = 0.98$). Differences between transport across Faroes–Wick and Faroes–Shetland sections are due to net flow between Shetland and Wick (Fig. 9a). The sign convention of this residual transport as plotted is positive into the North Sea. This residual flow alternates between typically positive values in winter and negative values in summer. Across a section from 1.10 to 2.46° W at 59.27° N (the western JONSIS line), we obtain transports representative of the Fair Isle Current (Fig. 9b). This transport is almost always positive, i.e. into the North Sea, peaking in winter with seasonal amplitude very similar to that seen in Fig. 9a. Given the varying sign of residual flow and persistent southward transport in the Fair Isle Current, we infer a steady recirculation around Shetland of ~ 1 Sv.

To emphasize the dominant contribution of Fair Isle Current fluctuations to the variability of Atlantic inflow, we co-plot 30-day running means of anomalies in inferred Atlantic inflow and Fair Isle Current transport (Fig. 9c). These time series are highly correlated: $r = 0.81$; significant at 99 % confidence level. To evaluate the realism of this part of the Atlantic inflow in ORCA12, Fig. 10a shows monthly Atlantic inflow estimates from the ICES Report on Ocean Climate alongside a 30-day running mean of the 5-day transports in Fig. 9b. Close correspondence between the two time series is again consistent with dominance of Atlantic inflow by the Fair Isle Current. In Fig. 10b and c, we show annual-mean transport anomalies from the ICES Atlantic inflow and ORCA12 Fair Isle Current transport, both relative to the 1988–2007 mean. It appears that transports declined by 0.3–

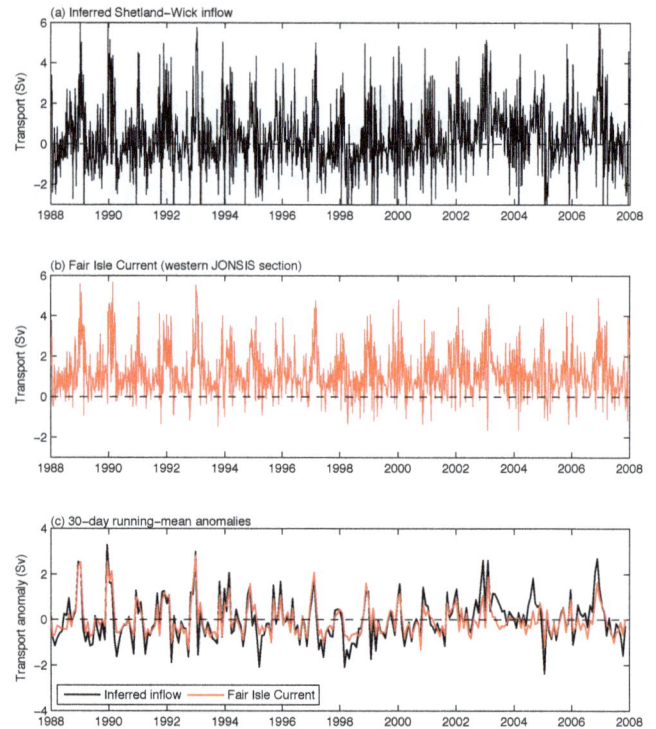

Figure 9. Transports related to Atlantic inflow into the northwest North Sea: **(a)** inferred inflow between Shetland and Wick, 5-day averages; **(b)** transport in the Fair Isle Current between 1.10 and 2.46° W at 59.27° N (along the western JONSIS line), 5-day averages; **(c)** 30-day running mean of anomalies in inferred Atlantic inflow and Fair Isle Current transport.

0.4 Sv over the 1990s, corresponding to around 20 % of the mean flow.

Returning to the Lagrangian analysis, and noting the tendency for particles to separately branch into the northwest and northeast North Sea (see Fig. 2), we consider the percentage of particle counts in "NW North Sea" and "NE North Sea" sub-regions (see Sect. 2.7 for definitions). For each year over 1988–2007, we average these statistics for both January and July releases. Figure 11 shows histograms of this annual mean percentage, for each sub-region (Fig. 11a and b) and both combined (Fig. 11c). In terms of a combined presence in the North Sea, the percentage of particles released at the FASTNEt section declines from near 40 % in the early 1990s to around 15 % in the mid-2000s. This long-term decline is suggestive of a reduced influence of Slope Current water of Atlantic origin in the northern North Sea, in the ORCA12 hindcast. The decline is similarly evident in the separate January and July releases (not shown), indicating a year-round character.

Finally, we consider salinity as a tracer of Atlantic Water, evaluating salinity variations in the hindcast alongside available observations. Sampling salinity along each model trajectory, we obtain averages per 0.5° × 0.5° grid cell (see

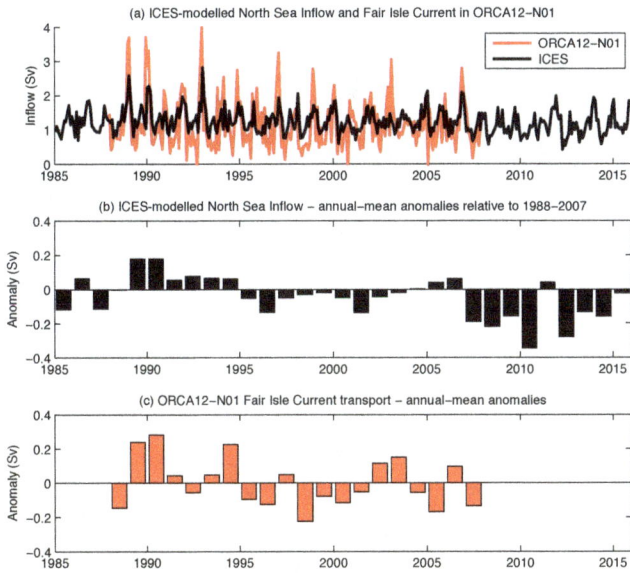

Figure 10. (a) Monthly and depth-averaged inflow to North Sea (source: ICES Report on Ocean Climate (thick line)), and 30-day running average of 5-day averaged Fair Isle Current transport in ORCA12-N01 (thin line); **(b)** annual-mean transport anomalies from ICES transport estimates of Atlantic inflow (relative to 1988–2007 mean); **(c)** annual-mean transport anomalies of Fair Isle Current transport in ORC12-N01.

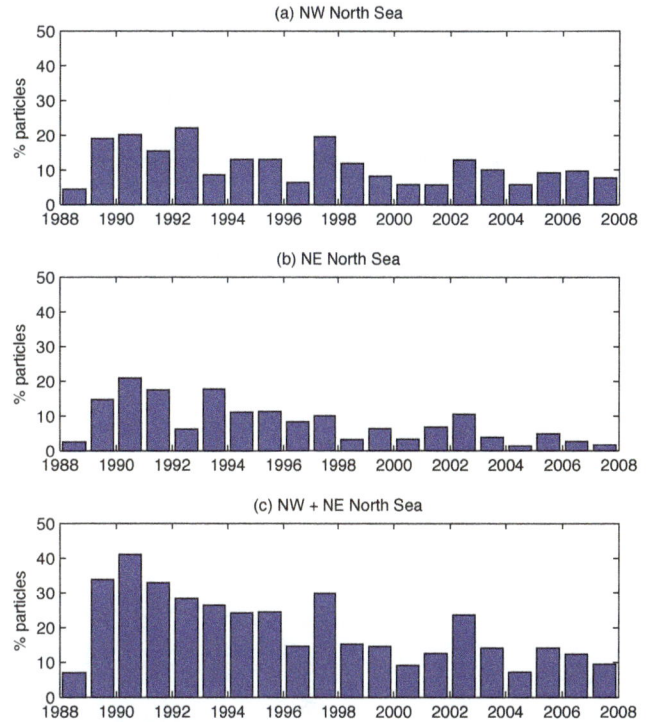

Figure 11. Histograms (per year) of mean fraction of (combined) January and July released particles residing: **(a)** in the northwest North Sea (4.5° W to 1.5° E, south of 59° N); **(b)** in the northeast North Sea (east of 1.5° E, south of 62° N); **(c)** the combined total.

Fig. S3) specific to time-varying flows at our four selected sections. We sample the NCEP Global GODAS analysis fields at locations central to each section (see Sect. 2.1), to obtain monthly mean salinity in that part of the water column most influenced by Atlantic Water (see Fig. S10), subtracting climatological seasonal cycles to obtain time series of salinity anomalies (see Fig. S11). We also consider direct observations, synthesized by ICES, where the Slope Current and Atlantic inflow have been monitored since 1950 and 1960 respectively: annual-mean observed salinity of Atlantic Water for the FSC Shetland Shelf; and annually and vertically averaged salinity in the Fair Isle Current (see Sect. 2.1). These salinity data are presented in Fig. 12.

At the FASTNEt and EEL sections (Fig. 12a and b), there is an overall decline of salinity from the early 1990s to the mid-2000s, in both ORCA12 and GODAS data. Superimposed on these declines is notable interannual variability, more dominant in the GODAS data, which also indicate a reversion to increasing trends from the mid-2000s onwards. At the Shetland Slope, long-term observed increases of salinity from 1980 to the early 2000s are evident in both GODAS and ICES data, while a declining trend persists in ORCA12 (Fig. 12c). We suspect that increasing trends observed in the FSC are associated with an additional influence from oceanic Atlantic Water (separate from Slope Current traversing FASTNEt and EEL sections) that is not well represented in the hindcast. At the western JONSIS line, a slight freshen-

ing trend is evident from the mid-1990s to the mid-2000s, in the direct observations, the GODAS analysis and ORCA12, although interannual variability is considerable in the observations but much reduced in ORCA12. The observed variability may be associated with local processes that are underrepresented in ORCA12. We note a remarkable increase of salinity around 2008 in the GODAS data (see also Fig. S11d), with positive anomalies being sustained up to 2016.

3.5 Sea level differences as proxies for Slope Current transport and Atlantic inflow to the North Sea

As Slope Current transport is strongly barotropic, we expect a strong correlation with the sea surface difference across the current. Given the available tide gauge data at Wick, Lerwick and Tórshavn (Sect. 2.5), we consider the differences in relative sea surface height (SSH) between Shetland and the Faroes (Lerwick–Tórshavn), and between mainland Scotland and Shetland (Wick–Lerwick), in ORCA12, presented in Fig. 13. For the 1988–2007 hindcast, Fig. 13a shows 5-day averages of SSH at the nearest ocean grid cells to Wick (green curve), Lerwick (red curve) and Tórshavn (blue curve), and the differences, Lerwick minus Tórshavn (Fig. 13b) and Wick minus Lerwick (Fig. 13c). There is a clear seasonal cycle in SSH at all three locations, with higher (lower) SSH in summer (winter), largely due to the

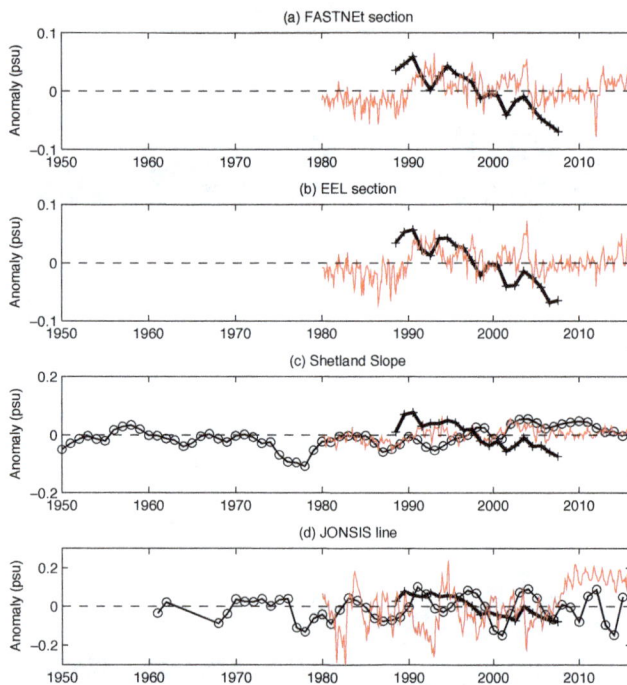

Figure 12. Annual and monthly mean salinity anomalies (relative to the length of each time series), from observations (thin lines) and in ORCA12-N01 (thick lines), at each section: **(a)** FASTNEt; **(b)** EEL; **(c)** Shetland Slope; and **(d)** JONSIS. The thin red lines are obtained by subtracting climatological seasonal cycles from salinity in the NCEP Global Ocean Data Assimilation System (GODAS) analysis fields, sampled at the following locations: 9.5° W, 55.83° N at 100 m (FASTNEt); 9.5° W, 57.17° N at 100 m (EEL); 2.5° W, 60.83° N at 100 m (Shetland Slope); 1.5° W, 59.17° N at 50 m (JONSIS) – see Figs. S10, S11. The thin black curves in **(c, d)** are observations for the FSC Shetland Shelf and the Fair Isle Current (see Sect. 2.1).

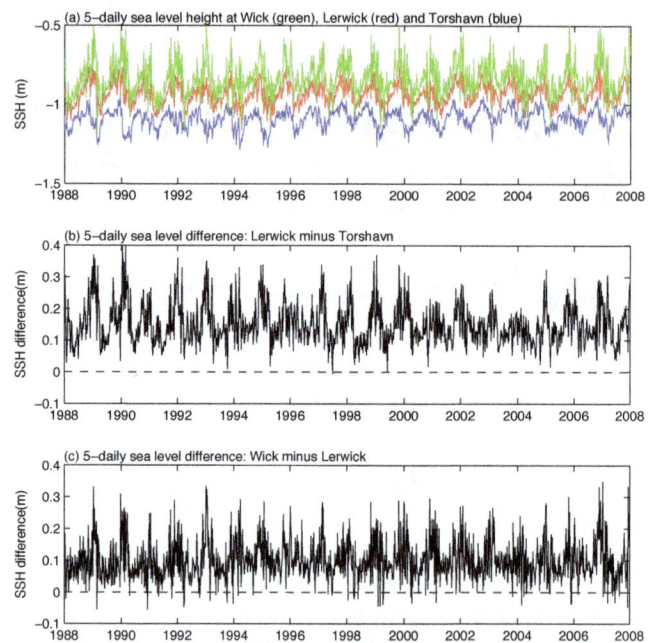

Figure 13. Relative sea surface height at Wick, Lerwick and Tórshavn, and the differences, 5-day averages over 1988–2007 in ORCA12-N01.

thermosteric effect of winter cooling (summer warming) and contraction (expansion) of water columns. The seasonal cycle increases in amplitude from Tórshavn to Wick; hence there is also a seasonal cycle in the SSH difference, with larger differences in winter. Comparing Figs. 13b and 4c, seasonal cycles of Shetland Slope transport and Lerwick–Tórshavn SSH difference are clearly in phase, as are Fair Isle Current transports (Fig. 9b) and Wick–Lerwick SSH differences (Fig. 13c).

Removing mean seasonal cycles in transports and SSH differences from 5-day averaged data, we find that transport anomalies are strongly correlated with SSH difference anomalies: $r = 0.68$ for Shetland Slope transports and Lerwick–Tórshavn SSH differences; $r = 0.85$ for Fair Isle Current transports and Wick–Lerwick SSH differences; both significant at 99 % confidence level. Linear regressions indicate a transport sensitivity of ~ 0.25 Sv cm^{-1}. Illustrating this sensitivity, Fig. 14 shows 30-day running averages of anomalies, relative to seasonal cycles over 1988–2007, for

Shetland Slope transport and Lerwick–Tórshavn SSH difference (Fig. 14a), and for Fair Isle Current transport and Wick–Lerwick SSH difference (Fig. 14b). SSH differences may therefore be useful proxies for Slope Current transport and Atlantic inflow to the North Sea.

To relate changes in Atlantic inflow to the North Sea with changes in Slope Current transport and SSH difference, we correlate the "combined" percentage of North Sea particles (Fig. 11c) with annual-mean anomalies for Shetland Slope transport and Lerwick–Tórshavn sea level difference. Correlation coefficients of 0.52 (with transports) and 0.70 (with SSH differences), both significant at 99 % confidence level, indicate that larger SSH differences and stronger transports are indeed associated with more Atlantic Water reaching the North Sea. The stronger correlation of the percentage of North Sea particles with SSH difference (compared to transport) indicates that this metric more completely captures transport variability than the short and fixed Shetland Shelf section. Clearly then, declining Slope Current transport at the Shetland Slope over 1988–2007 is broadly representative of changes already evident in the wider (upstream) Slope Current system, and consistent with the declining percentage of Slope Current water particles reaching the North Sea.

Evidence for variable Slope Current transport over a longer time period is now considered, using historical tide gauge data and the longer ORCA12-N06 hindcast. Figure 15 shows sea level at Wick, Lerwick and Tórshavn from 1957 onwards in tide gauge records, monthly averaged after correcting for glacial isostatic adjustment (Fig. 15a), 30-day

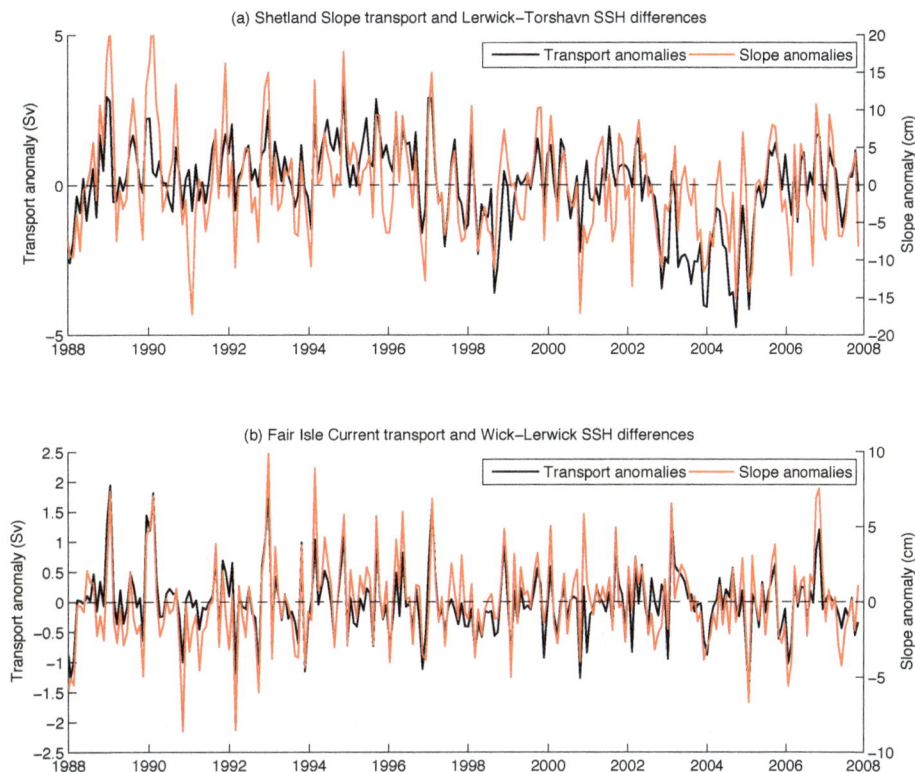

Figure 14. 30-day running averages of anomalies (relative to seasonal cycles over 1988–2007) of (**a**) Shetland Slope transports and Lerwick–Tórshavn SSH differences, and (**b**) Fair Isle Current transports and Wick–Lerwick SSH differences.

running means of SSH over 1958–2013 in the ORCA12-N06 hindcast (Fig. 15b), and the corresponding differences (Fig. 15c and d). Considering the sea level time series (Fig. 15a and b), there is general agreement between the tide gauges and ORCA12 in terms of a seasonal cycle and long-term sea level rise primarily associated with thermal expansion, although irreconcilable differences currently remain between the datum levels for the two tide gauge records (therefore Fig. 15a and b should not be directly compared).

Considering Lerwick–Tórshavn sea level differences in the model (Fig. 15c), ORCA12-N06 indicates a degree of low-frequency variability, with smaller differences in the last ~ 15 years that are in close agreement with ORCA12-N01 over the period of overlap (see Fig. S12): averaged over 1989–1999 and 1999–2012, sea level differences in the hindcast are 15.0 and 13.1 cm respectively, equating to a transport reduction of around 0.5 Sv in the latter period. This reduction in the sea level differences is seen to an extent in the tide gauge records whenever the data is available. However, substantially higher differences over 1957–1964, apparent from the tide gauge records, are not seen in the hindcast. Considering Wick–Lerwick sea level differences, higher amplitude variability on decadal timescales is apparent in the tide gauge data, compared to the hindcast. Considering the period 1965 onwards, when Wick tide gauge data are available, correlation coefficients between monthly tide gauge sea

level differences and model SSH differences are strong for Lerwick–Tórshavn ($r = 0.61$, significant at 99 % confidence level), but weak for Wick–Lerwick ($r = 0.11$, significant at 95 % confidence level).

As related to variability in sea level differences, Atlantic inflow to the North Sea is strongly correlated with FSC Slope Current transport in the hindcast ($r = 0.71$ for Lerwick–Tórshavn and Wick–Lerwick SSH differences, significant at 99 % confidence level). The corresponding tide gauge differences are, however, not correlated ($r = -0.05$, not significant), indicating that other factors influence the majority of observed sea level variability at Wick in particular. These preliminary findings help to validate the Slope Current variability simulated in ORCA12 hindcasts, while indicating the limited extent to which variable Slope Current transport and Atlantic inflow to the North Sea may be reconstructed with tide gauge data over the longer historical era. To obtain a useful proxy for Atlantic inflow, it will be necessary to first remove that part of the variability in the Wick tide gauge record that is not associated with a dynamical signal.

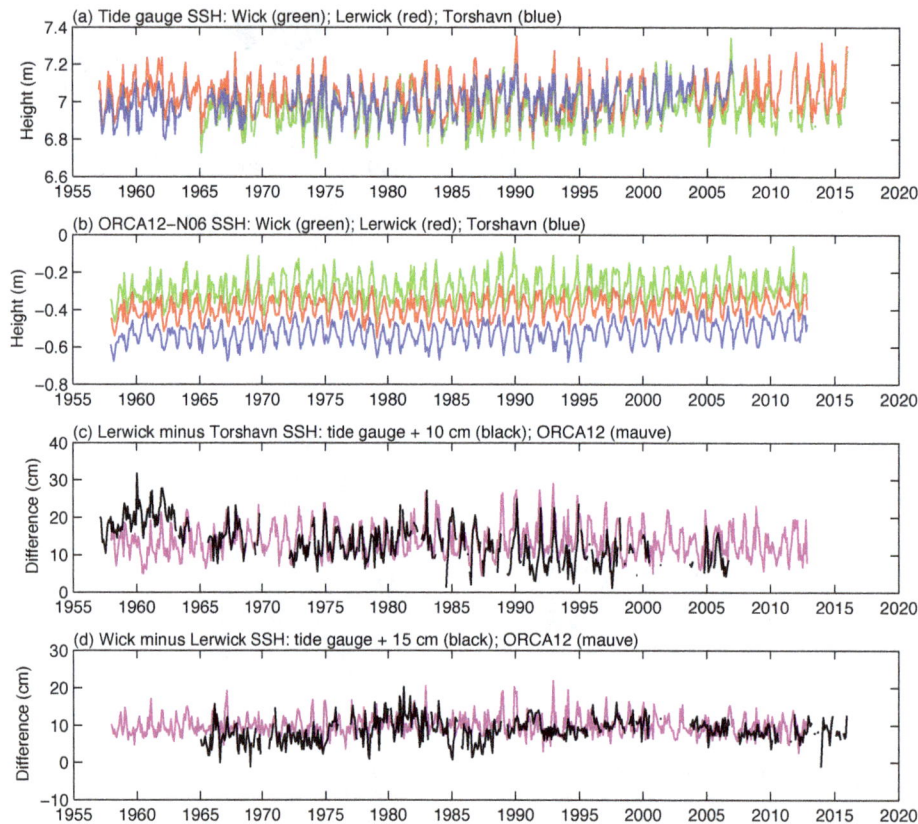

Figure 15. Sea surface height at Wick, Lerwick and Tórshavn: **(a)** relative sea level (monthly means) from tide gauge records (1957 onwards); **(b)** relative sea surface height (30-day running means) in ORCA12-N06 (1958–2012); **(c)** Lerwick minus Tórshavn differences; and **(d)** Wick minus Lerwick differences.

4 Discussion and conclusions

The Slope Current system that is observed to follow the shelf break to the west and north of Scotland is investigated using a range of observations and an eddy-resolving ocean model (ORCA12) hindcast spanning 1988–2007. Deployments of drogued drifters over 1995–1997 reveal a variety of pathways and timescales in the Slope Current system, hinting at seasonal to interannual variations. To further explore this variability, offline particle trajectories are calculated with model currents. Particles are tracked both forwards and backwards in time, for 183 days, from a section across the Slope Current (9.46–9.28° W at ∼ 55.82° N, in the upper 371 m) where floats have been deployed as part of the UK NERC project FASTNEt (http://www.sams.ac.uk/fastnet). Tracked backwards, particle trajectories reveal a major source of Slope Current water in the eastern subpolar gyre, with a smaller proportion advecting with the Slope Current from more southern latitudes.

Variable pathways are related to both seasonal and interannual variability in Slope Current transport. The latter variability is related to large-scale forcing mechanisms. Downward trends in Slope Current transport, similar to those inferred from altimetry (Xu et al., 2015), are principally re-

lated to basin-scale changes in the subpolar North Atlantic. Across the northeast Atlantic over 1988–2007, we identify 25–50 % reductions of meridional density gradients in the depth range 500–1000 m representative of a layer that supports geostrophic inflow to the Slope Current, which can be considered as "fed from the west". In particular, we find abrupt reductions of density gradients over 1995–1997, coincident with weakening of the Slope Current at the FAST-NEt and EEL sections. The reductions in meridional density gradients are primarily due to warming in the eastern subpolar gyre, coincident with weakening of the gyre (Johnson et al., 2013). The new OSNAP monitoring array (http://www.o-snap.org/), spanning the subpolar gyre and incorporating the EEL section, should provide the observations needed to further investigate large-scale drivers of variability in Slope Current transport.

Using a common framework, we find that changes in annual-mean wind forcing contribute around 20 % to the density gradient variability. Changes in wind forcing, specifically the along-slope component of wind stress, are associated with the transition of the North Atlantic Oscillation (NAO) from a positive to a neutral phase during the mid-1990s, weakening wind forcing of the Slope Current at that time. To summarize forcing mechanisms, the schematics in Fig. 16 indicate the density gradients, wind forcing, Ekman transports and sea surface slopes associated with weak and strong Slope Current transport. Strong (weak) transport is associated with a strong (weak) subpolar gyre, and the NAO in a positive (negative) phase. Emphasizing the conditions for strong Slope Current transport: in the deep ocean, colder water to the north sets up a stronger northward density gradient, while the downward sea surface slope to the north steepens; stronger eastward geostrophic flow is supported by the combined effect of density gradient and strengthened along-slope (northward) winds and onshore Ekman transports; northward steepening of the cross-slope gradient in sea surface height becomes more pronounced in proportion to inflow recruited to the barotropic Slope Current.

Downstream consequences of changes in Slope Current transport have also been investigated. Tracked forwards in the ORCA12 hindcast, a substantial number of particle trajectories reach the northern North Sea, and we accordingly diagnose the percentage of particle locations in the northwest and northeast North Sea, as metrics for cumulative Atlantic inflow in these regions. Over the 1988–2007 hindcast, we thus identify a decline from $\sim 40\%$ in the early 1990s to $\sim 15\%$ in the mid-2000s, accompanied by the reductions in Slope Current transport. Around half of the Atlantic inflow is mixed with fresher North Sea water (including Baltic outflow) before outflow in the Norwegian Coastal Current (Winther and Johannessen, 2006). Mean salinities along particle trajectories indicate a reduction of 0.2–0.3 psu from inflow to outflow (see Fig. S3). Variable Atlantic and Baltic inflows must contribute to salinity variability in the North Sea. There is a climatological seasonal variation of North Sea freshwater content by $\sim 20\%$, with peak values in July–August that lag by 2–3 months the net fresh inflow from the Baltic, which in turn varies by a factor of ~ 3 over the seasonal cycle (see Sündermann and Pohlmann, 2011 – their Fig. 17). While Baltic inflow dominates the seasonal cycle of salinity, Atlantic inflow is thought to dominate mean salinity of the North Sea (Sündermann and Pohlmann, 2011).

Changes in salinity of the Atlantic inflow, on interannual and longer timescales, are also likely to impact North Sea salinity. While observed increases of Atlantic Water salinity in the FSC over 1988–2007 (Holliday et al., 2008) are not reproduced in the hindcast, observed salinity in the Fair Isle Current is highly variable, and salinity remained relatively invariant in the northern North Sea (Larsen et al., 2016). At the same time, Atlantic inflow to the North Sea weakened to an extent, in both our hindcast and an independent model simulation (Larsen et al., 2016). Partitioning the influence of Atlantic inflow on North Sea salinity between changes in volume transport and changes in salinity, we may distinguish between "anomalous volume transport of mean salinity" and "mean volume transport of anomalous salinity". Increasing salinity in the Atlantic Water may have thus broadly compensated for declining Atlantic inflow during the 1990s, explain-

(a) Weak Slope Current

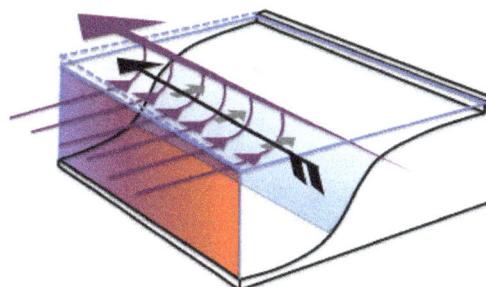

(b) Strong Slope Current

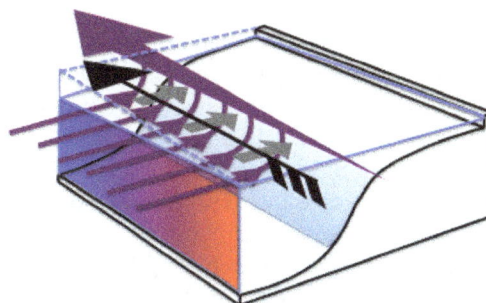

Figure 16. Schematics showing density gradients (shaded red to pale or dark blue), eastward geostrophic inflow, wind forcing (black alongshore arrow), Ekman transports (grey onshore arrows) and sea surface slopes at an idealized eastern boundary, associated with **(a)** weak and **(b)** strong Slope Current transport.

ing the absence of an observed salinity trend in the northern North Sea during this period.

Variable Atlantic inflow to the North Sea has a likely impact on North Sea ecosystems via hydrographic changes, as previously suggested by Reid et al. (2001). The northern North Sea undergoes seasonal stratification, with associated patterns and timings of productivity (Sharples et al., 2006), which may be sensitive to the relative influence of Atlantic Water. Changes of inflow prior to our study period may also help to explain a widely documented ecosystem regime shift in the early 1980s that was observed in phytoplankton and zooplankton populations (Beaugrand, 2004). Existing EEL observations additionally indicate warming and declining nutrient concentrations in the Rockall trough from 1996 to the mid-2000s (Johnson et al., 2013), which may have further influenced North Sea ecosystems that previously underwent a regime shift over 1982–1988, from a "cold dynamic equilibrium" (1962–1983) to a "warm dynamic equilibrium" (1984–1999) (Beaugrand, 2004).

Looking back over a longer period, we evaluate sea level differences as proxies for Slope Current transport (since 1957) and Atlantic inflow to the North Sea (since 1965). Slope Current transport variability is identified with sea level differences between Lerwick (Shetland) and Tórshavn (Faroes), while Atlantic inflow to the North

Sea is identified with differences between Wick (Scottish mainland) and Lerwick, in both the tide gauge records and in a longer ORCA12 hindcast spanning 1958–2012. In the shorter ORCA12 hindcast that provided the basis for in-depth analysis, Slope Current transport at the Shetland Shelf section is highly correlated with Lerwick–Tórshavn sea level differences. Looking to the longer periods, variability of Slope Current transport on a wide range of timescales, from seasonal to multi-decadal, is implicit in Lerwick–Tórshavn sea level differences. Wick–Lerwick sea level differences in tide gauge records indicate considerable decadal variability in the Fair Isle Current transport that dominates Atlantic inflow to the northwest North Sea, while sea level differences in the hindcast are dominated by strong seasonal variability. With locally strong isostacy, differences in local datums, and seasonal steric effects, there are considerable challenges in extracting from tide gauge records the signals that are associated with Slope Current transport and Atlantic inflow. We also recognize that contributions to variability in Lerwick–Tórshavn sea level differences may be associated with variability in (1) the recirculating Faroe branch of Atlantic inflow and (2) flow that negotiates the Faroe Bank and Wyville Thomson Ridge (Berx et al., 2013). Nevertheless, hindcast Lerwick–Tórshavn sea level differences are highly correlated with both the tide gauge equivalent and the hindcast Wick–Lerwick differences. However, the Wick–Lerwick and Lerwick–Tórshavn tide gauge differences are not significantly correlated. This suggests that the sea level variability recorded by the tide gauge at Wick is either not capturing the dynamical signal, or is dominated by other influences. Prospects for using sea level records to reconstruct or monitor Atlantic inflow thus depend on refined use of the tide gauge record at Wick.

The larger-scale context for long-term changes in the meridional density gradients that support the Slope Current, and Atlantic inflow to the North Sea, likely involves the basin-scale ocean circulation. Previous studies provide evidence for a decline of the Atlantic Meridional Overturning Circulation (AMOC) in mid-latitudes (at 48° N) between the early 1990s and the mid-2000s (Balmaseda et al., 2007; Grist et al., 2009), while Josey et al. (2009) show this decline to be representative across ∼ 48–60° N, a zone encompassing Slope Current inflow. The striking shift to weaker Slope Current transport at the EEL section over 1996–1998 coincides with a major warming of the subpolar gyre at this time (Robson et al., 2012). More recently, there has been a major reversal of temperature in the eastern subpolar gyre along with formation of a particularly dense mode of Subpolar Mode Water, associated with extreme cooling in the winter of 2013–2014 (Grist et al., 2015), reinforced through further cooling during 2015 (Duchez et al., 2016). These events may have restored strong meridional density gradients and re-strengthened the Slope Current, bringing more high-salinity Atlantic Water to the shelf break. Evidence for such a response in the Slope Current is found in an increase by around

0.05 psu of salinity at around 800 m over much of 2014–2016 (see Fig. S11a and b), while the thermal wind relation predicts an approximate doubling of eastward geostrophic transport in mid-latitudes to the west of the shelf break, associated with increased meridional density gradients due to subpolar cooling.

Author contributions. Robert Marsh designed the study and undertook analysis of the ORCA12-N01 hindcast, including the Lagrangian diagnostics and evaluation with observations (ICES, GODAS). Stuart A. Cunningham analysed the observations of mean absolute dynamic topography and climatological density. Mark E. Inall developed the wind forcing metric. Marie Porter analysed drifter observations and calculated the wind forcing metric. Ivan D. Haigh analysed tide gauge records at Wick, Lerwick and Tórshavn. Ben I. Moat diagnosed the ORCA12-N06 hindcast. Robert Marsh prepared the paper with contributions from all co-authors.

Competing interests. The authors declare that they have no conflict of interest.

Acknowledgements. Robert Marsh acknowledges the support of a 2013 Research Bursary awarded by the Scottish Association for Marine Science. Mark E. Inall and Marie Porter acknowledge the UK National Environment Research Council (NERC) programme FASTNEt (NERC Ref. NE/I030224/1). Stuart A. Cunningham acknowledges the NERC project UK-OSNAP (NERC Ref. NE/K010700/1) and the EU-funded project NACLIM. Ben I. Moat acknowledges funding from NERC through the RAPID-AMOC Climate Change (RAPID) programme. The ORCA12 simulations were undertaken at the National Oceanography Centre, using the NEMO framework. NEMO is a state-of-the-art, portable modelling framework developed by a consortium of European institutions, namely the National Centre for Scientific Research (CNRS) in Paris, the UK Met Office (UKMO), Mercator Ocean, and NERC. GODAS data are provided by the NOAA/OAR/ESRL PSD, Boulder, Colorado, USA, available from their Web site at http://www.esrl.noaa.gov/psd/. We thank two anonymous reviewers for many insightful comments that helped us to substantially improve the paper. This study is in memory of Kate Stansfield.

Edited by: M. Hecht

References

Amante, C. and Eakins, B. W.: ETOPO1 1 arc-minute global relief model: procedures, data sources and analysis, US Department of Commerce, National Oceanic and Atmospheric Administration, National Environmental Satellite, Data, and Information Service, National Geophysical Data Center, Marine Geology and Geophysics Division, Colorado, 2009.

AVISO+: SSALTO/DUACS User Handbook: MSLA and (M)ADT Near-Real Time and Delayed Time Products, SALP-MU-P-EA-21065-CLS, CLS-DPS-NT-06-034 (5rev0, 20/08/2016), available at: http://www.aviso.altimetry.fr/fileadmin/documents/data/tools/hdbk_duacs.pdf (last access: 10 April 2017), 2016.

Balmaseda, M. A., Smith, G. C., Haines, K., Anderson, D., Palmer, T. N., and Vitard, A.: Historical reconstruction of the Atlantic meridional overturning circulation from the ECMWF operational ocean reanalysis, Geophys. Res. Lett., 34, L23615, doi:10.1029/2007GL031645, 2007.

Beaugrand, G.: The North Sea regime shift: evidence, causes, mechanisms and consequences, Progr. Oceanogr., 60, 245–262, 2004.

Berx, B., Hansen, B., Østerhus, S., Larsen, K. M., Sherwin, T., and Jochumsen, K.: Combining in situ measurements and altimetry to estimate volume, heat and salt transport variability through the Faroe-Shetland Channel, Ocean Sci., 9, 639–654, doi:10.5194/os-9-639-2013, 2013.

Blaker, A. T., Hirschi, J. J.-M., McCarthy, G., Sinha, B., Taws, S., Marsh, R., de Cuevas, B. A., Alderson, S. G., and Coward, A. C.: Historical analogues of the recent extreme minima observed in the Atlantic meridional overturning circulation at 26° N, Clim. Dynam., 44, 457–473, doi:10.1007/s00382-014-2274-6, 2015.

Blanke, B. and Raynaud, S.: Kinematics of the Pacific Equatorial Undercurrent: a Eulerian and Lagrangian approach from GCM results, J. Phys. Oceanogr., 27, 1038–1053, 1997 (data available at: http://stockage.univ-brest.fr/~grima/Ariane/doc.html).

Brodeau, L., Barnier, B., Treguier, A.-M., Penduff, T., and Gulev, S.: An ERA40-based atmospheric forcing for global ocean circulation models, Ocean Model., 31, 88–104, 2010.

Burrows, M. and Thorpe, S. A.: Drifter observations of the Hebrides slope current and nearby circulation patterns, Ann. Geophys., 17, 280–302, doi:10.1007/s00585-999-0280-5, 1999.

Burrows, M., Thorpe, S. A., and Meldrum, D. T.: Dispersion over the Hebridean and Shetland shelves and slopesm Cont. Shelf Res., 19, 49–55, 1999.

Dee, D. P., Uppala, S. M., Simmons, A. J., Berrisford, P., Poli, P., Kobayashi, S., Andrae, U., Balmaseda, M. A., Balsamo, G., Bauer, P., Bechtold, P., Beljaars, A. C. M., van de Berg, L., Bidlot, J., Bormann, N., Delsol, C., Dragani, R., Fuentes, M., Geer, A. J., Haimberger, L., Healy, S. B., Hersbach, H., Hólm, E. V., Isaksen, L., Kållberg, P., Köhler, M., Matricardi, M., McNally, A. P., Monge-Sanz, B. M., Morcrette, J.-J., Park, B.-K., Peubey, C., de Rosnay, P., Tavolato, C., Thépaut, J.-N., and Vitart, F.: The ERA-Interim reanalysis: Configuration and performance of the data assimilation system, Q. J. Roy. Meteorol. Soc., 137, 553–597, doi:10.1002/qj.828, 2011 (data available at: http://www.ecmwf.int/en/research/climate-reanalysis/era-interim).

Dooley, H. D.: Hypothesis concerning the circulation of the North Sea, Journal du Conseil International pour l'Exploration de la Mer, 36, 64–61, 1974.

Duchez, A., Frajka-Williams, E., Josey, S. A., Evans, D., Grist, J. P., Marsh, R., McCarthy, G. D., Sinha, B., Berry, D. I., and Hirschi, J. J.-M.: Drivers of exceptionally cold North Atlantic Ocean temperatures and their link to the 2015 European heat wave, Environ. Res. Lett., 11, 074004, doi:10.1088/1748-9326/11/7/074004, 2016.

GODAS (NCEP Global Ocean Data Assimilation System): Salinity Data, available at: https://www.esrl.noaa.gov/psd/data/gridded/data.godas.html, last access 3 February 2017.

Good, S. A., Martin, M. J., and Rayner, N. A.: EN4: quality controlled ocean temperature and salinity profiles and monthly objective analyses with uncertainty estimates, J. Geophys. Res., 118, 6704–6716, doi:10.1002/2013JC009067, 2013 (data available at: http://www.metoffice.gov.uk/hadobs/en4/).

Grist, J. P., Marsh, R., and Josey, S. A.: On the relationship between the North Atlantic meridional overturning circulation and the surface-forced overturning stream function, J. Climate, 22, 4989–5002, doi:10.1175/2009JCLI2574.1, 2009.

Grist, J. P., Josey, S. A., Jacobs, Z. L., Marsh, R., Sinha, B., and van Sebille, E.: Extreme air-sea interaction over the North Atlantic subpolar gyre during the winter of 2013-14 and its sub-surface legacy, Clim. Dynam., 46, 4027–4045, doi:10.1007/s00382-015-2819-3, 2015.

Hecht, M. W. and Smith, R. D.: Toward a Physical Understanding of the North Atlantic: A Review of Model Studies in an Eddying Regime, Geophysical Monograph Series, Am. Geophys. Un., 177, 213–239, 2013.

Hjøllo, S. S., Skogen, M. D., and Svendsen, E.: Exploring currents and heat within the North Sea using a numerical model, J. Mar. Syst., 78, 180–192, 2009.

Holgate, S. J., Matthews, A., Woodworth, P. L., Rickards, K. J., Tamisiea, M. E., Bradshaw, E., Foden, P. R., Gordon, K., Jevrejeva, S., and Pugh, J.: New data systems and products at the Permanent Service for Mean Sea Level, J. Coast. Res., 29, 493–504, 2013.

Holliday, N. P., Hughes, S. L., Bacon, S., Beszczynska-Möller, A., Hansen, B., Lavín, A., Loeng, H., Mork, K. A., Østerhus, S., Sherwin, T., and Walczowski, W.: Reversal of the 1960s to 1990s Freshening Trend in the Northeast North Atlantic and Nordic Seas, Geophys. Res. Lett., 35, L03614, doi:10.1029/2007GL032675, 2008.

Huthnance, J. M.: Slope Currents and "JEBAR", J. Phys. Oceanogr., 14, 795–810, 1984.

Huthnance, J. M., Holt, J. T., and Wakelin, S. L.: Deep ocean exchange with west-European shelf seas, Ocean Sci., 5, 621–634, doi:10.5194/os-5-621-2009, 2009.

Inall, M. E., Gillibrand, P. A., Griffiths, C. R., MacDougal, N., and Blackwell, K.: On the oceanographic variability of the North-West European Shelf to the West of Scotland, J. Mar. Syst., 77, 210–226, doi:10.1016/j.jmarsys.2007.12.012, 2009.

Johnson, C., Inall, M., and Häkkinen, S.: Declining nutrient concentrations in the northeast Atlantic as a result of a weakening Subpolar gyre, Deep-Sea Res. Pt. I, 82, 95–107, 2013.

Josey, S. A., Grist, J. P., and Marsh, R.: Estimates of meridional overturning circulation variability in the North Atlantic from surface density flux fields, J. Geophys. Res., 114, C09022, doi:10.1029/2008JC005230, 2009.

Large, W. and Pond, S.: Open ocean momentum flux measurements in moderate to strong winds, J. Phys. Oceanogr., 11, 324–336, 1981.

Larsen, K. M. H., Gonzalez-Pola, C., Fratantoni, P., Beszczynska-Möller, A., and Hughes, S. L. (Eds.): ICES Report on Ocean Climate 2015, ICES Cooperative Research Report No. 331, 79 pp., Data sets: Salinity time series for Faroe Shetland Channel – Shetland Shelf (North Atlantic Water) and Fair Isle Current Water (Waters entering North Sea from Atlantic), Data Provider – Marine Scotland Science, Aberdeen, UK, Modelled North Sea Inflow, Data Provider – Institute of Marine Research, Norway, available at: http://ocean.ices.dk/iroc/, last access: 5 December 2016.

MacKenzie, B. R. and Schiedek, D.: Daily ocean monitoring since the 1860s shows record warming of northern European seas, Global Change Biol., 13, 1335–1347, doi:10.1111/j.1365-2486.2007.01360.x, 2007.

Madec, G.: NEMO ocean engine, in: Vol. 27, Institut Pierre-Simon Laplace, France, 2008.

McClean, J. L., Poulain, P.-M., Pelton, J. W., and Maltrud, M. E.: Eulerian and Lagrangian statistics from surface drifters and a high-resolution POP simulation in the North Atlantic, J. Phys. Oceanogr., 32, 2472–2491, 2002.

McKenna, C., Berx, B., and Austin, W. E. N.: The decomposition of the Faroe-Shetland Channel water masses using Parametric Optimum Multi-Parameter analysis, Deep-Sea Res. Pt. I, 107, 9–21, 2016.

Moat, B. I., Josey, S. A., Sinha, B., Blaker, A. T., Smeed, D. A., McCarthy, G., Johns, W. E., Hirschi, J.-M., Frajka-Williams, E., Rayner, D., Duchez, A., and Coward, A. C.: Major variations in subtropical North Atlantic heat transport at short (5 day) timescales and their causes, J. Geophys. Res., 121, 3237–3249, doi:10.1002/2016JC011660, 2016.

Petersen, M. R., Williams, S. J., Maltrud, M. E., Hecht, M. W., and Hamann, B.: A three-dimensional eddy census of a high-resolution global ocean simulation, J. Geophys. Res., 118, 1759–1774, 2013.

PSMSL (Permanent Service for Mean Sea Level): Tide Gauge Data, available at: http://www.psmsl.org/data/obtaining/, last access: 22 January 2017.

Reid, P. C., Holliday, N. P., and Smyth, T. J.: Pulses in eastern margin current with higher temperatures and North Sea ecosystem changes, Mar. Ecol.-Prog. Ser., 215, 283–287, 2001.

Reid, P. C., Edwards, M., Beaugrand, G., Skogen, M., and Stevens, D.: Periodic changes in the zooplankton of the North Sea during the twentieth century linked to oceanic inflow, Fish. Oceanogr., 12, 260–269, 2003.

Richter, K., Segtnan, O. H., and Furevik, T.: Variability of the Atlantic inflow to the Nordic Seas and its causes inferred from observations of sea surface height, J. Geophys. Res., 117, C04004, doi:10.1029/2011JC007719, 2012.

Rio, M. H., Guinehut, S., and Larnicol, G.: New CNES-CLS09 global mean dynamic topography computed from the combination of GRACE data, altimetry, and in situ measurements, J. Geophys. Res., 116, C07018, doi:10.1029/2010JC006505, 2011.

Robson, J., Sutton, R., Lohmann, K., Smith, D., and Palmer, M. D.: Causes of the rapid warming of the North Atlantic Ocean in the mid 1990s, J. Climate, 25, 4116–4134, doi:10.1175/JCLI-D-11-00443.1, 2012.

Sharples, J., Ross, O. N., Scott, B. E., Greenstreet, S., and Fraser, H.: Interannual variability in the timing of stratification and the spring bloom in the north-western North Sea, Cont. Shelf Res., 26, 733–751, 2006.

Sherwin, T. J., Hughes, S. L., Turrell, W. R., Hansen, B., and Østerhus, S.: Wind-driven monthly variations in transport and the flow field in the Faroe–Shetland Channel, Polar Res., 27, 7–22, 2008.

Simpson, J. H. and Sharples, J.: Introduction to the Physical and Biological Oceanography of Shelf Seas, Cambridge University Press, Cambridge, 2012.

Souza, A. J., Simpson, J. H., Harikrishnan, M., and Malarkey, J.: Flow structure and seasonality in the Hebridean slope current, Oceanol. Acta, 24, S63–S76, 2001.

Sündermann, J. and Pohlmann, T.: A brief analysis of North Sea physics, Oceanologia, 53, 663–689, doi:10.5697/oc.53-3.663, 2011.

Winther, N. G. and Johannessen, J. A.: North Sea circulation: Atlantic inflow and its destination, J. Geophys. Res., 111, C12018, doi:10.1029/2005JC003310, 2006.

Xu, W., Miller, P. I., Quartly, G. D., and Pingree, R. D.: Seasonality and interannual variability of the European Slope Current from 20 years of altimeter data with in situ measurement comparisons, Remote Sens. Environ., 162, 196–207, doi:10.1016/j.rse.2015.02.008, 2015.

11

Numerical investigation of the Arctic ice–ocean boundary layer and implications for air–sea gas fluxes

Arash Bigdeli[1], **Brice Loose**[1], **An T. Nguyen**[2], **and Sylvia T. Cole**[3]

[1]Graduate School of Oceanography, University of Rhode Island, Rhode Island, 02882, USA
[2]Institute of Computational Engineering and Sciences, University of Texas at Austin, Austin, Texas, 78712, USA
[3]Woods Hole Oceanographic Institution, Woods Hole, Massachusetts, 02543, USA

Correspondence to: Arash Bigdeli (arash_bigdeli@uri.edu)

Abstract. In ice-covered regions it is challenging to determine constituent budgets – for heat and momentum, but also for biologically and climatically active gases like carbon dioxide and methane. The harsh environment and relative data scarcity make it difficult to characterize even the physical properties of the ocean surface. Here, we sought to evaluate if numerical model output helps us to better estimate the physical forcing that drives the air–sea gas exchange rate (k) in sea ice zones. We used the budget of radioactive ^{222}Rn in the mixed layer to illustrate the effect that sea ice forcing has on gas budgets and air–sea gas exchange. Appropriate constraint of the ^{222}Rn budget requires estimates of sea ice velocity, concentration, mixed-layer depth, and water velocities, as well as their evolution in time and space along the Lagrangian drift track of a mixed-layer water parcel. We used 36, 9 and 2 km horizontal resolution of regional Massachusetts Institute of Technology general circulation model (MITgcm) configuration with fine vertical spacing to evaluate the capability of the model to reproduce these parameters. We then compared the model results to existing field data including satellite, moorings and ice-tethered profilers. We found that mode sea ice coverage agrees with satellite-derived observation 88 to 98 % of the time when averaged over the Beaufort Gyre, and model sea ice speeds have 82 % correlation with observations. The model demonstrated the capacity to capture the broad trends in the mixed layer, although with a significant bias. Model water velocities showed only 29 % correlation with point-wise in situ data. This correlation remained low in all three model resolution simulations and we argued that is largely due to the quality of the input atmospheric forcing. Overall, we found that even

the coarse-resolution model can make a modest contribution to gas exchange parameterization, by resolving the time variation of parameters that drive the ^{222}Rn budget, including rate of mixed-layer change and sea ice forcings.

1 Introduction

The ocean surface is a dynamic region where momentum, heat and salt, as well as biogeochemical compounds, are exchanged with the atmosphere and with the deep ocean. At the sea–air interface, gases of biogenic origin and geochemical significance are exchanged with the atmosphere. Theory indicates that the aqueous viscous sublayer, which has a length scale of 20 to 200 µm (Jähne and Haubecker, 1998), is the primary bottleneck for air–water exchange. Limitations in measurement at this critical scale have led to approximations of sea–air gas exchange based on indirect measurements. Four approaches involving data are typically used (Bender et al., 2011): (1) parametrization of the turbulent kinetic energy (TKE) at the base of the viscous sublayer, (2) tracing purposefully injected gases (Ho et al., 2006; Nightingale et al., 2000), (3) micro-meteorological methods (Zemmelink et al., 2006, 2008; Blomquist et al., 2010; Salter et al., 2011), and (4) radon deficit method. Here, we examine the radon deficit method (4), together with a parameterization of the TKE forcing (1) that theoretically leads to the observed deficit in mixed-layer radon.

When the ocean surface is not restricted by fetch, TKE is mostly dominated by wind speed and waves (Wanninkhof, 1992; Zemmelink et al., 2006; Wanninkhof and McGillis,

1999; Nightingale et al., 2000; Sweeney et al., 2007; Takahashi et al., 2009). In the polar oceans, wind energy and atmospheric forcing are transferred in a more complex manner as a result of sea ice cover (Loose et al., 2009, 2014; Legge et al., 2015). Sea ice drift due to Ekman flow (McPhee and Martinson, 1992), freezing and melting of ice leads on the surface ocean (Morison et al., 1992) and short period waves (Wadhams et al., 1986; Kohout and Meylan, 2008) all constitute important sources of momentum transfer. Considering the scarcity of data on marginally covered sea ice zones (Johnson et al., 2007; Gerdes and Köberle, 2007), especially during Arctic winter time, the environment is too poorly sampled to constrain these processes through direct measurement or empirical relationships.

Lacking sufficient data to constrain these processes, we wonder whether it is possible for a numerical model to adequately capture forcing of air–sea gas exchange in the sea ice zone and consequently improve predictions of air–sea flux. The parameters of interest are sea ice concentration (or fraction of open water), sea ice velocity, mixed-layer depth (MLD), and water current speed and direction in the ice–ocean boundary layer (IOBL) (Loose et al., 2014). Here we use the budget of ^{222}Rn gas in the IOBL as an example, because the radon deficit method has emerged as one of the principle methods to estimate gas exchange velocity in ice-covered waters (Rutgers Van Der Loeff et al., 2014; Loose et al., 2016).

The radon deficit method involves sampling ^{222}Rn and ^{226}Ra in the mixed layer to examine any difference in the concentration or (radio) activity of the two species. Radon is a gas, radium is a cation; in the absence of gas exchange ^{222}Rn and ^{226}Ra enter secular equilibrium meaning the amount of ^{222}Rn produced is equal to decay rate of ^{226}Ra. Any missing ^{222}Rn in the mixed layer is attributed to exchange with atmosphere (Peng et al., 1979).

Since the ^{222}Rn concentration in air is very low (less than 5 %, Smethie et al., 1985) and considering that concentration is proportional to activity and/or decay rate A, we can use Eq. (1) to determine gas exchange. Where k gas transfer velocity in $(\mathrm{m\,d^{-1}})$, A_E is the activity or decay rate of ^{222}Rn which in secular equilibrium is equal to ^{226}Ra activity, A_M is ^{222}Rn measured decay rate in the mixed layer, λ is decay constant of ^{222}Rn $(0.181\,\mathrm{d^{-1}})$ and h is the MLD.

$$k = \left[A_E / A_M - 1 \right] \lambda h \qquad (1)$$

The MLD, h, is calculated from the measurements performed at the hydrographic stations during ^{222}Rn sampling process. Gas transfer velocities from Eq. (1) reflect the memory of ^{222}Rn for a period of 2 to 4 weeks (Bender et al., 2011), which is 4 to 8 times the half-life of ^{222}Rn (3.8 days).

This memory integrates the physical oceanography properties of the IOBL, including sea ice cover, MLD and water current speed. These processes are likely to vary significantly during this period and it is important to consider them as a source of uncertainty in Eq. (1). To illustrate this uncertainty,

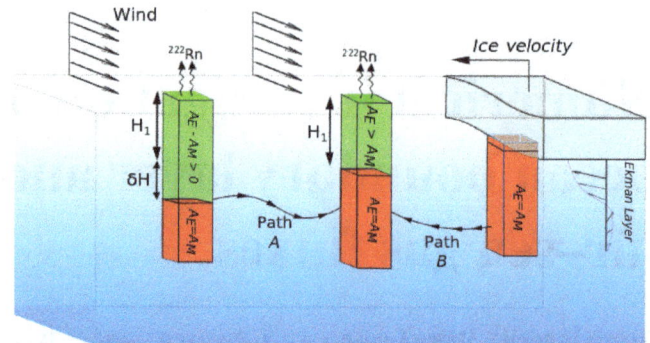

Figure 1. A graphic illustration of two possible back trajectories for a single sampling station.

consider a mixed layer that rapidly changes by a factor of 2 just prior to sampling for radon. If the mixed-layer becomes shallower by stratification, h will be smaller by factor of 0.5 while A_E / A_M in the mixed layer remains the same. Based on Eq. (1), this causes k to be half of its true value. That is, prior to stratification, TKE forcing was sufficient to ventilate the ocean to a depth greater than the apparent h (Bender et al., 2011).

Conversely, if the mixed layer deepens due to mixing, h increases and a new parcel of water with $A_E / A_M = 1$ is added to the mixed layer, causing the activity ratio to come closer to unity. These two influences on Eq. (1) (increasing h and A_E / A_M approaching unity) work against each other, but the net effect is to cause k to appear larger. The change of factor of 2 or higher (in case of convection) in MLD in less than two weeks has been observed during several studies (Acreman and Jeffery, 2007; Ohno et al., 2008; Kara, 2003).

The "memory" of gas exchange forcing that radon experiences is further complicated by the presence of sea ice. Consider two alternate water parcel drift paths that lead to the ^{222}Rn sampling station in the sea ice zone (Fig. 1). Path B demonstrates a history in which water column spends most its back trajectory under sea ice. Path "A" shows a water column which experiences stratification and shoaling of MLD equal to δh when drifting through a region that is completely uncovered by ice. During most of Path "B" gas transfer happens in form of diffusion through sea ice and it will have a very low k (Crabeck et al., 2014; Loose et al., 2011), in contrast Path "A" will have a greater radon deficit, but a smaller h because of stratification. In either case, it is critical to take into account the time history of gas exchange forcing, including changes in the mixed-layer and ice cover, which has led to the apparent radon deficit at the time of measurement.

This observation about drift paths in the sea ice zone strongly implies that we must consider both time and space in estimating the forcing conditions that are recorded in the radon deficit. In other words, we require a Lagrangian back trajectory of water parcels to track the evolution of the mixed layer and its relative velocity 4 weeks prior to sampling.

Although satellite data, ice-tethered drifters (Krishfield et al., 2008) and moorings (Krishfield et al., 2014; Proshutinsky et al., 2009) have provided valuable seasonal and spatial information about the sea ice zone, they do not track individual water parcels and tend to convolve space and time variations. The spatial limitation of these data poses a challenge to producing a back trajectory of the water parcel.

To address the above mentioned challenges, we use a suite of the Estimation of the Circulation and Climate of the Ocean (ECCO) project's Arctic regional configurations to test the if a numerical model can be used to follow the back trajectory of a radon-labeled water parcel and the gas exchange forcing acting upon it and yield the missing information required for the Radon deficit method.

The variables and derived quantities of interest from the numerical model include MLD, sea ice concentration and speed (Loose et al., 2014), and the water velocity in the MLD. We note that as part of the Arctic Ocean Model Intercomparison Project (AOMIP), a number of Arctic ocean-ice models' capability to represent the main ice–ocean dynamics have been assessed (Proshutinsky et al., 2001; Lindsay and Rothrock, 1995, p. 995; Proshutinsky et al., 2008). Our reasons for choosing ECCO over other Arctic models stem from the higher correlation between the ECCO's regional Arctic simulated outputs to satellite-derived sea ice data (Johnson et al., 2012) and the feasibility in the Massachusetts Institute of Technology general circulation model (MITgcm) to adapt a high near-surface vertical resolution to existing configurations.

The remainder of the article is organized as follows: in Sect. 2 we provide the details of the ECCO ice–ocean models. Section 3.1 and 3.2 focus on model outputs of sea ice concentration and velocity and comparison with observations from satellite and ice-tethered profilers. Section 3.3 investigates the modeled output salinity and temperature structure and the resulting upper ocean density structure and mixed layer. Section 3.4 evaluates the correlation in near-surface water velocity. In Sect. 4 we discuss the results and sources of error and their impact on estimated gas exchange and lastly, Sect. 5 provides the summary of our results.

2 Method

2.1 ECCO model configurations

Three ECCO configurations are used, at horizontal grid spacings of 36, 9 and 2 km, respectively. The models are based on the MITgcm code and employ the z coordinate system described in Adcroft and Campin (2004). Our approach is to first assess the model outputs from the coarse-resolution model using model–data misfits, then to investigate if there

is quantitative reduction in model–data misfits with higher horizontal resolutions. Surface forcings are from the 25-year Japanese Reanalysis Project (JRA-25) (Onogi et al., 2007) for 36 and 9 km runs and the European Centre for Medium-Range Weather Forecasts (ECMWF) analysis for the 2 km run. Initial conditions are from World Ocean Atlas 2005 (Antonov et al., 2006; Locarnini et al., 2006) and initial sea ice conditions are from Zhang and Rothrock (2003) for 36 and 9 km, from which the models are allowed to spin up from 1992. The 2 km global run is initialized from a 4 km spin-up version of the ECCO adjoint-based state estimate for January 2011 and covers the period February 2011 to October 2012. The vertical mixing uses K profile parameterization (KPP) developed by Large et al. (1994) and 36 and 9 km runs utilize salt-plume parameterization (SPP) of Nguyen et al. (2009). The horizontal boundary condition for the 36 and 9 km configurations comes from existing global ECCO2 model outputs (Marshall et al., 1997; Menemenlis et al., 2008; Losch et al., 2010; Heimbach et al., 2010).

We introduced a set of new vertical grid spacings to allow us to capture near-surface small details which cannot be represented with the coarser grid system. In the 36 km (hereafter referred to as A1) and 9 km (called A2) models, the spacing is 2 m in the upper 50 m of the water column and gradually increases to a maximum of 650 m. In contrast, the 2 km model (called A3) has 25 layers in the top 100 m of water column, starting from 1 m and increasing to 15 m step. All the boundary conditions from ECCO2 have been interpolated to match the new vertical grid system.

2.2 Observations

Satellite-derived estimation of sea ice cover at 25 km horizontal resolution (Comiso, 2000) is interpolated to a horizontal grid system to facilitate model–data comparison. In addition, sea ice drift gathered by 28 ice-tethered profilers (ITPs) (Krishfield et al., 2008) which have more than 2 months of data in the Beaufort Sea between 2006 and 2013 have been used to do the ice velocity comparison.

We compared near-surface water velocity data from an ice-tethered profiler with velocity instruments (ITP-V) (Williams et al., 2010) to A1 and A2 and an upward-looking acoustic doppler current profiler installed on a McLane moored profiler (MMP) (McPhee et al., 2009; Cole et al., 2014) to A1, A2 and A3 in order to compute the accuracy and feasibility of calculating back trajectory of parcels located in the mixed layer. We limit our comparison of ITP-V, which runs from October 2009 to March 2010, to A1 and A2 since those models run from 2006 to 2013 and A3 runs from 2011 to 2013.

Using salinity and temperature profiles from ITPs (Krishfield et al., 2008) we calculated MLD and compared it to 2 m vertical resolution model output (A1, A2). Most of the observed data exist in the Beaufort Gyre, hence we mostly focus our comparison to that geographic perimeter. Figure 2

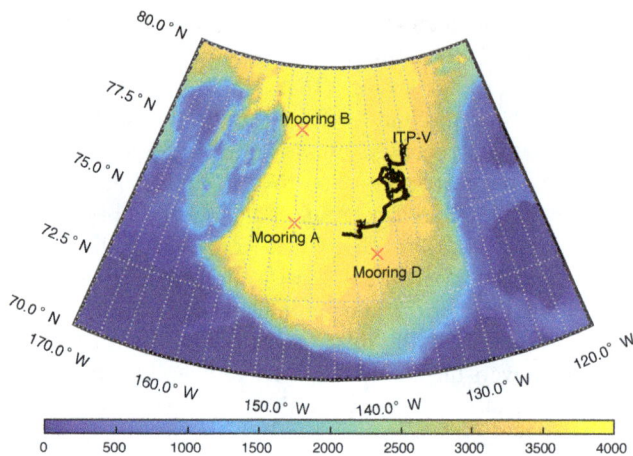

Figure 2. Bathymetry and location of ITP-V and mooring for data comparison.

depicts the bathymetry and location of most important observations we used to make the comparisons with the model.

3   Results

3.1   Sea ice concentration

For sea ice concentration analysis we introduced a grid system covering the Beaufort Gyre and interpolated the data from satellite (Comiso, 2000) and A1 onto the grid. The analysis grid extends from 70 to 80° N and 130 to 170° W, covering most of Beaufort Gyre (Fig. 3). Grid points can be divided into two main geographic zones that are marked out based on sea ice cover. The first zone contains grid points where the annual average sea ice cover is greater than 80 %. These sets of points are fully covered by sea ice most of the year. The second zone can be described as "marginally ice covered" wherein the ocean surface is free of ice for some fraction of the year. We chose three points within this sea ice geography to compare the seasonal and interannual behavior of the model with satellite ice cover. The points are located at 80° N, 131.82° W (P1), 70.82° N, 169.82° W (P2), and 74.76° N, 163.51° W (P3).

The ice cover at P1, P2 and P3 (Fig. 3) can be divided into 3 ice phases: (a) fully covered in ice, (b) open water and (c) a transition between (a) and (b). P3, which is the furthest south, has all three phases. In contrast P1 ice cover only dips below 60 % for two brief periods during the 7-year time series depicted in Fig. 3 – once in 2008 and again in 2012. These three points illustrate where and when the model has the greatest challenge reproducing the actual sea ice cover. At the extremities of the ice pack, where the water is predominantly covered by 100 or 0 % ice (P1 and P3), the model captures the seasonal advance and retreat and the percentage of ice cover itself is accurate. However, in the transition regions that are characterized by marginal ice for much of the year (P2), the

model has more difficulty reproducing the observed sea ice cover as well as the timing of the advance and retreat. This behavior is consistent with the description that has been explained by Johnson et al. (2007), that models have a higher accuracy predicting sea ice concentration in central Arctic and less accuracy near its periphery and the lower latitudes.

The spatial sensitivity of the model can be observed using root mean square (RMS) error (Hyndman and Koehler, 2006) Eq. (2), calculated over the 1992–2013 period (Fig. 3). The area with the highest misfits coincides with area between the 80 and 60 % contour lines (Fig. 3) and is concentrated primarily in the Western Beaufort. The RMSE error of 0.2 is the maximum value away from land, this same level of error can also be found near land which is caused by fast-ice generation. Fast ice in the model is replaced with packs of drifting sea ice; this error is common among numerical models and has been brought to attention during AOMIP (Johnson et al., 2012).

$$\mathrm{RMSE\,(point)} = \sqrt{\sum_{i=1}^{n} (C_{\mathrm{simulation}} - C_{\mathrm{satellite}})^2/n} \qquad (2)$$

If we compare the monthly climatology for sea ice cover over the 1992–2013 period, the RMS error between model and satellite data is least during the early winter months (e.g., January–March) when sea ice is close to its maximum extent. Comparing data and A1, Fig. 3 depicts an increase in RMSE during July, August, September and October and a minor decrease in May and November. The RMSE appears to be greater during the summer months of ice retreat, and slightly less during the autumn months of ice advance. Overall, the periods of transition (melt and freeze) coincide with the greatest RMSE.

An important source of errors in the model ice concentration comes from the reanalysis surface forcing. Fenty and Heimbach (2012) showed that adjustments in the air temperature that are within the uncertainties of this reanalysis field can help bring the model ice edge into agreement with the observations. Of note also is that the uncertainty in satellite-derived ice cover can be the highest in the marginal ice zone due to tracking algorithms that are sensitive to cloud liquid water or cannot distinguish thin ice from open water (Ivanova et al., 2015); this error also manifests itself in quantification of model–data misfits.

3.2   Sea ice velocity

Ekman turning causes ice and water to move at divergent angles with respect to each other. Ice moves the fastest, with mean values of 0.09 m s^{-1} (Cole et al., 2014), and the water column progressively winds down in velocity, along the Ekman spiral. Stratification in the Arctic leads to a confinement of the shear stress closer to the air–sea interface and also produces greater divergent flow vectors between ice and water (McPhee, 2012). In the marginal ice zone or in regions

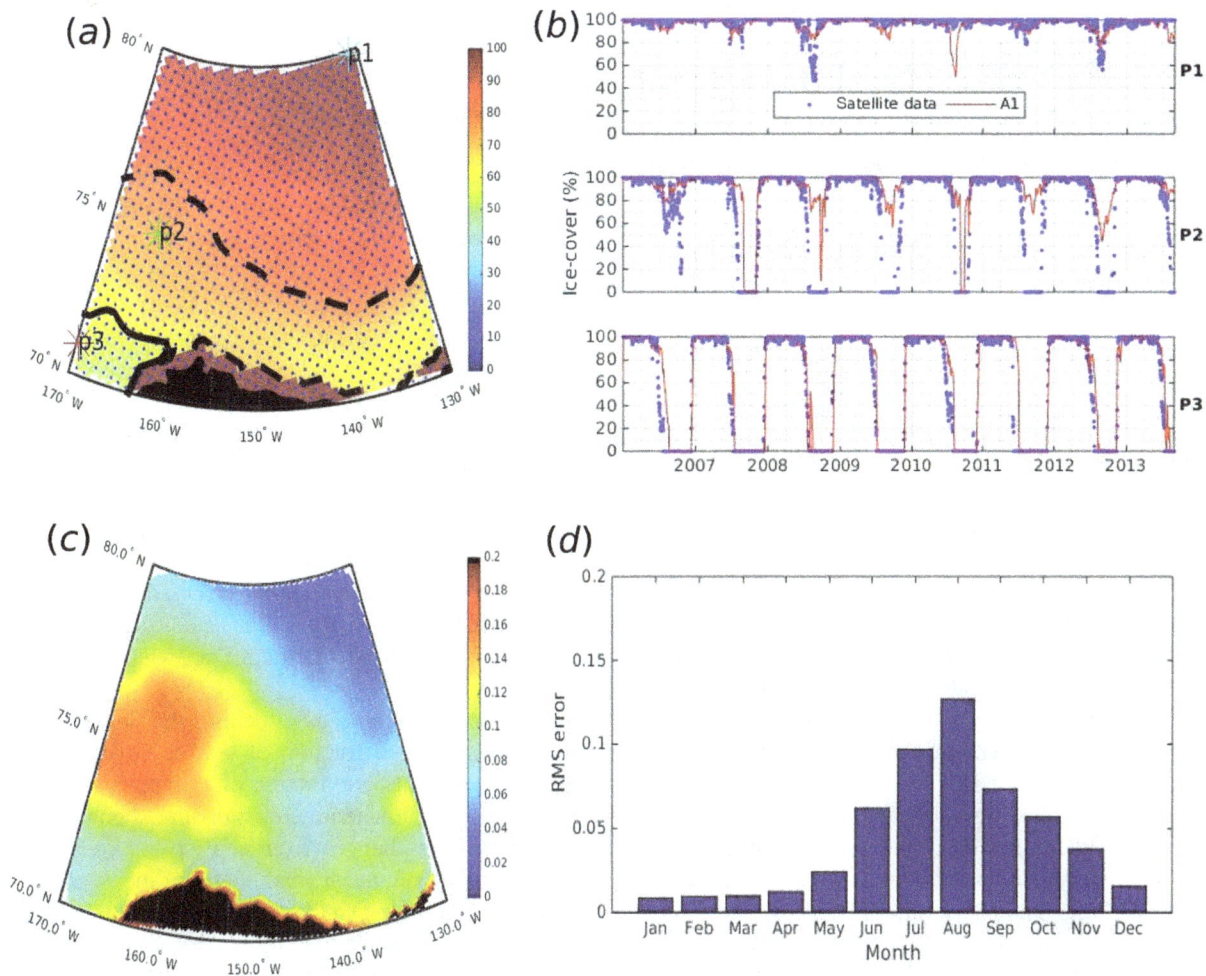

Figure 3. (a) Averaged satellite sea ice cover from 2006 to 2013, with the solid black line marking 60 % cover and dashed black line marking 80 %; blue dots show the analysis grid, stars show the location of the three points Cyan P1, Green P2 and Red P3 where time-series data is graphed in **(b)**. **(b)** Time history of sea ice fraction from top P1, P2 and P3, with satellite data represented by blue dots, compared with A1. **(c)** Horizontal distribution of RMS error of A1 sea ice concentration averaged over time from 2006 to 2013; black mask covers the grid points on the land. **(d)** Spatially averaged annual RMS error of A1 sea ice concentration.

where ice is converging or diverging, these motions, relative to the motion of the water column, can produce significant changes in the water column momentum budget as well as air–sea fluxes. Thankfully, the ITPs can provide us with a measure of the real ice drift.

To generate a more quantitative comparison between the results, we utilized the same method introduced by Timmermans et al. (2011), to compare ice velocity components (eastward–northward) of A1 to ITP velocity and compute the correlation coefficient of each experiment with the daily-averaged actual drift velocity from the ITPs (Fig. 4).

When averaged over all the ITPs operating in Beaufort Gyre during 2006 to 2013, A1 had correlations of 0.8 with actual velocity components and 0.82 correlation with speed magnitude. RMSE calculated for A1 based on Eq. (2) shows an error of 0.043 ms^{-1} and no significant bias.

3.3 Temperature, salinity, density and MLD

3.3.1 Vertical salinity and temperature profiles

We chose four hydrographic profiles in the Beaufort Sea to assess the simulated vertical salinity and temperature. The first two sets of profiles are from ITP-1 winter and summer 2006; the third set is from ITP-43 during winter 2010 and the fourth is from ITP-13 during summer 2008 (Fig. 5). For visualization we linearly extrapolated the profiles from the first layer of the model up to the surface, which occurs over the top 1 m of the water column.

During winter time, the model temperature and salinity profiles show a well mixed layer that extends below 15 m, followed by a very large gradient. The mixed-layer temperature is close to the local freezing point in a condition called "ice bath" (Shaw et al., 2009). The ITP profiles are similar;

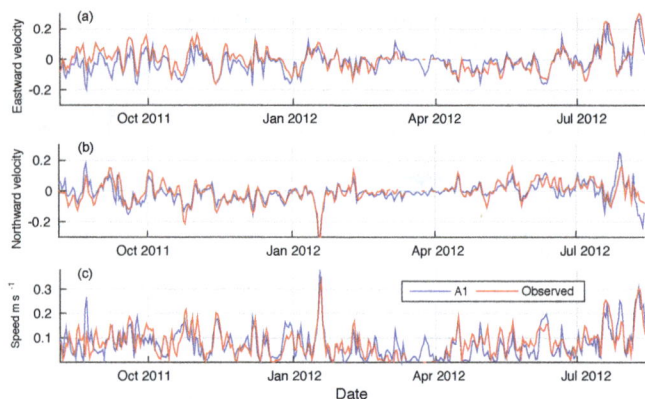

Figure 4. Time series of sea ice velocity components and speed of ITP 53 vs. 36 km horizontal resolution of MITgcm (A1). The correlations between eastward, northward and magnitude of velocity between ITP 53 data and A1 are 78, 75 and 80 %, respectively.

however the ITP MLD is deeper by nearly 10 m, indicating more ice formation and convective heat loss over this water column, as compared to the model water column. In summer the model mixed-layer shoals to approximately 5 m depth following two local temperature extrema; the bigger maximum is at $\sim$ 35 m, generated by intrusion of the Pacific Summer Water (PSW) which is a dominant feature in the Canada Basin. The second smaller maximum happens around 10 m called the summer mixed layer (Shimada et al., 2001, p. 201) or near-surface temperature maximum (NSTM) (Jackson et al., 2010) which is a seasonal feature generated by shortwave solar heat diffusion (Perovich and Maykut, 1990). These two well-defined phenomena are broadly descriptive of the summer surface layer in the Beaufort Gyre. They are, however, absent from the ITP data at this location, indicating a different ice and heat budget time history.

Data and model profiles in Fig. 5b show better agreement in the shape and the absolute value of the T and S profiles. Both model and ITP data have a 20 m deep mixed layer during 2010 winter. The model in this case does not show as much change in vertical temperature structure compared to actual data. In the profile from ITP-13 (Fig. 5) the model again over-estimates the temperature beneath the mixed layer, although certain features including the NSTM can still be found near 10 m, yet not as pronounced since it is very close to PSW. Bearing in mind that density in the Arctic is dominated by changes in salinity, we move forward to density profiles from this point on.

In addition, we note that recent studies show that eddies with diameters of 30 km or less (Nguyen et al., 2012; Spall et al., 2008; Zhao et al., 2014; Zhao and Timmermans, 2015; Zhao et al., 2016) play an important role in transporting Pacific water from the shelf break into the Canada Basin. Adequate representation of ocean eddies and investigation into their roles in setting the water column stratification require a model with finer horizontal resolution. Hence, moving for-

ward, in addition to A1, we utilize the 9 km model (A2) to investigate the density profiles as well as study the MLD.

3.3.2 Density profiles

We compared the 36 and 9 km model outputs of density to the time series of density profiles from ITP-35 (Fig. 6) from October 2009 to March 2010. A black mask indicates locations where there is no data from ITP-35 – particularly in the upper 7 m of the water column. As ITP-35 transited through the Canada Basin, density profiles contain both temporal and spatial changes.

We are able to discern some broad similarities between the model and ITP density profiles. From November through January, both ITP and model density profiles remain relatively constant. Between February and March, ITP-35 appears to drift through a zone of convection, likely caused by ice formation, with a sudden increase of density near the surface. The same feature can be observed in both A1 and A2 density. However, on a smaller scale, there is significantly more variation in the ITP data than what the model represents.

For exploring the reason behind the density signals, we used the simulated fraction of sea ice cover and ice thickness (Fig. 6). The dominating effect appears to result from a sea ice fraction when there is an almost continuously covered area. The changes from sea ice thickness can be observed in the volume of fresh water in the water column, as seen by outcropping of the 1022.5 isopycnal coinciding with the increase of sea ice thickness. An increase in near-surface density can be seen in late January and early February accompanied by an increase in ice thickness and insertion of brine in the water column. The second peak, which is not as pronounced, happens in late February when ice fraction decreases from 100 to 95 % and exposes the surface water to cold atmosphere, leading to the production of newly formed sea ice. We further examine these signals in the MLD section below.

3.3.3 Mixed-layer depth

There are many different methods in the literature for calculating MLD (Brainerd and Gregg, 1995; Wijesekera and Gregg, 1996; Thomson and Fine, 2003; de Boyer Montégut et al., 2004; Lorbacher et al., 2006; Shaw et al., 2009). The methods can be divided into two main types (Dong et al., 2008): the first type of algorithm looks for the depth (z_{MLD}) at which there has been a density increase of $\delta\rho$ between the ocean surface and z_{MLD}. A typical range of values for $\delta\rho$ are 0.005 to 0.125 kg m^{-3} (Brainerd and Gregg, 1995; de Boyer Montégut et al., 2004). The second type uses slightly different criteria, where the base of the mixed layer is determined as the depth where the gradient of density ($\partial\rho\,/\,\partial z$) equals or exceeds a threshold; typical numbers for $\partial\rho\,/\,\partial z$ are 0.005 to 0.05 kg m^{-4} (Brainerd and Gregg, 1995; Lor-

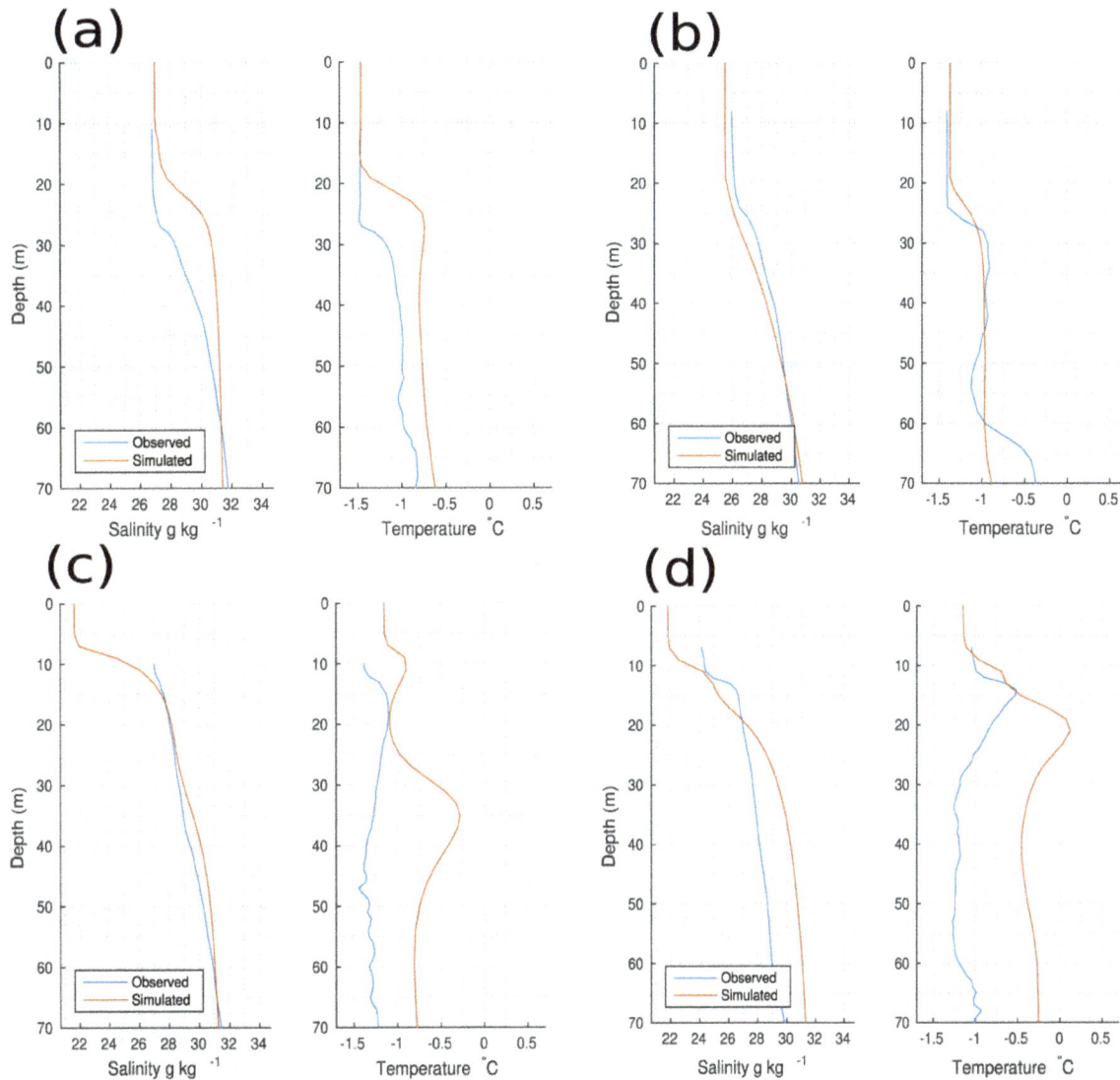

Figure 5. Salinity and temperature of the top 70 m based on ITPs and A1. **(a)** ITP 1 on 13 December 2006 at 74.80° N and 131.44° W. **(b)** ITP-43 on 27 November 2010 at 75.41° N and 143.09° W. **(c)** ITP 1 on 28 August 2006 at 76.96° N and 133.32° W. **(d)** ITP-13 on 30 July 2008 at 75.00° N and 132.78° W.

bacher et al., 2006, p. 200). A more sophisticated approach to type 1 of these criteria is to utilize a differential between $\rho_{100\,m} - \rho_{\text{surface}}$ as the cut of point (instead of using a fixed $\delta\rho$) to account for the effects of surface ρ changes during winter and summer (Shaw et al., 2009). Here, we have implemented two of these methods M1 and M2, with M1 using $\delta\rho$ equal to 0.2 of $\rho_{100\,m} - \rho_{\text{surface}}$ (Shaw et al., 2009) and M2 with a gradient $(\partial\rho \,/\, \partial z)$ cut off point equal to 0.02 $\text{kg\,m}^{-4}$, which matches innate model parameterization of MLD (Nguyen et al., 2009).

We compare these two methods by applying them to the profiles from Fig. 5, and the results are shown in Fig. 7. In case (a) and (b) M1 produces a MLD that is 8 to 12 m deeper, compared to the other method. A visual examination of pro-

files indicates that the M1 criteria may be too flexible of a criteria. The results from M1 appear to be intermittently "realistic", whereas M2 can be difficult to implement for data sampled at high vertical resolution as a result of greater small-scale variability. In practice, we find M1 is the most straight-forward to implement.

It should be mentioned that it is difficult to consistently compare performance of the M1($\delta\rho$) and M2 methods on ITP and model data, because the model data extends to the top 1 m of water column, whereas the ITP data stops at 7 m depth (Peralta-Ferriz and Woodgate, 2015). Furthermore, it has been shown that the summer mixed layer in the Canada Basin can be less than 12 m (Toole et al., 2010). To account for this effect, we apply an additional restriction wherein any

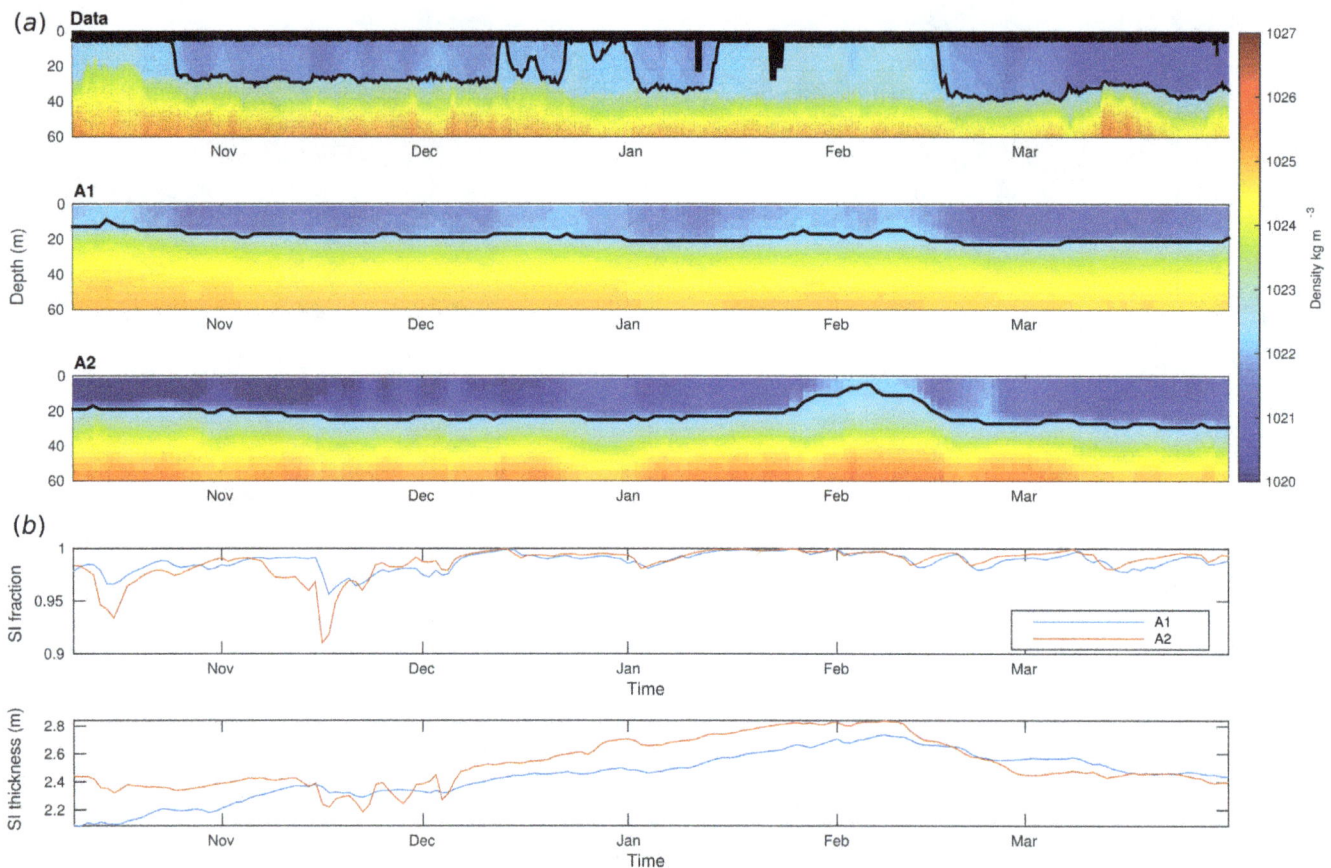

Figure 6. (a) Observed upper-ocean density vs. 36 km (A1) and 9 km (A2) resolution MITgcm density along the path of ITP drift; the black mask covers areas where no ITP data is available and solid black line shows isopycnal of $1022.5 \, \mathrm{kg \, m^{-3}}$. **(b)** Simulated sea ice fraction and thickness on top of the water column.

profile whose MLD is less than 2 m below the shallowest ITP measurement is discarded. This restriction effectively removes any MLDs shallower than 10 m due to the ITP sampler not resolving the upper 8 m of water column. In some cases, a remnant mixed-layer from the previous winter may exist in the water column. In this case, the methods incorrectly identify the remnant mixed layer (ML) as actual MLD.

To compare the methods over a longer time period, we calculated the MLD from model data and ITP-35 data along the ITP-35 drift track. We used M1 to determine the MLD for A1, A2 and for ITP-35 data (Fig. 8). Both model results show a shallower ML compared to the ITP data; the most prominent feature in late January corresponds to a sudden change in density found in Fig. 6. Beside the above-mentioned peak, A1 fails to capture any variability in MLD whereas A2 shows that the ML deepens by about 10 m in mid February corresponding to ice opening occurring during the same time span (Fig. 6). The difference between A1 and A2, and their ability to capture MLD change, can be explained by the capability of a higher-resolution model to capture small-scale fractures in the ice cover (Fig. 8), and conversely, the inability of the coarser resolution to do so is due to averaging over a larger

grid. The wind appears to be the primary driving mechanism for the divergence in ice cover, which in turn exposes the ocean to the cold atmosphere and leads to a loss of buoyancy and an increase in MLD. With higher resolution these openings can be captured, leading to a better agreement with data in marginal ice zones. The changes in MLD are of first-order importance to the calculation of gas budgets such as the radon deficit. In this regard a fine-scale grid resolution has real advantages through its ability to capture both the ice advection and openings in ice cover that lead to MLD change. Coarser resolution would be justified when the point of interest is sufficiently far away from leads and marginal ice zones where the effect of sea ice dynamics on MLD is important, so the effects of area averaging would be small enough to omit.

One last important note is the effect of the SPP on MLD. Nguyen et al. (2009) demonstrated the need to remove the artificial excessive vertical mixing in coarse horizontal resolution models. To rule out the dependency of this parameterization to vertical resolution as a source in MLD bias, we performed a suite of 1-D tests, with and without the SPP, on a variety of vertical resolutions (not shown here) and sea ice melting/freezing scenarios and confirmed that SPP is not de-

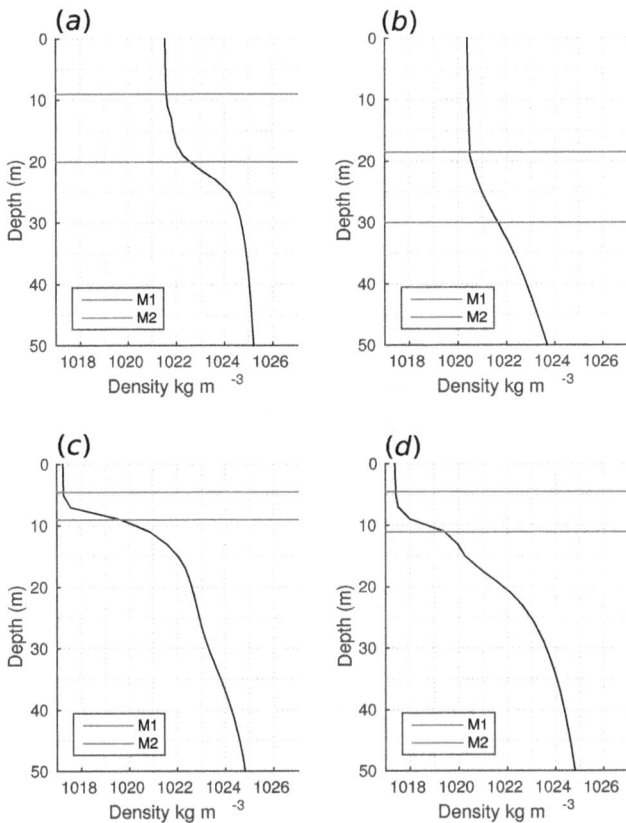

Figure 7. Methods M1 and M2 applied to selected ITP profiles, **(a)** ITP 1 on 13 December 2006 at 74.80° N, 131.44° W. **(b)** ITP-43 on 27 November 2010 at 75.41° N, 143.09° W. **(c)** ITP 1 on 28 August 2006 at 76.96° N, 133.32° W. **(d)** ITP-13 on 30 July 2008 at 75.00° N, 132.78° W.

(Fig. 9). The ITP data has been daily averaged to remove higher frequency information which we do not expect the model to capture due to the low frequency (6-hourly) wind forcing. Both A1 and A2 show less than 0.3 correlations with data with no improvement in respect to resolution.

We further add A3 to our comparison for moorings velocities (Fig. 9), and compared velocities at 25 m, which is the level that is shared between all our models and removes the necessity of any interpolation. The simulation results show RMSE normalized by data of higher than 5 and correlations of less than 0.3 over three moorings and almost 2 years of data. This result indicates that ocean currents are not well captured in the model irrespective of horizontal grid resolution. We must therefore look into the atmospheric forcing as a likely source of error on high frequency water velocities near the surface. As noted above, the wind inputs into the model from the reanalyses are available at a 6-hourly frequency. Chaudhuri et al. (2014) and Lindsay et al. (2014) have compared various available reanalysis products over the Arctic which we used to force our model, along with multiple other reanalysis products with available ship-based and weather station data, and found out that wind products in all of those have low correlation, i.e., less than 0.2. To investigate we compared JRA-55 (Onogi et al., 2007) and NCEP (Kalnay et al., 1996) to shipboard data gathered during 2014 in a time span of 2 months in the Arctic and found that JRA-55 had −0.20 correlation, RMSE of 7.36 and bias of −1.3, and NCEP had a correlation of 0.10, RMSE of 5.73 and bias of −1.40 when compared with high-frequency data on each cruise, reinforcing our suspicion of high-frequency wind as a source of error in water currents.

4 Gas exchange estimation

Up to this point we have spent extensive effort assessing the skill of the MITgcm to reproduce the key forcing parameters listed in our introduction. This effort is motivated by the potential for using the MITgcm model output as a tool to improve our ability to model gas budgets in the IOBL and to improve our estimates of k in the sea ice zone, both of which depend on sea ice processes in the IOBL. To illustrate the potential impact that IOBL properties can have on the estimate of k, we perform a simple experiment, using estimates of k over the range of variation in model output at three locations in the sea ice zone. The intention is to illustrate the variability in k and in the radon deficit that can arise as a result of sea ice processes.

4.1 Constraining gas exchange forcings

Utilizing the results from Sect. 3.1, 3.2 and 3.3, we calculated gas exchange velocities at P1, P2 and P3 (Fig. 10), over the course of the model simulation (i.e., $n = 2557$ days $\times 3 = 7671$) introduced in Sect. 3.1. The MIT-

pendent on vertical grid spacing. We also investigated MLD in A3 (no SPP) run compared to A2, and confirmed that the average MLD is the same between these two runs.

3.4 Velocities in the water column

We have very little information from direct observations that permit us to track a water parcel, especially beneath sea ice. This is one area where model output could be critical as there are not obvious alternatives. To assess the consistency of the model water current field, we compared 2-D model water velocity to data gathered from two sources: (1) from ADCPs mounted on moorings that were deployed starting in 2008 in Beaufort Gyre (Proshutinsky et al., 2009) and (2) the ITP-V sensor equipped with MAVSs (Modular Acoustic Velocity Sensors) (Williams et al., 2010), which was the only operating ITP before 2013 which had an acoustic sensor mounted on it.

We compared the velocity components averaged from 5 to 50 m to account for flow direction that is moving the water parcels in the mixed layer over the duration of ITP-V working days, which was from 9 October 2009 to 31 March 2010

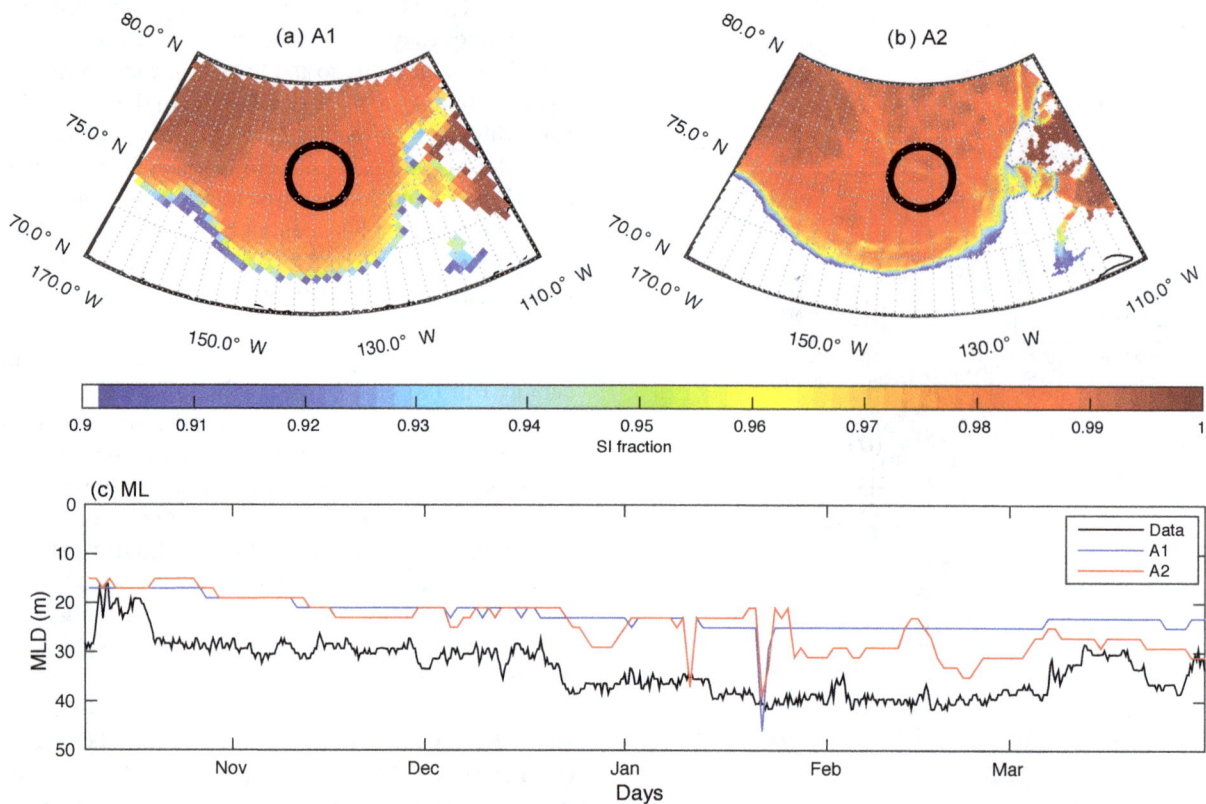

Figure 8. Sea ice cover higher than 0.9 with gray circle marking the area of ITP operation for **(a)** 36 km (A1) and **(b)** 9 km(A2) horizontal resolution of the model. A2 captured the ice opening and resulting mixed-layer change while this phenomenon has been averaged out by a **(c)** coarse-resolution model observed and simulated evolution of mixed-layer depth on the path of ITP.

gcm IOBL properties are fed to the estimator of k, considering sea ice processes (Loose et al., 2014). Our selected points have the mean sea ice concentrations of 96.1, 87.62 and 61.69 %, sea ice speeds of 0.05, 0.086 and 0.10 ms^{-1}, wind speeds of 8.73, 5.87 and 4.11 ms^{-1}.

The result yields a point cloud of values that varies depending primarily on the range of ice velocity, wind speed and sea ice cover. The values of k range between 0.1 and 14.0, with a mean of 2.4 and standard deviation of 1.55. This exercise demonstrates the sensitivity of k to the IOBL forcing parameters. In the event that we can trust the majority of the model outputs, such as the case here with high fidelity in the simulated SI concentration and SI velocities in A1, we conclude that a numerical model, even a coarse-resolution one, can make significant improvement to the estimate of k. The question of constraining the radon budget within a Lagrangian water parcel is somewhat more complicated.

4.2 Application of forcings on radon budget

The results in Sect. 3.4 showed that the difference between model and data water trajectories accumulated too much error to be useful, and indicate that for a regional GCM to be useful for reconstructing the back trajectories of radon-

labeled water parcels, we will need improved wind-forcing fields. With current reanalysis products, finding the back trajectory of radon-labeled water parcels is not feasible. When improved wind fields are available, the Green's functions approach (Menemenlis et al., 2005; Nguyen et al., 2011) or adjoint method (Forget et al., 2015, p. 4; Wunsch and Heimbach, 2013) can be used to reduce misfits between modeled and observed MLD velocity and likely make the model a valuable tool for tracking back trajectories, either in a smaller domain or full Arctic regional configuration. A possible source of wind data can be from shipboard measurements, assuming the measurements persist over 10 days in the given sampling station.

However, it may be possible to improve on the existing approach. When the drift trajectory is not known, one solution is to resort to averaging IOBL properties within a radius that is equal to the 30-day drift track (e.g., as done by Rutgers Van Der Loeff et al., 2014). The averages within this circle are treated as the representative IOBL properties. The radius of spatial averaging should be restricted by the average magnitude of the water parcel's velocity multiplied by the time span of interest. When applying a spatial averaging, if the timescale of changes in forcings is smaller than time span of interest, the time dependency of forcings should be ac-

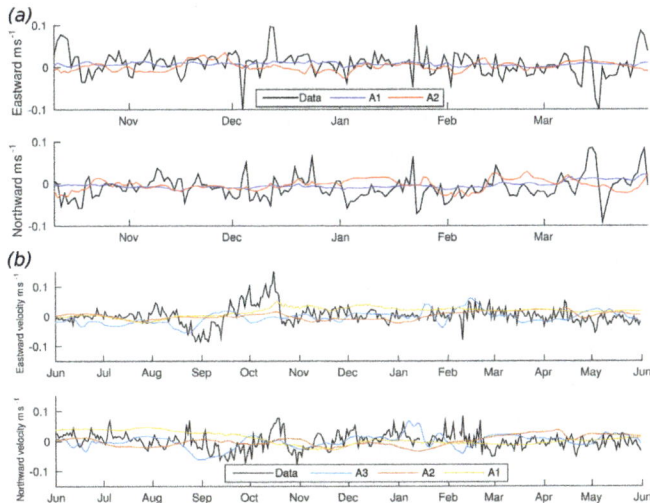

Figure 9. (a) Daily-averaged velocity components from 5 to 50 m observed by ITP-V vs. those simulated by A1 and A2. **(b)** Daily-averaged velocity components at 25 m observed by mooring D vs. A1, A2 and A3.

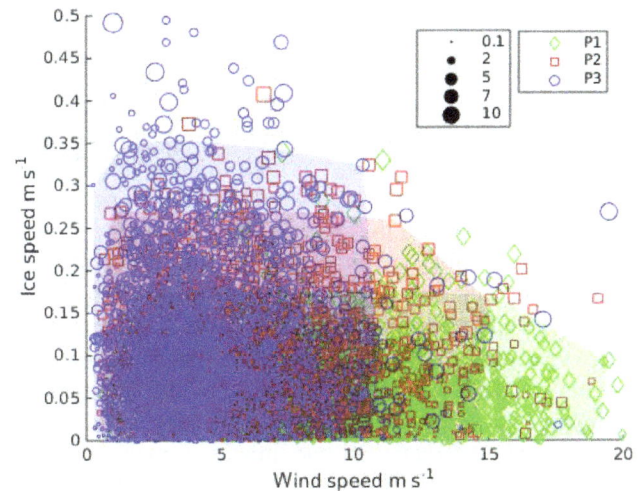

Figure 10. Gas exchange estimated model outputs of wind and sea ice speed at locations P1: 77.4° N, 143.6° W, P2: 74.8° N, 163.5° W and P3: 70.59° N, 159.4° W from January 2006 to December 2012, Areas enclose the outputs around the mean and two standard deviation. The size of the points demonstrate the magnitude of the gas exchange velocities normalized by sea ice cover.

counted for. Typically sea ice velocity is ∼5 times greater than vertically averaged water velocity in the mixed layer (Cole et al., 2014). In this regard, it may be acceptable to assume that the water parcel is stationary as long as ice advection is accounted for. Hence, spatial averaging should account for ice drift over the point of radon and radium sampling. The same logic also applies to the changes in the MLD and sea ice concentration. For example, gas exchange calculated (Eq. 1) based on assumption of constant MLD of 27.5 m with limits of 5 to 50 m (Peralta-Ferriz and Woodgate 2015) would have limits of ±80 %, whereas gas exchange calculated based on model MLD would have ±50 % error and accounts for time variability. With the current level of uncertainty in reanalysis products and inherent heterogeneity of marginal sea ice zones, we suggest a mixed weighted combination of model outputs and shipboard data to be the way forward for constraining gas budget in sea ice zones.

5 Summary

We have used 36, 9 and 2 km versions of the ECCO ocean–sea ice coupled models based on the MITgcm to investigate whether numerical model outputs can be used to compensate for lack of data in constraining air–sea gas exchange rate in the Arctic. The goal is to understand if model outputs can improve estimation of gas exchange velocity calculation and to evaluate the capability of the model to fill in the missing information in the radon deficiency method. This systematic comparison of upper-ocean processes has revealed the following.

The coarse-resolution model showed a good fidelity in regard to reproducing sea ice concentration. Depending on the

location and/or season, the error of simulated ice concentration varied between 0.02 and 0.2. Away from ice fronts or active melting/freezing zones the model tended to have higher accuracy. Even in the marginal ice zone, due to the potentially high error in the satellite-derived ice concentration, the model can still be used to quantify the air–sea gas exchange rate, though with an expected higher uncertainty due to the combination of model and data errors. In addition to sea ice concentration, we also found good correlation (82 %) between model ice speed and ITP drift.

The estimation of MLD is challenging due to its dependence on unconstrained density anomaly or density gradient thresholds. No MLD algorithm performs well in all situations. In addition, CTD profiles from drifting buoys often do not include the top 7–10 m of the surface ocean where stratification can be important. Adding to the challenge is the dependence of the ocean density structure on vertical fluxes. In these model–data comparisons we found model MLD to be consistently biased on the shallower side in all model resolutions. We note however that this result can partly be due to the missing upper 7 m in moored drifters such as ITPs, thus resulting in a one-sided bias in the observed MLD. The evolution of the mixing events showed that MLD correlates to sea ice fraction: in areas of nearly-full ice cover, small openings may result in exposure of water to the cold atmosphere and the resulting freezing events would deepen the mixed layer via brine rejection. The higher the resolution, the higher the capability of the model to capture these openings and the resulting deepening effects. The usage of the SPP does not play an important role in determining the MLD.

The A1, A2 and A3 experiments consistently could not capture the water velocity observed in ITPs or mooring. We speculate that this discrepancy may be the result of the quality of the reanalysis wind products that are forcing these models. The wind products have been shown to have poor correlation with observed data at high frequencies. Considering that the response of near-surface water is occurs almost simultaneously to the wind forcing, low correlation in wind velocity would have direct impact on the modeled near-surface water velocities and likely yield low correlations between modeled and observed ocean currents. Conversely, the same wind fields at lower frequencies and on broader spatial scale have higher accuracy, as evidenced by the high correlation between the modeled and observed sea ice velocity.

Taking into accounts all the misfits through detailed model–data comparisons, we were able to quantify the usefulness of a numerical model to improve gas exchange rate and parameterization methods. We showed an example of how the sea ice concentration, velocity and MLD can affect the gas exchange rate by up to 200 % in marginal sea ice zones and that the model outputs can help constrain this rate. By finding the low correlation in near-surface ocean velocities, irrespective of model horizontal resolution, we concluded that finding the back trajectory of radon-labeled water parcels is currently not feasible. Furthermore, we speculate as to the source for the common errors in our models, namely the high frequency and under-constrained atmospheric forcing fields, and identify alternative approaches to enable the use of a model to achieve the back trajectory calculation task. The alternative approach includes using the MITgcm Green's functions and adjoint capability to help constrain the model ocean velocity to observations, and performing the simulations in a smaller dedicated domain based on the specific spatial distribution of data for both atmospheric winds and ocean currents in the mixed layer.

Acknowledgement. We gratefully acknowledge computational resources and support from the NASA Advanced Supercomputing (NAS) Division and from the JPL Supercomputing and Visualization Facility (SVF) and high-performance computing support from Yellowstone (ark:/85065/d7wd3xhc) provided by NCAR's Computational and Information Systems Laboratory, sponsored by the National Science Foundation. The ice-tethered profiler data were collected and made available by the Ice-Tethered Profiler Program (Toole et al., 2011; Krishfield et al., 2008) based at the Woods Hole Oceanographic Institution (http://www.whoi.edu/itp). Funding for this research was provided by the NSF Arctic Natural Sciences program through Award # 1203558.

Edited by: D. Stevens

References

Acreman, D. M. and Jeffery, C. D.: The Use of Argo for Validation and Tuning of Mixed Layer Models, Ocean Model., 19, 53–69, doi:10.1016/j.ocemod.2007.06.005, 2007.

Adcroft, A. and Campin, J.-M.: Rescaled Height Coordinates for Accurate Representation of Free-Surface Flows in Ocean Circulation Models, Ocean Model., 7, 269–284, doi:10.1016/j.ocemod.2003.09.003, 2004.

Antonov, J., Locarnini, R., Boyer, T., Mishonov, A., and Garcia, H.: World Ocean Atlas 2005 Vol. 2 Salinity, NOAA Atlas NESDIS, 62, NOAA, Silver Spring, Md, 2006.

Bender, M. L., Kinter, S., Cassar, N., and Wanninkhof, R.: Evaluating Gas Transfer Velocity Parameterizations Using Upper Ocean Radon Distributions, J. Geophys. Res., 116, C02010, doi:10.1029/2009JC005805, 2011.

Blomquist, B. W., Huebert, B. J., Fairall, C. W., and Faloona, I. C.: Determining the sea-air flux of dimethylsulfide by eddy correlation using mass spectrometry, Atmos. Meas. Tech., 3, 1–20, doi:10.5194/amt-3-1-2010, 2010.

Brainerd, K. E. and Michael, C. G.: Surface Mixed and Mixing Layer Depths, Deep-Sea Res. Pt. I, 42, 1521–1543, doi:10.1016/0967-0637(95)00068-H, 1995.

Chaudhuri, A. H., Ponte, R. M., and Nguyen, A. T.: A Comparison of Atmospheric Reanalysis Products for the Arctic Ocean and Implications for Uncertainties in Air–Sea Fluxes, J. Climate, 27, 5411–1521, doi:10.1175/JCLI-D-13-00424.1, 2014.

Cole, S. T., Timmermans, M.-L., Toole, J. M., Krishfield, R. A., and Thwaites, F. T.: Ekman Veering, Internal Waves, and Turbulence Observed under Arctic Sea Ice, J. Phys. Oceanogr., 44, 1306–1328, doi:10.1175/JPO-D-12-0191.1, 2014.

Comiso, J.: Bootstrap Sea Ice Concentrations from Nimbus-7 SMMR and DMSP SSM/I-SSMIS. Boulder, Colorado USA, NASA DAAC at the National Snow and Ice Data Center, 2000.

Crabeck, O., Delille, B., Rysgaard, S., Thomas, D. N., Geilfus, N.-X., Else, B., and Tison, J.-L.: First "in Situ" Determination of Gas Transport Coefficients (DO_2, DAr, and DN_2) from Bulk Gas Concentration Measurements (O_2, N_2, Ar) in Natural Sea Ice, J. Geophys. Res.-Oceans, 119, 6655–6668, doi:10.1002/2014JC009849, 2014.

de Boyer Montégut, C., Madec, G., Fischer, A. S., Lazar, A., and Iudicone, D.: Mixed Layer Depth over the Global Ocean: An Examination of Profile Data and a Profile-Based Climatology, J. Geophys. Res.-Oceans, 109, C12003, doi:10.1029/2004JC002378, 2004.

Dong, S., Sprintall, J., Gille, S. T., and Talley, L.: Southern Ocean Mixed-Layer Depth from Argo Float Profiles, J. Geophys. Res.-Oceans, 113, C06013, doi:10.1029/2006JC004051, 2008.

Fenty, I. and Heimbach, P.: Coupled Sea Ice–Ocean-State Estimation in the Labrador Sea and Baffin Bay, J. Phys. Oceanogr., 43, 884–904, doi:10.1175/JPO-D-12-065.1, 2012.

Forget, G., Campin, J.-M., Heimbach, P., Hill, C. N., Ponte, R. M., and Wunsch, C.: ECCO version 4: an integrated framework for non-linear inverse modeling and global ocean state estimation, Geosci. Model Dev., 8, 3071–3104, doi:10.5194/gmd-8-3071-2015, 2015.

Gerdes, R. and KöBerle, C.: Comparison of Arctic Sea Ice Thickness Variability in IPCC Climate of the 20th Century Experiments and in Ocean-Sea Ice Hindcasts, J. Geophys. Res.-Oceans, 112, C04S13, doi:10.1029/2006JC003616, 2007.

Heimbach, P., Menemenlis, D., Losch, M., Campin, J.-M., and Hill, C.: On the Formulation of Sea-Ice Models. Part 2: Lessons from Multi-Year Adjoint Sea-Ice Export Sensitivities through the Canadian Arctic Archipelago, Ocean Model., 33, 145–158, doi:10.1016/j.ocemod.2010.02.002, 2010.

Ho, D. T., Law, C. S., Smithh, M. J., Schlosser, P., Harvey, M., and Hill, P.: Measurements of Air-Sea Gas Exchange at High Wind Speeds in the Southern Ocean: Implications for Global Parameterizations, Geophys. Res. Lett., 33, L16611, doi:10.1029/2006GL026817, 2006.

Hyndman, R. J. and Koehler, A. B.: Another Look at Measures of Forecast Accuracy, Int. J. Forecasting, 22, 679–688, doi:10.1016/j.ijforecast.2006.03.001, 2006.

Ivanova, N., Pedersen, L. T., Tonboe, R. T., Kern, S., Heygster, G., Lavergne, T., Sørensen, A., Saldo, R., Dybkjær, G., Brucker, L., and Shokr, M.: Inter-comparison and evaluation of sea ice algorithms: towards further identification of challenges and optimal approach using passive microwave observations, The Cryosphere, 9, 1797–1817, doi:10.5194/tc-9-1797-2015, 2015.

Jackson, J. M., Carmack, E. C., McLaughlin, F. A., Allen, S. E., and Ingram, R. G.: Identification, Characterization, and Change of the near-Surface Temperature Maximum in the Canada Basin, 1993–2008, J. Geophys. Res., 115, C05021, doi:10.1029/2009JC005265, 2010.

Jähne, B. and Haubecker, H.: Air-Water Gas Exchange, Annu. Rev. Fluid Mech., 30, 443–448, doi:10.1146/annurev.fluid.30.1.443, 1998.

Johnson, M., Gaffigan, S., Hunke, E., and Gerdes, R.: A Comparison of Arctic Ocean Sea Ice Concentration among the Coordinated AOMIP Model Experiments, J. Geophys. Res.-Oceans, 112, C04S11, doi:10.1029/2006JC003690, 2007.

Johnson, M., Proshutinsky, A., Aksenov, Y., Nguyen, A. T., Lindsay, R., Haas, C., Zhang, J., Diansky, N., Kwok, R., Maslowski, W., and Haekkinen, S.: Evaluation of Arctic Sea Ice Thickness Simulated by Arctic Ocean Model Intercomparison Project Models, J. Geophys. Res., 117, C00D13, doi:10.1029/2011JC007257, 2012.

Kalnay, E., Kanamitsu, M., Kistler, R., Collins, W., Deaven, D., Gandin, L., Iredell, M., Saha, S., White, G., Woollen, J., and Zhu, Y.: The NCEP/NCAR 40-year reanalysis project, B. Am. Meteorol. Soc., 77, 437–471, 1996.

Kara, A. B.: Mixed Layer Depth Variability over the Global Ocean, J. Geophys. Res., 108, 3079, doi:10.1029/2000JC000736, 2003.

Kohout, A. L. and Meylan, M. H.: An Elastic Plate Model for Wave Attenuation and Ice Floe Breaking in the Marginal Ice Zone, J. Geophys. Res., 113, C09016, doi:10.1029/2007JC004434, 2008.

Krishfield, R., Toole, J., Proshutinsky, A., and Timmermans, M.-L.: Automated Ice-Tethered Profilers for Seawater Observations under Pack Ice in All Seasons, J. Atmos. Ocean. Tech., 25, 2091–2105, doi:10.1175/2008JTECHO587.1, 2008.

Krishfield, R. A., Proshutinsky, A., Tateyama, K., Williams, W. J., Carmack, E. C., McLaughlin, F. A., and Timmermans, M.-L.: Deterioration of Perennial Sea Ice in the Beaufort Gyre from 2003 to 2012 and Its Impact on the Oceanic Freshwater Cycle, J. Geophys. Res.-Oceans, 119, 1271–1305, doi:10.1002/2013JC008999, 2014.

Large, W. G., McWilliams, J. C., and Doney, S. C.: Oceanic Vertical Mixing: A Review and a Model with a Nonlocal Boundary Layer Parameterization, Rev. Geophys., 32, 363–403, doi:10.1029/94RG01872, 1994.

Legge, O. J., Bakker, D. C. E., Johnson, M. T., Meredith, M. P., Venables, H. J., Brown, P. J., and Lee, G. A.: The Seasonal Cycle of Ocean-Atmosphere CO_2 Flux in Ryder Bay, West Antarctic Peninsula, Geophys. Res. Lett., 42, 2934–2942, doi:10.1002/2015GL063796, 2015.

Lindsay, R., Wensnahan, M., Schweiger, A., and Zhang, J.: Evaluation of Seven Different Atmospheric Reanalysis Products in the Arctic, J. Climate, 27, 2588–2606, 2014.

Lindsay, R. W. and Rothrock, D. A.: Arctic Sea Ice Leads from Advanced Very High Resolution Radiometer Images, J. Geophys. Res., 100, 4533–4544, 1995.

Locarnini, R., Mishonov, J., Boyer, T., Antonov, J. I., and Garcia, H. E.: World Ocean Atlas 2005, Vol. 1 Temreature, NOAA Atlas NESDIS 62 (NOAA, Silver Spring, Md.), 2006.

Loose, B., McGillis, W. R., Schlosser, P., Perovich, D., and Takahashi, T.: Effects of Freezing, Growth, and Ice Cover on Gas Transport Processes in Laboratory Seawater Experiments, Geophys. Res. Lett., 36, L05603, doi:10.1029/2008GL036318, 2009.

Loose, B., Schlosser, P., Perovich, D., Ringelberg, D., Ho, D. T., Takahashi, T., Richter-Menge, J., Reynolds, C. M., Mcgillis, W. R., and Tison, J.-L.: Gas Diffusion through Columnar Laboratory Sea Ice: Implications for Mixed-Layer Ventilation of CO2 in the Seasonal Ice Zone, Tellus B, 63, 23–39, doi:10.1111/j.1600-0889.2010.00506.x, 2011.

Loose, B., McGillis, W. R., Perovich, D., Zappa, C. J., and Schlosser, P.: A Parameter Model of Gas Exchange for the Seasonal Sea Ice Zone, Ocean Sci., 10, 1–16, doi:10.5194/os-10-1-2014, 2014.

Loose, B., Kelly, R., Bigdeli, A., Krishfield, R. A., Rutgers Van Der Loeff, M., and Moran, B.: How Well Does Wind Speed Predict Air-Sea Gas Transfer in the Sea Ice Zone? A Synthesis of Radon Deficit Profiles in the Upper Water Column of the Arctic Ocean, J. Geophys. Res., accepted, 2016.

Lorbacher, K., Dommenget, D., Niiler, P. P., and Köhl, A.: Ocean Mixed Layer Depth: A Subsurface Proxy of Ocean-Atmosphere Variability, J. Geophys. Res.-Oceans, 111, C07010, doi:10.1029/2003JC002157, 2006.

Losch, M., Menemenlis, D., Campin, J.-M., Heimbach, P., and Hill, C.: On the Formulation of Sea-Ice Models. Part 1: Effects of Different Solver Implementations and Parameterizations, Ocean Model., 33, 129–144, doi:10.1016/j.ocemod.2009.12.008, 2010.

Marshall, J., Adcroft, A., Hill, C., Perelman, L., and Heisey, C.: A Finite-Volume, Incompressible Navier Stokes Model for Studies of the Ocean on Parallel Computers, J. Geophys. Res.-Oceans, 102, 5753–5766, doi:10.1029/96JC02775, 1997.

McPhee, M. G.: Advances in Understanding Ice–ocean Stress during and since AIDJEX, Cold Reg. Sci. Technol., 76–77, 24–36, doi:10.1016/j.coldregions.2011.05.001, 2012.

McPhee, M. G., Proshutinsky, A., Morison, J. H., Steele, M., and Alkire, M. B.: Rapid Change in Freshwater Content of the Arctic Ocean, Geophys. Res. Lett., 36, L04606, doi:10.1029/2008GL036587, 2009.

McPhee, M. and Martinson, D. G.: Turbulent Mixing under Drifting Pack Ice in the Weddell Sea, Science, 263, 218–220, 1992.

Menemenlis, D., Fukumori, I., and Lee, T.: Using Green's Functions to Calibrate an Ocean General Circulation Model, Mon. Weather Rev., 133, 1224–1240, doi:10.1175/MWR2912.1, 2005.

Menemenlis, D., Campin, J.-M., Heimbach, P., Hill, C., Lee, T., Nguyen, A., Schodlok, M., and Zhang, H.: ECCO2: High Resolution Global Ocean and Sea Ice Data Synthesis, Mercator Ocean Quarterly Newsletter, 31, 13–21, 2008.

Morison, J. H., McPhee, M. G., Curtin, T. B., and Paulson, C. A.: The oceanography of winter leads, J. Geophys. Res., 97, 199–11, 1992.

Nguyen, A. T., Menemenlis, D., and Kwok, R.: Improved Modeling of the Arctic Halocline with a Subgrid-Scale Brine Rejection Parameterization, J. Geophys. Res., 114, C11014, doi:10.1029/2008JC005121, 2009.

Nguyen, A. T., Menemenlis, D., and Kwok, R.: Arctic Ice-Ocean Simulation with Optimized Model Parameters: Approach and Assessment, J. Geophys. Res.-Oceans, 116, C04025, doi:10.1029/2010JC006573, 2011.

Nguyen, A. T., Kwok, R., and Menemenlis, D.: Source and Pathway of the Western Arctic Upper Halocline in a Data-Constrained Coupled Ocean and Sea Ice Model, J. Phys. Oceanogr., 42, 802–823, doi:10.1175/JPO-D-11-040.1, 2012.

Nightingale, P. D., Malin, G. M., Law, C., Watson, A., Liss, P. S., Liddicoat, M. I., Boutin, J., and Upstill-Goddard, R. C.: In Situ Evaluation of Air-Sea Gas Exchange Parameterizations Using Novel Conservative and Volatile Tracers, Global Biogeochem. Cy., 14, 373–387, 2000.

Ohno, Y., Iwasaka, N., Kobashi, F., and Sato, Y.: Mixed Layer Depth Climatology of the North Pacific Based on Argo Observations, J. Oceanogr., 65, 1–16, doi:10.1007/s10872-009-0001-4, 2008.

Onogi, K., Tsutsui, J., Koide, H., Sakamoto, M., Kobayashi, S., Hatsushika, H., Matsumoto, T. et al.: The JRA-25 Reanalysis, J. Meteorol. Soc. Jpn. Ser. II, 85, 369–432, doi:10.2151/jmsj.85.369, 2007.

Peng, T.-H., Broecker, W. S., Mathieu, G. G., and Li, Y.-H.: Radon Evasion Rates in the Atlantic and Pacific Oceans as Determined during the Geosecs Program, J. Geophys. Res., 84, 2471–2486, 1979.

Peralta-Ferriz, C. and Woodgate, R. A.: Seasonal and Interannual Variability of Pan-Arctic Surface Mixed Layer Properties from 1979 to 2012 from Hydrographic Data, and the Dominance of Stratification for Multiyear Mixed Layer Depth Shoaling, Prog. Oceanogr., 134, 19–53, doi:10.1016/j.pocean.2014.12.005, 2015.

Perovich, D. and Maykut, G. A.: Solar Heating of a Stratified Ocean in the Presence of a Static Ice Cover, J. Geophys. Res., 95, 18233–18245, doi:10.1029/JC095iC10p18233, 1990.

Proshutinsky, A., Steele, M., Zhang, J., Holloway, G., Steiner, N., Hakkinen, S., Holland, D., Gerdes, R., Koeberle, C., Karcher, M., and Johnson, M.: Multinational Effort Studies Differences among Arctic Ocean Models, EOS Transact American Geophysical Union, 82, 637–644, doi:10.1029/01EO00365, 2001.

Proshutinsky, A., Gerdes, R., Holland, D., Holloway, G. and Steele, M.: AOMIP: Coordinated activities to improve models and model predictions, CLIVAR Exchanges, 13 (1), (Exchanges; 44), 17, http://eprints.soton.ac.uk/50120/01/Exch_44.pdf (last access: 20 September 2015), 2008.

Proshutinsky, A., Krishfield, R., Timmermans, M.-L., Toole, J., Carmack, E., McLaughlin, F., Williams, W. J., Zimmermann, S., Itoh, M., and Shimada, K.: Beaufort Gyre Freshwater Reservoir: State and Variability from Observations, J. Geophys. Res., 114, C00A10, doi:10.1029/2008JC005104, 2009.

Rutgers Van Der Loeff, M., Cassar, N., Nicolaus, M., Rabe, B., and Stimac, I.: The Influence of Sea Ice Cover on Air-Sea Gas Exchange Estimated with Radon-222 Profiles, J. Geophys. Res.-Oceans, 119, 2735–2751, doi:10.1002/2013JC009321, 2014.

Salter, M. E., Upstill-Goddard, R. C., Nightingale, P. D., Archer, S. D., Blomquist, B., Ho, D. T., Huebert, B., Schlosser, P., and Yang, M.: Impact of an Artificial Surfactant Release on Air-Sea Gas Fluxes during Deep Ocean Gas Exchange Experiment II, J. Geophys. Res., 116, C11016, doi:10.1029/2011JC007023, 2011.

Shaw, W. J., Stanton, T. P., McPhee, M. G., Morison, J. H., and Martinson, D. G.: Role of the Upper Ocean in the Energy Budget of Arctic Sea Ice during SHEBA, J. Geophys. Res., 114, C06012, doi:10.1029/2008JC004991, 2009.

Shimada, K., Carmack, E. C., Hatakeyama, K., and Takizawa, T.: Varieties of Shallow Temperature Maximum Waters in the Western Canadian Basin of the Arctic Ocean, Geophys. Res. Lett., 28, 3441–3444, 2001.

Smethie, W. M., Takahashi, T., Chipman, D. W., and Ledwell, J. R.: Gas Exchange and CO_2 Flux in the Tropical Atlantic Ocean Determined from 222Rn and pCO_2 Measurements, J. Geophys. Res.-Oceans, 90, 7005–7022, 1985.

Spall, M. A., Pickart, R. S., Fratantoni, P. S., and Plueddemann, A. J.: Western Arctic Shelfbreak Eddies: Formation and Transport, J. Phys. Oceanogr., 38, 1644–1668, doi:10.1175/2007JPO3829.1, 2008.

Sweeney, C., Gloor, E., Jacobson, A. R., Key, R. M., McKinley, G., Sarmiento, J.-L., and Wanninkhof, R.: Constraining Global Air-Sea Gas Exchange for CO_2 with Recent Bomb ^{14}C Measurements, Global Biogeochem. Cy., 21, GB2015, doi:10.1029/2006GB002784, 2007.

Takahashi, T., Sutherland, S. C., Wanninkhof, R., Sweeney, C., Feely, R. A., Chipman, D. W., Hales, B., Friederich, G., Chavez, F., Sabine, C., and Watson, A.: Climatological Mean and Decadal Change in Surface Ocean pCO_2, and Net Sea-Air CO_2 Flux over the Global Oceans, Deep-Sea Res. Pt. II, 56, 554–577, 2009.

Thomson, R. E. and Fine, I. V.: Estimating Mixed Layer Depth from Oceanic Profile Data, J. Atmos. Ocean. Tech., 20, 319–329, doi:10.1175/1520-0426(2003)020<0319:EMLDFO>2.0.CO;2, 2003.

Timmermans, M.-L., Proshutinsky, A., Krishfield, R. A., Perovich, D. K., Richter-Menge, J. A., Stanton, T. P., and Toole, J. M.: Surface Freshening in the Arctic Ocean's Eurasian Basin: An Apparent Consequence of Recent Change in the Wind-Driven Circulation, J. Geophys. Res., 116, C00D03, doi:10.1029/2011JC006975, 2011.

Toole, J. M., Timmermans, M. L., Perovich, D. K., Krishfield, R. A., Proshutinsky, A., and Richter-Menge, J. A.: Nfluences of the Ocean Surface Mixed Layer and Thermohaline Stratification on Arctic Sea Ice in the Central Canada Basin, J. Geophys. Res., 115, C10018, doi:10.1029/2009JC005660, 2010.

Wadhams, P., Squire, V. A., Ewing, J. A., and Pascal, R. W.: The Effect of the Marginal Ice Zone on the Directional Wave Spectrum of the Ocean, J. Phys. Oceanogr., 16, 358–376, 1986.

Wanninkhof, R.: Relationship between Wind Speed and Gas Exchange over the Ocean, J. Geophys. Res.-Oceans, 97, 7373–7382, doi:10.1029/92JC00188, 1992.

Wanninkhof, R. and McGillis, W. R.: A Cubic Relationship between

Air-Sea CO_2 Exchangeand Wind Speed, Geophys. Res. Lett., 26, 1889–1892, 1999.

Wijesekera, H. W. and Gregg, M. C.: Surface Layer Response to Weak Winds, Westerly Bursts, and Rain Squalls in the Western Pacific Warm Pool, J. Geophys. Res.-Oceans, 101, 977–997, doi:10.1029/95JC02553, 1996.

Williams, A. J., Thwaites, F. T., Morrison, A. T., Toole, J. M., and Krishfield, R. A.: Motion Tracking in an Acoustic Point-Measurement Current Meter, IEEE, doi:10.1109/OCEANSSYD.2010.5603862, 2010.

Wunsch, C. and Heimbach, P.: Dynamically and Kinematically Consistent Global Ocean Circulation and Ice State Estimates, in: In Ocean Circulation and Climate: A 21 Century Perspective, Elsevier, BV, https://dash.harvard.edu/handle/1/12136112 (last access: 14 January 2017), 2013.

Zemmelink, H. J., Delille, B., Tison, J. L., Hintsa, E. J., Houghton, L., and Dacey, J. W. H.: CO_2 Deposition over Multi-Year Ice of the Western Weddell Sea, Geophys. Res. Lett., 33, L13606, doi:10.1029/2006GL026320, 2006.

Zemmelink, H. J., Dacey, J. W. H., Houghton, L., Hintsa, E. J., and Liss, P. S.: Dimethylsulfide Emissions over the Multi-Year Ice of the Western Weddell Sea, Geophys. Res. Lett., 35, L06603, doi:10.1029/2007GL031847, 2008.

Zhang, J. and Rothrock, D. A.: Modeling Global Sea Ice with a Thickness and Enthalpy Distribution Model in Generalized Curvilinear Coordinates, Mon. Weather Rev., 131, 845–861, doi:10.1175/1520-0493(2003)131<0845:MGSIWA>2.0.CO;2, 2003.

Zhao, M. and Timmermans, M.-L.: Vertical Scales and Dynamics of Eddies in the Arctic Ocean's Canada Basin, J. Geophys. Res.-Oceans, 120, 8195–8209, doi:10.1002/2015JC011251, 2015.

Zhao, M., Timmermans, M.-L., Cole, S., Krishfield, R., Proshutinsky, A., and Toole, J.: Characterizing the Eddy Field in the Arctic Ocean Halocline, J. Geophys. Res.-Oceans, 119, 8800–8817, doi:10.1002/2014JC010488, 2014.

Zhao, M., Timmermans, M. L., Cole, S., Krishfield, R., and Toole, J.: Evolution of the eddy field in the Arctic Ocean's Canada Basin, 2005–2015, Geophys. Res. Lett., 43, 8106–8114, 2016.

12

Shelf–Basin interaction along the East Siberian Sea

Leif G. Anderson[1], **Göran Björk**[1], **Ola Holby**[2], **Sara Jutterström**[3], **Carl Magnus Mörth**[4], **Matt O'Regan**[4], **Christof Pearce**[4,5], **Igor Semiletov**[6,7,8], **Christian Stranne**[4,10], **Tim Stöven**[1,9], **Toste Tanhua**[9], **Adam Ulfsbo**[1,11], and **Martin Jakobsson**[4]

[1]Department of Marine Sciences, University of Gothenburg, P.O. Box 461, 40530 Gothenburg, Sweden
[2]Department of Environmental and Energy Systems, Karlstad University, 651 88 Karlstad, Sweden
[3]IVL Swedish Environmental Research Institute, Box 530 21, 400 14 Gothenburg, Sweden
[4]Department of Geological Sciences, Stockholm University, 106 91 Stockholm, Sweden
[5]Department of Geoscience, Aarhus University, Aarhus, Denmark
[6]International Arctic Research Center, University Alaska Fairbanks, Fairbanks, AK 99775, USA
[7]Pacific Oceanological Institute, Russian Academy of Sciences Far Eastern Branch, Vladivostok 690041, Russia
[8]The National Research Tomsk Polytechnic University, Tomsk, Russia
[9]Helmholtz Centre for Ocean Research Kiel, GEOMAR, Kiel, Germany
[10]Center for Coastal and Ocean Mapping/Joint Hydrographic Center, Durham, NH 03824, USA
[11]Division of Earth and Ocean Sciences, Nicholas School of the Environment, Duke University, Durham, NC 27704, USA

Correspondence to: Leif G. Anderson (leif.anderson@marine.gu.se)

Abstract. Extensive biogeochemical transformation of organic matter takes place in the shallow continental shelf seas of Siberia. This, in combination with brine production from sea-ice formation, results in cold bottom waters with relatively high salinity and nutrient concentrations, as well as low oxygen and pH levels. Data from the SWERUS-C3 expedition with icebreaker *Oden*, from July to September 2014, show the distribution of such nutrient-rich, cold bottom waters along the continental margin from about 140 to 180° E. The water with maximum nutrient concentration, classically named the upper halocline, is absent over the Lomonosov Ridge at 140° E, while it appears in the Makarov Basin at 150° E and intensifies further eastwards. At the intercept between the Mendeleev Ridge and the East Siberian continental shelf slope, the nutrient maximum is still intense, but distributed across a larger depth interval. The nutrient-rich water is found here at salinities of up to ~ 34.5, i.e. in the water classically named lower halocline. East of 170° E transient tracers show significantly less ventilated waters below about 150 m water depth. This likely results from a local isolation of waters over the Chukchi Abyssal Plain as the boundary current from the west is steered away from this area by the bathymetry of the Mendeleev Ridge. The water with salini-

ties of ~ 34.5 has high nutrients and low oxygen concentrations as well as low pH, typically indicating decay of organic matter. A deficit in nitrate relative to phosphate suggests that this process partly occurs under hypoxia. We conclude that the high nutrient water with salinity ~ 34.5 are formed on the shelf slope in the Mendeleev Ridge region from interior basin water that is trapped for enough time to attain its signature through interaction with the sediment.

1 Introduction

The extensive, flat, and shallow shelf areas of the Laptev and East Siberian seas are particularly influenced by the changing climate in the Arctic. Coastal erosion from wave action becomes widespread when the summer sea-ice cover shrinks and river discharge increases in warmer humid conditions, both affecting organic matter and nutrient supply (Charkin et al., 2011). At the same time, the decrease in summer sea-ice coverage changes the dynamics of the ocean by increasing vertical mixing and brine production in the fall when sea ice again starts to form over areas that in the past used to be sea-ice covered. The changes may impact shelf basin exchange (e.g. Dethleff, 2010; Nishino et al., 2013).

Here we assess data collected in 2014 along the continental shelf break of northern Siberia. Acquired oceanographic and bottom sediment data add to our understanding of water mass modification in the central Arctic Ocean basin. The objectives are to describe the spreading of shelf waters, including those richest in nutrients, from the East Siberian Sea and assess their sources, as well as to evaluate potential effects of diminishing sea-ice coverage under a warmer climate.

The Arctic Ocean has an area of about $9.5 \times 10^{12} \, \mathrm{m}^2$ of which more than half is comprised of shallow continental shelf seas (Jakobsson, 2002). The deep central part consists of several basins; the Nansen and Amundsen basins are together denoted the Eurasian Basin, and the Canada and Makarov basins constitute the Amerasian Basin. The Lomonosov Ridge stretches from the continental slope of the Laptev Sea to the slope off northern Greenland and separates the Eurasian Basin from the Amerasian Basin (Fig. 1). The deep waters of the Arctic Ocean are supplied from the Atlantic Ocean, entering either through the eastern Fram Strait (Fram Strait Branch, FSB) or over the Barents Sea (Barents Sea Branch, BSB). The latter water flows into the Kara Sea before exiting through the St Anna Trough along the continental margin where it covers a depth range down to about 1500 m (e.g. Schauer et al., 2002). Both branches flow to the east and follow the bathymetry in a cyclonic pattern around the basins (Rudels et al., 1994, Fig. 1), the difference being that the FSB takes the inner turn and is largely restricted to the Eurasian Basin. It is mainly the BSB that flows over the Lomonosov Ridge into the Makarov Basin north of the Laptev Sea.

The upper waters are entering from both the Pacific and Atlantic oceans, where the latter either pass over the Barents shelf or through Fram Strait. The upper waters have classically been divided into a surface mixed layer (SML) that varies seasonally, an upper halocline of mainly Pacific origin, and a lower halocline of Atlantic origin (e.g. Jones and Anderson, 1986; Rudels et al., 1996). The flow pattern of these waters differs. The lower halocline primarily follows the underlying Atlantic layer, while the upper halocline, and even more so the surface mixed layer circulation, is much impacted by the dominating wind field (e.g. Jones et al., 2008). The flow of the surface water is dominated by transport from the Laptev Sea towards Fram Strait, the Transpolar Drift, and one cyclonic circulation in the Canada Basin, the Beaufort Gyre. The size of the latter is determined by the atmospheric pressure field, where a negative Arctic Oscillation results in a larger Beaufort Gyre compared to a positive Arctic Oscillation (Proshutinsky et al., 2009).

The properties of the surface mixed layer and the upper halocline are modified over the shelves, and for the SML also in the central Arctic Ocean by, e.g. mixing with river runoff, sea-ice melt, and brine from sea-ice formation. Biogeochemical processes also modify the chemical signature, e.g. low-ering the nutrient concentration of the SML through primary production and increasing the nutrient concentration in the upper halocline through remineralization of organic matter (e.g. Jones and Anderson, 1986). The latter process has been reported to occur in the Chukchi Sea (Bates, 2006; Pipko et al., 2002), East Siberian Sea (Nishino et al., 2009; Anderson et al., 2011), and Laptev Sea (Semiletov et al., 2013, 2016).

One of the most pronounced signatures of the upper halocline of the central Arctic Ocean is a silicate maximum, which was first reported in 1968–1969 from observations made from the drifting T-3 ice island in the Canada Basin (Kinney et al., 1970). In 1979 the silicate maximum was observed during the LOREX study over the Lomonosov Ridge and into the fringe of the Amundsen Basin (Moore et al., 1983). In 1994 no silicate maximum was observed in the Makarov Basin along a section from the Chukchi Sea to the North Pole (Swift et al., 1997). It is clear that the distribution of the upper halocline with its prominent silicate signature has varied much in the past and with changing sea-ice coverage it might vary even more in the future. In this contribution we give some indications of the latter.

Furthermore, recently high silicate concentrations were found at salinities ~ 34.5 along the continental slope of the eastern East Siberian Sea (Nishino et al., 2009; Anderson et al., 2013). Nishino et al. (2009) suggested that this silicate maximum was produced by decomposition of opal-shelled organisms along the continental margin. Based on $\delta^{18}\mathrm{O}$ data collected in 2008, Anderson et al. (2013) reported a brine content of at least 4 % and a small temperature minimum signature associated with the high silicate concentration. The present study will expand on the formation process of this water

2   Methods

Water column data in this study were obtained along six oceanographic sections across the shelf break (A–F; Fig. 1) during the SWERUS-C3 (Swedish–Russian–US Arctic Ocean Investigation of Climate–Cryosphere–Carbon Interactions) expedition in 2014 with Swedish icebreaker *Oden*. SWERUS-C3 is a multi-disciplinary international program focusing on investigating the functioning of the Climate–Cryosphere–Carbon (C3) system of the East Siberian Arctic Ocean. The expedition consisted of two legs with the icebreaker *Oden*. Leg 1 started 5 July in Tromsø, Norway, and followed the Siberian continental shelf to end in Barrow, Alaska, on 21 August. Leg 2 took the return route from Barrow and ended in Tromsø on 3 October after concentrating the field program to the continental shelf break, slope, and the adjacent deep Arctic Ocean basin. Data from leg 2 focusing on the shelf break are discussed in this study.

Water samples were collected using a rosette system equipped with 24 bottles of the Niskin type, each having a volume of 7 L. The bottles were closed during the return of

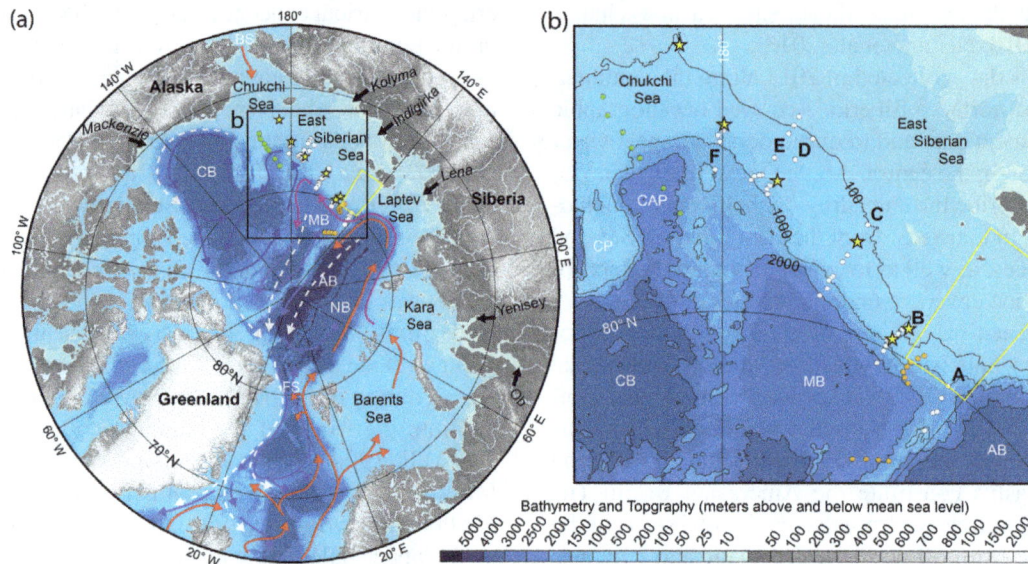

Figure 1. Map of the Arctic Ocean with general currents at intermediate depths over the deep basins and exchange with the surrounding oceans **(a)**. The black frame indicates the investigated area that is illustrated in panel **(b)** with the hydrographic station positions of sections A to F as white points and those of sediment cores in yellow stars. The Arctic Ocean Section 1994 stations are in green and the orange points show the positions of the ACSYS 96 stations, which are used as historic references. The yellow frame borders the area where the historic sea-ice coverage has been evaluated; see Fig. 10. Abbreviations: Fram Strait (FS), Bering Strait (BS) Nansen Basin (NB), Amundsen Basin (AB), Makarov Basin (MB), Canada Basin (CB), Chukchi Plateau (CP), and Chukchi Abyssal Plain (CAP).

the CTD rosette package from the bottom to the surface and water samples for all constituents were drawn soon after the rosette was secured in the sampling container.

The following constituents are used here: bottle practical salinity, dissolved inorganic carbon (DIC), total alkalinity (TA), pH, oxygen, nutrients ($NO_3^- + NO_2^-$, PO_4^{3-}, SiO_2), and the transient tracer sulfur hexafluoride (SF_6). The order of sampling was determined by the risk of contamination meaning that transient tracer samples were collected first followed by oxygen, the carbon system parameters, nutrients, and salinity.

Salinity and temperature data were collected using a SeaBird 911+ CTD with dual SeaBird temperature (SBE 3), conductivity (SBE 04C), and oxygen sensors (SBE 43) attached to a 24 bottle rosette for water sampling. Salinity data were calibrated against deep water samples analysed onboard using an Autosal 8400B lab salinometer. The salinometer was calibrated using one standard seawater ampule (IAPSO standard seawater from OSIL Environmental Instruments and Systems) before each batch of 24 samples. The accuracy of the Autosal salinities and CTD salinities should both be within ±0.003 and the accuracy for temperature ±0.002 °C. Water samples for salinity were analysed for more than 90 % of the depth and when no data were available the CTD salinity was used in the evaluation.

The water samples for determination of the transient tracer SF_6 were directly drawn from the Niskin bottles using 250 mL glass syringes. The samples were stored in a cooling bath that was continuously rinsed with cold surface water

to prevent outgassing of the tracers. Measurements were directly performed on board, using a purge and trap GC-ECD system similar to the "PT3" set-up described in Stöven and Tanhua (2014). The column composition was as follows: the trap consisted of a 1/16" column packed with 70 cm Heysep D, the 1/8" precolumn was packed with 30 cm Porasil C and 60 cm Molsieve 5Å and the 1/8" main column with 200 cm Carbograph 1AC and 20 cm Molsieve 5Å. The precision for onboard measurements was ±0.02 fmol kg^{-1} for SF_6. In the evaluation of the data, SF_6 is given in partial pressure normalized to 1 atm, which is equal to the mixing ratio on the volume scale in parts per trillion (ppt). The advantage of using partial pressure instead of concentrations for dissolved gases in the ocean is that the partial pressure is not influence by temperature and salinity effects. The precision is given in concentration since it is related to the absolute value per kg seawater. Age modelling based on these transient tracers is complicated and erroneous at high latitudes due to ambiguous reasons (Stöven et al., 2015, 2016). Hence, we do not provide any statements about the ventilation timescale but rather the ventilation states of the water masses in the Arctic Ocean based on the concentration distribution.

An automated Winkler titration system was used for the oxygen measurements with a precision of ~ 1 µmol kg^{-1}. The accuracy was set by titrating known amounts of KIO_3 salts that were dissolved in sulfuric acid. As the amount was known to better than 0.1 % the accuracy should be significantly less than the precision.

DIC was determined by a coulometric titration method based on Johnson et al. (1987), having a precision of $2.0\,\mu mol\,kg^{-1}$, from duplicate sample analyses, with the accuracy set by calibration against certified reference materials (CRM; Batch nos. 123 and 136), supplied by A. Dickson, Scripps Institution of Oceanography (USA). TA was determined by an automated open-cell potentiometric titration (Haraldsson et al., 1997), with a precision better than $2.0\,\mu mol\,kg^{-1}$ and the accuracy ensured in the same way as for DIC. pH was determined by a spectrophotometric method, based on the absorption ratio of the sulfonephtalien dye, m-cresol purple (mCP) (Clayton and Byrne, 1993). Purified mCP was purchased from the laboratory of Robert H. Byrne, University of South Florida, USA. The accuracy was estimated to 0.006 from internal consistency calculations of analysed CRM samples and the precision, defined as the absolute mean difference of duplicate samples, was 0.001 pH units. The seawater pH is reported on the total scale and in situ temperature.

The partial pressure of carbon dioxide (pCO_2) was calculated from the combination of pH and TA, and pH and DIC, using CO2SYS version 1.1 (van Heuven et al., 2011) with the stoichiometric dissociation constants of carbonic acid (K_1^* and K_2^*) and bisulfate ($K_{HSO_4}^*$) given by Millero (2010) and Dickson (1990), respectively. Input data included salinity, temperature, PO_4, and SiO_2. The reported values are the average of the two calculated for each sample. The uncertainty was computed using a Monte Carlo approach (Legge et al., 2015) and is, expressed as double standard deviation, about 2.5 %.

Besides the extensive sampling and measuring of the water column, analyses were also performed on sediments. Sediment samples from six coring stations along the SWERUS leg 2 cruise track (Fig. 1) were taken from four different depths in the upper 16 cm (Table 1). Two different types of coring devices were used: a gravity corer (GC) and multicorer (MC). These 24 samples were analysed for biogenic silica (BSi) content, with the aim of investigating a possible sedimentary source of the silicate maximum observed in the water column. Biogenic silica was measured using a wet alkaline extraction technique (Conley and Schelske, 2001). Samples were freeze dried and approximately 30 mg of homogenized sediment was placed in a mild alkaline solution (1 % Na_2CO_3) at 85 °C and aliquots were taken at 3, 4, and 5 h during this leaching process. For each of these subsamples, dissolved Si was measured by Inductively Coupled Plasma Spectrometry, using a Thermo ICAP 6500 DUO. All BSi is assumed to have dissolved after 2 h leaching, after which only Si from minerals is being released. Based on this principle, the zero-hour intercept of the slope from the 3, 4, and 5 h Si concentrations is used to calculate the biogenic fraction. This method was validated by including blanks, and standards from a previous inter-laboratory comparison exercise (Conley, 1998). The relative uncertainties associated with this method are estimated to be ±20 % of the measured value and precision of the ICP is from certified standard measurements better than 5 %.

3 Results

The salinity distribution along the continental margin from the Lomonosov Ridge to the Chukchi Sea shows a similar general pattern, but with some significant variations especially in the top 50 m (Fig. 2). The thinnest layer of low-salinity surface water is found at the Lomonosov Ridge (section A), which increases in thickness eastward in the study area. In section B we find the lowest salinity of 24.55 at 10 m water depth, followed by a very sharp halocline with the salinity increasing from about 32 at 50 m to 34 at 100 m depth. Further to the east the halocline is less sharp with, e.g. the 34 salinity isoline deepening to a depth of more than 200 m. Here, also the > 33 isolines deepens from the shelf towards the deep basin, especially in sections D and E.

The silicate distribution is variable between the sections (Fig. 2). Over the Lomonosov Ridge (section A) the highest silicate concentration, reaching $15\,\mu mol\,L^{-1}$, is found in the surface. In section B the maximum is instead found at about 50 m depth and varies horizontally, with the highest concentration exceeding $30\,\mu mol\,L^{-1}$. At this depth the salinity is around 33. Further to the east at section C, the concentration in the silicate maximum is higher and is found somewhat deeper and also at a larger salinity range. It extends horizontally all over the shelf and slope, although with concentrations decreasing some 100–150 km seaward from the shelf break. At the station farthest out in the deep basin the concentration is close to the maximum in section B.

Sections D and E are fairly close to each other and both show a similar pattern. The maximum silicate concentration, above $50\,\mu mol\,L^{-1}$, is close to the bottom at 100–150 m depth (Fig. 2). From here the concentration decreases gradually away from the shelf break, to the lowest maximum at the outer station, around $30\,\mu mol\,L^{-1}$. Another specific characteristic of the silicate distribution at these sections are the wide depth range of concentrations more than $15\,\mu mol\,L^{-1}$. Here it spans the range of about 50 to 250 m, whereas in section C it only spans 50 to 150 m. To some degree this is attributed to the more gradual increase in salinity with depth, but there are also high concentrations at salinities above 34.5. In section F the silicate concentration is lower and also spans a narrower depth range. However, this section starts further away from the shelf break and may be difficult to compare with the other sections.

The waters of high silicate concentration have other distinct characters such as high concentrations of the other nutrients, phosphate and nitrate, high apparent oxygen utilization (AOU) and pCO_2, and low pH (Fig. 3). The top 100–150 m is colder than 0 °C and the nutrient maximum as represented by phosphate is largely confined to the coldest water. There are some small differences in the exact pattern of the differ-

Table 1. Geographic information of sediment cores and their biogenic silica content.

Core ID	Longitude (° E)	Latitude (° N)	Water depth (m)	Sediment depth (cm)	Biogenic silica (wt % SiO$_2$)	Average BSi (wt %)
L2-5-GC1	176.207	72.870	115.5	0–1	7.58	9.34
				3–4	6.04	
				8–9	10.24	
				13–14	13.51	
L2-7-GC1	179.820	74.993	391.5	0–1	1.99	0.80
				5–6	0	
				10–11	1.03	
				15–16	0.15	
L2-18-MC5	173.879	76.409	349	0–1	1.07	0.34
				5–6	0.06	
				10–11	0.19	
				15–16	0.04	
L2-21-MC6	163.308	77.579	153	0–1	0.19	0.25
				5–6	0.43	
				10–11	0.28	
				15–16	0.09	
L2-25-MC6	152.676	79.226	101	0–1	0.17	0.04
				5–6	0	
				10–11	0	
				15–16	0	
L2-27-MC6	154.126	79.665	276	0–1	0.30	0.30
				5–6	0.39	
				10–11	0.21	
				15–16	0.28	

ent parameters, e.g. the AOU maximum is located slightly deeper than that of phosphate farthest out in the deep basin.

Biogenic silica concentrations in the analysed sediments varied widely between the different sites. The full names of the cores include the prefix SWERUS-L2, which henceforth is omitted. The most western sites (coring stations 21MC1, 25MC1, 27MC1 $\sim$ water column sections A, B, C) had BSi levels of less than 0.5 % (Fig. 4). Values increased slightly to the east, reaching up to 1 % BSi in station 18MC1 ($\sim$ water column section D) and up to 2 % in station 7GC1 ($\sim$ section F). Concentrations in the most eastern station 5GC1 located on the western flank of Herald Canyon are, however, much higher and reach up to 13.5 %. The surface generally contains the highest concentration of BSi at all sites, except for station 5GC1 where the concentration increases down core. These subtler differences should however be treated with caution due to the large uncertainties associated with the measurement method.

The mean mixed layer partial pressure of SF$_6$ along all sections is $\sim$ 8.1 ppt (Fig. 5), which is slightly below the contemporaneous atmospheric value of 8.4 ppt. At all sections except A, a SF$_6$ minimum is associated with the maximum in AOU. Close to the shelf in section B this SF$_6$ minimum is 6.4 ppt at 80 m and shoals polewards to 50 m with increasing partial pressure to the range 6.7–7 ppt. The elevated AOU values are 75–138 µmol kg^{-1} at these depths. The SF$_6$ minimum becomes more significant at section C with partial pressures between 4.5–5.1 ppt at 95–130 m (135–183 µmol kg^{-1} AOU). The maximum deepens eastwards to about 200 m at sections D, E, and F with partial pressures between 2.5–3.4 ppt and 90–118 µmol kg^{-1} AOU.

The SF$_6$ partial pressure in the AW (Atlantic Water) layer between 250 and 600 m is homogeneously distributed with a mean value of about 6 ppt at sections B and C (Fig. 5). In contrast, sections D, E, and F show significant lower mean partial pressures of 4.1–3.4 ppt in the same depth interval with the lowest values at section F. Note that the deep SF$_6$ partial pressures at sections E and F are close to the values in the overlying minimum at 200 m and the minimum can thus not be defined by SF$_6$ data only. However, the minima can clearly be separated by the AOU values since the warm AW layer shows constant low values of about 50 µmol kg^{-1} along all sections.

The bottom water partial pressure of SF$_6$ has a general trend of decreasing values at a specific isobaths from the west to the east (Fig. 6). The highest partial pressures of 6.1–

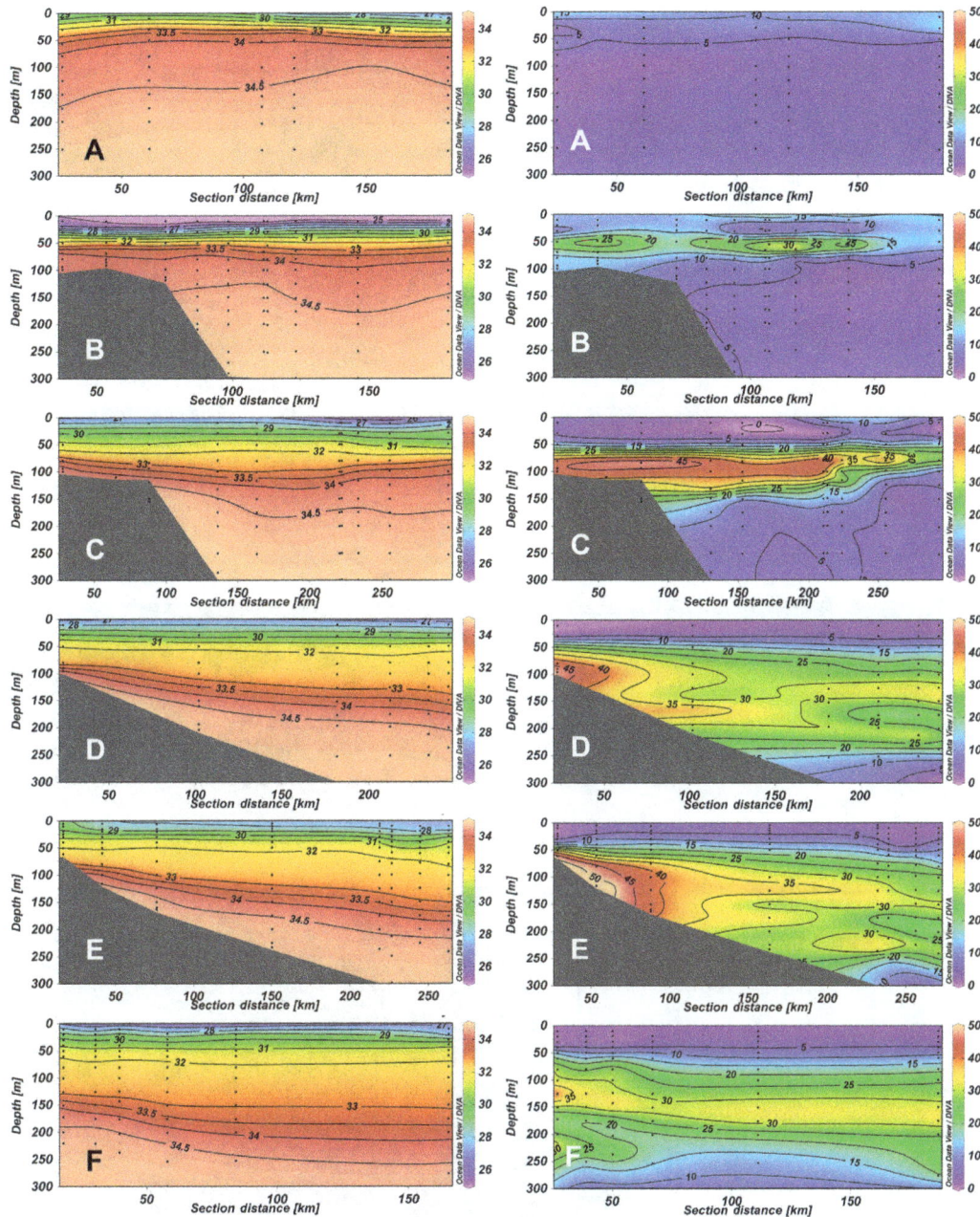

Figure 2. Sections of salinity (left) and silicate in µmol L^{-1} (right) of the upper 300 m of sections A to F; see Fig. 1 for location of sections. Sections drawn using Ocean Data View (Schlitzer, 2017).

6.9 ppt can be found at section B between 100 and 500 m. Section C shows increasing partial pressures with depth from 4.6 ppt at 100 m to 6.5 ppt at 500 m, with decreasing values deeper. A similar gradient of 4.4 to 5.7 ppt can be found at sections D, E, and F at the same depth range. Below ∼ 500 m the partial pressure decreases with increasing depth, reaching the detection limit at 1900–2000 m in the Makarov Basin (Fig. 6).

4   Discussion

In section A, the most western that is located over the Lomonosov Ridge, the highest silicate concentrations are found in the surface. Hence, no sub-surface maximum typical of the upper halocline water is present here. The surface silicate maximum is associated with low salinity, which is a typical signature of runoff from the Lena River. Thus, this surface water has a substantial fraction of freshwater from river, even if there also is a contribution of sea-ice melt. The

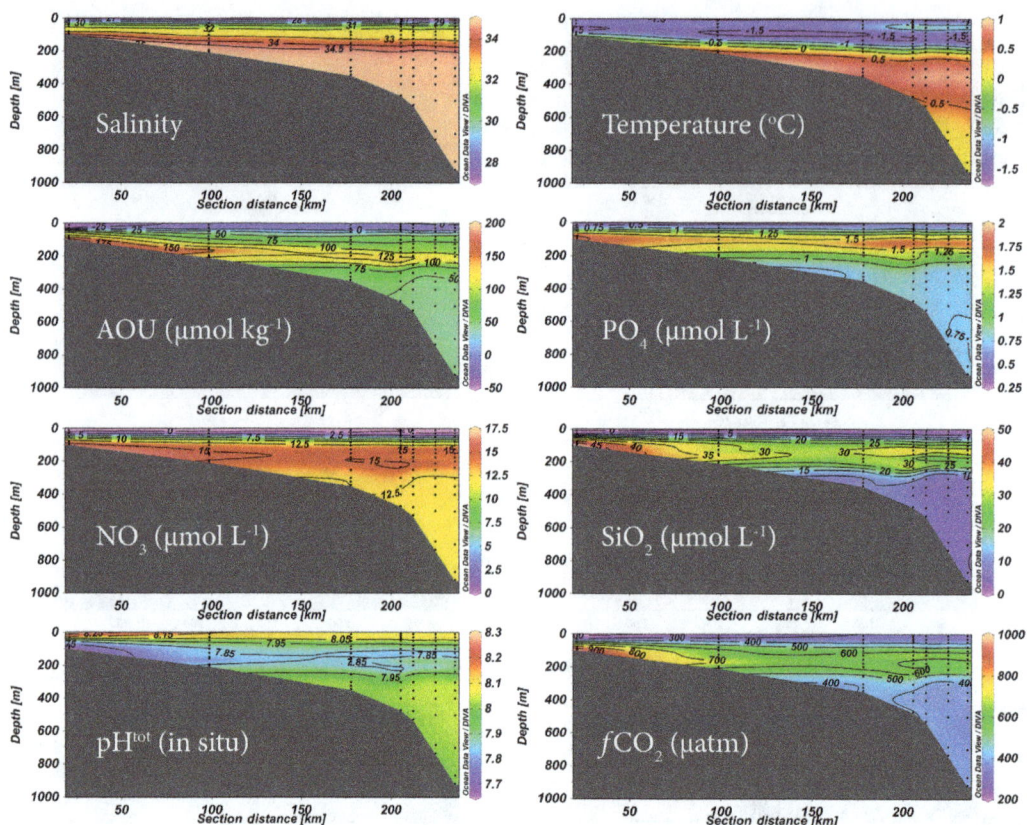

Figure 3. Properties in the upper 1000 m along section D of Fig. 1. Sections drawn using Ocean Data View (Schlitzer, 2017).

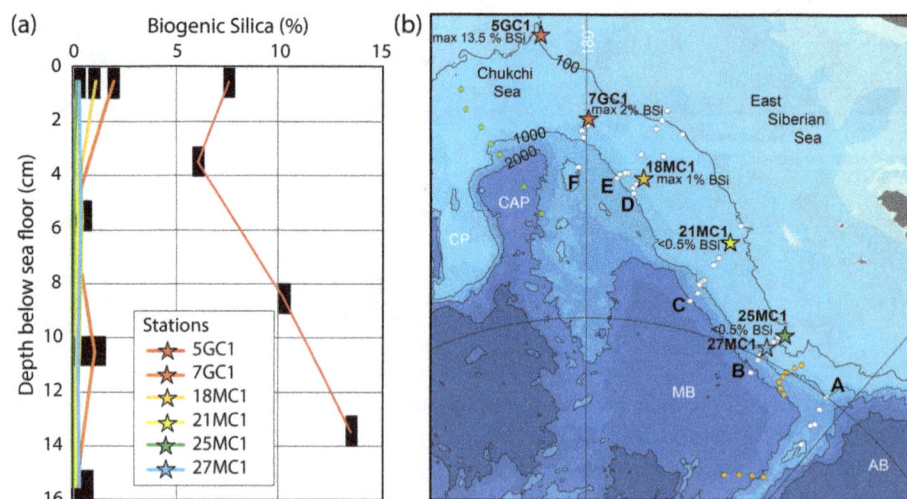

Figure 4. Biogenic silicate (BSi) concentrations (percent dry weight) in the upper 16 cm of the sediment **(a)** at coring sites shown in panel **(b)**. The symbols of the coring sites are colour coded after measured BSi concentration from the lowest concentrations in blue to the highest in red. The black bars represent the depth layer of the sediment that is analysed.

high silicate surface water is also seen in section B and partly in section C but not further to the east, indicating the limit of the river plume to this part of the deep Arctic Ocean.

The highest silicate concentrations are found at the shelf slope of sections D and E and this is also where the salin-ity isolines shoal (Fig. 2). The increase in salinity along the shelf slope at bottom depths less than 250 m is accompanied by an increase in temperature as illustrated in the depth pan-els (Fig. 7a and b) with no indication of mixing with another water mass, as all data from these sections have the same

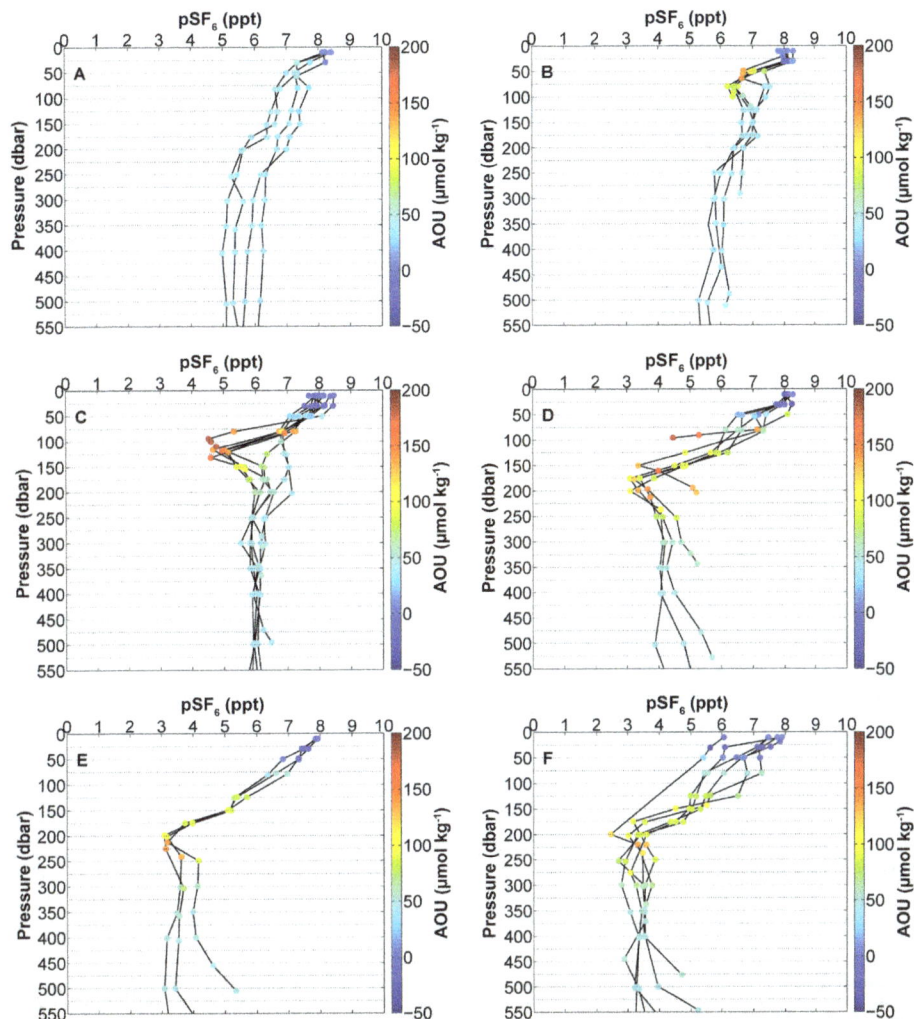

Figure 5. pSF$_6$ (ppt per volume) profiles from surface to 550 m depth, coloured by AOU (μmol kg^{-1}) of the stations in sections A to F.

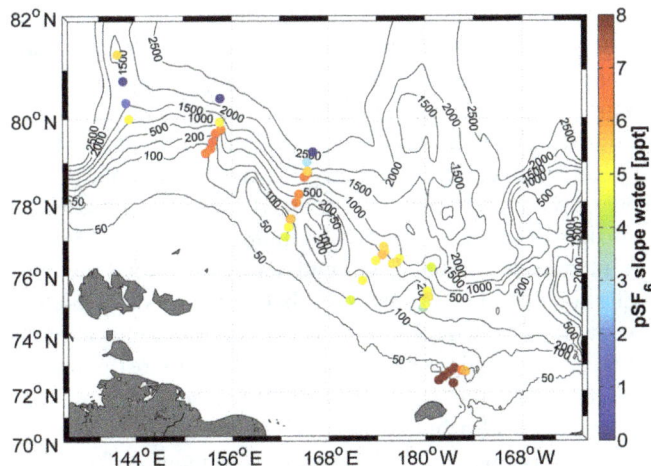

Figure 6. SF$_6$ partial pressure (ppt) in the sample collected closest to the bottom, typically 5–10 m above.

shape in a $T - S$ panel (Fig. 8a) for salinity > 32.5. The slope of the isolines infers a near-bottom increase of the current due to geostrophic shear, which is superimposed on the typical overall eastward current (e.g. Rudels, 2012). Although it is not possible to determine the absolute current velocity from just a geostrophic calculation, our data together with the known direction of the mean flow suggest that we have a bottom intensified flow in the eastward direction. The magnitude of this increase is about 3 cm s^{-1} (based on the density difference and distance between the two slope stations located at 164 and 241 m water depth in section E) over the depth range 100–150 m, which is not negligible.

At sections D and E is the maximum observed silicate concentration about 56 μmol L^{-1} and found at ~ 120 m (Fig. 7c) but elevated concentrations are found down to nearly 250 m depth. Plotting silicate concentrations against salinity (Fig. 8c) shows a clear pattern with a shallow maximum around 33 and a deep maximum at 34.5. These maxima are also evident in the section panels in Fig. 2.

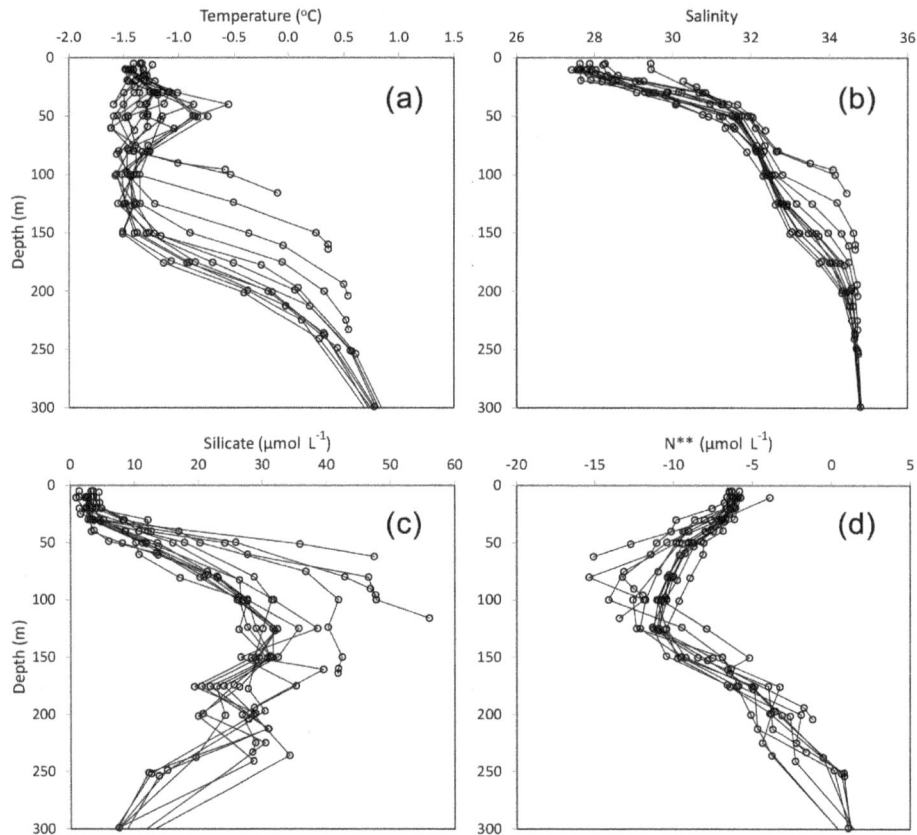

Figure 7. Depth profiles of temperature **(a)**, salinity **(b)**, silicate **(c)**, and N** **(d)** in the upper 300 m of all stations at sections D and E; see Fig. 1 for locations.

When the nutrient maximum is accompanied by an oxygen minimum (Fig. 3) it suggests organic matter mineralization, and if this occurs at low oxygen levels, nitrate is lost as electron acceptor via either denitrification or anammox. Such conditions can only be met close to, or in, the sediments of the shelves within the Arctic Ocean as all other waters observed in this region are well oxygenated. A deficit in nitrate is seen when computing the property $N^{**} = 0.87 \times [NO_3] - 16 \times [PO_4] + 2.9$ (Codispoti et al., 2005), which gives a constant value if the classical Redfield–Ketchum–Richard N : P ratio (Redfield et al., 1963) is followed. A low value indicates denitrification. The N** profiles (Fig. 7d) show a broad minimum focused at depths around 100 m, strongly indicating that this water has had its signature influenced by hypoxic conditions, i.e. loss of nitrate when used as electron acceptor during mineralization of organic matter at low oxygen concentration.

More information on the formation history of the high salinity silicate maximum water can be obtained from property versus salinity panels (Fig. 8). The $T - S$ curve show a typical shape for the halocline, with a warmer low-salinity water at $S \approx 31$, followed by a temperature minimum at $32 < S < 33$ and then increasing temperature with salinity to a maximum in the Atlantic Layer, followed by colder water towards the highest salinity in the deep water (Fig. 8a). The temperature minimum has historically been attributed to winter water, often with a signature of brine contribution (e.g. Aagaard et al., 1981; Anderson et al., 2013). This brine enhanced water follows the shelf bottom and gets enriched in organic matter decay products during its flow towards the deep basin. A nearly strait mixing line can be seen in the salinity range from about $S = 34$ to that of the temperature maximum (Fig. 8a), i.e. no T_{min} at the high salinity silicate maximum water. Oxygen profiles, on the other hand, show a clear minimum at $S \approx 34.5$ (Fig. 8b) indicating organic matter remineralization. Comparing the oxygen signature with those of silicate and N** (Fig. 8c and d) reveals interesting information. The broad silicate maximum around $S \approx 33$ has a minimum in N** but no minimum in oxygen even if the concentration is some 100 µmol kg^{-1} below saturation, while the silicate maximum at $S \approx 34.5$ has a clear oxygen minimum but with only a slight minimum in N**. The most plausible explanation for this pattern is that the nutrient maximum at low salinity had a higher oxygen concentration before exposure to organic matter decay at the sediment surface. The waters with $S > 34$ at some stations with lower oxygen and higher silicate concentrations also have lower N** (Fig. 8b,

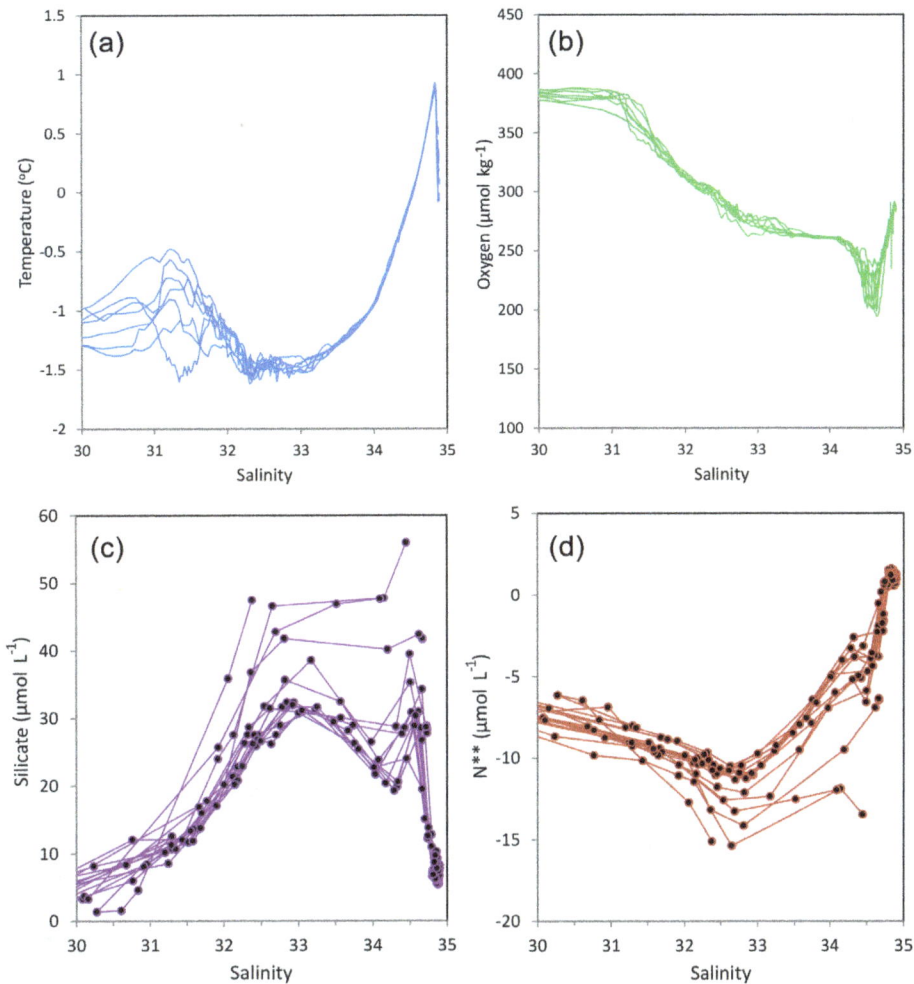

Figure 8. Temperature (**a**), oxygen (**b**), silicate (**c**) and N** (**d**) versus salinity for the stations at sections D and E. The panels (**a**) and (**b**) are from the CTD output, while (**c**) and (**d**) are from water samples analysed.

c and d), indicating less mixing and thus potentially more recent contact with the sediment surface.

The conclusion is that both nutrient maxima are formed in contact with hypoxic sediments, with one maxima at salinity around $S \approx 33$ mainly being formed on the shelf where the preformed water is well oxygenated by interaction with the atmosphere during ice-free periods and ice formation periods with cooling and convection, while the nutrient maximum at $S \approx 34.5$ is formed at the shelf break of more than 100 m depth. Such a scenario is consistent with the SF_6 partial pressure of the silicate maximum at $S \approx 34.5$ being close to those in the deep basin, while that around $S \approx 33$ has a significantly higher level of around 7 ppt (Fig. 5). At section B the maximum AOU is associated with $S \approx 33$ and found at about 50 m depth (Fig. 5b) and at C it is found at around 100 m depth (Fig. 5c). At the latter section the maximum AOU is found at $S \approx 34.5$ associated with the SF_6 partial pressure minimum of around 5 ppt. At the same salinity there is also a weaker minimum in section B at about 75 m depth. In sec-

tions D, E, and F the SF_6 minimum is also found at $S \approx 34.5$ but at a deeper depth of 200 m, all associated with the AOU maximum. However, at these sections the AOU maximum has a SF_6 partial pressure close to that of the water deeper, indicating that the basin water in the Chukchi Abyssal Plain is the source of this high salinity nutrient maximum water. The presence of the high salinity SF_6 minimum at section C, and to a lesser degree at section B, points to the existence of a westward penetration of water at the shelf break. However, this does not need to be a persistent flow, but can be something that occurs intermittently. Strengthening of these concussions are seen in Fig. 9 where the pSF_6 interpolated to a salinity of 34.5 has a strong gradient with increasing partial pressure towards the west and significant higher silicate concentrations at lower pSF_6.

The formation of the silicate maximum at $S \approx 34.5$ on the shelf break is in line with Nishino et al. (2009), who suggested that the silicate maximum at this high salinity was produced by decomposition of opal-shelled organisms along the continental margin. Anderson et al. (2013) showed that this high salinity silicate maximum had a brine content of at least 4 % and that the CTD record had a small temperature minimum signature. This was not the situation in 2014, illustrating that the conditions likely are not constant with time. This water with a salinity of just over 34 has historically been named lower halocline water (Jones and Anderson, 1986), without a signature of any nutrient maximum. Hence, this more recent finding of high silicate concentrations along the continental margin of the East Siberian Sea is either a local or a new phenomenon. The sediment record (Fig. 4) clearly shows that the BSi content is low in the slope off the western part of East Siberian Sea (sections B and C; coring stations 27MC1, 25MC1, 21MC1), making local decomposition in this region unlikely. Further to the east the BSi content increases slightly to the location of sections D and F, with a large increase at the most eastern station in the Herald Canyon (13.5 % BSi at site 5GC1) where opal-shelled organism, primarily diatoms, are abundant in the bottom sediments. This strongly supports a Pacific Ocean source of silicate, but does not exclude that some of the silicate-rich water enters the eastern East Siberian Sea before transformation and escape to the slope and deep central basins. Such a scenario is consistent with the decreasing silicate concentrations to the west in the salinity range 34.3 to 34.7 (Fig. 9b). However, it is not possible to fully elucidate the transport and transformation of silicate from these few sediment profiles, especially since they are also from variable bottom depths (Table 1). Nevertheless, these sediment observations do not contradict occasional westward flow along the shelf break, as suggested by the SF_6 signature.

Variability is also seen in a comparison with historic data. Our section F is on the border to the Chukchi Abyssal Plain, where the Arctic Ocean Section hydrographic program collected a section of data in 1994. In Fig. 10 we compare these two data sets and it is clear that the silicate maximum at $S \approx 34.5$ was more or less absent in 1994. However, at stations with bottom depths ~ 180 m the silicate concentration was close to 20 μmol L^{-1} towards the seafloor, and at the station with bottom depth ~ 250 m, the concentration decreased to 18 μmol L^{-1} towards the seafloor. These are the stations where the salinity does not reach the maximum salinity in Fig. 10a. Hence, there seems to be a signal from the shelf slope that did not penetrate deep into the Chukchi Abyssal Plain. Also N^{**} had relative to the deepest data higher values at $S \approx 34$ except for at the shallowest stations with elevated silicate concentrations (Fig. 10b). Centred at $S \approx 33$ the silicate maximum is higher and the N^{**} minimum is lower in 1994, indicating a stronger contribution of organic matter decay at low oxygen levels from the shelf. There is also an indication of a wider salinity interval of the silicate maximum in 2014 compared to 1994, especially towards the high salinity end (Fig. 10a).

Historically the extent of the nutrient maximum has varied but few studies along the Siberian continental margin

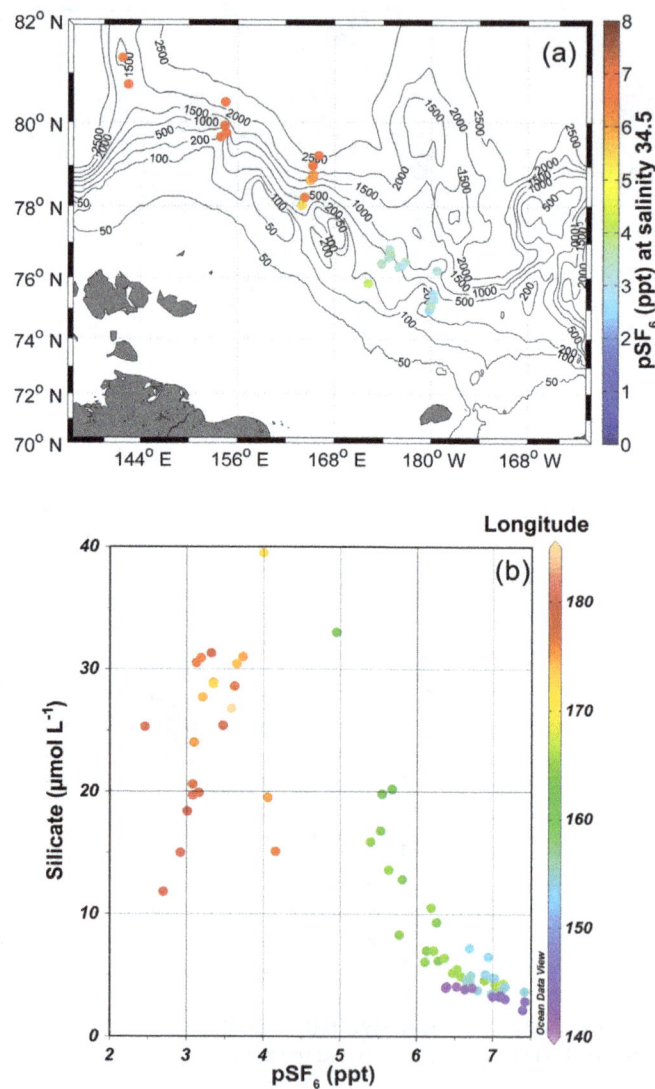

Figure 9. SF_6 partial pressure (ppt) interpolated to the salinity 34.5 (a) and silicate versus pSF_6 in the salinity range 34.3 to 34.7 colour coded by longitude (b).

have been reported. Data from east of 175° E for years between 2001 and 2010 were compiled by Nishino et al. (2013) showing the presence of the nutrient maximum but with some variability in both the maximum concentration and the vertical extent between the years. Our 2014 data show a clear maximum in the Makarov Basin, at section B some 150 nautical miles east of the Lomonosov Ridge at about longitude 153° E, with higher concentrations in the sections further to the east (Fig. 2). As the concentration of silicate generally increases towards the shelf in sections B and C and also increases from section B to C (Fig. 2) it is likely that the source is the local shelf area. Data obtained from RV *Polarstern* in the same area as section B (see Fig. 1b for station locations) during the summer of 1996 (ACSYS 96) did not show any elevated silicate concentrations in the halocline (Fig. 11). This

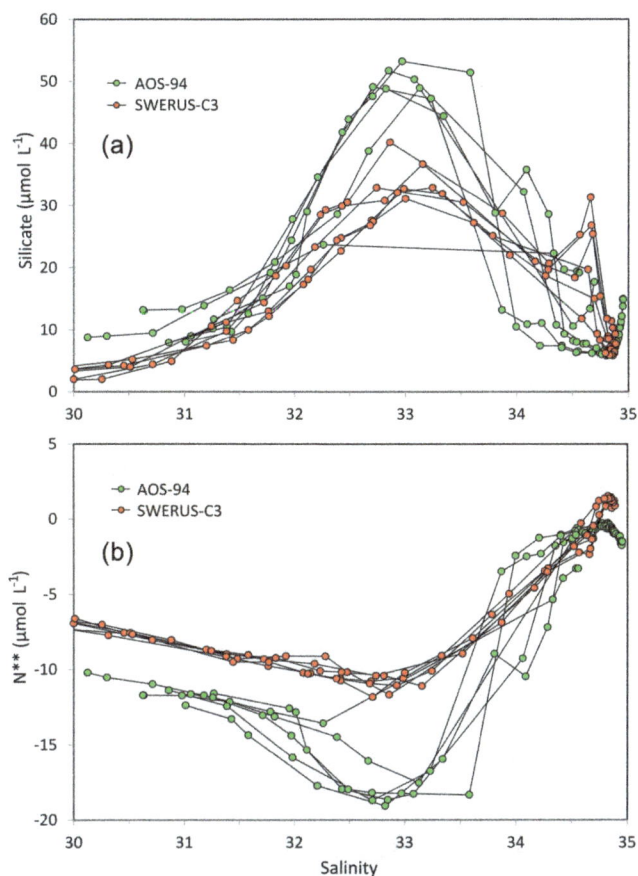

Figure 10. Silicate **(a)** and N** **(b)** versus salinity in the Chukchi Abyssal Plain area; data from the Arctic Ocean Section in 1994 in green (stations marked by green points in Fig. 1b) and from SWERUS-C3 in red (section F in Fig. 1b).

Figure 11. Silicate versus depth for data collected in 1996 (green) and from section B in 2014 (red). The positions of the stations in 1996 are shown by orange points in Fig. 1b.

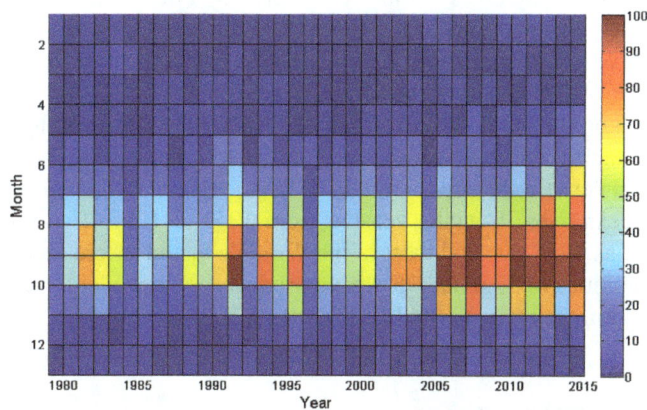

Figure 12. Percentage of ice-free area in the region: latitude 76 to 80° N, longitude 140 to 150° E (framed yellow in Fig. 1b), for each month from 1980 to 2014, evaluated from the passive microwave data of NSIDC (Cavalieri et al., 1996).

could be an effect of either that in 1996 the nutrient-rich water was confined closer to the shelf on its transport to the east, or that the production site of this water has extended further to the west since 1996. We find it most plausible that the latter is the cause as the sea-ice climate has changed significantly over the shelf south of these sections during the last 20 to 30 years (Fig. 12) that potentially have moved the production areas further to the west. Up to about the year 2000 most summers had more than 50 % sea-ice coverage, with a few years with less than 10 %. During the last 10 years the typical conditions for the month of September is more than 90 % open water. All through the record the area is more or less ice covered in November, a situation that lasts until April. Consequently, there has been more sea-ice formation and, thus, brine production in this region during the last 10 years compared to the 1980s and 1990s. When this sea-ice formation is further away from the coast line the initial salinity is probably also higher and thus also the resulting brine.

Indications of shelf plumes penetrating all the way to the deep basin are seen in sections D and F where salinity, silicate, and SF_6 levels increases towards the bottom, a signa-

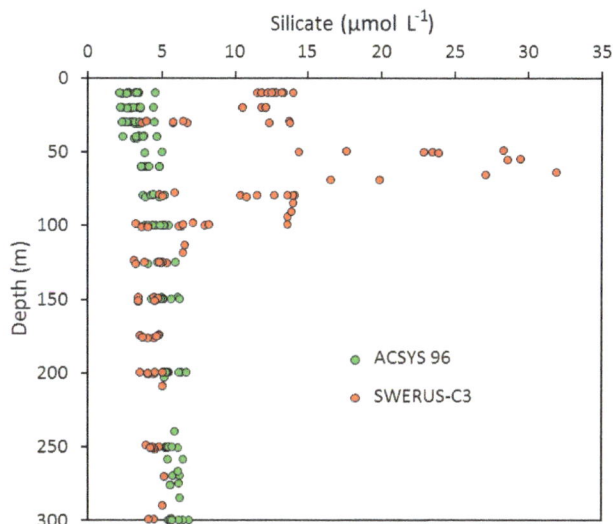

ture that generally fades away down the shelf slope (Fig. 13). This is not observed in the more western sections and is consistent with shelf plumes penetrating down into this eastern region. A rough computation of the fraction of shelf water can be done as follows. With an increase in SF_6 partial pressure from the intermediate to the bottom water of about 0.5 ppt, and the intermediate water and shelf water partial pressures being 1.5 and 8 ppt, respectively, a little less than 10 % of shelf water is needed. This is quite substantial but not unrealistic if matched with the other properties. The shelf water silicate concentration should be $\sim 25\ \mu mol\ L^{-1}$ in order to achieve the observed $2\ \mu mol\ L^{-1}$ increase, and the shelf water salinity would be 36 to get an increase of 0.15. These computations completely ignore mixing and entrain-

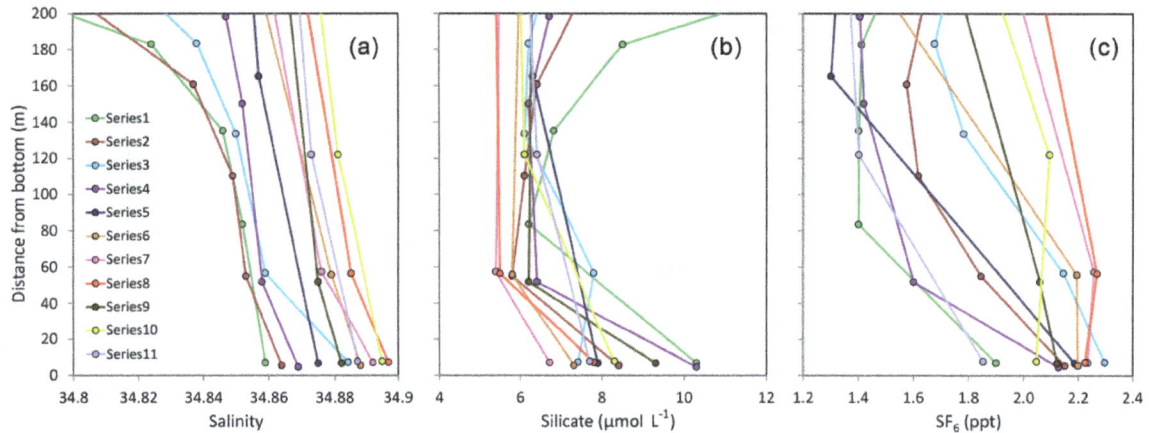

Figure 13. Salinity (**a**), silicate (**b**), and SF$_6$ (**c**) as a function of distance from the bottom for all stations deeper than 400 m in sections D and F. Series 1 to 11 represent the bottom depth, increasing from 483 to 1120 m.

ment, but provide some indication that a shelf water contribution to the deep water of the Chukchi Abyssal Plain is realistic. The realism of the shelf source concentrations is supported by observations and modelling. For instance, silicate concentrations in the range 40 to 60 µmol L^{-1} was observed on the western flank of Herald Canyon in the summer of 2008, even if the salinity was well below 34 (Anderson et al., 2013). Windsor and Björk (2000) used a polynya model driven by atmospheric forcing to compute ice, salt, and dense water production in different regions of the Arctic Ocean over 39 winter seasons from 1959 to 1997. Two regions were east and west of the Wrangle Island where mean salinities of 37.0 and 35.9 were produced, respectively, well within the range needed.

5 Conclusions

We have showed that this region of the Arctic Ocean is much more dynamic and variable than previously reported. Our data collected in the summer of 2014 are consistent with a shelf–basin exchange scenario as summarized in Fig. 14. A boundary current of Atlantic Layer water follows the shelf break from the west to the east, where some of the water crosses the Lomonosov Ridge into the Makarov Basin. This boundary current follows the shelf break to the Alpha Ridge where it turns towards north at its western flank. The water at the corresponding depth in the Chukchi Abyssal Plain has a substantially lower partial pressure of SF$_6$, consistent with a more isolated circulation in this region.

Surface water with substantial input of river water exits the shelf north of the New Siberian Islands to follow the Lomonosov Ridge out into the central basins. High nutrient water with salinity centred at 33 exits the East Siberian Sea from its western end and contributes to the cold halocline of the central Arctic Ocean. Compared to historic data the high nutrient water is found outside the shelf break further to the

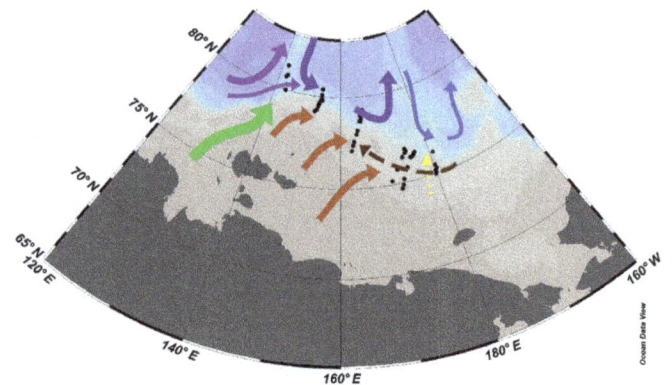

Figure 14. Summary of deduced circulation. Green arrow shows the runoff spreading in the surface out north of the New Siberian Islands. The light brown illustrates the export of nutrient-rich water from the shelf into the deep basin at a salinity of around 33 and the dark brown interrupted line the nutrient-rich water of a salinity around 34.5. The dark blue arrows in the deep basin show the intermediate deep (500–1500 m depth) boundary currents. The yellow dotted line illustrates the deep water plumes off the shelf break. Map drawn using Ocean Data View (Schlitzer, 2017).

west in 2014, which is associated with a lower degree of ice cover during the summer north of the New Siberian Islands. Where the Mendeleev Ridge connects to the shelf slope a water body with salinity around 34.5, elevated nutrient concentrations and low SF$_6$ partial pressure hugs the shelf slope. Water of such property is also found further to the west. As the source of the low SF$_6$ partial pressure most likely is in the Chukchi Abyssal Plain, at least an occasional flow to the west follows, a conclusion that is supported by the surface sediment biogenic silicate (BSi) content. In the eastern study region plumes of high salinity, silicate, and SF$_6$ levels flow off the shelf into the deep basin.

Competing interests. The authors declare that they have no conflict of interest.

Acknowledgements. We thank the supporting crew and Master of I/B *Oden* as well as the support of the Swedish Polar Secretariat.

This research was supported by grants from the Swedish Research Council (contract 621-2013-5105); the Swedish Research Council Formas (project reference 214-2008-1383, A.U. 214-2014-1165); the Swedish Knut and Alice Wallenberg Foundation (KAW); the European Union FP7 project CarboChange (under grant agreement no. 264879). I. Semiletov acknowledge support from the Russian Government (grant no. 14.Z50.31.0012/03/19.2014), the Far Eastern Branch of the Russian Academy of Sciences and the ICE-ARC-EU FP7 project. The transient tracer measurements were supported by the Deutsche Forschungsgemeinschaft in the framework of the "Antarctic Research with comparative investigations in Arctic ice areas" priority program by a grant to T. Tanhua and M. Hoppema; Carbon and transient tracers dynamics: a bi-polar view on Southern Ocean eddies and the changing Arctic Ocean (TA 317 = 5, HO 4680 = 1).

Edited by: T. Tesi

References

Aagaard, K., Coachman, L. K., and Carmack, E. C.: On the halocline of the Arctic Ocean, Deep-Sea Res., 28, 529–545, 1981.

Anderson, L. G., Björk, G., Jutterström, S., Pipko, I., Shakhova, N., Semiletov, I., and Wåhlström, I.: East Siberian Sea, an Arctic region of very high biogeochemical activity, Biogeosciences, 8, 1745–1754, doi:10.5194/bg-8-1745-2011, 2011.

Anderson, L. G., Andersson, P., Björk, G., Jones, E. P., Jutterström, S., and Wåhlström, I.: Source and formation of the upper halocline of the Arctic Ocean, J. Geophys. Res., 118, 410–421, doi:10.1029/2012JC008291, 2013.

Bates, N. R.: Air-sea CO_2 fluxes and the continental shelf pump of carbon in the Chukchi Sea adjacent to the Arctic Ocean, J. Geophys. Res., 111, C10013, doi:10.1029/2005JC003083, 2006.

Cavalieri, D., Parkinson, C., Gloersen, P., and Zwally, H.J.: Sea Ice Concentrations from Nimbus-7 SMMR and DMSP SSM/I-SSMIS Passive Microwave Data, [1996-01-01; 2010-12-31], National Snow and Ice Data Center, Boulder, Colorado USA, Digital media, updated yearly, 1996.

Charkin, A. N., Dudarev, O. V., Semiletov, I. P., Kruhmalev, A. V., Vonk, J. E., Sánchez-García, L., Karlsson, E., and Gustafsson, Ö.: Seasonal and interannual variability of sedimentation and organic matter distribution in the Buor-Khaya Gulf: the primary recipient of input from Lena River and coastal erosion in the southeast Laptev Sea, Biogeosciences, 8, 2581–2594, doi:10.5194/bg-8-2581-2011, 2011.

Clayton, T. D. and Byrne, R. H.: Spectrophotometric seawater pH measurements: total hydrogen ion concentration scale calibration of m-cresol purple and at-sea results, Deep-Sea Res. Pt. I, 40, 2115–2129, 1993.

Codispoti, L. A., Flagg, C., Kelly V., and Swift, J. H.: Hydrographic conditions during the 2002 SBI process experiments, Deep-Sea Res. Pt. II, 52, 3199–3226, 2005.

Conley, D. J.: An interlaboratory comparison for the measurement of biogenic silica in sediments, Mar. Chem., 63, 39–48, doi:10.1016/S0304-4203(98)00049-8, 1998.

Conley, D. J. and Schelske, C. L.: Biogenic Silica, in: Tracking Environmental Change Using Lake Sediments, edited by: Smol, J. P., Birks, H. J. B., Last, W. M., Bradley, R. S., and Alverson, K., Kluwer Academic Publishers, Dordrecht, the Netherlands, 281–293, 2001.

Dethleff, D.: Dense water formation in the Laptev Sea flaw lead, J. Geophys. Res., 115, C12022, doi:10.1029/2009JC006080, 2010.

Dickson, A. G.: Standard potential of the reaction: $AgCl(s) + 1/2H_2(g) = Ag(s) + HCl(aq)$, and the standard acidity constant of the ion HSO_4^- in synthetic sea water from 273.15 to 318.15 K, J. Chem. Thermodyn., 22, 113–127, doi:10.1016/0021-9614(90)90074-Z, 1990.

Haraldsson, C., Anderson, L. G., Hassellöv, M., Hulth, S., and Olsson, K.: Rapid, high-precision potentiometric titration of alkalinity in the ocean and sediment pore waters, Deep-Sea Res. Pt. I, 44, 2031–2044, 1997.

Jakobsson, M.: Hypsometry and volume of the Arctic Ocean and its constituent seas, Geochem. Geophys. Geosys., 3, 1–18, 2002.

Johnson, K. M., Sieburth, J. M., Williams, P. J., and Brändström, L.: Coulometric total carbon dioxide analysis for marine studies: automation and calibration, Mar. Chem., 21, 117–133, 1987.

Jones, E. P. and Anderson, L. G.: On the Origin of the Chemical Properties of the Arctic Ocean Halocline, J. Geophys. Res., 91, 10759–10767, 1986.

Jones, E. P., Anderson, L. G., Jutterström, S., Mintrop, L., and Swift, J. H.: Pacific freshwater, river water and sea ice meltwater across Arctic Ocean basins: Results from the 2005 Beringia Expedition, J. Geophys. Res., 113, C08012, doi:10.1029/2007JC004124, 2008.

Kinney, P., Arhelger, M. E., and Burrell, D. C.: Chemical characteristics of water masses in the Amerasian Basin of the Arctic Ocean, J. Geophys. Res., 75, 4097–4104, 1970.

Legge, O. J., Bakker, D. C. E., Johnson, M. T., Meredith, M. P., Venables, H. J., Brown, P. J., and Lee, G. A.: The seasonal cycle of ocean-atmosphere CO_2 flux in Ryder Bay, west Antarctic Peninsula, Geophys. Res. Lett., 42, 2934–2942, doi:10.1002/2015GL063796, 2015.

Millero, F.: Carbonate constants for estuarine waters, Mar. Freshwater Res., 61, 139–142, doi:10.1071/MF09254, 2010.

Moore, R. M., Lowings, M. G. and Tan, F. C.: Geochemical profiles in the central Arctic Ocean: Their relation to freezing and shallow circulation, J. Geophys. Res., 88, 2667–2674, 1983.

Nishino, S., Shimada, K., Itoh, M., and Chiba, S.: Vertical Double Silicate Maxima in the Sea-Ice Reduction Region of the Western Arctic Ocean: Implications for an Enhanced Biological Pump due to Sea-Ice Reduction, J. Oceanogr., 65, 871–883, 2009.

Nishino, S., Itoh, M., Williams, W. J., and Semiletov, I.: Shoaling of the nutricline with an increase in near-freezing temperature water in the Makarov Basin, J. Geophys. Res.-Oceans, 118, 635–649, doi:10.1029/2012JC008234, 2013.

Pipko, I. I., Semiletov, I. P., Tishchenko, P. Ya., Pugach, S. P., and Christensen, J. P.: Carbonate chemistry dynamics in Bering Strait and the Chukchi Sea, Prog. Ocean., 55, 77–94, 2002.

Proshutinsky, A., Krishfield, R., Timmermans, M.-L., Toole, J., Carmack, E., McLaughlin, F., Williams, W. J., Zimmermann, S., Itoh, M., and Shimada, K.: Beaufort Gyre freshwater reservoir: State and variability from observations, J. Geophys. Res., 114, C00A10, doi:10.1029/2008JC005104, 2009.

Redfield, A. C., Ketchum, B. H., and Richards, F. A.: The influence of organisms on the composition of sea water, in: The Sea, Vol. 2., edited by: Hill, M. N., Wiley, New York, USA, 26–77, 1963.

Rudels, B.: Arctic Ocean circulation and variability – advection and external forcing encounter constraints and local processes, Ocean Sci., 8, 261–286, doi:10.5194/os-8-261-2012, 2012.

Rudels, B., Jones, E. P., Anderson L. G., and Kattner, G.: On the intermediate depth waters of the Arctic Ocean, in: The Polar Oceans and Their Role in Shaping the Global Environment, edited by: Johannessen, O. M., Muench, R. D., and Overland, J. E., American Geophysical Union, Washington, D.C., USA, 33–46, 1994.

Rudels, B., Anderson, L. G., and Jones, E. P.: Formation and evolution of the surface mixed layer and halocline of the Arctic Ocean, J. Geophys. Res., 101, 8807–8821, 1996.

Schauer, U., Rudels, B., Jones, E. P., Anderson, L. G., Muench, R. D., Björk, G., Swift, J. H., Ivanov, V., and Larsson, A.-M.: Confluence and redistribution of Atlantic water in the Nansen, Amundsen and Makarov basins, Ann. Geophys., 20, 257–273, doi:10.5194/angeo-20-257-2002, 2002.

Schlitzer, R.: Ocean Data View, available at: http://odv.awi.de, last access: February 2017.

Semiletov, I., Pipko, I., Gustafsson, O., Anderson, L. G., Sergienko, V., Pugach, S., Dudarev, O., Charkin, A., Gukov, A., Broder, L., Andersson, A., Spivak, E., and Shakhova, N.: Acidification of East Siberian Arctic Shelf waters through addition of freshwater and terrestrial carbon, Nat. Geosci., 9, 361–365, doi:10.1038/ngeo2695, 2016.

Semiletov, I. P., Shakhova, N. E., Pipko, I. I., Pugach, S. P., Charkin, A. N., Dudarev, O. V., Kosmach, D. A., and Nishino, S.: Space-time dynamics of carbon and environmental parameters related to carbon dioxide emissions in the Buor-Khaya Bay and adjacent part of the Laptev Sea, Biogeosciences, 10, 5977–5996, doi:10.5194/bg-10-5977-2013, 2013.

Stöven, T. and Tanhua, T.: Ventilation of the Mediterranean Sea constrained by multiple transient tracer measurements, Ocean Sci., 10, 439–457, doi:10.5194/os-10-439-2014, 2014.

Stöven, T., Tanhua, T., Hoppema, M., and Bullister, J. L.: Perspectives of transient tracer applications and limiting cases, Ocean Sci., 11, 699–718, doi:10.5194/os-11-699-2015, 2015.

Stöven, T., Tanhua, T., Hoppema, M., and von Appen, W.-J.: Transient tracer distributions in the Fram Strait in 2012 and inferred anthropogenic carbon content and transport, Ocean Sci., 12, 319–333, doi:10.5194/os-12-319-2016, 2016.

Swift, J. H., Jones, E. P., Carmack, E. C., Hingston, M., Macdonald, R. W., McLaughlin, F. A., and Perkin, R. G.: Waters of the Makarov and Canada Basins, Deep-Sea Res. Pt. II, 44, 1503–1529, 1997.

van Heuven, S., Pierrot, D., Lewis, E., and Wallace, D. W. R.: MATLAB Program developed for CO_2 system calculations, version 1.1 2011, ORNL/CDIAC-105b. Carbon dioxide information analysis center, Oak Ridge National Laboratory, U.S. Department of Energy, Oak Ridge, USA, 2011.

Winsor, P. and Björk, G.: Polynya activity in the Arctic Ocean from 1958 to 1997, J. Geophys. Res., 105, 8789–8803, 2000.

Seabirds as samplers of the marine environment – a case study of northern gannets

Stefan Garthe[1], Verena Peschko[1], Ulrike Kubetzki[1,2], and Anna-Marie Corman[1]

[1]Research & Technology Centre (FTZ), Kiel University, Hafentörn 1, 25761 Büsum, Germany
[2]Department of Animal Ecology and Conservation, Biocentre Grindel, Hamburg University, Martin-Luther-King Platz 3, 20146 Hamburg, Germany

Correspondence to: Stefan Garthe (garthe@ftz-west.uni-kiel.de)

Abstract. Understanding distribution patterns, activities, and foraging behaviours of seabirds requires interdisciplinary approaches. In this paper, we provide examples of the data and analytical procedures from a new study in the German Bight (North Sea) tracking northern gannets (*Morus bassanus*) at their breeding colony on the island of Heligoland. Individual adult northern gannets were equipped with different types of data loggers for several weeks, measuring geographic positions and other parameters mostly at 3–5 min intervals. Birds flew in all directions from the island to search for food, but most flights targeted areas to the (N)NW (north–northwest) of Heligoland. Foraging trips were remarkably variable in duration and distance; most trips lasted 1–15 h and extended from 3 to 80 km from the breeding colony on Heligoland. Dives of gannets were generally shallow, with more than half of the dives only reaching depths of 1–3 m. The maximum dive depth was 11.4 m. Gannets showed a clear diurnal rhythm in their diving activity, with dives being almost completely restricted to the daylight period. Most flight activity at sea occurred at an altitude between the sea surface and 40 m. Gannets mostly stayed away from the wind farms and passed around them much more frequently than flying through them. Detailed information on individual animals may provide important insights into processes that are not detectable at a community level.

1 Introduction

Seabirds are marine animals that live mostly at or near the air–water interface. The dynamics of both these media may consequently have a strong influence on the ecology of seabirds (Schneider, 1991). Many studies in the world's oceans have shown that the physical environment has a substantial influence on seabird distributions (e.g. Briggs et al., 1987; Hunt Jr., 1990). Physical processes are particularly relevant to seabirds when they cause predictable prey aggregations, either regular or irregular. However, other opportunities (such as fisheries' discards; e.g. Ryan and Moloney, 1988; Garthe et al., 1996) and constraints (such as the need to breed on land; e.g. Schneider and Hunt Jr., 1984; Wilson et al., 1995a) may also influence seabird distributions and related behaviours. An essential feature in the marine system is "scale". Quantitative relations between abiotic and biotic variables are strongly influenced by the scale at which they are measured (Schneider, 1994). Thus, general seabird distribution patterns often correspond best with physical phenomena at large scales, whereas smaller-scale patterns are associated with biological features such as foraging range, social interactions, and prey availability (Schneider and Duffy, 1985; Hunt Jr. and Schneider, 1987). Most behaviours at sea are directly related to foraging (e.g. searching, feeding) or the result of foraging-related constraints (waiting for food to become available, digesting, commuting). Several external and internal characteristics and limitations influence foraging activities, e.g. diurnal rhythms, flight manoeuvrability, feeding techniques, prey-detection capabilities, social attractions, learning and age-dependent skills, foraging ranges, and dietary preferences (Furness and Monaghan, 1987; Shealer, 2002).

For decades, studies of seabird biology were mainly land based, with a particular focus on the breeding period. Although there were understandable logistic reasons for this, it has led to severe biases in our understanding of seabird ecology. Two subsequent approaches focusing on the behaviours of seabirds at sea have allowed significant progress in our understanding of seabird ecology. One such approach involved studying seabird distributions at sea from boats. Whereas early work was targeted towards establishing the distribution patterns of seabirds (e.g. Brown, 1986; Tasker et al., 1987), later studies concentrated on improving our understanding of the underlying factors, including habitat parameters, mainly hydrographic features measured synoptically at sea or by remote techniques, and food availability, assessed by detecting and possibly quantifying prey at sea (e.g. Hunt Jr. et al., 1998; Davoren et al., 2003; Jahncke et al., 2005). The second approach was to equip seabirds with telemetric devices and/or data-logging units (e.g. Jouventin and Weimerskirch, 1990; Wilson et al., 2002; Wilson and Vandenabeele, 2012). These devices record the bird's position and/or other parameters, such as temperature and depth, while the bird is at sea. Because seabirds are fast-moving and wide-ranging animals, this approach also enables us to study them in logistically inaccessible areas. Furthermore, it allows information on individual birds to be collected, in contrast to boat-based observations, which involve larger samples of birds but where individuals cannot be tracked over larger areas or time spans.

Understanding patterns in distributions, activities, and foraging behaviours of seabirds requires interdisciplinary approaches. The physical properties of the sea establish the basic habitat parameters with which both the seabirds and their prey have to cope, while biological conditions influence the birds' food supply (e.g. by prey behaviour) and foraging behaviours. Furthermore, anthropogenic activities may also influence different aspects of the marine environment, both directly on individual seabirds, and indirectly by affecting habitat conditions and prey availability. A combination of these methodological and conceptual approaches will further improve our understanding of the ecology of seabirds within the study area.

In this paper, we provide an overview of a new study in the German Bight (North Sea) connected to the Coastal Observing System for Northern and Arctic Seas (COSYNA) network. We started tracking northern gannets (*Morus bassanus*) at their breeding colony on the island of Heligoland in 2014 (Garthe et al., 2017). Gannets were selected as they have the largest foraging ranges of all abundant seabird species on Heligoland and are large animals that can carry various types of data loggers. Here, we provide examples of the data and analytical procedures based on selected data sets from 2015, and explain the value and perspectives of such studies, especially in relation to coastal observation systems such as COSYNA.

Figure 1. Breeding colony of northern gannets on the island of Heligoland. Photo: S. Garthe.

Figure 2. Flying northern gannets with a Bird Solar GPS logger attached to the tail feathers. Photo: K. Borkenhagen.

2 Methods

2.1 Field work

Field work was conducted on the island of Heligoland (54°11′ N, 7°55′ E) in the southeastern North Sea (Fig. 1). A total of 14 adult northern gannets that were either incubating or rearing chicks were caught on 12–13 May, 17–18 June, or 22–23 July 2015. All birds were equipped with data loggers. A total of 10 gannets each received a Bird Solar GPS logger (e-obs GmbH, Munich, Germany) and the other four birds were equipped with both a CatLog-S GPS logger (Catnip Technologies, Hong Kong SAR, China) and a precision temperature–depth (PTD) logger (Earth and Ocean Technologies, Kiel, Germany). All loggers were attached to the base of the four central tail feathers using TESA® tape (Beiersdorf AG GmbH, Hamburg, Germany; Fig. 2). Data obtained from these loggers covered durations of 0.4–10.9 weeks.

The total masses of the attached devices (including sealing, base plate, and tape) were about 48 g (Bird Solar) and 64 g (CatLog-S plus PTD), representing 1.5 and 1.9 %, respectively, of the mean gannet body mass of 3286 g (Wanless and Okill, 1994). This is well below the potential threshold of 3 % (Phillips et al., 2003; see Vandenabeele et al., 2012). Although attachments to the tail may have a negative influence on flight behaviour (Vandenabeele et al., 2014), most pairs successfully incubated their eggs and/or raised their chicks, similar to non-handled nests, with no visible effects on bird behaviour.

2.2 Technology

Bird Solar GPS logger

These loggers recorded date, time, position (latitude, longitude), ground speed, heading and acceleration. The sampling interval was mostly set to 3–5 min, and the triaxial accelerometer to 0.25–3 min. The onboard memories were either 32 or 64 MB. The outer diameters of the devices were $63 \times 22 \times 16$ mm, plus base plates and an antenna of 76 mm for data transfer. Data could be downloaded remotely using a hand-held device when approaching the birds in the colony.

2.3 CatLog-S GPS logger

These devices recorded date, time, and position (latitude, longitude) and were set at an interval of 5 min. Dimensions varied slightly according to battery type but were about $50 \times 35 \times 8$ mm. The plate was encased by a heatable plastic housing. Data were retrieved by recapturing the bird and downloading data from the device.

PTD loggers

These loggers had 2 MB onboard memory and measured date, time, pressure, and internal and external temperatures (Earth and Ocean Technologies). Temperature measurements were obtained from an external, fast-responding, temperature sensor that allowed sampling of the water column with minimal time lag in thermal signals (temperature-response time $T0.9$ (i.e. time to reach 90 % ΔT, following a temperature change) of approximately 1.8 s (Daunt et al., 2003). The streamlined lightweight carbon-fibre composite casing (outer diameter 19 mm, length 80 mm) weighed about 23 g. Recording intervals for temperature and pressure were set at 3 s. Data could be retrieved by recapturing the bird and downloading the data from the device.

2.4 Northern gannets

The northern gannet is the largest seabird species in the North Atlantic. It has a body mass of 2.3–3.6 kg and breeds in colonies of up to several tens of thousands of pairs. Northern gannets spend their entire life at sea, except for breeding on land. They usually start breeding at 5–6 years old, and may live to 20 years or older. They lay one egg that is incubated for 6 weeks, followed by a chick-rearing period of about 13 weeks. Only one adult of the pair is usually at the nest at any one time during incubation or chick guarding, while the other is at sea (Nelson, 2002; Bauer et al., 2005). Apart from short flights to collect nesting material or due to disturbance/interactions at the nest site, gannets carry out foraging trips to collect food for themselves and their offspring. They usually forage using so-called plunge dives, which are initiated when flying (and searching) at a few to several tens of metres above the sea surface (Nelson, 2002; Garthe et al., 2014). Two different dive types can be distinguished in this species (Garthe et al., 2000). U-shaped dives occur when the birds remain at a largely constant depth for a period after plunging into the water, with little vertical movement, before returning to the sea surface. In contrast, V-shaped dives are usually short and shallow, with the ascent almost immediately following the descent.

Northern gannets have recently been studied intensively by satellite telemetry and data loggers in various places (e.g. Hamer et al., 2001; Pettex et al., 2012; Wakefield et al., 2013), thus allowing comparisons among regions and populations.

3 Products and analyses

3.1 Flight patterns

Figure 3a and b show the flight patterns of two adult northern gannets that were typical of 13 of the 14 individuals tracked in 2015. Birds flew in all directions from the island to search for food, but most flights targeted areas to the (N)NW (north–northwest) of Heligoland. Foraging trips (defined in this paper as absences from the nest site of at least 20 min and of at least 2.0 km direct distance) were remarkably variable in duration and distance; most trips lasted 1–15 h and extended from 3 to 80 km from the breeding colony on Heligoland. One individual's behaviour differed from that of the other gannets by repeatedly flying far north to forage in the Skagerrak (Fig. 3c). These long-distance foraging trips were almost identical in their structures ($n = 3$; duration of 44.3–59.3 h, most distant location 375–388 km away, total distance of 971–1019 km flown) and were interspersed with "normal" foraging trips into the German Bight.

3.2 Diving behaviour

Dives of gannets breeding on Heligoland and foraging in the (south) eastern North Sea were generally shallow, with more than half of the dives only reaching depths of 1–3 m (Fig. 4). The maximum dive depth was 11.4 m, and the median dive depth was 2.2 m ($n = 4$ individuals, $n = 2577$ dives). Most dives were V-shaped, though the measuring interval of 3 s did not allow a precise determination of the proportions of

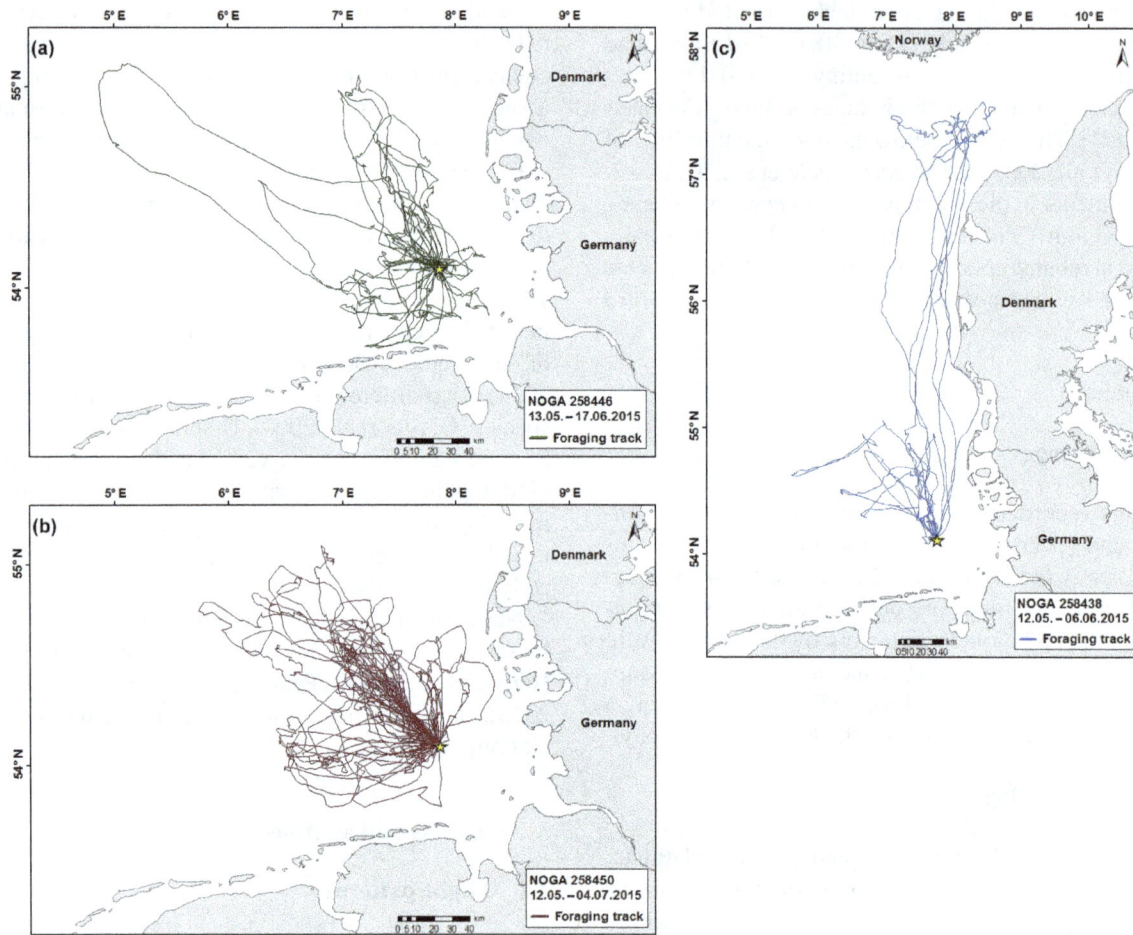

Figure 3. Flight patterns of three northern gannets (NOGAs) breeding on Heligoland in 2015. Birds were tracked over 8 (**a**), 5 (**b**), and 3.5 (**c**) weeks, respectively.

U- and V-shaped dives (see Garthe et al., 2000). The measuring interval of 3 s might also mask the deepest parts of some dives and may thus underestimate dive depth in general. We therefore re-analysed a random sample of 100 dives from Garthe et al. (2014), where gannets exhibited a similar high percentage of V-shaped dives, using both 1 and 3 s intervals (10 individuals, 10 dives each). Scaling down to 3 s missed 10 % of the dives, while the median-detected dive depth was only slightly smaller (4.3 vs. 4.5 m). These subtle differences demonstrate the validity of 3 s measuring intervals to determine the dive-depth pattern.

These dives were shallow compared with previous studies conducted in the northwestern North Sea (Lewis et al., 2002), the English Channel (Grémillet et al., 2006), the northwest Atlantic (Garthe et al., 2000), and the Gulf of Saint Lawrence, Canada (Garthe et al., 2007). Although the sample sizes of individuals are small, this does not hold true for the number of days the birds were tagged and the number of dives. Dive depths recorded from Heligoland gannets in 2015 remained much shallower even when subsampling small data

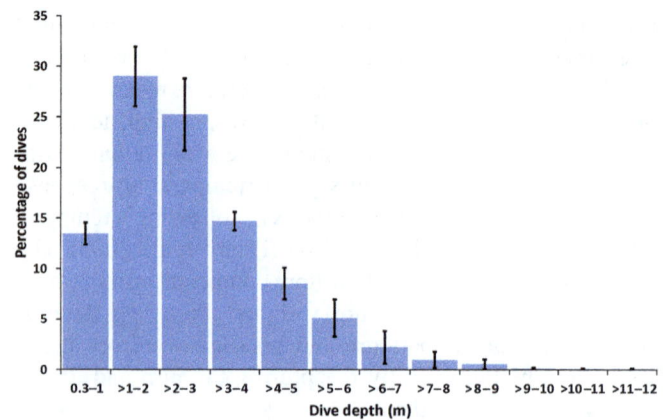

Figure 4. Dive depths of northern gannets in 2015. Data are based on 2557 dives recorded from four adults breeding on Heligoland. Vertical bars show values averaged over the four individuals; extended lines show standard errors. Immersions of < 0.3 m were excluded as potentially indicating bathing and other behaviours.

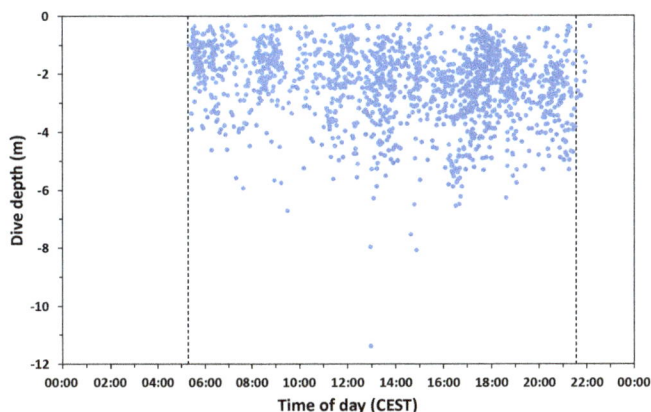

Figure 5. Diurnal rhythm in diving activity of northern gannets in the German Bight in 2015. Data are for the period 12–31 May 2015 and are based on dive recordings of three adults breeding on Heligoland. Each dot represents one dive, showing the maximum depth during the dive. Vertical dashed lines indicate sunrise and sunset for the median day of the period covered.

sets from a large database from eastern Canada (S. Garthe et al., unpublished data).

Gannets showed a clear diurnal rhythm in their diving activity (Fig. 5). Dives were almost completely restricted to the daylight period, with the remaining dives occurring around dawn and dusk. No dives were recorded between 22:09 and 05:21 Central European Summer Time (CEST). This pattern fits well previous studies (Garthe et al., 2000, 2003).

3.3 Habitat analyses

Dive positions were analysed for the fixed habitat parameters including distance to colony (Heligoland), water depth, and distance to nearest land (except for Heligoland; Fig. 6). Almost two-thirds of the dives were carried out at a distance of less than 50 km from the colony, with proportions declining further away from the colony. However, at the largest distances, proportions of dives increased again, strongly indicating that gannets may have specifically targeted such distant foraging areas (see also Sect. 3.5, Fig. 9). As related to water depth, gannets from Heligoland were diving most often in waters of 20–40 m depths, less often in shallower, and rarely in deeper waters (Fig. 6). For foraging, gannets mostly stayed away from the coast, with the highest proportions at a distance of 40–60 km. This pattern differs completely from studies in eastern Canada where gannets were found to concentrate their diving efforts on the coastal zone (Garthe et al., 2007). Both Heligoland (located approximately 43 km north of the East Frisian Islands) and Funk Island (Newfoundland, Canada; located approximately 50 km away from the coast) have a similar placement and, in consequence, the location of the colonies cannot explain the observed difference in coastal focus. However, the near-coastal waters in the Canadian study sites are characterised by much more marine con-

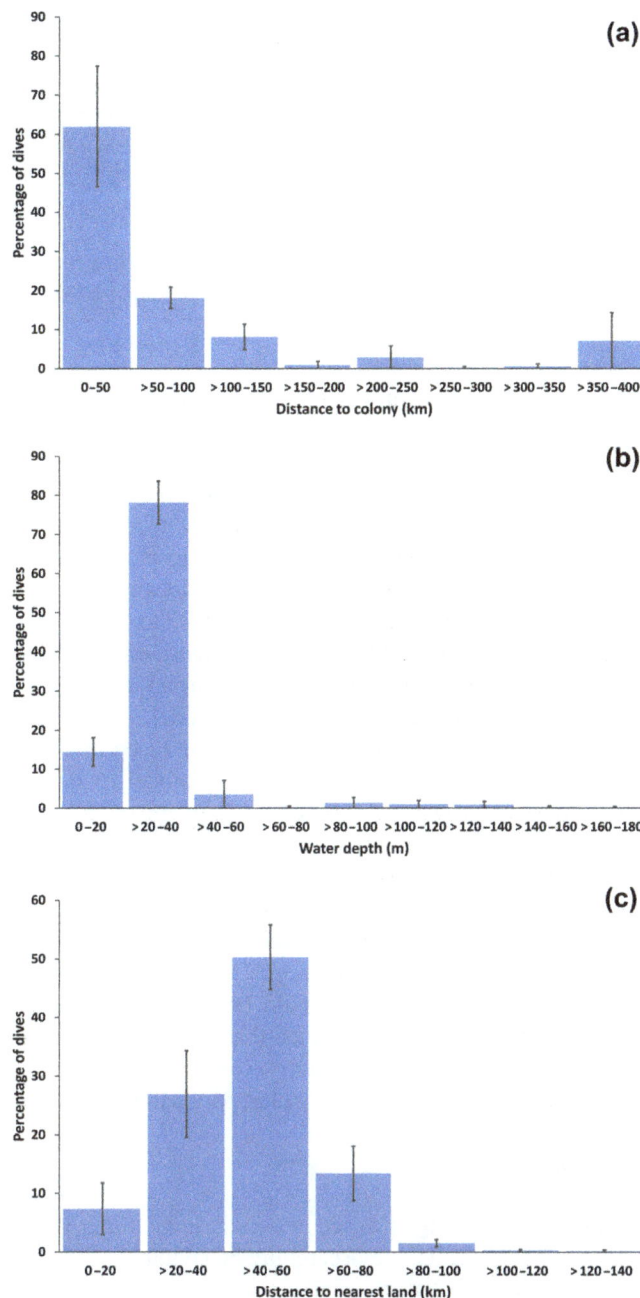

Figure 6. Habitat relationships (upper graph: distance to colony; middle graph: water depth; lower graph: distance to nearest land) of diving northern gannets in 2015. Data are based on 2557 dives recorded from four adults breeding on Heligoland. Vertical bars show mean values averaged over the four individuals; extended lines show standard errors.

ditions and larger water depths compared to the Wadden Sea coast with extended shallow waters in the German Bight.

Two Gaussian linear mixed models were used to analyse the impact of the habitat parameters including distance to colony, water depth, and distance to nearest land on (1) the dive depth and (2) the dive duration of the tagged birds (using

Table 1. Linear mixed model of two key dive parameters (dive depth and dive duration) and their possible explanation by the three fixed habitat variables (distance to colony, water depth, and distance to nearest land). AIC indicates Akaike's information criterion. LRT indicates the likelihood ratio test. Significant results are shown in bold.

	Dive depth			Dive duration		
	AIC	LRT	p	AIC	LRT	p
Full model	4314.1		531.2			
Distance to colony	4312.4	0.265	0.606	531.9	2.747	0.097
Water depth	4312.9	0.760	0.383	537.4	8.174	**0.004**
Distance to nearest land	4315.2	3.102	0.078	530.5	1.342	0.247

R version 3.3.2; R Development Core Team, 2016; package "lme4", function "lmer" by Bates et al., 2015). Both response variables were log transformed to approach normality. The three habitat parameters were used as numeric explanatory variables. Dive ID nested within bird ID was taken as a random factor to avoid pseudo-replication due to multiple measurements per bird. As the function lmer does not provide p values, we further used the function "drop1" to find the relevant habitat parameter explaining the variance of dive depth and/or dive duration. This function tests every term in the model as if it was the last entering the model. In turn, every term in the model is omitted, and the reduced model is then (automatically) compared to the full model by a likelihood ratio test under 1 degree of freedom (i.e. a so-called marginal frequentist F test; Korner-Nievergelt et al., 2015). Though the sample size of individuals was low, the temporal coverage (12–18 days) and the number of dives per individual (381–773) were high. In this data set, dive depth could not be explained statistically by any of the three fixed habitat variables, while dive duration could be explained by water depth (Table 1). It is to be expected that larger data sets that will be collected in the future may exhibit more significant relationships to these and other habitat parameters.

3.4 Flight altitudes

The altitudes of flying birds are important in relation to their migratory movements, prey-searching behaviour, and potential overlap with technical installations at sea.

The Bird Solar GPS loggers calculated height above the ellipsoid when connecting with satellites during positional fixes, and altitude measurements were therefore corrected for geoid height (39.1 m at colony location). Altitude estimates are improved when connection time to the satellite is increased (e.g. Corman and Garthe, 2014), and we therefore used pulses of GPS fixes over 11–15 s and analysed the last and assumed best altitude measurement from each pulse. Figure 7 shows non-smoothed altitude measurements for one foraging trip of 22.7 h. Colony attendance was derived from positional fixes and known nest position; on-water periods were determined from ground speed (<3 km h^{-1}) and positional fixes. Although values fluctuated slightly even for fixed places, such as the nest site in the colony and the sea surface,

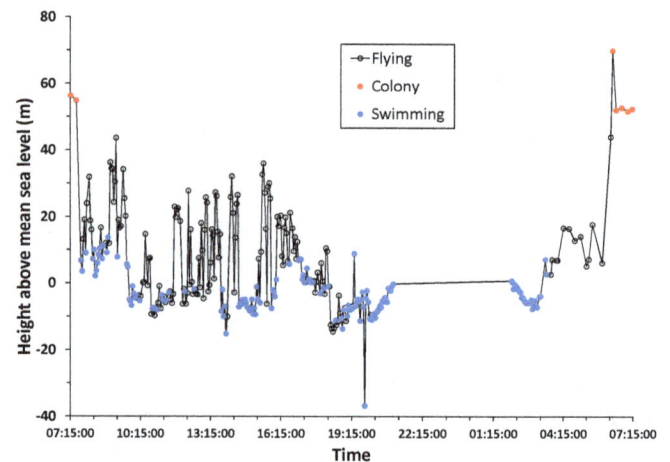

Figure 7. Altitude measurements for a northern gannet tracked in summer 2015 on Heligoland. Altitude measurements are related to activities "in colony" (before and after the 22.7 h foraging trip), swimming, and flying. For details, see text. Please note that this device was switched off during the core darkness hours to save energy, and because birds are known to either stay at the nest site or rest at the sea surface during this period (as shown here).

the measurements appeared reasonable and showed that most flight activity occurred at an altitude between the sea surface and 40 m, with maximum values in this study for when birds were commuting to/from the colony. Measurements in the colony and on water can be used to calibrate altitude measurements because of their relatively well-known heights.

Flight heights of gannets have also been determined by radar measurements (e.g. Krijgsveld et al., 2011), visual observations (e.g. Johnston et al., 2014), and pressure sensors (Garthe et al., 2014; Cleasby et al., 2015). Overall, flight heights of gannets tend to be low, with relatively few flights above 50 m and very few recorded above 100 m, though no comprehensive analysis has yet been published.

3.5 Behavioural patterns

Animal movements can be tracked using motion sensors. Many data loggers contain accelerometers that ideally cover all three axes (x, y, z), and frequent measurements allow be-

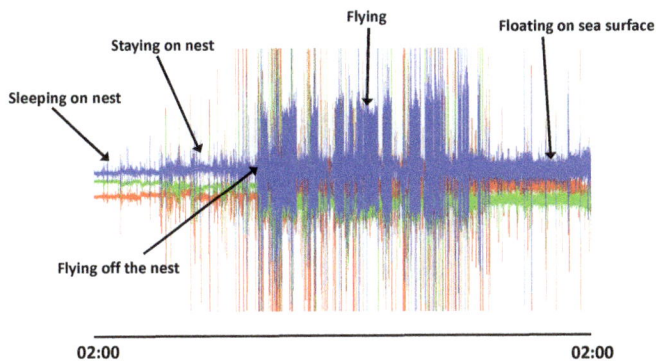

Figure 8. Accelerometer measurements for a northern gannet breeding on Heligoland. This example shows the values for the three different axes (red indicates x, green indicates y, blue indicates z) over 24 h, from 02:00 to 02:00 CEST on the next day. Different activities are indicated by arrows; higher peaks indicate greater movement.

havioural differentiation at a fine scale (e.g. Sakamoto et al., 2009).

Figure 8 shows an example of accelerometer measurements of a northern gannet at a 0.25 min interval over 24 h. Recordings start when the bird is on its nest, with very little activity, obviously sleeping. After about 3 h, the bird remains on its nest but its activity increases, coinciding with dawn. A few hours later, the bird leaves the nest and flies off, followed by a period of about 10 h of mostly flying, interrupted by a few shorter swimming periods. Towards the end of the recording period, the bird settles down on the sea surface and remains floating there overnight (Fig. 8). Such information is important in many ways. It may help identifying the relevance of certain sea areas, i.e. whether areas are used for foraging or just for resting, or for long(er)-distance movements. Quantifying birds' activities is a well-established tool to measure energy expenditure. Such energy budgets may, e.g. help unravelling seabird movement strategies as has been shown by Garthe et al. (2012) for northern gannets wintering in different regions of the northeast Atlantic. Analyses of these kinds will be done for more birds in a separate study.

To determine the relevance of certain sea areas and to improve our understanding of the flight patterns of the birds, it is necessary to know when and where the birds are feeding. Because gannets almost always obtain food by plunge diving, observing dives provides a good proxy for determining feeding areas. Figure 9 shows the flight tracks and dive locations of the gannet that flew repeatedly towards the Skagerrak. It shows that the gannet was foraging intensively in the Skagerrak, while longer passages on outbound and inbound flights were long-distance flights without much foraging activity.

3.6 Overlap with human pressures

The ability to track seabird movements at small spatial and temporal scales makes it possible to study the potential im-

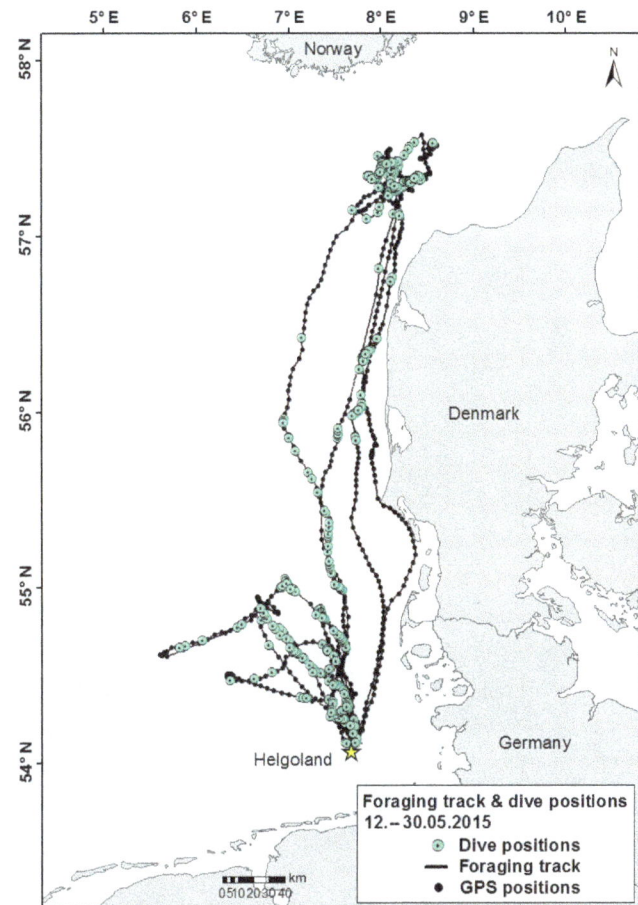

Figure 9. Foraging tracks and dive positions for the northern gannet shown in Fig. 3c. Please note that the data set is smaller than in Fig. 1c because only the synoptic GPS and pressure data are shown (the memory of the PTD logger was full after about 18 days).

pacts of human activities at sea comprehensively. A total of 12 offshore wind farms have been built and became operational in the German Bight between 2008 and November 2016, and a further 5 are currently under construction. Another 15 wind farms have been given consent, and several tens more have been applied for. The impact of wind farms on seabirds, which is currently a hot topic in conservation biology and environmental policy (e.g. Furness et al., 2013; Masden et al., 2015), can thus be studied comprehensively in German North Sea waters. In 2014, gannets were tracked near existing wind farms for the first time. All three individuals largely avoided the wind farm area north of Heligoland (Garthe et al., 2017).

The flight tracks of the gannets shown in Fig. 3 were projected on top of the wind farms that were operational or under construction during the tracking period. The three gannets mostly stayed away from the wind farms and passed around them much more frequently than flying through them (Fig. 10). Wind farms further from Heligoland were not entered, but gannets visited the areas around them. Focusing on

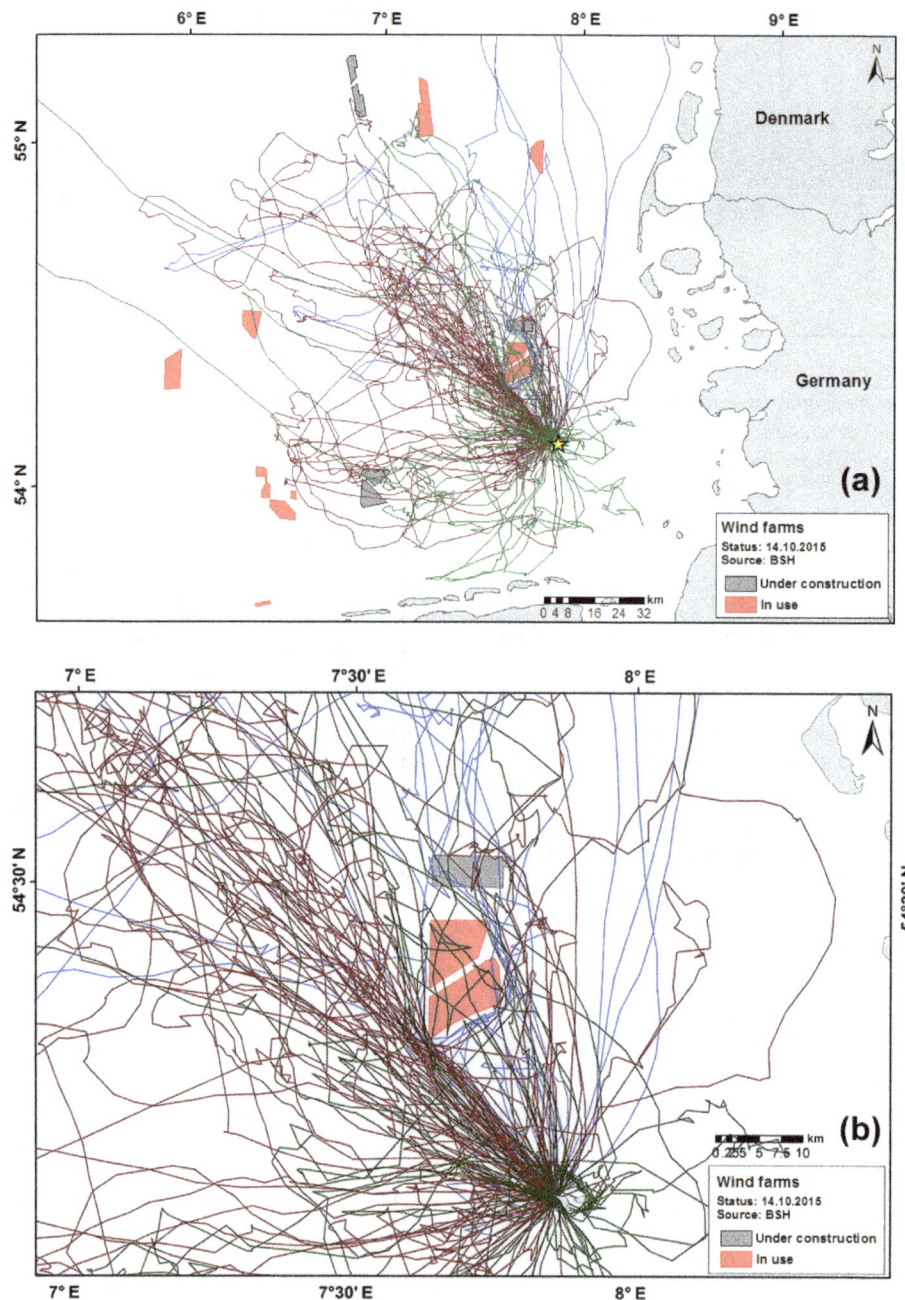

Figure 10. Overlap of flight patterns for the three northern gannets shown in Fig. 3 with the locations of wind farms in the German Bight. Information on the location and status of wind farms was collated from the Federal Maritime and Hydrographic Agency (BSH, personal communication) and the Global Offshore Wind Farms Database (http://www.4coffshore.com/offshorewind/). The upper graph shows the whole German Bight; the lower graph shows the area with the three wind farms near Heligoland only.

the three wind farms north of Heligoland, 5 of the 14 gannets tracked in 2015 did not enter them, 4 only flew into the wind farms once, while 4 visited them occasionally and 1 frequently.

4 Conclusions and perspectives

Tracking free-living animals such as seabirds can open up new dimensions in biological, ecological, and environmental research (Kays et al., 2015). The latest developments in microelectronics can even provide real-time data transfer through mobile-phone networks (e.g. Gilbert et al., 2016). Information collected by data loggers can be used for various purposes, including applied topics, such as assessing the possible effects of wind farms, as well as fundamental research. In a review of offshore wind-farm studies, Bailey et al. (2014) concluded that traditional visual surveys of birds and mammals from ships and aircraft were unlikely to have enough power to detect changes in behaviour or fine-scale spatial or temporal shifts in distribution, given that observers can only be in one place at a time and can only reliably survey in calm sea conditions during daylight hours. Other techniques such as GPS tracking are thus likely to provide more useful data in many cases (Bailey et al., 2014).

Substantial added-value information can be retrieved by combining geographic-position information with other parameters. For birds feeding under water, pressure sensors are essential to characterise foraging areas, allowing diving activity to be described comprehensively (e.g. Boyd, 1997; Ronconi and Burger, 2011). Pressure sensors and/or high-rate GPS measurements can also be used to estimate flight heights (Corman and Garthe, 2014; Scales et al., 2014). Further detailed behavioural and energetic information can be derived from three-dimensional accelerometer measurements (Gómez Laich et al., 2008; Sakamoto et al., 2009), making this a topical research interest.

To understand the distributions of food-searching seabirds and their variation over time, analysing the birds' habitat choice is an important and promising approach. While some habitat parameters may be collected by the loggers on the birds directly (e.g. sea surface temperatures; Wilson et al., 1995b), a full set of variables can only be derived from a combination of remote-sensing and in situ measurements. In most studies tracking seabirds, sea surface temperature and chlorophyll have been analysed and compared to bird distributions, often with limited success (e.g. Grémillet et al., 2008). In future activities of our study, we will make use of the project consortium COSYNA (Baschek et al., 2016) which provides comprehensive and relevant information on important habitat variables. We expect that fixed-point measurements may be particularly valuable for studying seasonal and/or annual variability of the foraging behaviour and distributions of northern gannets, while moving and remote-sensing platforms may be best used to unravel the spatial distribution of the birds at any time. The advantage of COSYNA in this context will be the variety of measured variables as well as the three-dimensional measurements so that stratification can be assessed which would not be available from remote-sensing sources (Baschek et al., 2016). Furthermore, the generation of models may prove particularly valuable for analysing and possibly predicting the distribution of seabirds (Breitbach et al., 2016; Stanev et al., 2016). Finally, information on the marine environment may also be generated through the study of foraging seabirds directly, as their distinct prey-search behaviours may also inform physical oceanographers on the location of physical features, especially small-scale features such as fronts (e.g. Sabarros et al., 2014; Scales et al., 2014).

Competing interests. The authors declare that they have no conflict of interest.

Acknowledgements. This study was funded by the Federal Ministry for Economic Affairs and Energy according to the decision of the German Bundestag (HELBIRD, 0325751). The e-obs loggers used in this study were provided through the COSYNA project led by the Helmholtz Zentrum Geesthacht (HZG). Logistic support on Heligoland was provided by J. Dierschke and the Institute of Avian Research on Heligoland. K. Borkenhagen, R. M. Borrmann, L. Enners, K. Fließbach, N. Guse, J. Jeglinski, K. Lehmann-Muriithi, B. Mendel, S. Müller, G. Schultheiß, H. Schwemmer, S. Vandenabeele, and S. Weiel helped with fieldwork. S. Furness provided linguistic support. All institutional and national guidelines for the handling and equipping of birds were followed. Birds were equipped under a licence issued by the Ministry of Energy Transition, Agriculture, Environment and Rural Areas Schleswig-Holstein, Germany (file no. V 312-7224.121-37 (80-6/13)). All animals were handled in strict accordance with good animal practice to minimise handling times and stress.

Edited by: H. Brix

References

Bailey, H., Brookes, K. L., and Thompson, P. M.: Assessing environmental impacts of offshore wind farms: lessons learned and recommendations for the future, Aquat. Biosyst., 10, doi:10.1186/2046-9063-10-8, 2014.

Baschek, B., Schroeder, F., Brix, H., Riethmüller, R., Badewien, T. H., Breitbach, G., Brügge, B., Colijn, F., Doerffer, R., Eschenbach, C., Friedrich, J., Fischer, P., Garthe, S., Horstmann, J., Krasemann, H., Metfies, K., Ohle, N., Petersen, W., Pröfrock, D., Röttgers, R., Schlüter, M., Schulz, J., Schulz-Stellenfleth, J., Stanev, E., Winter, C., Wirtz, K., Wollschläger, J., Zielinski, O., and Ziemer, F.: The Coastal Observing System for Northern and Arctic Seas (COSYNA), Ocean Sci. Discuss., doi:10.5194/os-2016-31, in review, 2016.

Bates, D., Maechler, M., Bolker, B., and Walker, S.: Fitting Linear Mixed-Effects Models Using lme4, J. Stat. Software, 67, 1–48, 2015.

Bauer, H.-G., Bezzel, E., and Fiedler, W.: Das Kompendium der Vögel Mitteleuropas. Alles über Biologie, Gefährdung und Schutz. Vol. 1: Nonpasseriformes – Nichtsperlingsvögel, Aula-Verlag, Wiebelsheim, Germany, 2005.

Boyd, I. L.: The behavioural and physiological ecology of diving, Trends Ecol. Evol., 12, 213–217, 1997.

Breitbach, G., Krasemann, H., Behr, D., Beringer, S., Lange, U., Vo, N., and Schroeder, F.: Accessing diverse data comprehensively – CODM, the COSYNA data portal, Ocean Sci., 12, 909–923, doi:10.5194/os-12-909-2016, 2016.

Briggs, K. T., Tyler, W. B., Lewis, D. B., and Carlson, D. R.: Bird communities at sea off California: 1975 to 1983, Stud. Avian Biol., 11, 1–74, 1987.

Brown, R. G. B.: Revised Atlas of Eastern Canadian Seabirds. I. Shipboard surveys, Canadian Wildlife Service, Dartmouth, Canada, 1986.

Cleasby, I. R., Wakefield, E. D., Bearhop, S., Bodey, T. W., Votier, S. C., and Hamer, K. C.: Three-dimensional tracking of a wide-ranging marine predator: flight heights and vulnerability to offshore wind farms, J. Appl. Ecol., 52, 1474–1482, 2015.

Corman, A.-M. and Garthe, S.: What flight heights tell us about foraging and potential conflicts with wind farms: a case study in lesser black-backed gulls (Larus fuscus), J. Ornithol., 155, 1037–1043, 2014.

Daunt, F., Peters, G., Scott, B., Grémillet, D., and Wanless, S.: Rapid-response recorders reveal interplay between marine physics and seabird behaviour, Mar. Ecol.-Prog. Ser., 255, 283–288, 2003.

Davoren, G. K., Montevecchi, W. A., and Anderson, J. T.: Distributional patterns of a marine bird and its prey: habitat selection based on prey and conspecific behaviour, Mar. Ecol.-Progr. Ser., 256, 229–242, 2003.

Furness, R. W. and Monaghan, P.: Seabird Ecology, Blackie, Glasgow, 1987.

Furness, R. W., Wade, H., and Masden E. A.: Assessing vulnerability of seabird populations to offshore wind farms, J. Environ. Manage., 119, 56–66, 2013.

Garthe, S., Camphuysen, C. J., and Furness, R. W.: Amounts of discards by commercial fisheries and their significance as food for seabirds in the North Sea, Mar. Ecol.-Prog. Ser., 136, 1–11, 1996.

Garthe, S., Benvenuti, S., and Montevecchi, W. A.: Pursuit-plunging by Northern Gannets (Sula bassana) feeding on Capelin (Mallotus villosus), Proc. R. Soc. London Biol. Sci., 267, 1717–1722, 2000.

Garthe, S., Benvenuti, S., and Montevecchi, W. A.: Temporal patterns of foraging activities of northern gannets, Morus bassanus, in the northwest Atlantic Ocean, Can. J. Zool., 81, 453–461, 2003.

Garthe, S., Montevecchi, W. A., Chapdelaine, G., Rail, J.-F., and Hedd, A.: Contrasting foraging tactics by northern gannets (Sula bassana) breeding in different oceanographic domains with different prey fields, Mar. Biol., 151, 687–694, 2007.

Garthe, S., Ludynia, K., Hüppop, O., Kubetzki, U., Meraz, J. F., and Furness, R. W.: Energy budgets reveal equal benefits of varied migration strategies in northern gannets, Mar. Biol., 159, 1907–1915, 2012.

Garthe, S., Guse, N., Montevecchi, W. A., Rail, J.-F., and Grégoire, F.: The daily catch: Flight altitude and diving behavior of northern gannets feeding on Atlantic mackerel, J. Sea Res., 85, 456–462, 2014.

Garthe, S., Markones, N., and Corman, A.-M.: Possible impacts of offshore wind farms on seabirds: a pilot study in Northern Gannets in the southern North Sea, J. Ornithol., 158, 345–349, 2017.

Gilbert, N. I., Correia, R. A., Silva, J. P., Pacheco, C., Catry, I., Atkinson, P. W., Gill, J. A., and Franco, A. M. A.: Are white storks addicted to junk food? Impacts of landfill use on the movement and behaviour of resident white storks (Ciconia ciconia) from a partially migratory population, Movement Ecol., 4, doi:10.1186/s40462-016-0070-0, 2016.

Gómez Laich, A., Wilson, R. P., Quintana, F., and Shepard, E. L. C.: Identification of imperial cormorant Phalacrocorax atriceps behaviour using accelerometers, Endang. Spec. Res., 10, 29–27, 2008.

Grémillet, D., Pichegru, L., Siorat, F., and Georges, J.-Y.: Conservation implications of the apparent mismatch between population dynamics and foraging effort in French Northern Gannets from the English Channel, Mar. Ecol.-Prog. Ser., 319, 15–25, 2006.

Grémillet, D., Lewis, S., Drapeau, L., van der Lingen, C. D., Huggett, J. A., Coetzee, J. C., Verheye, H. M., Daunt, F., Wanless, S., and Ryan, P. G.: Spatial match–mismatch in the Benguela upwelling zone: should we expect chlorophyll and sea-surface temperature to predict marine predator distributions?, J. Appl. Ecol., 45, 610–621, 2008.

Hamer, K. C., Phillips, R. A., Hill, J. K., Wanless, S., and Wood, A. G.: Contrasting foraging strategies of gannets Morus bassanus at two North Atlantic colonies: foraging trip duration and foraging area fidelity, Mar. Ecol.-Prog. Ser., 224, 283–290, 2001.

Hunt Jr., G. L.: The pelagic distribution of marine birds in a heterogeneous environment, Polar Res., 8, 43–54, 1990.

Hunt Jr., G. L. and Schneider, D. C.: Scale-dependent processes in the physical and biological environment of marine birds, in: Seabirds: Feeding Ecology and Role in Marine Ecosystems, edited by: Croxall, J. P., Cambridge University Press, Cambridge, 7–41, 1987.

Hunt Jr., G. L., Russell, R. W., Coyle, K. O., and Weingartner, T.: Comparative foraging ecology of planktivorous auklets in relation to ocean physics and prey availability, Mar. Ecol.-Prog. Ser., 167, 241–259, 1998.

Jahncke, J., Coyle, K. O., and Hunt Jr., G. L.: Seabird distribution, abundance and diets in the eastern and central Aleutian Islands, Fish. Oceanogr., 14 (Suppl. 1), 160–177, 2005.

Johnston, A., Cook, A. S. C. P., Wright, L. J., Humphreys, E. M., and Burton, N. H. K.: Modelling flight heights of marine birds to more accurately assess collision risk with offshore wind turbines, J. Appl. Ecol., 51, 31–41, 2014.

Jouventin, P. and Weimerskirch, H.: Satellite tracking of Wandering Albatrosses, Nature, 343, 746–748, 1990.

Kays, R., Crofoot, M. C., Jetz, W., and Wikelski, M.: Terrestrial animal tracking as an eye on life and planet, Science, 348, aaa2478, doi:10.1126/science.aaa2478, 2015.

Korner-Nievergelt, F., von Felten, S., Roth, T., Guélat, J., Almasi, B., and Korner-Nievergelt, P.: Bayesian data analysis in ecology using linear models with R, Bugs, and Stan, Academic Press Inc, Waltham, MA, 2015.

Krijgsveld, K. L., Fijn, R. C., Japink, M., van Horssen, P. W., Heunks, C., Collier, M. P., Poot, M. J. M., Beuker, D., and Dirksen, S.: Effect Studies Offshore Wind Farm Egmond aan Zee: Final report on fluxes, flight altitudes and behaviour of flying birds, Bureau Waardenburg Report, 10–219, 2011.

Lewis, S., Benvenuti, S., Dall'Antonia, L., Griffiths, R., Money, L., Sherratt, T. N., Wanless, S., and Hamer, K. C.: Sex-specific foraging behaviour in a monomorphic seabird, Proc. R. Soc. London B, 269, 1687–1693, 2002.

Masden, E. A., McCluskie, A., Owen, E., and Langston, R. H. W.: Renewable energy developments in an uncertain world: The case of offshore wind and birds in the UK, Mar. Policy, 51, 169–172, 2015.

Nelson, J. B.: The Atlantic Gannet, Fenix Books, Norfolk, 2002.

Pettex, E., Lorentsen, S.-H., Grémillet, D., Gimenez, O., Barrett, R. T., Pons, J.-B., Le Bohec, C., and Bonadonna, F.: Multi-scale foraging variability in Northern gannet (*Morus bassanus*) fuels potential foraging plasticity, Mar. Biol., 159, 2743–2756, 2012.

Phillips, R. A., Xavier, J. C., and Croxall, J. P.: Effects of satellite transmitters on albatrosses and petrels, Auk, 120, 1082–1090, 2003.

R Development Core Team: R: A language and environment for statistical computing. R Foundation for Statistical Computing, Vienna, Austria, https://www.R-project.org/, 2016.

re3data.org: COSYNA Data web portal; editing status 2016-04-27; re3data.org – Registry of Research Data Repositories, doi:10.17616/R3K02T, last access: 24 April 2017.

Ronconi, R. A. and Burger, A. E.: Foraging space as a limited resource: inter- and intra-specific competition among sympatric pursuit-diving seabirds, Can. J. Zool., 89, 356–368, 2011.

Ryan, P. G. and Moloney, C. L.: Effect of trawling on bird and seal distributions in the southern Benguela region, Mar. Ecol.-Prog. Ser., 45, 1–11, 1988.

Sabarros, P. S., Grémillet, D., Demarcq, H., Moseley, H., Pichegru, L., Mullers, R. H. E., Stenseth, N. C., and Machu, E.: Fine-scale recognition and use of mesoscale fronts by foraging Cape gannets in the Benguela upwelling region, Deep-Sea Res. Pt. II, 107, 77–84, 2014.

Sakamoto, K. Q., Sato, K., Ishizuka, M., Watanuki, Y., Takahashi, A., Daunt, F., and Wanless, S.: Can ethograms be automatically generated using body acceleration data from free-ranging birds?, PLoS ONE, 4, e5379, doi:10.1371/journal.pone.0005379, 2009.

Scales, K. L., Miller, P. I., Embling, C. B., Ingram, S. N., Pirotta, E., and Votier, S. C.: Mesoscale fronts as foraging habitats: composite front mapping reveals oceanographic drivers of habitat use for a pelagic seabird, J. R. Soc. Interface, 11, 20140679, doi:10.1098/rsif.2014.0679, 2014.

Schneider, D. C.: The role of fluid dynamics in the ecology of marine birds, Oceanogr. Mar. Biol. Annu. Rev., 29, 487–521, 1991.

Schneider, D. C.: Quantitative Ecology. Spatial and Temporal Scaling, Academic Press, San Diego, 1994.

Schneider, D. C. and Duffy, D. C.: Scale-dependent variability in seabird abundance, Mar. Ecol.-Prog. Ser., 25, 211–218, 1985.

Schneider, D. and Hunt Jr., G. L.: A comparison of seabird diets and foraging distribution around the Pribilof Islands, Alaska, in: Marine Birds: their Feeding Ecology and Commercial Fisheries Relationships, edited by: Nettleship, D. N., Sanger, G. A., and Springer, P. F., Can. Wildl. Serv. Spec. Publ., Canadian Wildlife Service, 86–95, 1984.

Shealer, D. A.: Foraging behavior and food of seabirds, in: Biology of Marine Birds, edited by: Schreiber, E. A. and Burger, J., CRC Press, Boca Raton, 137–177, 2002.

Stanev, E. V., Schulz-Stellenfleth, J., Staneva, J., Grayek, S., Grashorn, S., Behrens, A., Koch, W., and Pein, J.: Ocean forecasting for the German Bight: from regional to coastal scales, Ocean Sci., 12, 1105–1136, doi:10.5194/os-12-1105-2016, 2016.

Tasker, M. L., Webb, A., Hall, A. J., Pienkowski, M. W., and Langslow, D. R.: Seabirds in the North Sea. Final report of phase 2 of the Nature Conservancy Council Seabirds at Sea Project November 1983–October 1986, Nature Conservancy Council, Peterborough, 1987.

Vandenabeele, S. P., Shepard, E. L. C., Grogan, A., and Wilson, R. P.: When three per cent may not be three per cent; device-equipped seabirds experience variable flight constraints, Mar. Biol., 159, 1–14, 2012.

Vandenabeele, S. P., Grundy, E., Friswell, M. I., Grogan, A., Votier, S. C., and Wilson, R. P.: Excess baggage for birds: Inappropriate placement of tags on gannets changes flight patterns, PLoS ONE, 9, e92657, doi:10.1371/journal.pone.0092657, 2014.

Wakefield, E. D., Bodey, T. W., Bearhop, S., Blackburn, J., Colhoun, K., Davies, R., Dwyer, R. G., Green, J., Grémillet, D., Jackson, A. L., Jessopp, M. J., Kane, A., Langston, R. H. W., Lescroël, A., Murray, S., Le Nuz, M., Patrick, S. C., Péron, C., Soanes, L., Wanless, S., Votier, S. C., and Hamer, K. C.: Space partitioning without territoriality in gannets, Science, 341, 68–70, 2013.

Wanless, S. and Okill, J. D.: Body measurements and flight performance of adult and juvenile Gannets *Morus bassanus*, Ring. Migr., 15, 101–103, 1994.

Wilson, R. P. and Vandenabeele, S. P.: Technological innovation in archival tags used in seabird research, Mar. Ecol.-Prog. Ser., 451, 245–262, 2012.

Wilson, R. P., Scolaro, J. A., Peters, G., Laurenti, S., Kierspel, M., Gallelli, H., and Upton, J.: Foraging areas of Magellanic Penguins *Spheniscus magellanicus* breeding at San Lorenzo, Argentina, during the incubation period, Mar. Ecol.-Prog. Ser., 129, 1–6, 1995a.

Wilson, R. P., Weimerskirch, H., and Lys, P.: A device for measuring seabird activity at sea, J. Avian Biol., 26, 172–175, 1995b.

Wilson, R. P., Grémillet, D., Syder, J., Kierspel, M. A. M., Garthe, S., Weimerskirch, H., Schäfer-Neth, C., Scolaro, J. A., Bost, C.-A., Plötz, J., and Nel, D.: Remote-sensing systems and seabirds: their use, abuse and potential for measuring marine environmental variables, Mar. Ecol.-Prog. Ser., 228, 241–261, 2002.

Permissions

All chapters in this book were first published in OS, by Copernicus Publications; hereby published with permission under the Creative Commons Attribution License or equivalent. Every chapter published in this book has been scrutinized by our experts. Their significance has been extensively debated. The topics covered herein carry significant findings which will fuel the growth of the discipline. They may even be implemented as practical applications or may be referred to as a beginning point for another development.

The contributors of this book come from diverse backgrounds, making this book a truly international effort. This book will bring forth new frontiers with its revolutionizing research information and detailed analysis of the nascent developments around the world.

We would like to thank all the contributing authors for lending their expertise to make the book truly unique. They have played a crucial role in the development of this book. Without their invaluable contributions this book wouldn't have been possible. They have made vital efforts to compile up to date information on the varied aspects of this subject to make this book a valuable addition to the collection of many professionals and students.

This book was conceptualized with the vision of imparting up-to-date information and advanced data in this field. To ensure the same, a matchless editorial board was set up. Every individual on the board went through rigorous rounds of assessment to prove their worth. After which they invested a large part of their time researching and compiling the most relevant data for our readers.

The editorial board has been involved in producing this book since its inception. They have spent rigorous hours researching and exploring the diverse topics which have resulted in the successful publishing of this book. They have passed on their knowledge of decades through this book. To expedite this challenging task, the publisher supported the team at every step. A small team of assistant editors was also appointed to further simplify the editing procedure and attain best results for the readers.

Apart from the editorial board, the designing team has also invested a significant amount of their time in understanding the subject and creating the most relevant covers. They scrutinized every image to scout for the most suitable representation of the subject and create an appropriate cover for the book.

The publishing team has been an ardent support to the editorial, designing and production team. Their endless efforts to recruit the best for this project, has resulted in the accomplishment of this book. They are a veteran in the field of academics and their pool of knowledge is as vast as their experience in printing. Their expertise and guidance has proved useful at every step. Their uncompromising quality standards have made this book an exceptional effort. Their encouragement from time to time has been an inspiration for everyone.

The publisher and the editorial board hope that this book will prove to be a valuable piece of knowledge for researchers, students, practitioners and scholars across the globe.

List of Contributors

H. A. Dijkstra and M. A. Kliphuis
Institute of Marine and Atmospheric Research Utrecht, Utrecht University, Princetonplein 5, 3584 CC Utrecht, the Netherlands

S.-E. Brunnabend
Institute of Marine and Atmospheric Research Utrecht, Utrecht University, Princetonplein 5, 3584 CC Utrecht, the Netherlands
Leibniz Institute for Baltic Sea Research Warnemünde, Seestrasse 15, 18119 Rostock, Germany

H. E. Bal
Department of Computer Science, VU University Amsterdam, 1081 HV Amsterdam, the Netherlands

F. Seinstra, B. van Werkhoven, J. Maassen, and M. van Meersbergen
Netherlands eScience Center, 1098 XG Amsterdam, the Netherlands

Jan Kaiser, Karen J. Heywood, Dorothee C.E. Bakker, Gareth Lee and Oliver Legge
Centre for Ocean and Atmospheric Sciences, School of Environmental Sciences, University of East Anglia, Norwich Research Park, Norwich NR4 7TJ, UK

Jacqueline Boutin
Laboratoire d'Océanographie et du Climat, 4 Place Jussieu, 75005 Paris, France

Michael P. Hemming
Centre for Ocean and Atmospheric Sciences, School of Environmental Sciences, University of East Anglia, Norwich Research Park, Norwich NR4 7TJ, UK
Laboratoire d'Océanographie et du Climat, 4 Place Jussieu, 75005 Paris, France

Kiminori Shitashima
Tokyo University of Marine Science and Technology, 4-5-7 Konan, Minato, Tokyo 108-0075, Japan

Reiner Onken
Helmholtz-Zentrum Geesthacht, Max-Planck-Straße 1, 21502 Geesthacht, Germany

Philipp Fischer, Max Schwanitz and Markus Brand
Alfred-Wegener-Institut Helmholtz Centre for Polar and Marine Research, Centre for Scientific Diving at the Biological Station Helgoland, Kurpromenade 211, 27498 Helgoland, Germany

Reiner Loth
loth-engineering GmbH, Lochmühle 1, 65527 Niedernhausen, Germany

Uwe Posner
-4H-JENA engineering GmbH, Mühlenstr. 126, 07745 Jena, Germany

Friedhelm Schröder
Helmholtz-Zentrum Geesthacht, Institut für Material- und Küstenforschung, Max-Planck-Straße 1, 21502 Geesthacht, Germany

Ming Xi Yang, Frances E. Hopkins, John Stephens, and Thomas G. Bell
Plymouth Marine Laboratory, Plymouth, UK

Richard P. Sims
Plymouth Marine Laboratory, Plymouth, UK
University of Exeter, Exeter, UK

Ute Schuster and Andrew J. Watson
University of Exeter, Exeter, UK

Sergey V. Prants, Maxim V. Budyansky, and Michael Y. Uleysky
Laboratory of Nonlinear Dynamical Systems, Pacific Oceanological Institute of the Russian Academy of Sciences, 43 Baltiyskaya st., 690041 Vladivostok, Russia

Jiping Xie, Laurent Bertino, François Counillon and Knut A. Lisæter
Nansen Environmental and Remote Sensing Center, Bergen 5006, Norway

Pavel Sakov
Bureau of Meteorology, Melbourne VIC3001, Australia

Wenjun Yao and Jiuxin Shi
Physical Oceanography Laboratory/CIMST, Ocean University of China and Qingdao National Laboratory for Marine Science and Technology, Qingdao, China

Xiaolong Zhao
North China Sea Marine Forecasting Center, State Oceanic Administration, Qingdao, 266061, Shandong, China

Burkard Baschek, Friedhelm Schroeder, Holger Brix, Rolf Riethmüller, Gisbert Breitbach, Franciscus Colijn, Roland Doerffer, Christiane Eschenbach, Jana Friedrich, Jochen Horstmann, Hajo Krasemann, Lucas Merckelbach, Wilhelm Petersen, Daniel Pröfrock, Rüdiger Röttgers Johannes Schulz-Stellenfleth, Emil Stanev, Joanna Staneva, Jochen Wollschläger, Kai Wirtz and Friedwart Ziemer
Institute of Coastal Research, Helmholtz-Zentrum Geesthacht, Geesthacht, Germany

Thomas H. Badewien, Jan Schulz and Oliver Zielinski
Institute for Chemistry and Biology of the Marine Environment, University of Oldenburg, Oldenburg, Germany

Bernd Brügge
Federal Maritime and Hydrographic Agency, Hamburg, Germany

Philipp Fischer, Katja Metfies and Michael Schlüter
Alfred Wegener Institute, Helmholtz Center for Polar and Marine Research, Center for Polar and Marine Research, Bremerhaven, Germany

Stefan Garthe
Research and Technology Centre (FTZ), University of Kiel, Büsum, Germany

Nino Ohle
Hamburg Port Authority, Hamburg, Germany

ChristianWinter
MARUM, Center for Marine Environmental Sciences, Bremen University, Bremen, Germany

Henk W. van den Brink
KNMI, Utrechtseweg 297, De Bilt, the Netherlands

Sacha de Goederen
Rijkswaterstaat, Boompjes 200, Rotterdam, the Netherlands

Robert Marsh and Ivan D. Haigh
Ocean and Earth Science, University of Southampton, National Oceanography Centre, Southampton, European Way, Southampton SO14 3ZH, UK

Stuart A. Cunningham, Mark E. Inall and Marie Porter
Scottish Association for Marine Science, Scottish Marine Institute, Oban, Argyll PA37 1QA, UK

Ben I. Moat
National Oceanography Centre, European Way, Southampton SO14 3ZH, UK

Arash Bigdeli and Brice Loose
Graduate School of Oceanography, University of Rhode Island, Rhode Island, 02882, USA

An T. Nguyen
Institute of Computational Engineering and Sciences, University of Texas at Austin, Austin, Texas, 78712, USA

Sylvia T. Cole
Woods Hole Oceanographic Institution, Woods Hole, Massachusetts, 02543, USA

Leif G. Anderson and Göran Björk
Department of Marine Sciences, University of Gothenburg, P.O. Box 461, 40530 Gothenburg, Sweden

Ola Holby
Department of Environmental and Energy Systems, Karlstad University, 651 88 Karlstad, Sweden

Sara Jutterström
IVL Swedish Environmental Research Institute, Box 530 21, 400 14 Gothenburg, Sweden

Carl Magnus Mörth, Matt O'Regan and Martin Jakobsson
Department of Geological Sciences, Stockholm University, 106 91 Stockholm, Sweden

Christof Pearce
Department of Geological Sciences, Stockholm University, 106 91 Stockholm, Sweden
Department of Geoscience, Aarhus University, Aarhus, Denmark

Igor Semiletov
International Arctic Research Center, University Alaska Fairbanks, Fairbanks, AK 99775, USA
Pacific Oceanological Institute, Russian Academy of Sciences Far Eastern Branch, Vladivostok 690041, Russia
The National Research Tomsk Polytechnic University, Tomsk, Russia

Toste Tanhua
Helmholtz Centre for Ocean Research Kiel, GEOMAR, Kiel, Germany

Christian Stranne
Department of Geological Sciences, Stockholm University, 106 91 Stockholm, Sweden
Center for Coastal and Ocean Mapping/Joint Hydrographic Center, Durham, NH 03824, USA

Tim Stöven
Department of Marine Sciences, University of Gothenburg, P.O. Box 461, 40530 Gothenburg, Sweden
Helmholtz Centre for Ocean Research Kiel, GEOMAR, Kiel, Germany

Adam Ulfsbo
Department of Marine Sciences, University of Gothenburg, P.O. Box 461, 40530 Gothenburg, Sweden
Division of Earth and Ocean Sciences, Nicholas School of the Environment, Duke University, Durham, NC 27704, USA

Stefan Garthe, Verena Peschko, and Anna-Marie Corman
Research & Technology Centre (FTZ), Kiel University, Hafentörn 1, 25761 Büsum, Germany

Ulrike Kubetzki
Research & Technology Centre (FTZ), Kiel University, Hafentörn 1, 25761 Büsum, Germany
Department of Animal Ecology and Conservation, Biocentre Grindel, Hamburg University, Martin-Luther-King Platz 3, 20146 Hamburg, Germany

Index